Data Analysis with Microsoft® Excel

Updated for Office 2007®

Kenneth N. Berk
Illinois State University

Patrick Carey
Carey Associates, Inc.

Australia • Brazil • Japan • Korea • Mexico • Singapore • Spain • United Kingdom • United States

Data Analysis with Microsoft® Excel: Updated for Office 2007®, Third Edition
Berk, Carey

Publisher: Richard Stratton
Senior Sponsoring Editor: Molly Taylor
Associate Editor: Daniel Seibert
Editorial Assistant: Shaylin Walsh
Associate Media Editor: Catie Ronquillo
Senior Marketing Manager: Greta Kleinert
Marketing Coordinator: Erica O'Connell
Marketing Communications Manager: Mary Anne Payumo
Content Project Manager: Jessica Rasile
Art Director: Linda Helcher
Print Buyer: Linda Hsu
Permissions Editor: Margaret Chamberlain-Gaston
Production Service/Compositor: PrePress PMG
Photo Manager: John Hill
Cover Designer: Blue Bungalow Design
Cover Image: ©Fotolia

Library of Congress Control Number: 2009928574

ISBN-13: 978-0-495-39178-4

ISBN-10: 0-495-39178-6

Brooks/Cole
20 Channel Center Street
Boston, MA 02210
USA

Cengage Learning products are represented in Canada by Nelson Education, Ltd.

For your course and learning solutions, visit **www.cengage.com**

Purchase any of our products at your local college store or at our preferred online store **www.ichapters.com**

Printed in the United States of America
1 2 3 4 5 6 7 13 12 11 10 09

About the Authors

Kenneth N. Berk

Kenneth N. Berk (Ph.D., University of Minnesota) is an emeritus professor of mathematics at Illinois State University and a Fellow of the American Statistical Association. Berk was editor of Software Reviews for the *American Statistician* for six years. He served as chair of the Statistical Computing Section of the American Statistical Association. He has twice co-chaired the annual Symposium on the Interface between Computing Science and Statistics.

Patrick Carey

Patrick Carey received his M.S. in biostatistics from the University of Wisconsin where he worked as a researcher in the General Clinical Research Center designing and analyzing clinical studies. He coauthored his first textbook with Ken Berk on using Excel as a statistical tool. He and his wife Joan founded *Carey Associates, Inc.*, a software textbook development company. He has since authored or coauthored over 20 academic and trade texts for the software industry. Besides books on data analysis, Carey has written on the Windows® operating system, Web page design, database management, the Internet, browsers, and presentation graphics software. Patrick, Joan, and their six children live in Wisconsin.

I thank my wife Laura for her advice, because here she is the one who knows about publishing books.

—*Kenneth N. Berk*

Thanks to my wife, Joan, and my children, John Paul, Thomas, Peter, Michael, Stephen, and Catherine, for their love and support.

—*Patrick M. Carey*

Introduction

Data Analysis with Microsoft® Excel: Updated for Office 2007® harnesses the power of Excel and transforms it into a tool for learning basic statistical analysis. Students learn statistics in the context of analyzing data. We feel that it is important for students to work with real data, analyzing real-world problems, so that they understand the subtleties and complexities of analysis that make statistics such an integral part of understanding our world. The data set topics range from business examples to physiological studies on NASA astronauts. Because students work with real data, they can appreciate that in statistics no answers are completely final and that intuition and creativity are as much a part of data analysis as is plugging numbers into a software package. This text can serve as the core text for an introductory statistics course or as a supplemental text. It also allows nontraditional students outside of the classroom setting to teach themselves how to use Excel to analyze sets of real data so they can make informed business forecasts and decisions.

Users of this book need not have any experience with Excel, although previous experience would be helpful. The first three chapters of the book cover basic concepts of mouse and Windows operation, data entry, formulas and functions, charts, and editing and saving workbooks. Chapters 4 through 12 emphasize teaching statistics with Excel as the instrument.

Using Excel in a Statistics Course

Spreadsheets have become one of the most popular forms of computer software, second only to word processors. Spreadsheet software allows the user to combine data, mathematical formulas, text, and graphics together in a single report or workbook. For this reason, spreadsheets have become indispensable tools for business, as they have also become popular in scientific research. Excel in particular has won a great deal of acclaim for its ease of use and power.

As spreadsheets have expanded in power and ease of use, there has been increased interest in using them in the classroom. There are many advantages to using Excel in an introductory statistics course. An important advantage is that students, particularly business students, are more likely to be familiar with spreadsheets and are more comfortable working with data entered into a spreadsheet. Since spreadsheet software is very common at colleges and universities, a statistics instructor can teach a course without requiring students to purchase an additional software package.

Having identified the strengths of Excel for teaching basic statistics, it would be unfair not to include a few warnings. Spreadsheets are not statistics packages, and there are limits to what they can do in replacing a full-featured statistics package. This is why we have included our own downloadable add-in, StatPlus™. It expands some of Excel's statistical capabilities. (We explain the use of StatPlus where appropriate throughout the text.) Using Excel for anything other than an introductory statistics course would probably not be appropriate due to its limitations. For example, Excel can easily perform balanced two-way analysis of variance but not unbalanced two-way analysis of variance. Spreadsheets are also limited in handling data with missing values. While we recommend Excel for a basic statistics course, we feel it is not appropriate for more advanced analysis.

System Information

You will need the following hardware and software to use *Data Analysis with Microsoft® Excel: Updated for Office 2007®:*

- A Windows-based PC.
- Windows XP or Windows Vista.
- Excel 2007. If you are using an earlier edition of Excel, you will have to use an earlier edition of *Data Analysis with Microsoft® Excel.*
- Internet access for downloading the software files accompanying the text.

The *Data Analysis with Microsoft® Excel* package includes:

- The text, which includes 12 chapters, a reference section for Excel's statistical functions, Analysis ToolPak commands, StatPlus Add-In commands, and a bibliography.
- The companion website at **www.cengage.com/statistics/berk** contains 92 different data sets from real-life situations plus a summary of what the data set files cover, ten interactive Concept Tutorials, and installation files for StatPlus—our statistical application. Chapter 1 of the text includes instructions for installing the files.
- An Instructor's Manual with solutions to all the exercises in the text is available, password-protected on the companion website, to adopting instructors.

Excel's Statistical Tools

Excel comes with 81 statistical functions and 59 mathematical functions. There are also functions devoted to business and engineering problems. The statistical functions that basic Excel provides include descriptive statistics such as means, standard deviations, and rank statistics. There are also cumulative distribution and probability density functions for a variety of distributions, both continuous and discrete.

The Analysis ToolPak is an add-in that is included with Excel. If you have not loaded the Analysis ToolPak, you will have to install it from your original Excel installation.

The Analysis ToolPak adds the following capabilities to Excel:

- Analysis of variance, including one-way, two-way without replication, and two-way balanced with replication
- Correlation and covariance matrices
- Tables of descriptive statistics
- One-parameter exponential smoothing
- Histograms with user-defined bin values
- Moving averages
- Random number generation for a variety of distributions
- Rank and percentile scores
- Multiple linear regression
- Random sampling
- *t* tests, including paired and two sample, assuming equal and unequal variances
- *z* tests

In this book we make extensive use of the Analysis ToolPak for multiple linear regression problems and analysis of variance.

StatPlus™

Since the Analysis ToolPak does not do everything that an introductory statistics course requires, this textbook comes with an additional add-in called the **StatPlus™ Add-In** that fills in some of the gaps left by basic Excel 2007 and the Analysis ToolPak.

Additional commands provided by the StatPlus Add-In give users the ability to:

- Create random sets of data
- Manipulate data columns
- Create random samples from large data sets
- Generate tables of univariate statistics

- Create statistical charts including boxplots, histograms, and normal probability plots
- Create quality control charts
- Perform one-sample and two-sample t tests and z tests
- Perform non-parametric analyses
- Perform time series analyses, including exponential and seasonal smoothing
- Manipulate charts by adding data labels and breaking charts down into categories
- Perform non parametric analyses
- Create and analyze tabular data

A full description of these commands is included in the Appendix's Reference section and through on-line help available with the application.

Concept Tutorials

Included with the StatPlus add-in are ten interactive Excel tutorials that provide students a visual and hands-on approach to learning statistical concepts.

These tutorials cover:

- Boxplots
- Probability
- Probability distributions
- Random samples
- Population statistics
- The Central Limit Theorem
- Confidence intervals
- Hypothesis tests
- Exponential smoothing
- Linear regression

Acknowledgments

We thank Mac Mendelsohn, Managing Editor at Course Technology, for his support and enthusiasm for the First Edition of this book. For this edition, our thanks to Jessica Rasile, Content Project Manager, Blue Bungalow Design for the cover design, and Carol A. Loomis, Copyeditor, for their professional attention to all the details of production.

Special thanks go to our reviewers, who gave us valuable insights into improving the book in each edition: Aaron S. Liswood, Sierra Nevada College; Abbot L. Packard, State University of West Georgia; Andrew E. Coop, US Air Force Academy; Barry Bombay, J. Sargeant Reynolds Community College; Beth Eschenback, Humboldt State University; Bruce Trumbo, California State University – Hayward; Carl Grafton, Auburn University; Carl R. Williams, University of Memphis; Cheryl Dale, William Carey College; Dang Tran, California State University – Los Angeles; Bruce Marsh, Texas A & M University – Kingsvile; Edward J. Williams, University of Michigan – Dearborn; Eric Zivot, University of Washington; Farrokh Alemi, George Mason University; Faye Teer, James Madison University; Gordon Dahl, University of Rochester; Ian Hardie, University of Maryland; Jack Harris, Hobart and William Smith Colleges; Ames E. Pratt, Cornell University; James Zumbrunnen, Colorado State University; John A. Austin, Jr., Louisiana State University – Shreveport; Kelwyn A. D'Souza, Hampton University; Kevin Griffin, Eastern Arizona College; Lea Cloninger, University of Illinois at Chicago; Lorrie Hoffman, University of Central Florida; Marion G. Sobol, Southern Methodist University, and Matthew C. Dixon, USAF Academy.

We thank Laura Berk, Peter Berk, Robert Beyer, David Booth, Orlyn Edge, Stephen Friedberg, Maria Gillett, Richard Goldstein, Glenn Hart, Lotus Hershberger, Les Montgomery, Joyce Nervades, Diane Warfield, and Kemp Wills for their assistance with the data sets in this book. We especially want to thank Dr. Jeff Steagall, who wrote some of the original material for Chapter 12, Quality Control. If we have missed anyone, please forgive the omission.

Kenneth N. Berk

Patrick M. Carey

Contents

Chapter 1

GETTING STARTED WITH EXCEL

Objectives

In this chapter you will learn to:

- Install StatPlus files
- Start Excel and recognize elements of the Excel workspace
- Work with Excel workbooks, worksheets, and chart sheets
- Scroll through the worksheet window
- Work with Excel cell references
- Print a worksheet
- Save a workbook
- Install and remove Excel add-ins
- Work with Excel add-ins
- Use the features of StatPlus

In this chapter you'll learn how to work with Excel 2007 in the Windows operating system. You'll be introduced to basic workbook concepts, including navigating through your worksheets and worksheet cells. This chapter also introduces StatPlus, an Excel add-in supplied with this book and designed to expand Excel's statistical capabilities.

Getting Started

This book does not require prior Excel 2007 experience, but familiarity with basic features of that program will reduce your start-up time. This section provides a quick overview of the features of Excel 2007. If you are using an earlier version of Excel, you should refer to the text *Data Analysis for Excel for Office XP*. There are many different versions of Windows. This text assumes that you'll be working with Windows Vista or Windows XP.

Special Files for This Book

This book includes additional files to help you learn statistics. There are three types of files you'll work with: StatPlus files, Explore workbooks, and Data (or Student) files.

Excel has many statistical functions and commands. However, there are some things that Excel does not do (or does not do easily) that you will need to do in order to perform a statistical analysis. To solve this problem, this book includes **StatPlus**, a software package that provides additional statistical commands accessible from within Excel.

The **Explore workbooks** are self-contained tutorials on various statistical concepts. Each workbook has one or more interactive tools that allow you to see these concepts in action.

The **Data** or **Student files** contain sample data from real-life problems. In each chapter, you'll analyze the data in one or more Data file, employing various statistical techniques along the way. You'll use other Data files in the exercises provided at the end of each chapter.

Installing the StatPlus Files

The companion website at www.cengage.com/statistics/berk contains an installation program that you can use to install StatPlus on your computer. Install your files now.

To run the installation routine:

1. On the companion website click on the StatPlus link under the Book Resources section.
2. Download the ZIP file containing the StatPlus files to your hard drive.
3. Extract the ZIP file, which will contain a folder called **StatPlus**.
4. Place the **StatPlus** folder in the desired location on your hard drive. If you want, you may rename this folder to a different name of your choice.

The installation folder contains files arranged in three separate subfolders as shown in Figure 1-1.

Figure 1-1 The Stat Plus folders

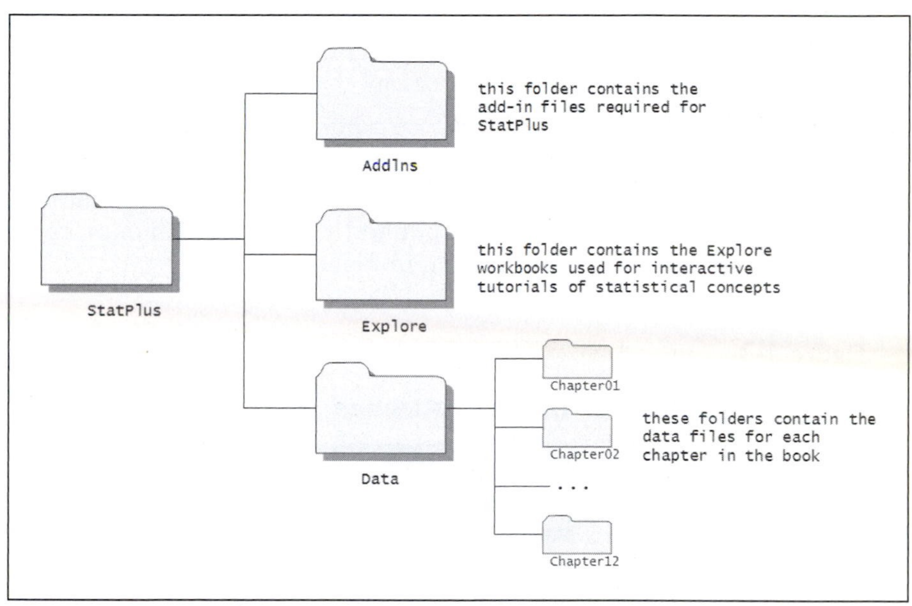

Later in this chapter, you'll learn how to access the StatPlus program from within Excel.

Excel and Spreadsheets

Excel is a software program designed to help you evaluate and present information in a spreadsheet format. **Spreadsheets** are most often used by business for cash-flow analysis, financial reports, and inventory management. Before the era of computers, a spreadsheet was simply a piece of paper with a grid of rows and columns to facilitate entering and displaying information as shown in Figure 1-2.

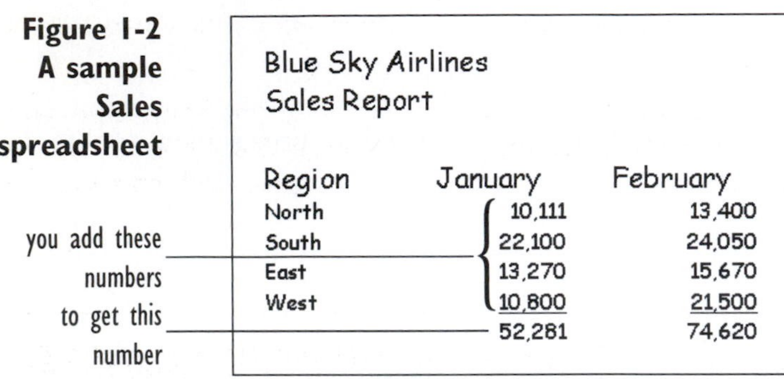

Figure 1-2 A sample Sales spreadsheet

Computer spreadsheet programs use the old hand-drawn spreadsheets as their visual model but add a few new elements, as you can see from the Excel worksheet shown in Figure 1-3.

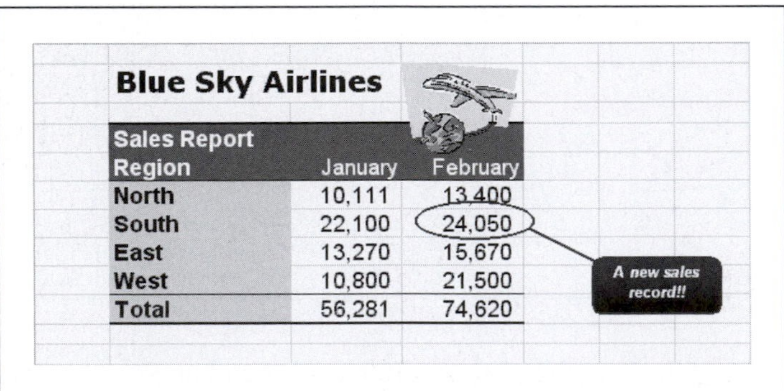

Figure 1-3 A sample spreadsheet as formatted within Excel

However, Excel is so flexible that its application can extend beyond traditional spreadsheets into the area of data analysis. You can use Excel to enter data, analyze the data with basic statistical tests and charts, and then create reports summarizing your findings.

Launching Excel

When Excel 2007 is installed on your computer, the installation program automatically inserts a shortcut icon to Excel 2007 in the Programs menu located under the Windows Start button. You can click this icon to launch Excel.

To start Excel:

1. Click the **Start** button on the Windows Taskbar and then click **All Programs**.

2. Click **Microsoft Office** and then click **Microsoft Office Excel 2007** as shown in Figure 1-4.

   Note: Depending on how Windows has been configured on your computer, your Start menu may look different from the one shown in Figure 1-4. Talk to your instructor if you have problems launching Excel 2007.

Figure 1-4 Starting Excel 2007

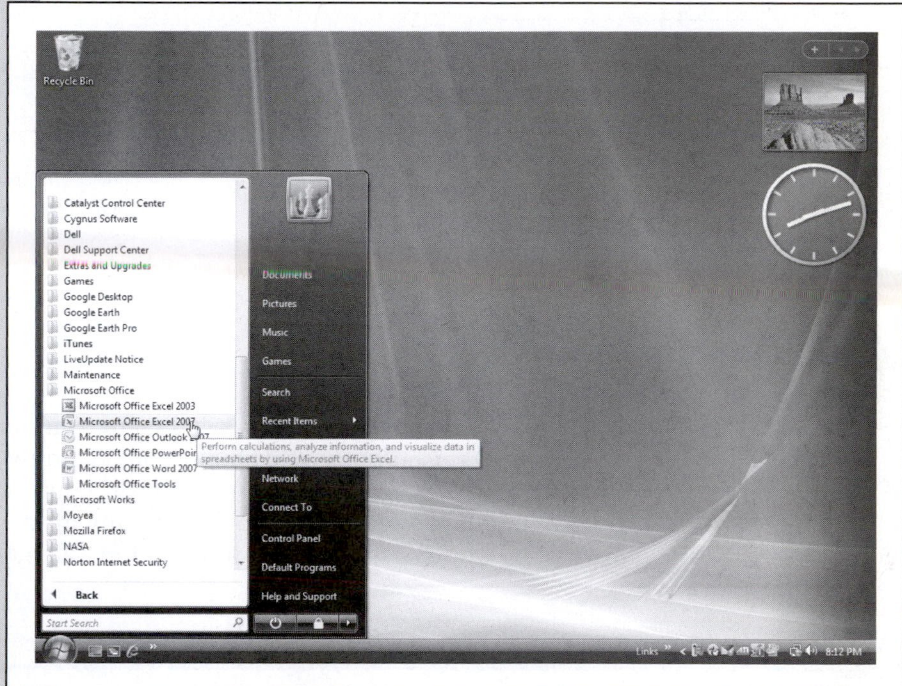

3. Excel starts up, displaying the window shown in Figure 1-5.

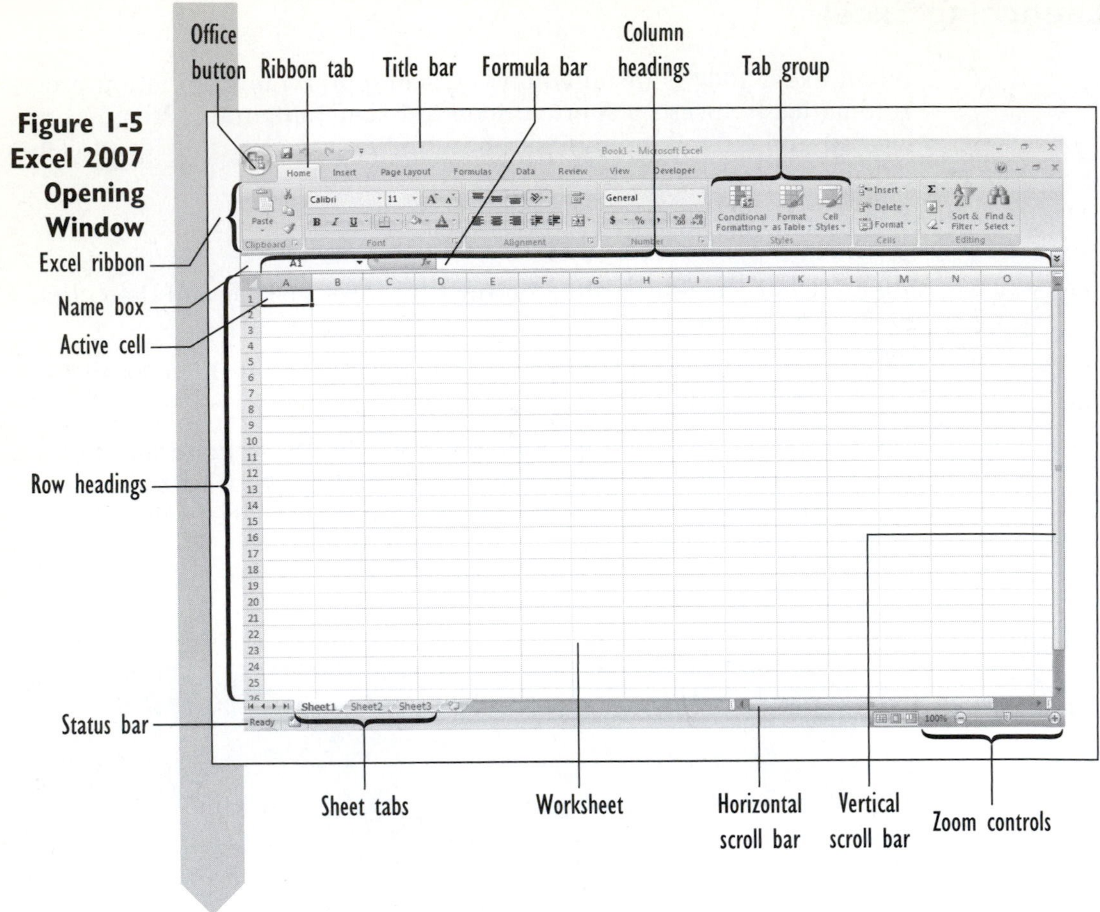

Figure 1-5 Excel 2007 Opening Window

Viewing the Excel Window

The Excel window shown in Figure 1-5 is the environment in which you'll analyze the data sets used in this textbook. Your window might look different depending on how Excel has been set up on your system. Before proceeding, take time to review the various elements of the Excel window. A quick description of these elements is provided in Table 1-1.

Table 1-1 Excel Elements

Excel Element	Purpose
Active cell	The cell currently selected in the worksheet
Cells	Stores individual text or numeric entries
Column headings	Organizes cells into lettered columns

(continued)

Excel ribbon	A toolbar containing Excel commands broken down into different topical tabs
Formula bar	Displays the formula or value entered into the currently selected cell
Horizontal scroll bar	Used to scroll through the contents of the worksheet in a horizontal direction
Name box	Displays the name or reference of the currently selected object or cell
Office button	Displays a menu of commands related to the operation and configuration of Excel and Excel documents
Ribbon tab	A tab containing Excel command buttons for a particular topical area
Row headings	Organizes cells into numeric rows
Sheet tabs	Click to display individual worksheets
Status bar	Displays messages about current Excel operations
Tab group	A group of command buttons within a ribbon tab containing commands focused on the same set of tasks
Title bar	Displays the name of the application and the current Excel document
Vertical scroll bar	Used to scroll through the contents of the worksheet in a vertical direction
Worksheet	A collection of cells laid out in a grid where each cell can contain a single text or numeric entry
Zoom controls	Controls used to increase or decrease the magnification applied to the worksheet

Running Excel Commands

You can run an Excel command either by clicking the icons found on the Excel ribbon or by clicking the Office button and then clicking one of the commands from the menu that appears. Figure 1-6 shows how you would open a file using the Open command available on the menu within the Office button. Note that some of the commands have **keyboard shortcuts**—key combinations that run a command or macro. For example, pressing the CTRL and keys simultaneously will also run the Open command.

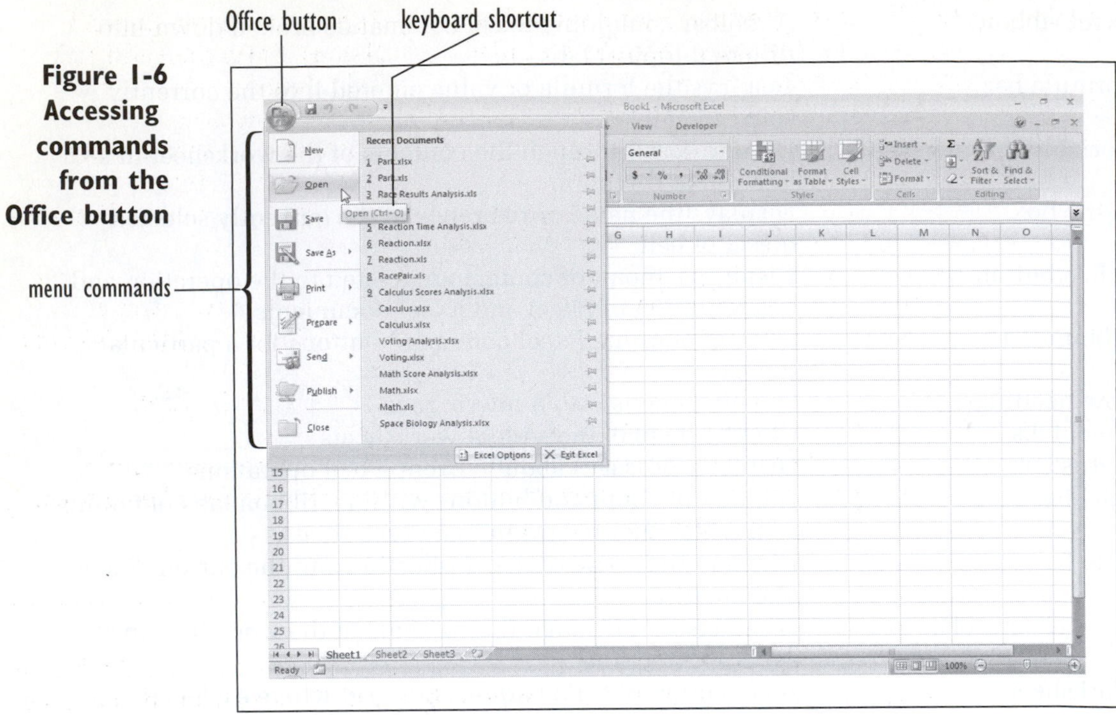

Figure 1-6 Accessing commands from the Office button

The menu commands below the Office button are used to set the properties of your Excel application and entire Excel documents. If you want to work with the contents of a document you work with the commands found on the Excel ribbon.

Each of the tabs on the Excel ribbon contains a rich collection of icons and buttons providing one-click access to Excel commands. Table 1-2 describes the different tabs available on the ribbon.

Note that this list of tabs and groups will change on the basis of how Excel is being used by you. Excel, like other Office 2007 products, is designed to show only the commands which are pertinent to your current task.

Table 1-2 Excel Ribbon Tabs

Ribbon tab	Description	Ribbon Groups
Home	Used to format the contents of worksheet cells	Clipboard, Font, Alignment, Number, Styles, Cells, Editing
Insert	Used to insert objects into an Excel workbook	Tables, Illustrations, Charts, Links, Text
Page Layout	Used to format the printed version of the Excel workbook and to control how each worksheet appears in the Excel window	Themes, Page Setup, Scale to Fit, Sheet Options, Arrange

(continued)

Formulas	Used to insert formulas into a worksheet and to audit the effects of your formulas on cells values	Function Library, Defined Names, Formula Auditing, Calculation
Data	Used to import data from different data sources and to group data values and perform what-if analysis on data	Get External Data, Connections, Sort & Filter, Data Tools, Outline
Review	Used to proof the contents of a workbook and to manage the document in a workgroup environment involving several users	Proofing, Comments, Changes
View	Controls the display of the Excel worksheet window including the ability to hide or display Excel elements	Workbook Views, Show/ Hide, Zoom, Window, Macros
Develop	Contains tools used to add macros and other features to extend the capabilities of Excel	Code, Controls, XML
Add-Ins	Contains user-define menus and tab groups created from add-ins (note that this tab will only appear when an add-in has been installed and activated.)	*various groups depending upon the add-ins being used.*

Each tab is broken up into different topical groups. For example the Home tab is broken into the following groups: Clipboard, Font, Alignment, Number, Styles, Cells, and Editing. When you are asked to run a command, you will be told which button to click from which tab group. For example, to copy the contents of a worksheet cell you would be given the following command:

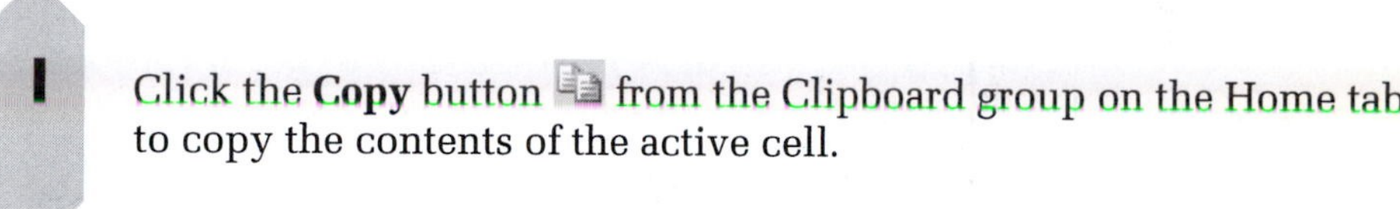

Click the **Copy** button from the Clipboard group on the Home tab to copy the contents of the active cell.

If you are asked to run a command using a keyboard shortcut, the keyboard combination will be shown in boldface with the keys joined by a plus sign to indicate that you should press these keys simultaneously. For example,

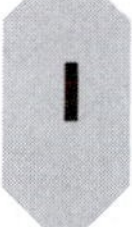

Press **CTRL+n** to create a new blank document.

In addition to the Excel ribbon, you may occasionally see **context-sensitive ribbons**. These ribbons only appear when certain items are selected in the Excel document. For example, when you select an Excel chart, Excel will display a Chart ribbon containing a collection of tabs and tab groups designed for use with charts.

Excel Workbooks and Worksheets

Excel documents are called **workbooks**. Each workbook is made up of individual spreadsheets called **worksheets** and sheets containing charts called **chart sheets**.

Opening a Workbook

To learn some basic workbook commands, you'll first look at an Excel workbook containing public-use data from Kenai Fjords National Park in Alaska. The data are stored in the Parks workbook, located in the Chapter01 subfolder of the Data folder. Open this workbook now.

To open the Park workbook:

1 Click the **Office** button and then click **Open** from the Office menu.

The Open dialog box appears as shown in Figure 1-7. Your dialog box will display a different folder and file list.

Figure 1-7 The Open dialog box

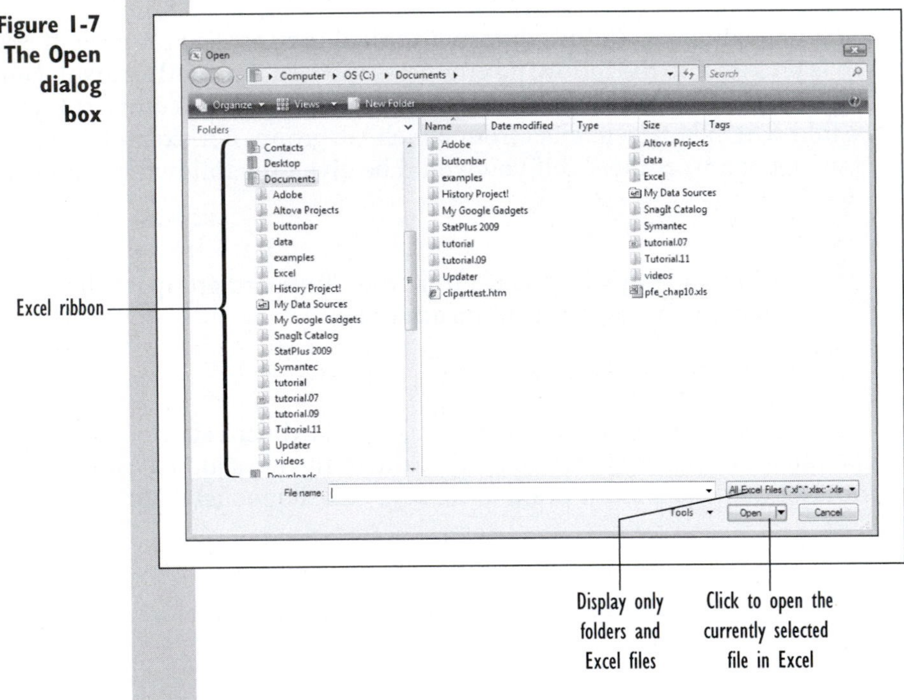

2 Locate the folder containing your Chapter01 data files.

3 Double-click the **Park** workbook.

Excel opens the workbook as shown in Figure 1-8.

Figure 1-8
The Park workbook

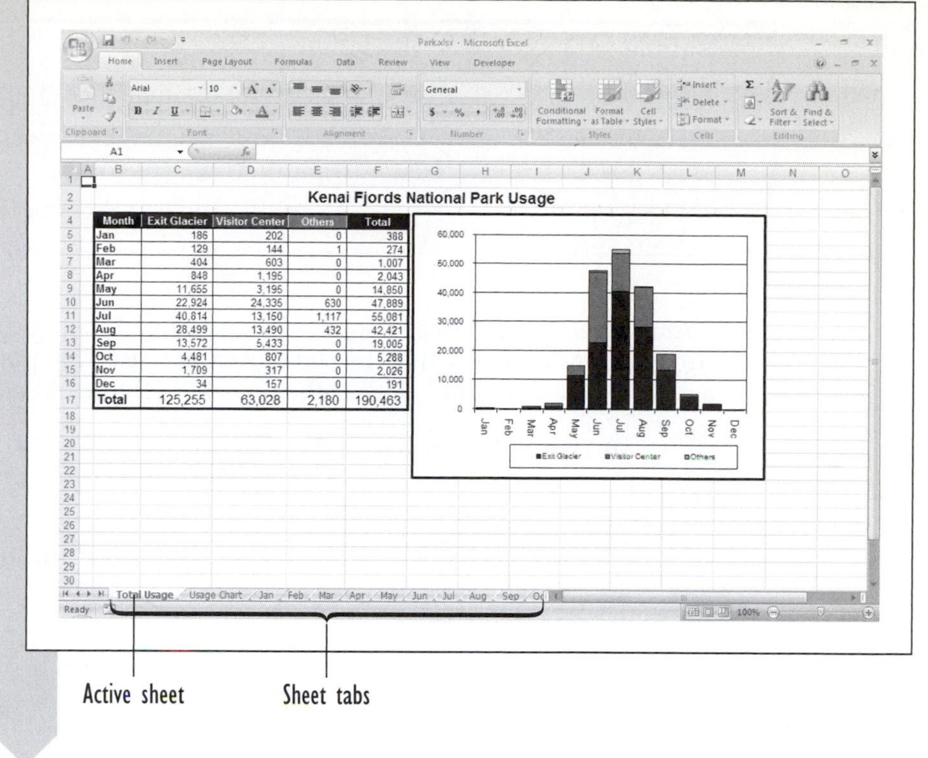

Month	Exit Glacier	Visitor Center	Others	Total
Jan	186	202	0	388
Feb	129	144	1	274
Mar	404	603	0	1,007
Apr	848	1,195	0	2,043
May	11,655	3,195	0	14,850
Jun	22,924	24,335	630	47,889
Jul	40,814	13,150	1,117	55,081
Aug	28,499	13,490	432	42,421
Sep	13,572	5,433	0	19,005
Oct	4,481	807	0	5,288
Nov	1,709	317	0	2,026
Dec	34	157	0	191
Total	125,255	63,028	2,180	190,463

A single workbook can have as many as 255 worksheets. The names of the sheets appear on tabs at the bottom of the workbook window. In the Park workbook, the first sheet is named Total Usage and contains information on the number of visitors at each location in the park over the previous year. The sheet shows both a table of visitor counts and a chart with the same information. Note that the chart has been placed within the worksheet. Placing an object like a chart on a worksheet is known as **embedding**. Glancing over the table and chart, we see that the peak-usage months were May through September.

The second tab is named Usage Chart and contains another chart of park usage. After the first two sheets are worksheets devoted to usage data from each month of the year. Your next task will be to move between the various sheets in the Park workbook.

Scrolling through a Workbook

To move from one sheet to another, you can either click the various sheet tabs in the workbook or use the navigational buttons located at the bottom of the workbook window. Table 1-3 provides a description of these buttons.

Table 1-3 Workbook Navigation Buttons

Button	Image	Purpose
First sheet	⏮	Scroll to the first sheet in the workbook
Previous sheet	◀	Scroll to the previous sheet
Next sheet	▶	Scroll to the next sheet
Last sheet	⏭	Scroll to the last sheet in the workbook

You can also move to a specific sheet by right clicking one of these navigation buttons and selecting the sheet from the resulting pop-up list of sheet names. Try viewing some of the other sheets in the workbook now.

To view other sheets:

1 Click the **Usage Chart** sheet tab.

2 Excel displays the chart. Click anywhere within the chart to select it. See Figure 1-9.

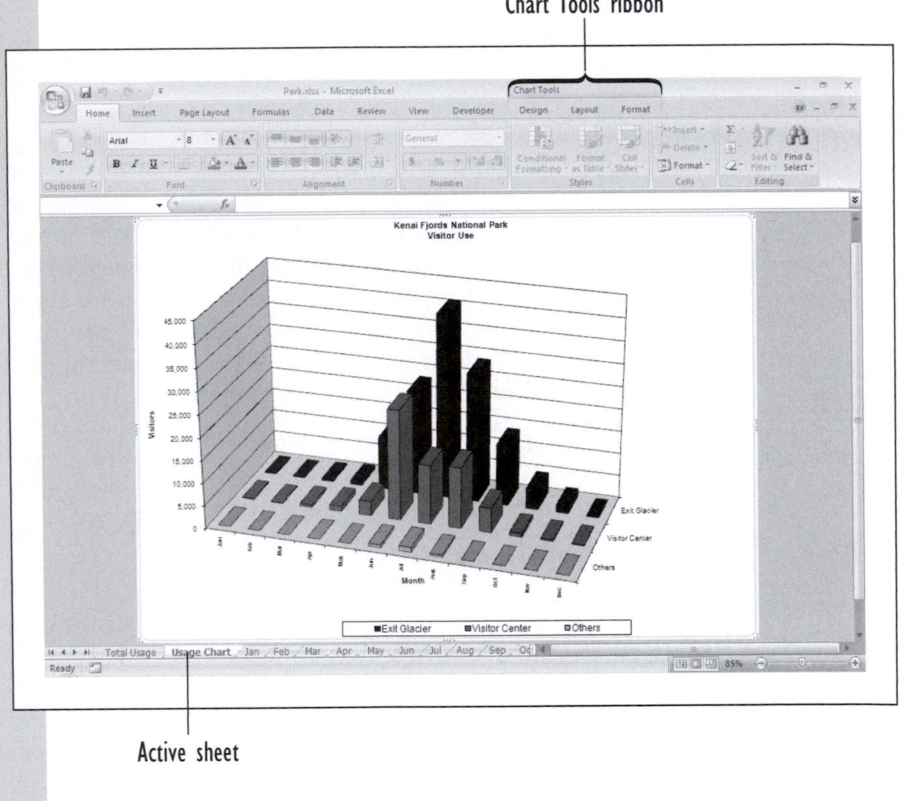

Figure 1-9 The Usage Chart sheet

Note that when you selected the chart, Excel displayed a new ribbon—the Chart Tools ribbon containing specific commands for working with charts. You'll learn more about Excel charts and working with this ribbon in Chapter 3.

3 Click the **Jan** sheet tab.

4 The worksheet for the month of January is displayed as shown in Figure 1-10.

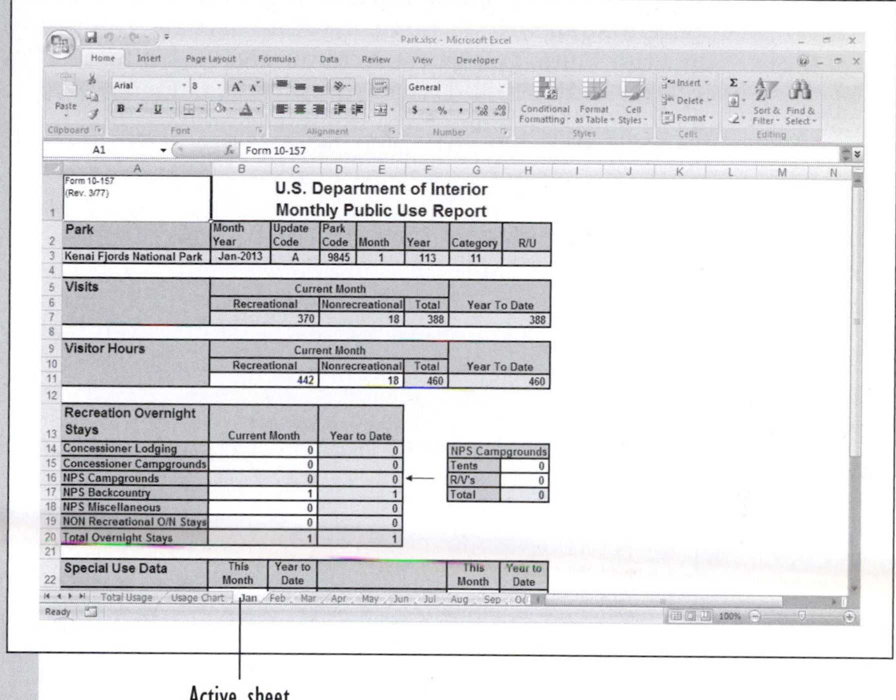

Figure 1-10 The Jan worksheet

The form that appears in this worksheet resembles the form used by the Kenai Fjords staff to record usage information. It contains information on the park, the number of visits each month, visitor hours, and other important data. Some of these data are hidden beyond the boundary of the worksheet window.

5 Drag the **Vertical scrollbar** down to move the worksheet down and view the rest of the January data.

Clearly, the Park workbook is complex. Its sheets contain many pieces of information, much of it interrelated. This book will not cover all the techniques used to create a workbook like this one, but you should be aware of the formatting possibilities that exist.

Worksheet Cells

Each worksheet can be thought of as a grid of **cells**, where each cell can contain a numeric or text entry. Cells are referenced by their location on the grid. For example, the total number of visitors at the park is shown in cell F17 of the Total Usage worksheet (see Figure 1-11.) As you'll see later in Chapter 2, if you were to use this value in a function or Excel command, you would use the cell reference F17.

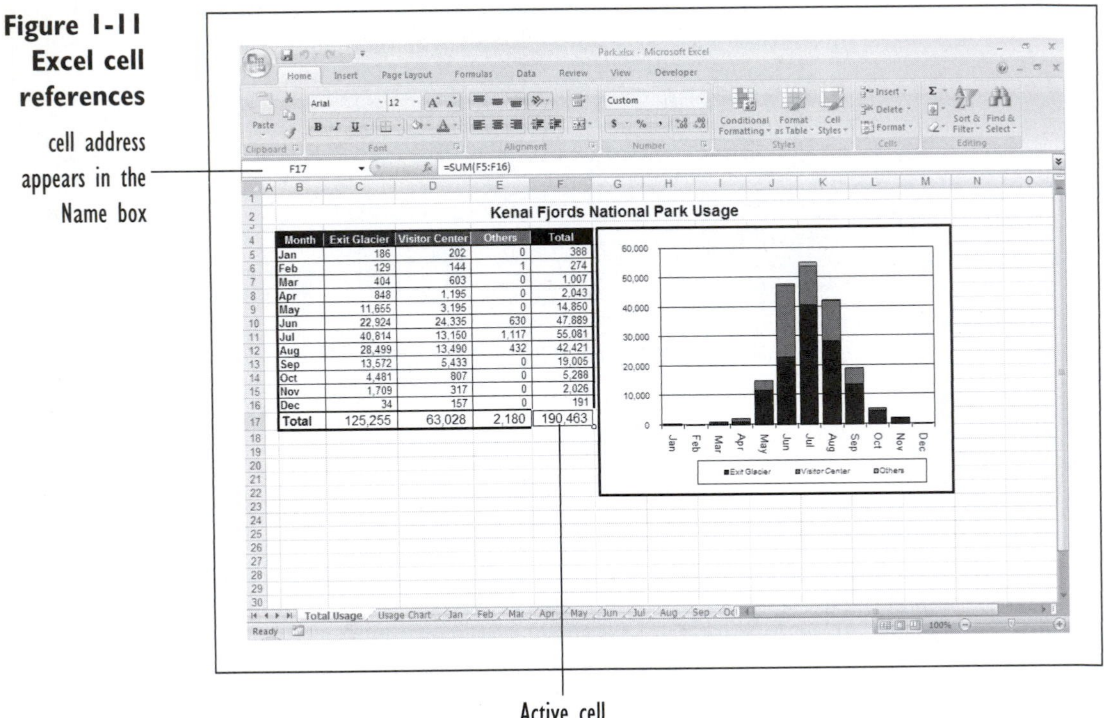

Figure 1-11
Excel cell references

Selecting a Cell

When you want to enter data or format a particular value, you must first select the cell containing the data or value. To do this, you simply click on the cell in the worksheet. Try this now with cell F17 in the Total Usage worksheet.

To select a cell from the worksheet:

1. Click the **Total Usage** sheet tab to move back to the front of the workbook.
2. Click **F17** in the worksheet grid.

Cell F17 now has a small box around it, indicating that it is the **active cell** (see Figure 1-11.) Moreover, when you selected cell F17, the Name box displays F17 indicating that this is the active cell. Also, the formula bar now displayed the formula =SUM(F5:F16). This formula calculates the sum of the values in cells F5 through F16. You'll learn more about formulas in Chapter 2.

If you want to select a group of cells, known as a **cell range** or **range**, you must select one corner of the range and then drag the mouse pointer over cells. To see how this works in practice, try selecting the usage table located in the cell range B4:F17 of the Total Usage worksheet.

To select a cell range:

1. Click **B4**.
2. With the mouse button still pressed, drag the mouse pointer over to cell **F17**.
3. Release the mouse button.

Now the range of cells from B4 down to F17 is selected. Observe that a selected cell range is highlighted to differentiate it from unselected cells. A cell range selected in this fashion is always rectangular in shape and contiguous. If you want to select a range that is not rectangular or contiguous, you must use the CTRL key on your keyboard and then select the separate distinct groups that make up the range. For example, if you want to select only the cells in the range B4:B17 and F4:F17, you must use this technique.

To select a noncontiguous range:

1. Select the range **B4:B17**.
2. Press the **CTRL** key on your keyboard.

3 With the CTRL key still pressed, select the range **F4:F17**.

The selected range is shown in Figure 1-12.

Figure 1-12
Noncontiguous cell range

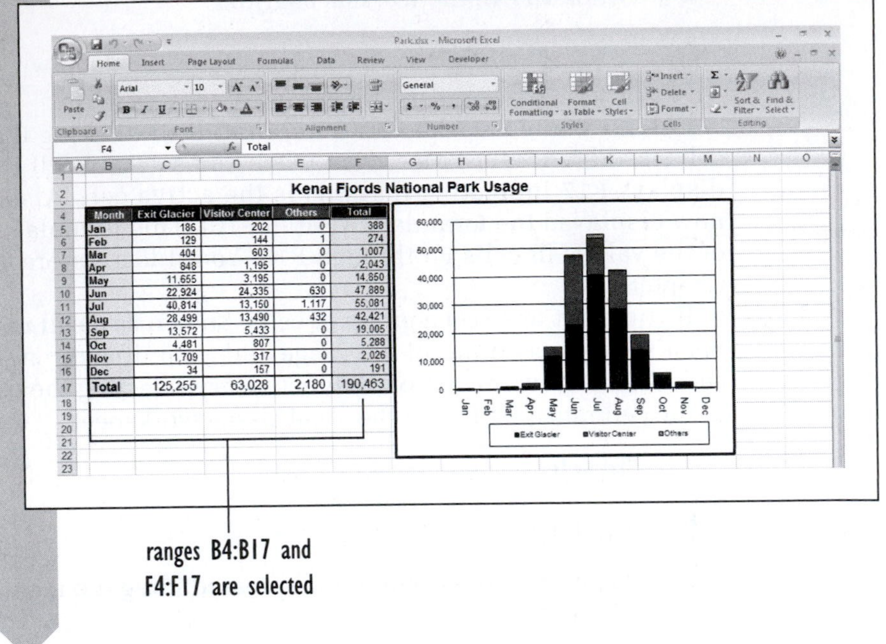

Month	Exit Glacier	Visitor Center	Others	Total
Jan	186	202	0	388
Feb	129	144	1	274
Mar	404	603	0	1,007
Apr	848	1,195	0	2,043
May	11,655	3,195	0	14,850
Jun	22,924	24,335	630	47,889
Jul	40,814	13,150	1,117	55,081
Aug	28,499	13,490	432	42,421
Sep	13,572	5,433	0	19,005
Oct	4,481	807	0	5,288
Nov	1,709	317	0	2,026
Dec	34	157	0	191
Total	125,255	63,028	2,180	190,463

ranges B4:B17 and F4:F17 are selected

The cell reference for this group of cells is B4:B17;F4:F17, where the semicolon indicates a joining of two distinct ranges.

Moving Cells

Excel allows you to move the contents of your cells around without affecting their values. This is a great help in formatting your worksheets. To move a cell or range of cells, simply select the cells and then drag the selection to a new location. Try this now with the table of usage data from the Total Usage worksheet.

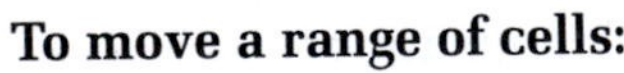

To move a range of cells:

1 Select the range **B4:F17**.

2 Move the mouse pointer to the border of the selected area so that the pointer changes from a ✥ to a ↖.

3 Drag the selected area down two cells, so that the new range is now B6:F19, and release the mouse button.

Note that as you moved the selected range, Excel displayed a screen tip with the new location of the range.

4 Click **F19** to deselect the cell range.

When you look at the formula bar for cell F19, note that the formula is now changed from =SUM(F4:F17) to =SUM(F7:F18). Excel will automatically update the cell references in your formulas to account for the fact that you moved the cell range.

You can also use the Cut, Copy, and Paste buttons to move a cell range. These buttons are essential if you want to move a cell range to a new workbook or worksheet (you can't use the drag and drop technique to perform that action). Try using the Cut and Paste method to move the table back to its original location.

To cut and paste a range of cells:

1 Select the range **B6:F19**.

2 Click the **Cut** button from the Clipboard group on the Home tab or press **CTRL+x**.

A flashing border appears around the cell range, indicating that it has been cut or copied from the worksheet.

3 Click **B4**.

4 Click the **Paste** button from the Clipboard group on the Home tab or press **CTRL+v**.

The table now appears back in the cell range, B4:F17.

5 Click cell **A1** to make A1 the active cell again.

If you want to copy a cell range rather than move it, you can use the Copy button in the above steps, or if you prefer the drag and drop technique, hold down the CTRL key while dragging the cell range to its new location; this will create a copy of the original cell range at the new location. You can refer to Excel's online Help for more information.

Printing from Excel

It would be useful for the chief of interpretation at Kenai Fjords National Park to have a hard copy of some of the worksheets and charts in the Park workbook. To do this, you can print out selected portions of the workbook.

Previewing the Print Job

Before sending a job to the printer, it's usually a good idea to preview the output. With Excel's Print Preview window, you can view your job before it's printed, as well as set up the page margins, orientation, and headers and footers. Try this now with the Total Usage worksheet.

To preview a print job:

1. Verify that Total Usage is still the active worksheet.
2. Click the **Office** button, then click **Print**, and then click **Print Preview**.

   The Print Preview opens as displayed in Figure 1-13.

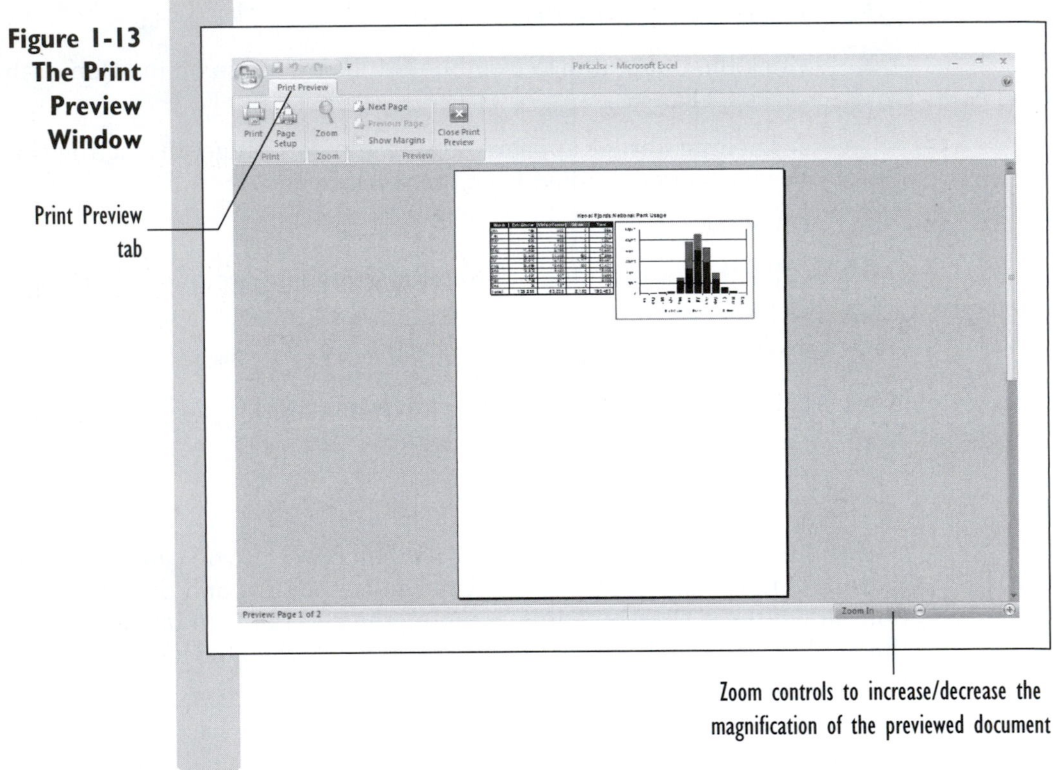

Figure 1-13 The Print Preview Window

Table 1-4 describes the variety of options available to you from the Print Preview tab in the Print Preview window.

Table 1-4 Print Preview Options

Button	Description
Print	Send the document to the printer
Page Setup	Set up the properties of the printed page
Zoom	Zoom in and out of the Preview window
Next Page	View the next page in the print job
Previous Page	View the previous page in the print job
Show Margins	Display margins in the Preview window
Close Print Preview	Close the Print Preview window

Setting Up the Page

The Preview window opens with the default print settings for the workbook. You can change these settings for each print job. You may add a header or footer to each page, change the orientation from portrait to landscape, and modify many other features. To see how this works, adjust the settings for the current print job by adding a header and changing the page layout.

To add a header to a print job:

1. Click the **Page Setup** button from the Print Preview tab.
2. Click the **Header/Footer** dialog sheet tab.
3. Excel provides a list of built-in headers that you can select from the Header drop-down list. You can also write your own; you'll do this now.
4. Click the **Custom Header** button.
5. Type **Yearly Usage Report** in the Center section of the Header dialog box as shown in Figure 1-14.

Figure 1-14
Adding a header to the printed page

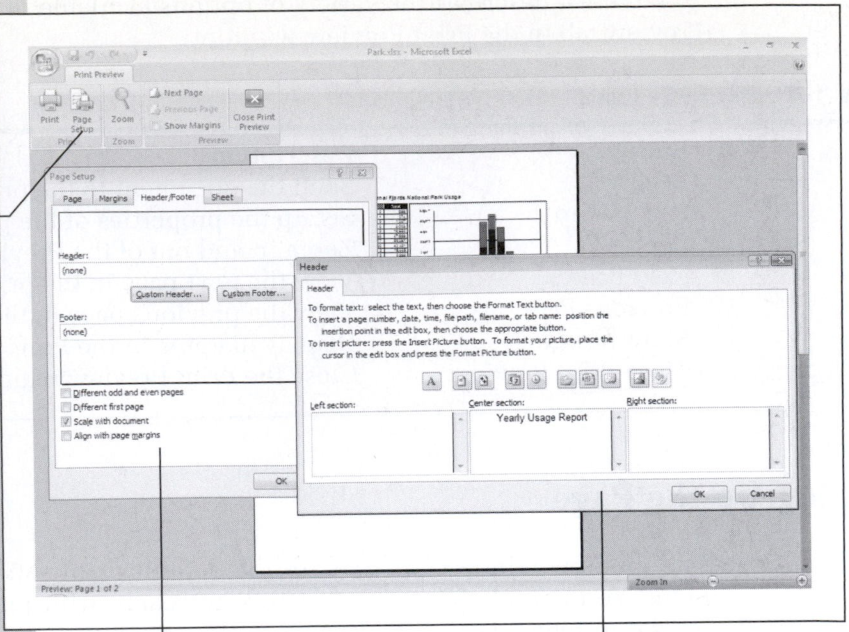

6 Click the **OK** button.

Because the print job is more horizontal than vertical, it would be a good idea to change the orientation from portrait to landscape.

To change the page orientation:

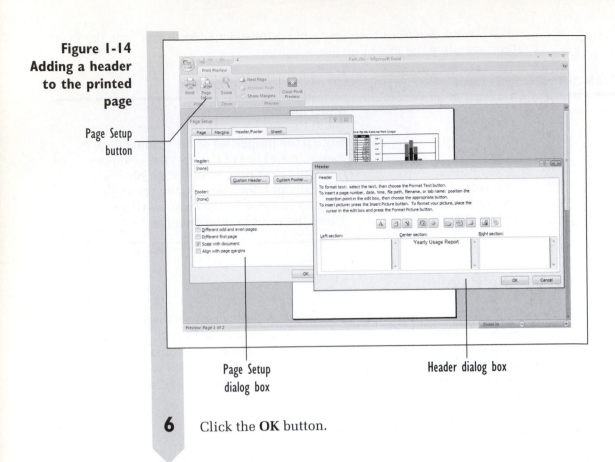

1 Click the **Page** dialog sheet tab within the Page Setup dialog box.

2 Click the **Landscape** option button.

3 Click the **OK** button.

Figure 1-15 shows the new layout of the print job with a header and landscape orientation.

Figure 1-15 Landscape orientation of the printed page

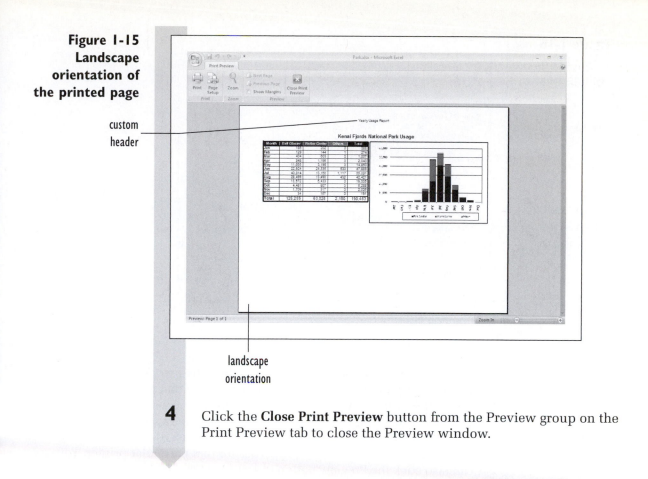

custom header

landscape orientation

4 Click the **Close Print Preview** button from the Preview group on the Print Preview tab to close the Preview window.

There are many other printing features available to you in Excel. Check the online Help for more information.

Printing the Page

To print your worksheet, you can select the Print command from the Office menu. Try printing the Total Usage worksheet now.

To print the Total Usage worksheet:

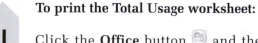

1 Click the **Office** button  and then click **Print** from the Office menu.

The Print Dialog box appears. See Figure 1-16.

Figure 1-16 Print dialog box

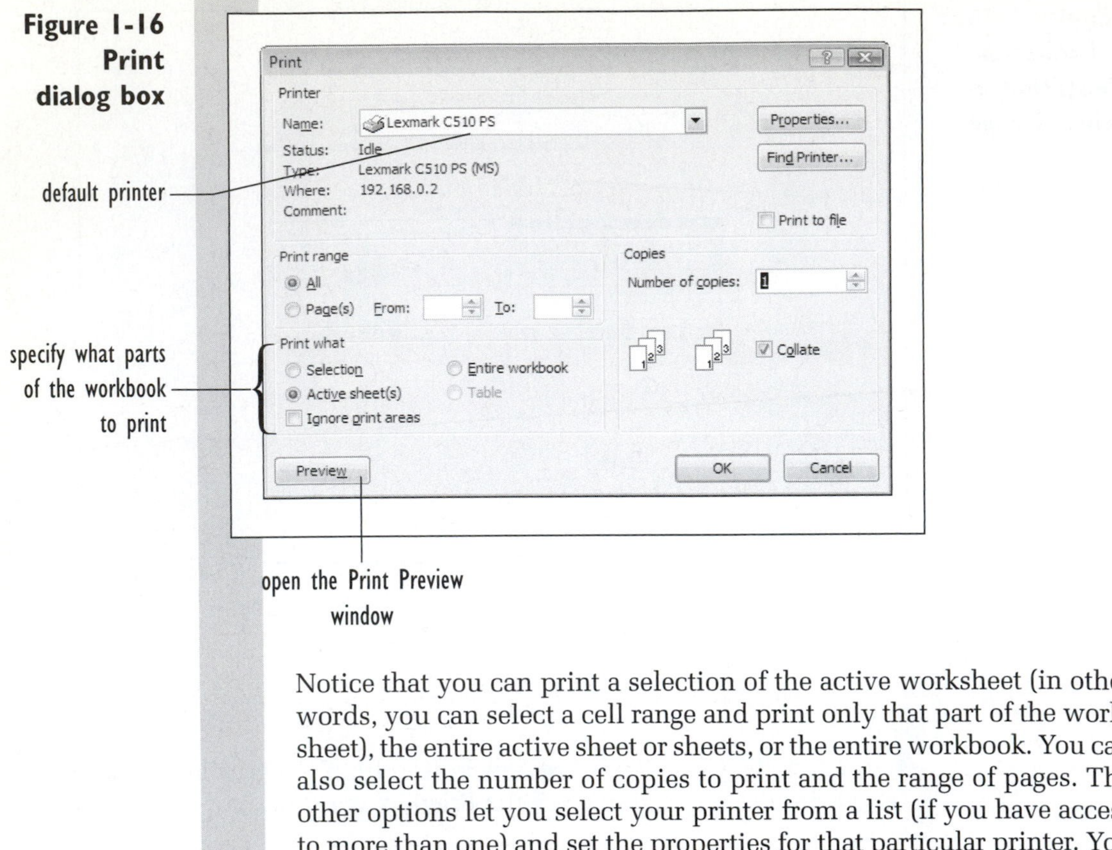

Notice that you can print a selection of the active worksheet (in other words, you can select a cell range and print only that part of the worksheet), the entire active sheet or sheets, or the entire workbook. You can also select the number of copies to print and the range of pages. The other options let you select your printer from a list (if you have access to more than one) and set the properties for that particular printer. You can also click the Preview button to go to the Print Preview window.

2 Click **OK** to start the print job.

Your printer should soon start printing the Total Usage worksheet.

If you were to hand this printout to the chief of interpretation of the park, he or she would be able to use the information contained in it to determine when to hire extra help at the various stations in the park.

Saving Your Work

You should periodically save your work when you make changes to a workbook or when you are entering a lot of data so that you won't lose much work if your computer or Excel crashes. Excel offers two options for saving your work: the Save command, which saves the file; and the Save As command, which allows you to save the file under a new name.

So that you do not change the original files (and can go through the chapters again with unchanged files if necessary), you'll be instructed throughout this book to save your work under new file names. To save the changes you made to the Park workbook, save the file as Park Usage Report. If using your own computer, you can save the workbook to your hard drive. If you are using a computer on the school network, you may be asked to save your work to your own floppy disk. This book assumes that you'll save your work to the same folder containing the original data workbook.

To save the Park workbook as Park Usage Report:

1. Click the **Office** button and then click **Save As** from the Office menu to open the Save As dialog box.
2. Navigate to and select the folder in which you want to save the file, or save the file in the same folder as the Park workbook.
3. Type **Park Usage Report** in the File name box. See Figure 1-17.

Figure 1-17 Save As dialog box

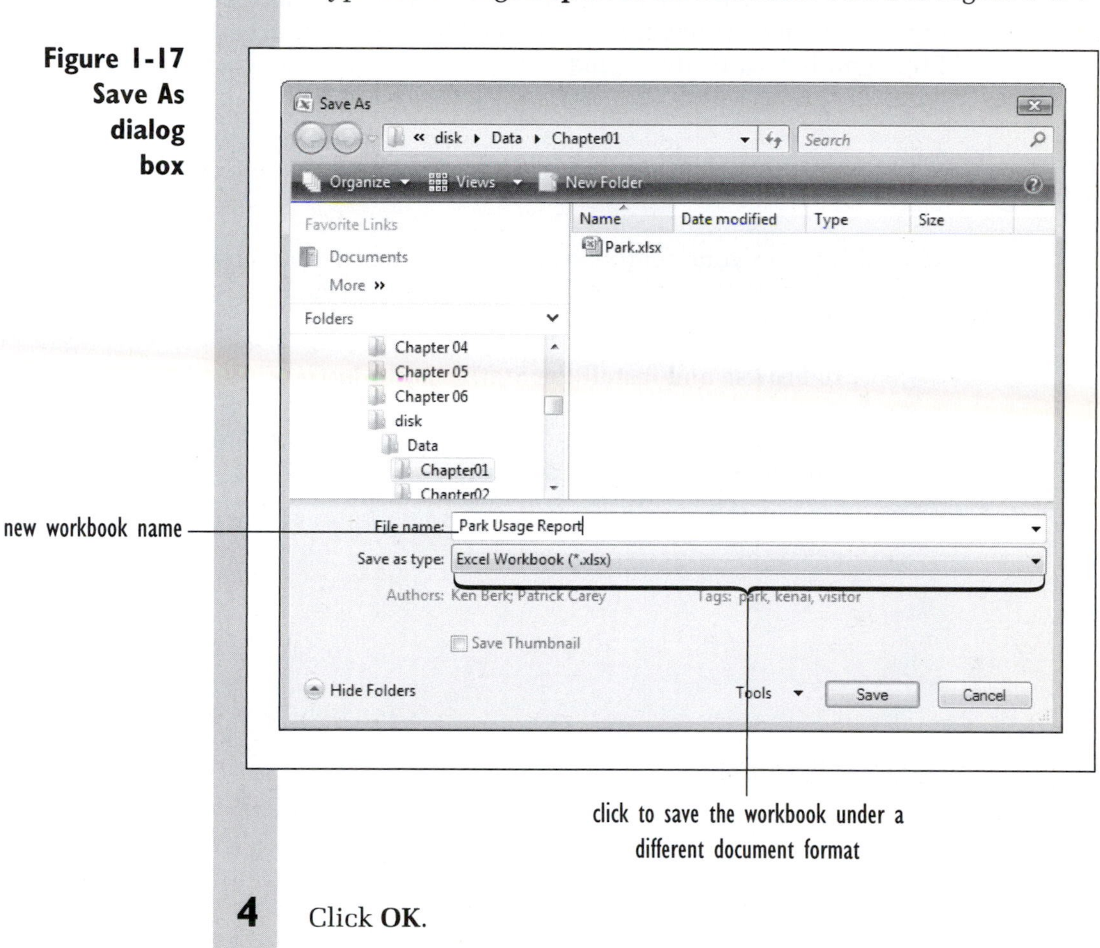

4. Click **OK**.

Excel then saves the workbook under the name Park Usage Report. Note that if you can save the workbook under a variety of formats by clicking the Save as type list box and choosing a file type.

Excel Add-Ins

Excel's capabilities can be expanded through the use of special programs called **add-ins**. These add-ins tie into Excel's special features, almost looking like a part of Excel itself. Various add-ins that allow you to easily generate reports, explore multiple scenarios, or access databases are supplied with Excel. To use these add-ins, you have to go through a process of saving the add-in files to a location on your computer, and then telling Excel where to find the add-in file.

Excel comes with an add-in called **Analysis ToolPak** that provides some of the statistical commands you'll need for this book. Another add-in, StatPlus, you have already copied to your hard disk. Now you will install the add-in in Excel.

Loading the StatPlus Add-In

The add-ins on your computer are stored in a list in Excel. From this list, you can activate the add-in or browse for new ones. First you'll browse for the StatPlus add-in.

To browse and install the StatPlus add-in:

1. Click the **Office** button and then click **Excel Options** located at the bottom of the pop-up menu.
2. Click **Add-Ins** from the list of Excel options as shown in Figure 1-18.

Figure 1-18 Excel Options dialog box

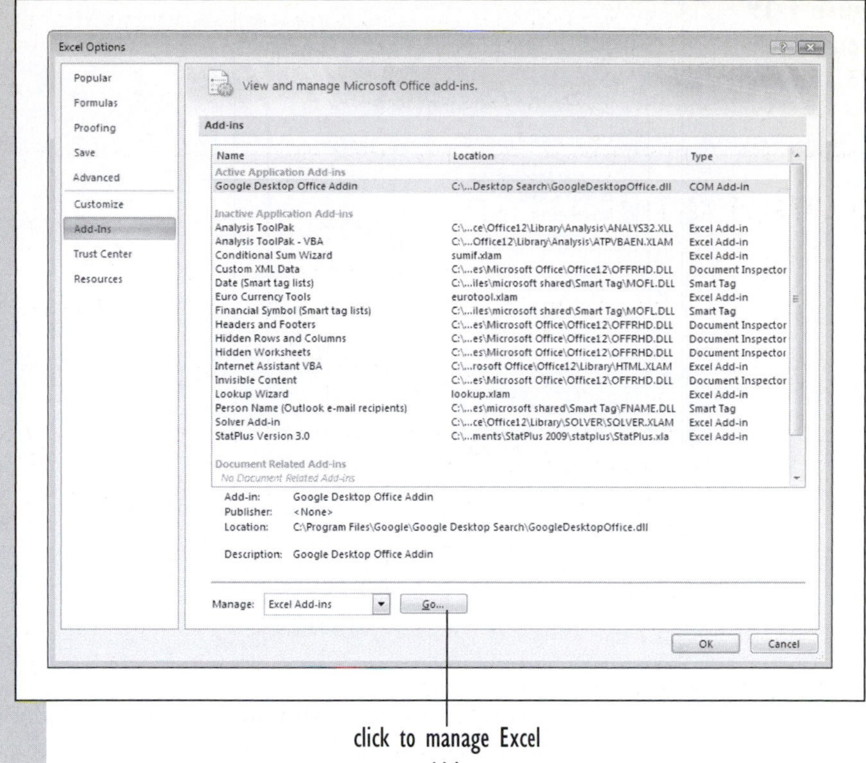

3 Click the **Manage** list box at the bottom of the window, select **Excel Add-Ins** and then click the **Go** button.

The Add-Ins dialog box opens as shown in Figure 1-19.

Figure 1-19
List of currently available add-ins

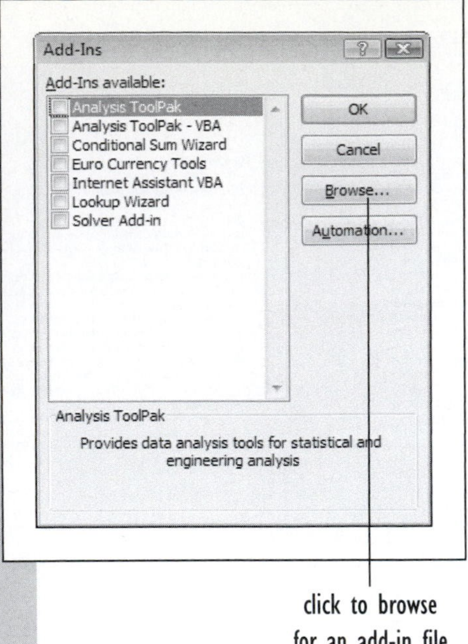

Each available add-in is shown in Figure 1-19 along with a checkbox indicating whether that add-in is currently loaded in Excel.

4 Click the **Browse** button.

5 Locate the installation folder on your hard drive where you placed the StatPlus files, and open the folder.

6 Open the **Addins** subfolder.

7 Click **StatPlus.xla** and click **OK**.

StatPlus Version 3.0 now appears in the Add-Ins dialog box. If it is not checked, click the checkbox. See Figure 1-20.

Figure 1-20
The StatPlus
add-in

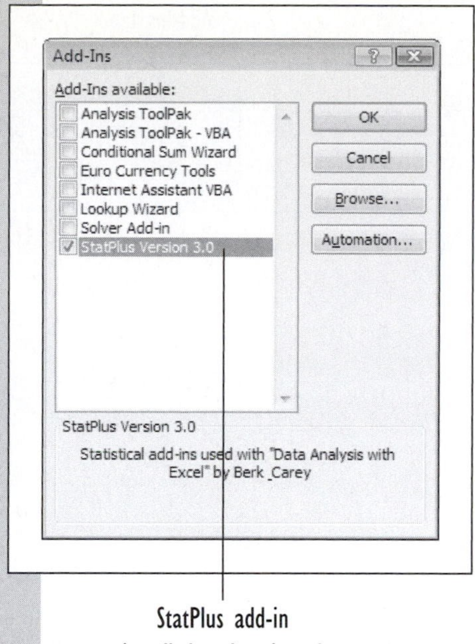

StatPlus add-in
installed and activated

8 Click the **OK** button.

After clicking the OK button, the Add-Ins dialog box closes and a new tab named Add-Ins should be added to the Excel ribbon.

9 Click the **Add-Ins** tab on the Excel ribbon and then click **StatPlus** from the Menu Commands group on the tab.

The menu commands offered by StatPlus are shown in Figure 1-21. You'll have a chance to work with these commands later in the book.

Figure 1-21
The StatPlus menu

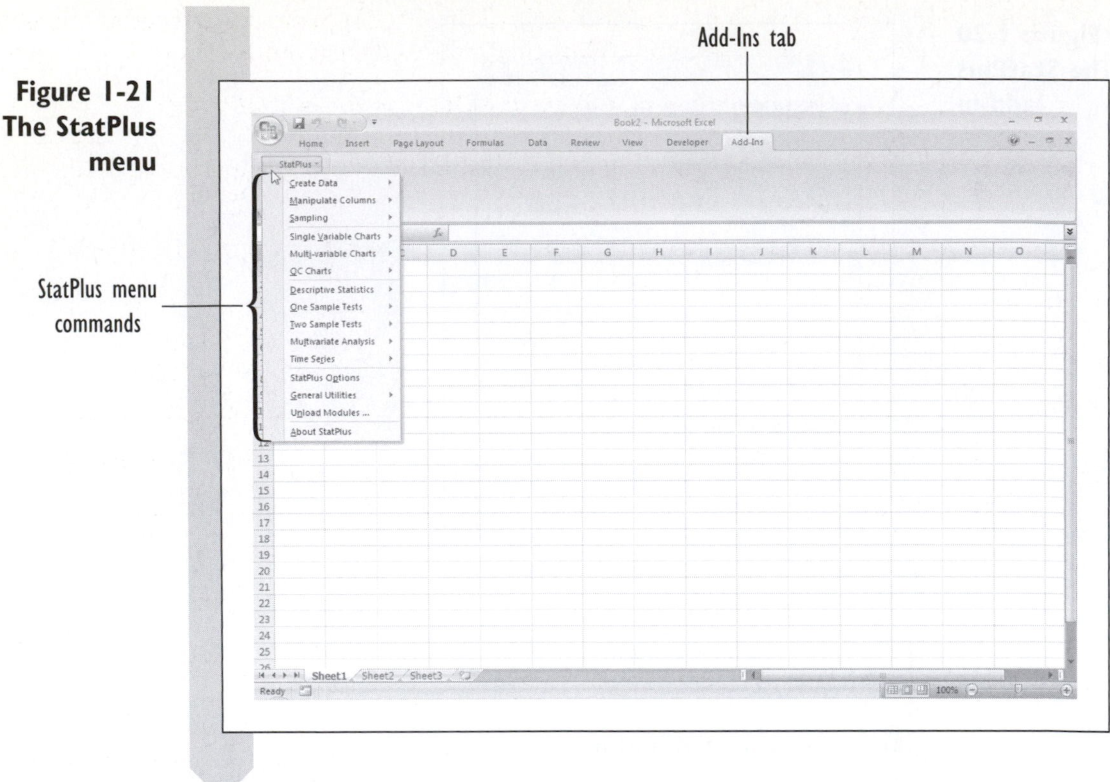

Loading the Data Analysis ToolPak

Now that you've seen how to load the StatPlus add-in, you can load the Data Analysis ToolPak. Since the Data Analysis ToolPak comes with Excel, you may need to have your Excel or Office installation disks handy.

To load the Data Analysis ToolPak:

1. Click the **Office** button and then click **Excel Options**. Click **Add-Ins** from the Excel options dialog box and then click **Go** next to the Manage Excel Add-Ins list box.
2. Click the checkbox for the **Analysis ToolPak** and click **OK**.
3. At this point, Excel may prompt you for the installation CD; if so, insert the CD into your CD-ROM drive and follow the installation instructions.

When activated, the Data Analysis ToolPak appears in a new group named Analysis on the Data tab. View the Data Analysis ToolPak now.

To access commands for the Data Analysis ToolPak:

1 Click the **Data** tab and then click **Data Analysis** from the Analysis group.

The Data Analysis dialog box appears as shown in Figure 1-22.

Figure 1-22 Viewing the Data Analysis ToolPak

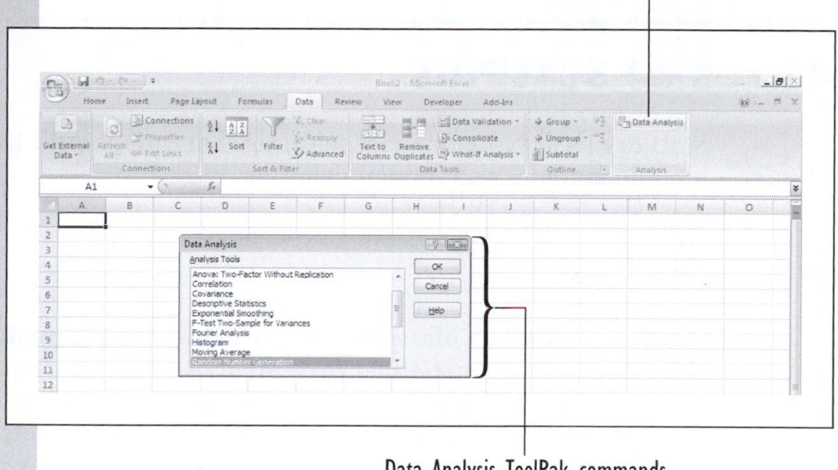

The list of commands available with the Data Analysis ToolPak is shown in the Analysis Tools list box. To run one of these commands, select it from the list box and then click the OK button. You'll have an opportunity to use some of the Data Analysis ToolPak's commands later in this book.

2 Click **Cancel** to close the Data Analysis dialog box.

Unloading an Add-In

If at any time, you want to unload the Data Analysis ToolPak or StatPlus, you can do so by returning to the list of available add-ins dialog box shown in Figure 1-19, and then deselecting the checkbox for the specific add-in. Unloading an add-in is like closing a workbook; it does not affect the add-in file. If you want to use the add-in again, simply reopen the Add-Ins dialog box and reselect the checkbox. If you exit Excel with an add-in loaded, Excel will assume that you want to run the add-in the next time you run Excel, so it will load it for you automatically.

Features of StatPlus

StatPlus has several special features that you should be aware of. These include modules and hidden data.

Using StatPlus Modules

StatPlus is made up of a series of add-in files, called **modules**. Each module handles a specific statistical task, such as creating a quality control chart or selecting a random sample of data. StatPlus will load the modules you need on demand (this way, you do not have to use up more system memory than needed). After using StatPlus for a while, you may have a great many modules loaded. If you want to reduce this number, you can view the list of currently opened modules and unload those you're no longer using.

To view a list of StatPlus modules:

1 Click **Unload Modules** from the StatPlus menu on the Add-Ins tab.

StatPlus displays a list of loaded modules. A sample list is shown in Figure 1-23. Yours will be different.

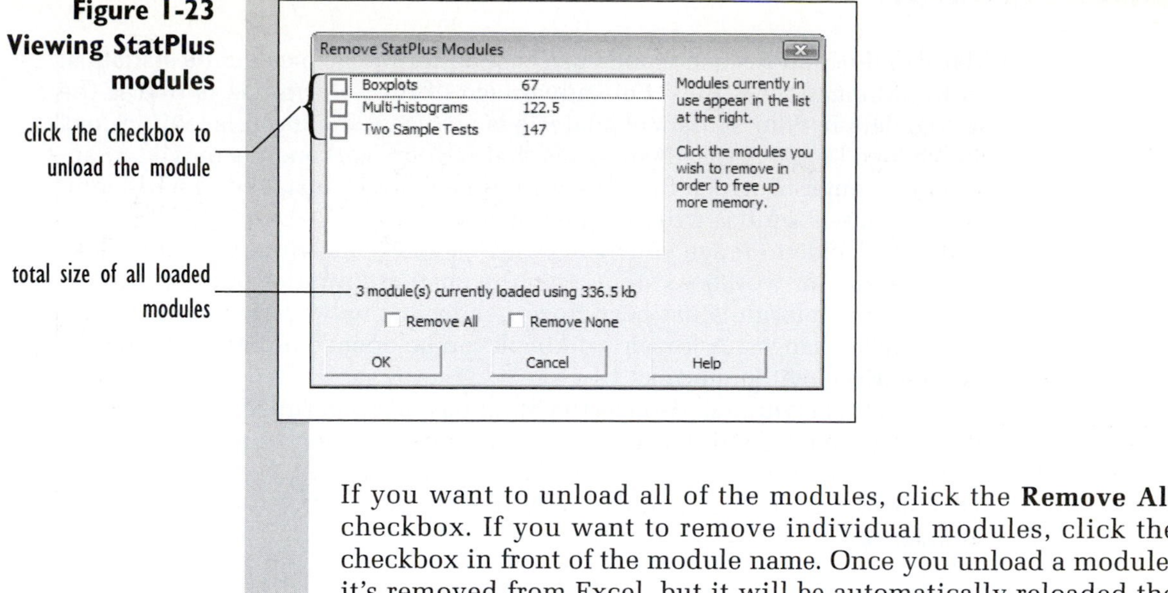

Figure 1-23 Viewing StatPlus modules

If you want to unload all of the modules, click the **Remove All** checkbox. If you want to remove individual modules, click the checkbox in front of the module name. Once you unload a module, it's removed from Excel, but it will be automatically reloaded the next time you try to use a command supported by the module.

2 Click **OK** to close the Remove StatPlus Modules dialog box.

Hidden Data

Several StatPlus commands employ hidden worksheets. A **hidden worksheet** is a worksheet in your workbook that is hidden from view. Hidden worksheets are used in creating histograms, boxplots, and normal probability plots (don't worry, you'll learn about these topics in later chapters). You can view these hidden worksheets if you need to troubleshoot a problem with one of these charts. There are three hidden worksheet commands in StatPlus (see Table 1-5; they are available in the General Utilities submenu).

Table 1-5 Hidden Worksheet Commands

Command	Description
View hidden data	Unhides a hidden StatPlus worksheet
Rehide hidden data	Rehides a StatPlus worksheet
Remove unlinked hidden data	Removes extraneous data, like hidden data for a deleted chart, from the hidden worksheet

Linked Formulas

Many of the StatPlus commands use custom formulas to calculate statistical and mathematical values. One advantage of these formulas is that if the source data in your statistical analysis is changed, the formulas will reflect the changed data. A disadvantage is that if your workbook is moved to another computer in which StatPlus is not installed (or installed in a different folder), those formulas will no longer work.

If you decide to move your workbook to a new location, you can freeze the data in your workbook, removing the custom formulas but keeping the values. Once a formula has been frozen, the value will not be updated if the source data changes. A frozen workbook can be opened on other computers running Excel without error.

If the other computer also has the StatPlus add-in installed but in a different folder, you can still use the custom formulas by pointing the workbook to the new location of the StatPlus add-in file. Table 1-6 describes the various linked formula commands (available in the StatPlus General Utilities submenu).

Table 1-6 Linked Formula Commands

Command	Description
Resolve StatPlus links	Find the location of the StatPlus add-in on the current computer and attach custom formulas to the new location
Freeze data in worksheet	Freeze all data on the current worksheet
Freeze hidden data	Freeze all data on hidden worksheets
Freeze data in workbook	Freeze all data in the current workbook

Setup Options

If you want to control how StatPlus operates in Excel, you can open the StatPlus Options dialog box from the StatPlus menu. The dialog box, shown in Figure 1-24 is divided into the four dialog sheets: Input, Output, Charts, Hidden Data.

Figure 1-24 StatPlus Options dialog box

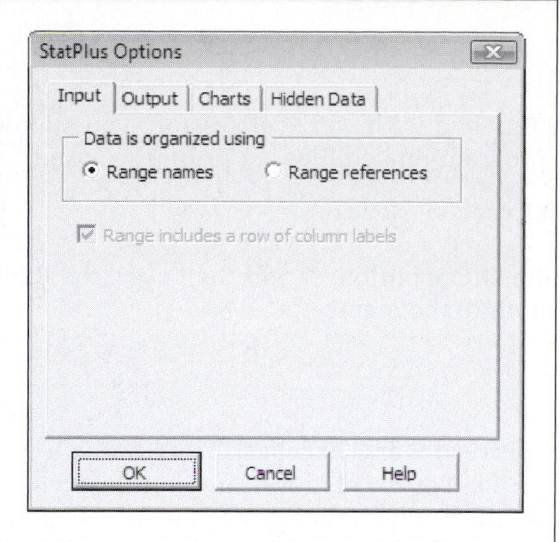

The **Input sheet** allows you to specify the default method used for referencing the data in your workbook. The two options are (1) using range names and (2) using range references. You'll learn about range names and range references in Chapter 2.

The **Output sheet** allows you to specify the default format for your output. You can choose between creating dynamic and static output. Dynamic output uses custom formulas which you'll have to adjust if you want to move your workbook to a new computer. Static output only displays the output values and does not update if the input data are changed. You can also choose the default location for your output, from among (1) a cell on the current worksheet, (2) a new worksheet in the current workbook, or (3) a new workbook.

The **Charts sheet** allows you to specify the default format for chart output. You can choose between creating charts as embedded objects in worksheets or as separate chart sheets. This will be discussed in Chapter 3.

The **Hidden Data** sheet allows you to specify whether to hide worksheets used for the background calculations involved in creating charts and statistical calculations.

All of the options specified in the StatPlus Options dialog box are default options. You can override any of these options in a specific dialog box as you perform your analysis.

You can learn more about StatPlus and its features by viewing the online Help file. Help buttons are included in every dialog box. You can also open the Help file by clicking **About StatPlus** from the StatPlus menu.

Exiting Excel

When you are finished with an Excel session, you should exit the program so that all the program-related files are properly closed.

To exit Excel:

Click the **Office** button and then click the **Exit Excel** button from the bottom of the menu.

If you have unsaved work, Excel asks whether you want to save it before exiting. If you click No, Excel closes and you lose your work. If you click Yes, Excel opens the Save As dialog box and allows you to save your work. Once you have closed Excel, you are returned to the Windows desktop or to another active application.

Chapter 2

WORKING WITH DATA

Objectives

In this chapter you will learn to:

- Enter data into Excel from the keyboard
- Work with Excel formulas and functions
- Work with cell references and range names
- Query and sort data using the AutoFilter and Advanced Filter
- Import data from text files and databases

In this chapter you'll learn how to enter data in Excel through the keyboard and by importing data from text files and databases. You'll learn how to create Excel formulas and functions to perform simple calculations.

You'll be introduced to cell references and learn how to refer to cell ranges using range names. Finally, you'll learn how to examine your data through the use of queries and sorting.

Data Entry

One of the many uses of Excel is to facilitate data entry. Error-free data entry is essential to accurate data analysis. Excel provides several methods for entering your data. Data sets can be entered manually from the keyboard or retrieved from a text file, database, or online Web sources. You can also have Excel automatically enter patterns of data for you, saving you the trouble of creating these data values yourself. You'll study all of these techniques in this chapter, but first you'll work on entering data from the keyboard.

Entering Data from the Keyboard

Table 2-1 displays average daily gasoline sales and other (nongasoline) sales for each of nine service station/convenience franchises in a store chain in a western city. There are three columns in this data set: Station, Gas, and Other. The Station column contains an id number for each of the nine stations.

The Gas column displays the gasoline sales for each station. The Other column displays sales for nongasoline items.

Table 2-1 Service Station Sales

Station	Gas	Other
1	$8,415	$7,211
2	$8,499	$7,500
3	$8,831	$7,899
4	$8,587	$7,488
5	$8,719	$7,111
6	$8,001	$6,281
7	$9,567	$13,712
8	$9,218	$12,056
9	$8,215	$7,508

To practice entering data, you'll insert this information into a blank worksheet. To enter data, you first select the cell corresponding to the upper left corner of the table, making it the active cell. You then type the value or text you want placed in the cell. To move between cells, you can either press the Tab key to move to the next column in the same row or press the Enter key to move to the next row in the same column. If you are entering data into several columns, the Enter key will move you to the next row in the first column of the data set.

To enter the first row of the service station data set:

1 Launch Excel as described in Chapter 1.

Excel shows an empty workbook with the name Book1 in the title bar.

2 Click cell **A1** to make it the active cell.

3 Type **Station** and then press **Tab**.

4 Type **Gas** in cell B1 and press **Tab**.

5 Type **Other** in cell C1 and press **Enter**.

Excel moves you to cell A2, making it the active cell.

6 Using the same technique, type the next two rows of the table, so that data for the first two stations are displayed. Your worksheet should appear as in Figure 2-1.

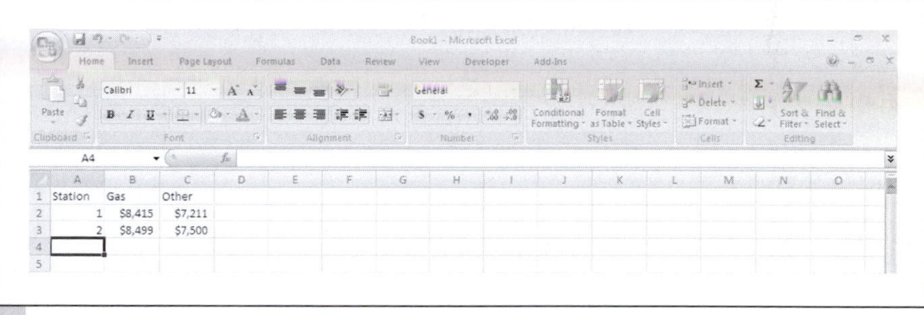

Figure 2-1 The first rows of the service station data set

Entering Data with Autofill

If you're inserting a column or row of values that follow some sequential pattern, you can save yourself time by using Excel's Autofill feature. The **Autofill** feature allows you to fill up a range of values with a series of numbers or dates. You can use Autofill to generate automatically columns containing data values such as:

1, 2, 3, 4, . . . 9, 10;
1, 2, 4, 8, . . . 128, 256;
Jan, Feb, Mar, Apr, . . . , Nov, Dec;

and so forth.

In the service station data, you have a sequence of numbers, 1–9, that represent the service stations. You could enter the values by hand, but this is also an opportunity to use the Autofill feature.

To use Autofill to fill in the rest of the service station numbers:

1. Select the range **A2:A3**.

   Notice the small black box at the lower right corner of the double border around the selected range. This is called a **fill handle**. To create a simple sequence of numbers, you'll drag this fill handle over a selected range of cells.

2. Move the mouse pointer over the fill handle until the pointer changes from a ✜ to a +. Click and hold down the mouse button.

3. Drag the fill handle down to cell **A10** and release the mouse button.

   Note that as you drag the fill handle down, a screen is displayed showing the value that will be placed in the active cell if you release the mouse button at that point.

4. Figure 2-2 shows the service station numbers placed in the cell range A2:A10.

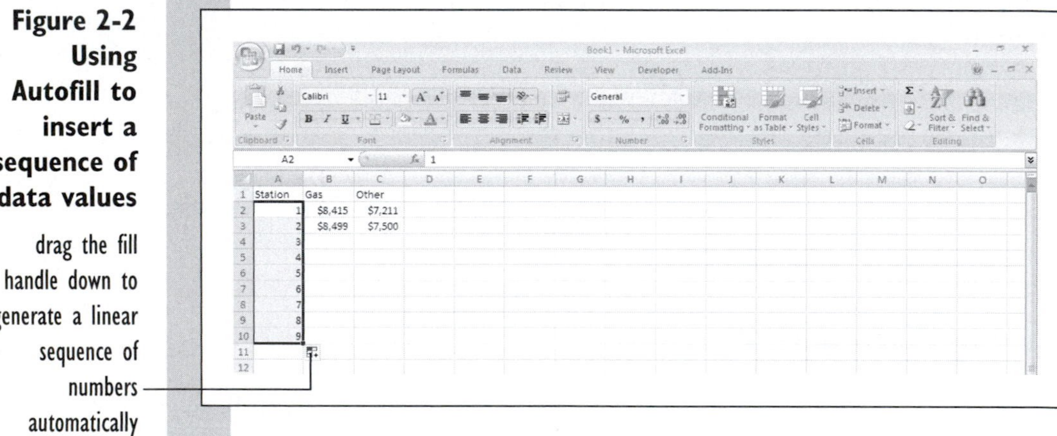

Figure 2-2 Using Autofill to insert a sequence of data values

drag the fill handle down to generate a linear sequence of numbers automatically

- If you want to create a geometric sequence of numbers, drag the fill handle with your right mouse button and then select Growth Trend from the pop-up menu.
- If you want to create a customized sequence of numbers or dates, drag the fill handle with your right mouse button and select Series… from the pop-up menu. Fill in details about your customized series in the Series dialog box.

With the service station numbers entered, you can add the rest of the sales figures to complete the data set.

To finish entering data:

1. Select the range **B4:C10**.
2. With B4 the active cell, start typing in the remaining values, using Table 2-1 as your guide.

   Note that when you're entering data into a selected range, pressing the Tab key at the end of the range moves you to the next row.
3. Click cell **A1** to remove the selection.

   The completed worksheet should appear as shown in Figure 2-3.

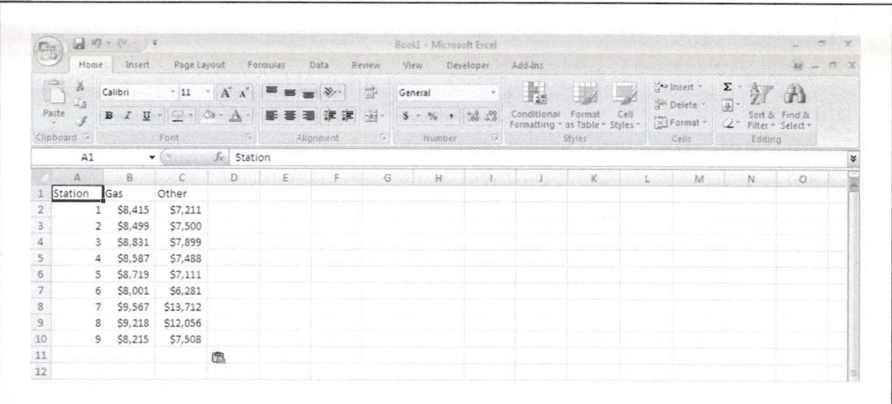

Figure 2-3 The completed service station data

Inserting New Data

Sometimes you will want to add new data to your data set. For example, you discover that there is a tenth service station with the following sales data:

Table 2-2 Additional Service Station Sales

Station	Gas	Other
0	$8,995	$6,938

You could simply append this information to the table you've already created, covering the cell range A11:C11. On the other hand, in order to maintain the sequential order of the station numbers, it might be better to place this information in the range A2:C2 and then have the other stations shifted down in the worksheet. You can accomplish this using Excel's Insert command.

To insert new data into your worksheet:

1 Select the cell range **A2:C2**.

2 Right-click the selected range and then click **Insert** from the pop-up menu. See Figure 2-4.

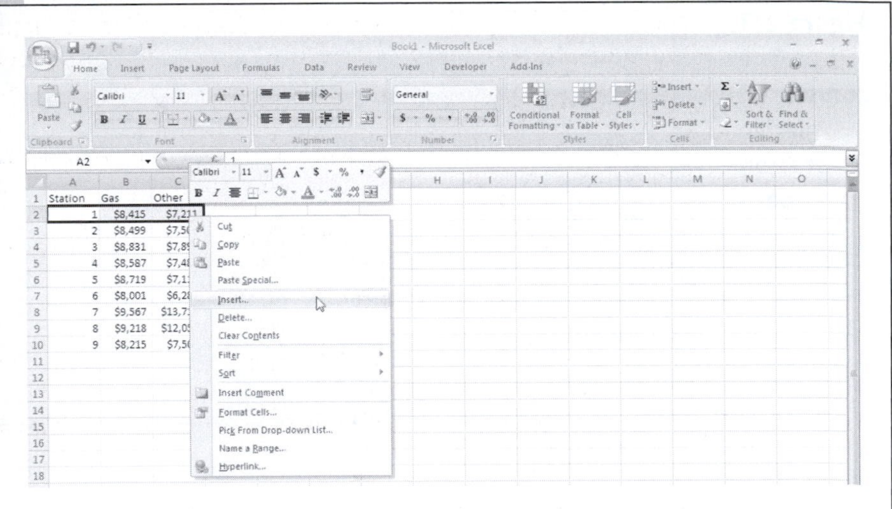

Figure 2-4 Running the Insert command from the shortcut menu

3 Verify that the **Shift cells down** option button is selected.

4 Click **OK**.

Excel shifts the values in cells A2:C10 down to A3:C11 and inserts a new blank row in the range **A2:C2**.

5 Enter the data for Station 0 from Table 2-2 in the cell range A2:C2.

6 Click **A1** to make it the active cell.

Data Formats

Now that you've entered your first data set, you're ready to work with data formats. **Data formats** are the fonts and styles that Excel applies to your data's appearance. Formats are applied to either text or numbers. Excel has already applied a currency format to the sales data you've entered. For example, if you click cell B2, note that the value in the formula bar is 8995, but the value displayed in the cell is $8,995. The extra dollar sign and comma separator are aspects of the currency format. You can modify this format if you wish by inserting additional digits to the value shown in the cell (for example, $8,995.00). You may do this if you want dollars and cents displayed to the user.

For the text displayed in the range A1:C1, Excel has applied a very basic format. The text is left justified within its cell and displayed in 11-point Calibri font (depending on how Excel has been configured on your system, a different font or font size may be used). You can modify this format as well.

Try it now, applying a boldface font to the column titles in A1:C1. In addition, center each column title within its cell.

To apply a boldface font and center the column titles:

1 Select the range **A1:C1**.

2 Click the **Bold** button B from the Font group on the Home tab.

3 Click the **Center** button from the Alignment group on the Home tab.

4 Click cell **A1** to remove the selection.

Your data set should look like Figure 2-5.

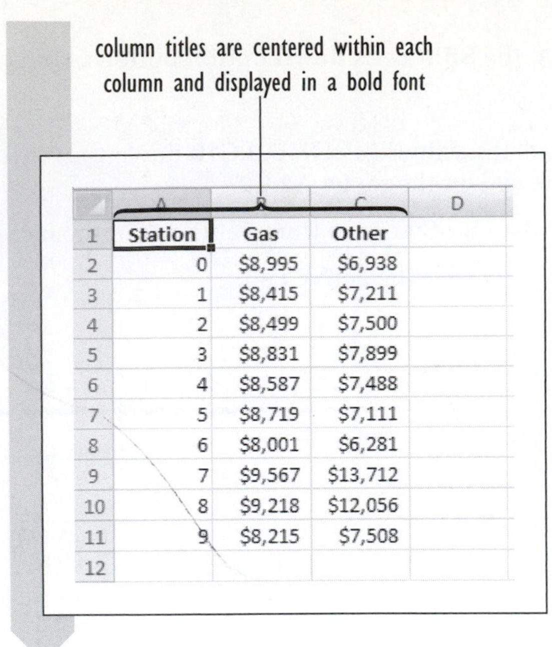

Figure 2-5 Applying a boldface font and centering the column titles

The Bold and Center buttons on the Home tab give you one-click access to two of Excel's popular formatting commands. Other format buttons are shown in Table 2-3.

Table 2-3 Data Format Buttons

Button	Icon	Purpose
Font	Calibri	Apply the font type.
Font Size	11	Change the size of the font (in points).
Bold	B	Apply a boldface font.
Italic	I	Apply an italic font.
Underline	U	Underline the selected text.
Align Left		Left-justify the text.
Align Center		Center the text.
Align Right		Right-justify the text.
Percent Style	%	Display values as percents (i.e., 0.05 = 5%).
Currency Style	$	Display values as currency (i.e., 5.25 = $5.25).

(continued)

Comma Style	,	Add comma separators to values (i.e., 43215 = 43,215).
Increase Decimal		Increase the number of decimal points (i.e., 4.3 = 4.300).
Decrease Decimal		Decrease the number of decimal points (i.e., 4.321 = 4.3).
Fill Color		Change the cell's background color.
Font Color		Change the color of the selected text.
Merge and Center		Merge the selected cells and center the text across the merged cells.

You can access all of the possible formatting options for a particular cell by opening the Format Cells dialog box. To see this feature of Excel, you'll use it to continue formatting the column titles, changing the font color to red.

To open the Format Cells dialog box:

1 Select the cell range, **A1:C1**.

2 Right-click the selection and click **Format Cells** from the shortcut menu.

The Format Cells dialog box contains six dialog sheets labeled

Number, Alignment, Font, Border, Patterns, and Protection. Each deals with a specific aspect of the cell's appearance or behavior in the workbook. You'll first change the font color to red. This option is located in the Font dialog sheet.

3 Click the **Font** tab.

4 Click the **Color** drop-down list box and click the **Red** checkbox (located as the second entry in the list of standard colors.) See Figure 2-6.

Figure 2-6
Changing the font color to red

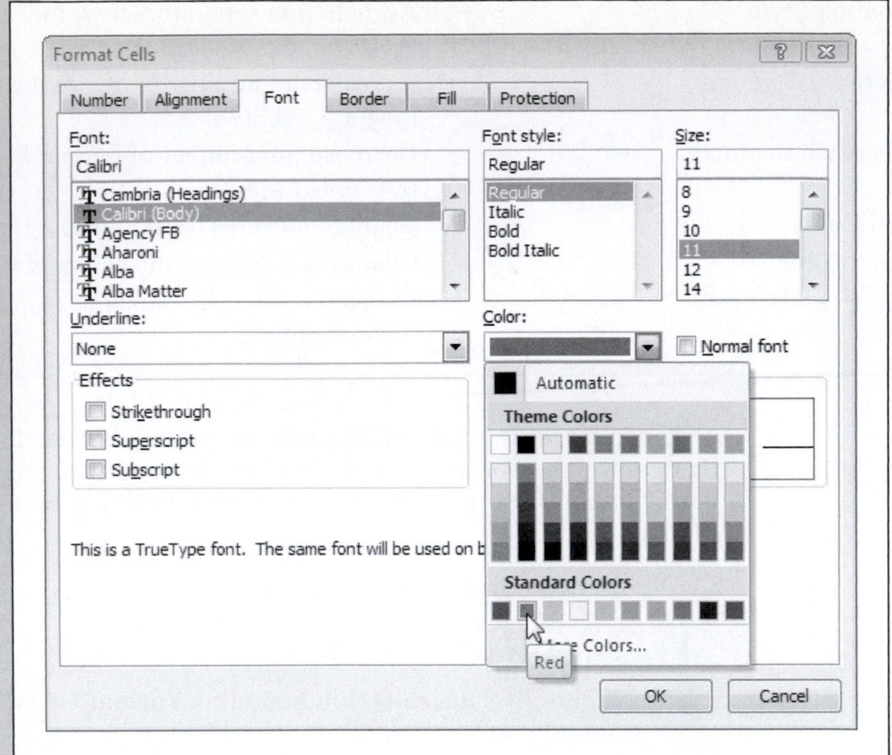

5 Click **OK** to close the Format Cells dialog box.

6 Click cell **D1** to unselect the cells.

Figure 2-7 displays the final format of the column titles.

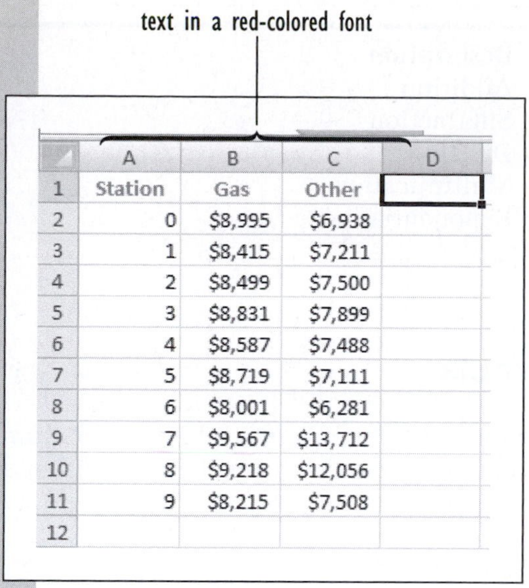

text in a red-colored font

Figure 2-7 Formatted column titles

	A	B	C	D
1	Station	Gas	Other	
2	0	$8,995	$6,938	
3	1	$8,415	$7,211	
4	2	$8,499	$7,500	
5	3	$8,831	$7,899	
6	4	$8,587	$7,488	
7	5	$8,719	$7,111	
8	6	$8,001	$6,281	
9	7	$9,567	$13,712	
10	8	$9,218	$12,056	
11	9	$8,215	$7,508	
12				

Before going further, this would be a good time to save your work.

To save your work:

1. Click the **Office** button and then click **Save**.
2. Save the workbook as **Gas Sales Data** in your student folder.

Formulas and Functions

Not all of the values displayed in a workbook come from data entry. Some values are calculated using formulas and functions. A formula always begins with an equals sign (=) followed by a function name, number, text string, or cell reference. Most functions contain mathematical operators such as + or −. A list of mathematical operators is shown in Table 2-4.

Table 2-4 Mathematical Operators

Operator	Description
+	Addition
–	Subtraction
/	Division
*	Multiplication
^	Exponentiation

Inserting a Simple Formula

To see how to enter a simple formula, add a new column to your data set displaying the total sales from both gasoline and other sources for each of the ten service stations.

To add a formula:

1. Type **Total** in cell D1 and press **Enter**.
2. Type =**b2+c2** in cell D2 and press **Enter**.

   Note that cells B2 and C2 contain the gas and other sales for Station 0.

   The value displayed in D2 is $15,933—the sum of these two values.

   At this point you could enter formulas for the remaining cells in the data set, but it's quicker to use Excel's Autofill capability to add those formulas for you.
3. Click cell **D2** to make it the active cell.
4. Click the fill handle and drag it down to cell **D11**. Release the mouse button.

   Excel automatically inserts the formulas for the cells in the range D3:D11. Thus, the formula in cell D11 is **=b11+c11**, to calculate the total sales for Station 9. Note that Excel has also applied the same currency format it used for the values in column B and C to values in column D. See Figure 2-8.

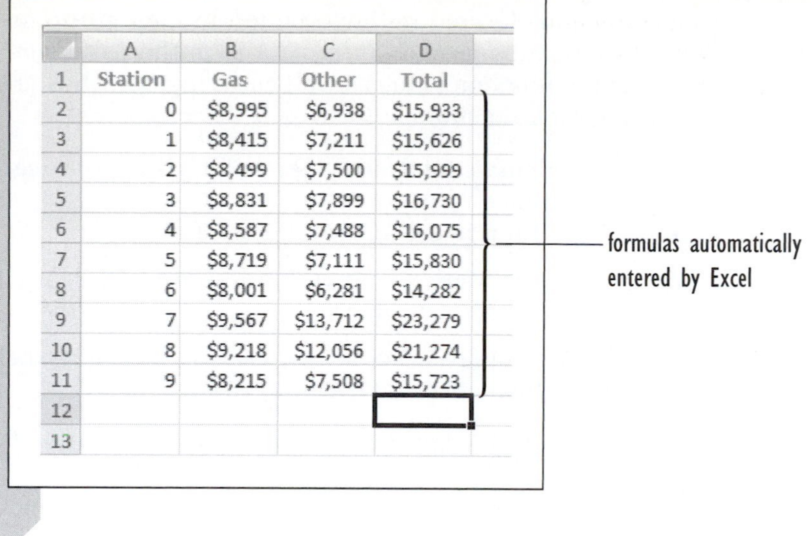

	A	B	C	D
1	Station	Gas	Other	Total
2	0	$8,995	$6,938	$15,933
3	1	$8,415	$7,211	$15,626
4	2	$8,499	$7,500	$15,999
5	3	$8,831	$7,899	$16,730
6	4	$8,587	$7,488	$16,075
7	5	$8,719	$7,111	$15,830
8	6	$8,001	$6,281	$14,282
9	7	$9,567	$13,712	$23,279
10	8	$9,218	$12,056	$21,274
11	9	$8,215	$7,508	$15,723
12				
13				

Figure 2-8 Adding new formulas with Autofill

This example illustrates a simple formula involving the addition of two numbers. What if you wanted to find the total gas and other sales for all ten of the service stations? In that case, you would be better off using one of Excel's built-in functions.

Inserting an Excel Function

Excel has a library containing hundreds of functions covering most financial, statistical, and mathematical needs. Users can also create their own custom functions using Excel's programming language. StatPlus contains its own library of functions, supplementing those offered by Excel. A list of statistics-related functions is included in the Appendix at the end of this book.

A function is composed of the **function name** and a list of **arguments**—values required by the function. For example to calculate the sum of a set of cells, you would use the SUM function. The general form or syntax of the SUM function is

= SUM(number1, number2, . . .)

where number1 and number2 are numbers or cell references. Note that the SUM function allows multiple numbers of cell references. Thus to calculate the sum of the cells in the range B2:B11, you could enter the formula

= SUM(B2:B11).

Although you can type in functions directly, you may find it easier to use the commands located in the Function Library group on the Formulas tab. These commands provide information on the parameters required for calculating the function value as well as giving one-click access to online help regarding each function.

To calculate total sales figures for all ten service stations:

1 Type **Total** in cell A13 and press **Tab**.

2 Click the **Math & Trig** button located in the Function Library group on the Formulas tab.

Excel displays a scroll box listing all of the Excel functions related to mathematics and trigonometry.

3 Scroll down the scroll box and click **SUM** from the list as shown in Figure 2-9.

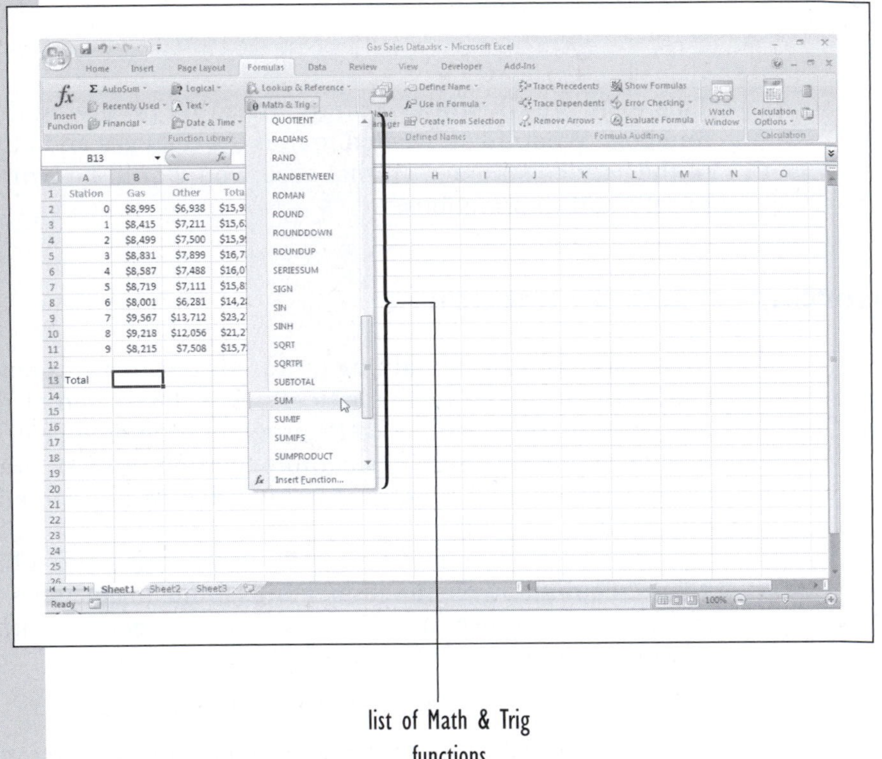

Figure 2-9 Accessing Math & Trig functions

Next, Excel displays a dialog box with the arguments for the **SUM** function. Excel has already inserted the cell reference B2:B12 for you in the first argument, but if this is not the reference you want, you can select a different one yourself. Try this now.

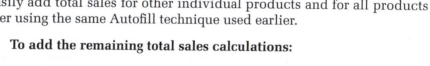

4. Click the **Collapse Dialog** button next to the number1 argument.
5. Drag your mouse pointer over the range **B2:B11**.
6. Click the **Restore Dialog** button .

   The cell range B2:B11 is entered into the number1 argument. See Figure 2-10.

Figure 2-10 Adding arguments to the SUM function

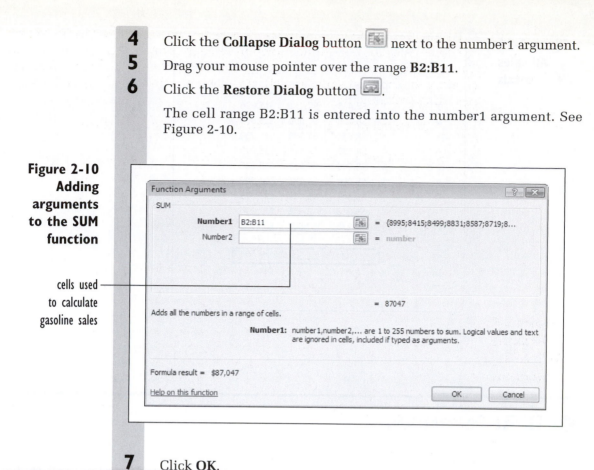

cells used to calculate gasoline sales

7. Click **OK**.

The total gasoline sales value of $87,047 is now displayed in cell B13. You can easily add total sales for other individual products and for all products together using the same Autofill technique used earlier.

To add the remaining total sales calculations:

1. Select **B13**.
2. Click the fill handle and drag it to cell **D13**.
3. Release the mouse button.

   Total sales figures are now shown in the range B13:D13. See Figure 2-11.

Figure 2-11
All sales totals

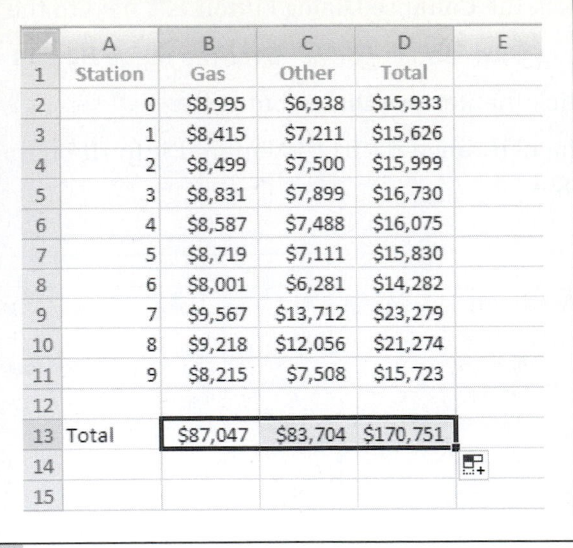

	A	B	C	D	E
1	Station	Gas	Other	Total	
2	0	$8,995	$6,938	$15,933	
3	1	$8,415	$7,211	$15,626	
4	2	$8,499	$7,500	$15,999	
5	3	$8,831	$7,899	$16,730	
6	4	$8,587	$7,488	$16,075	
7	5	$8,719	$7,111	$15,830	
8	6	$8,001	$6,281	$14,282	
9	7	$9,567	$13,712	$23,279	
10	8	$9,218	$12,056	$21,274	
11	9	$8,215	$7,508	$15,723	
12					
13	Total	$87,047	$83,704	$170,751	
14					
15					

Cell References

When Excel calculated the total sales for column C and column D on your worksheet, it inserted the following formulas into C13 and D13, respectively:

=SUM(C2:C11)

and

=SUM(D2:D11)

At this point you may wonder how Excel knew to copy everything except the cell reference from cell B13 and, in place of the original B2:B11 reference, to shift the cell reference one and two columns to the right. Excel does this automatically when you use relative references in your formulas. A **relative reference** identifies a cell range on the basis of its position relative to the cell containing the formula. One advantage of using relative references, as you've seen, is that you can fill up a row or column with a formula and the cell references in the new formulas will shift along with the cell.

Now what if you didn't want Excel to shift the cell reference when you copied the formula into other cells? What if you wanted the formula *always* to point to a specific cell in your worksheet? In that case you would need an **absolute reference**. In an absolute reference, the cell reference is prefixed with dollar signs. For example, the formula

=SUM($C$2:$C$11)

is an absolute reference to the range C2:C11. If you copied this formula into other cells, it would still point to C2:C11 and would not be shifted.

You can also create formulas that use **mixed references**, combining both absolute and relative references. For example, the formulas

=SUM($C2:$C11)

and

=SUM(C$2:C$11)

use mixed references. In the first example, the column is absolute but the row is relative, and in the second example, the column is relative but the row is absolute. This means that in the first example, Excel will shift the row references but not the column references, and in the second example, Excel will shift the column references but not the row references. You can learn more about reference types and how to use them in Excel's online Help. In most situations in this book, you'll use relative references, unless otherwise noted.

Range Names

Another way of referencing a cell in your workbook is with a range name. **Range names** are names given to specific cells or cells ranges. For example, you can define the range name Gas to refer to cells B2:B11 in your worksheet. To calculate the total gasoline sales, you could use the formula

=SUM(B2:B11)

or

=SUM(Gas).

Range names have the advantage of making your formulas easier to write and interpret. Without range names you would have to know something about the worksheet before you could determine what the formula =SUM(B2:B11) calculates.

Excel provides several tools to create range names. You'll find it easier to perform data analysis on your data set if you've defined range names for all of the columns. A simple way to create range names is to select the range of data including a row or column of titles. You can then use the titles from the worksheet to define the range name. Try it now with the service station data.

To create range names for the service station data:

1 Select the range **A1:D11**.

2 Click the **Create from Selection** button located in the Defined Names group on the Formulas tab.

Excel will create range names based on where you have entered the data labels. In this case, you'll use the labels you entered in the top row as the basis for the range names.

3 Verify that the **Top row** checkbox is selected as shown in Figure 2-12.

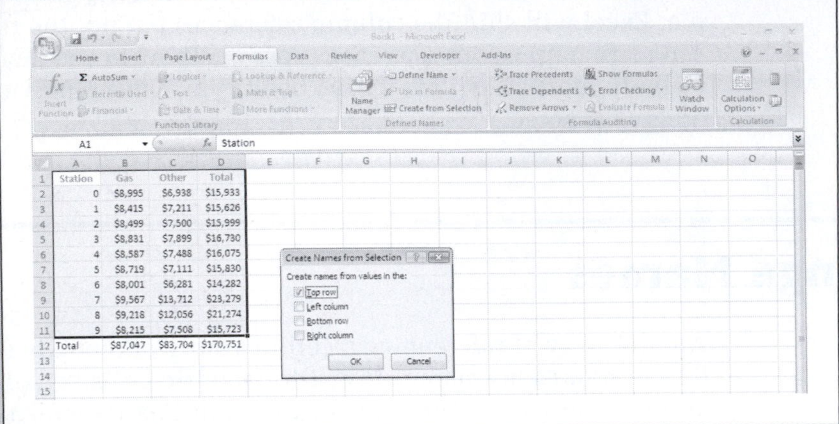

Figure 2-12 Creating range names from a selection

4 Click **OK**.

Four range names have been created for you: Station, Gas, Other, and Total. You can use Excel's Name Box to select those ranges automatically.

To select the Total range:

1 Click the **Name Box** (the drop-down list box) located directly above and to the left of the worksheet's row and column headers.

2 Click **Total** from the Name Box.

The cell range D2:D11 is automatically selected. See Figure 2-13.

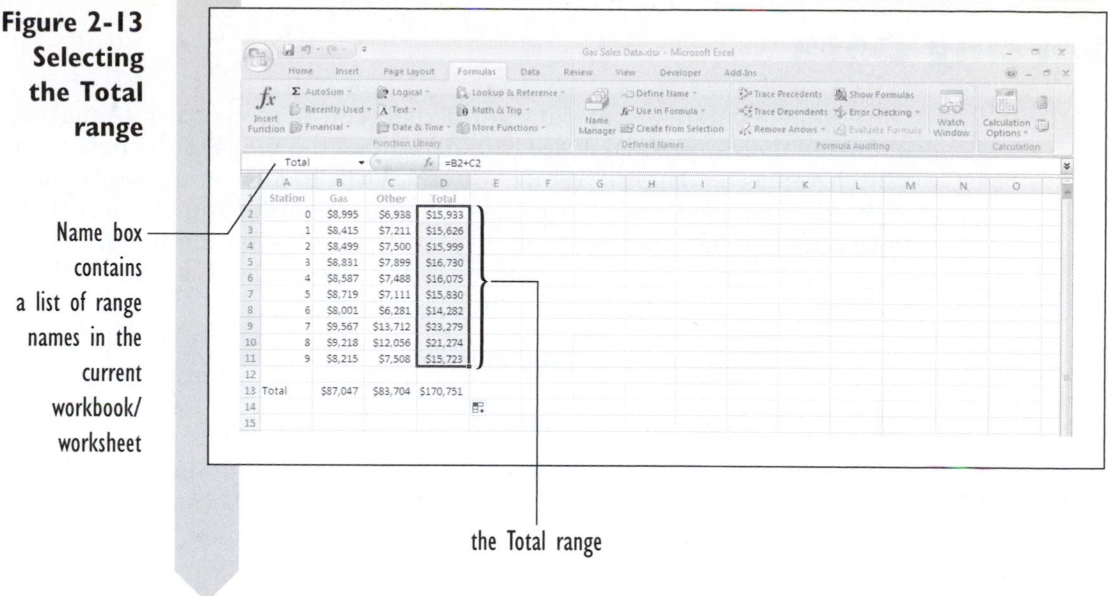

Figure 2-13 Selecting the Total range

All of the workbooks you'll use in this book will contain range names for each of their data columns.

EXCEL TIPS

- Another way to create range names is by first selecting the cell range and then typing the range name directly into the Name Box.
- You can view and organize all of the range names in the current workbook by clicking the Name Manager button located in the Defined Names group on the Formulas tab.
- Range names have a property called scope that determines where they are recognized in the workbook. Scope can be limited to the current worksheet only, allowing you to duplicate the same range name on different worksheets. If you wish to reference that a range name with a scope limited to a particular worksheet, you'll have to specify which worksheet you want to use. For example, you must use the reference 'Sheet 1'!Gas for the cell range "Gas" located on the Sheet 1 worksheet.
- You replace cell references with their range names by clicking the Define Name list button from the Defined Names group on the Formulas tab and then selecting Apply Names from the list box. You will then be prompted to apply the already-defined range names to formulas in the workbook.

Sorting Data

Once you've entered your data into Excel, you're ready to start analyzing it. One of the simplest analyses is to determine the range of the data values. Which values are largest? Which are smallest? To answer questions of this type, you can use Excel to sort the data. For example, you can sort the gas station data in descending order, displaying first the station that has shown the greatest total revenue down through the station that has had the lowest revenue. Try this now with the data you've entered.

To sort the data by Total amount:

1. Select the cell range **A1:D11**.

   The range A1:D11 contains the range you want to sort. Note that you do *not* include the cells in the range A13:D13, because these are the column totals and not individual service stations.

2. Click the **Sort & Filter** button located in the Editing group on the Home tab and then click **Custom Sort**.

   Excel opens the Sort dialog box. From this dialog box you can select multiple sorting levels. You can sort each level in an ascending or descending order.

3. Click the **Sort by** list box and select **Total** from the list range names found in the selected worksheet cells.

4. Click **Largest to Smallest** from the Order list box and shown in Figure 2-14.

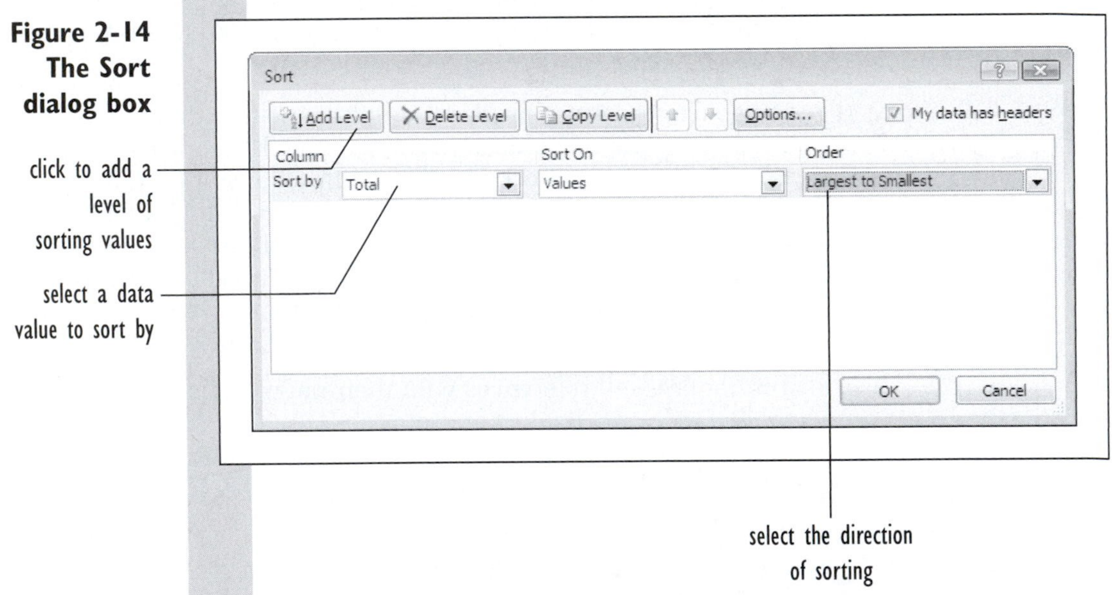

Figure 2-14 The Sort dialog box

5. Click the **OK** button.

6. Deselect the cell range by clicking cell **A1**.

   The stations are now sorted in order from Station 7, showing the largest total revenue, to Station 6 with the lowest revenue. See Figure 2-15.

Figure 2-15 The gas station data sorted in descending order to Total revenue

	A	B	C	D	E
1	Station	Gas	Other	Total	
2	7	$9,567	$13,712	$23,279	
3	8	$9,218	$12,056	$21,274	
4	3	$8,831	$7,899	$16,730	
5	4	$8,587	$7,488	$16,075	
6	2	$8,499	$7,500	$15,999	
7	0	$8,995	$6,938	$15,933	
8	5	$8,719	$7,111	$15,830	
9	9	$8,215	$7,508	$15,723	
10	1	$8,415	$7,211	$15,626	
11	6	$8,001	$6,281	$14,282	
12	Total	$87,047	$83,704	$170,751	
13					

EXCEL TIPS

- If you want to sort your list in nonnumeric order (in terms of days of the week or months of the year), click Custom List from the Order list box and then select one of the custom lists defined for your workbook.
- To sort your data by multiple levels, click the Add Level button in the Sort dialog box and then specify the data values corresponding to the next level.

Querying Data

In some cases you may be interested in a subset of your data rather than in the complete list. For instance, a manufacturing company trying to analyze quality control data from three work shifts might be interested in looking only at the night shift. A firm interested in salary data might want to consider just the subset of those making between $55,000 and $85,000. Excel allows you to specify the criteria for creating these subsets in the following two ways:

- **Comparison criteria**, which compares data values to specified values or constants
- **Calculated criteria**, which compares data values to a calculated value

For the gas station data, an example of a comparison criterion would be one that determines which service stations have gas sales exceeding $5,000. On the other hand, a calculated criterion would be one that determines which service stations have gas sales that exceed the average of gas sales of all stations in the data sample.

Once you have determined your criteria for creating a subset of the data values, you select worksheet cells that fulfill these criteria by **filtering** or **querying** the data. Excel provides two ways of filtering data. The first method, called the **AutoFilter**, is primarily used for simple queries employing comparison criteria. For more complicated queries and those involving calculated values, Excel provides the **Advanced Filter**. You'll have a chance to use both methods in exploring the gas station data.

Using the AutoFilter

Let's say the service station company plans a massive advertising campaign to boost sales for the service stations that are reporting gas sales of less than $8,500. You can construct a simple query using comparison criteria to have Excel display only service stations with gas sales <$8,500.

To query the service station list:

1 Click the **Sort & Filter** button from the Editing group on the Home tab and then click **Filter** from the drop-down menu.

Excel adds drop-down arrows to each of the column titles in the data list. By clicking these drop-down arrows you can filter the data list on the basis of the values in the selected column.

2 Click the Gas drop-down arrow to display the shortcut menu. Click Number Filters and then Less Than as shown in Figure 2-16.

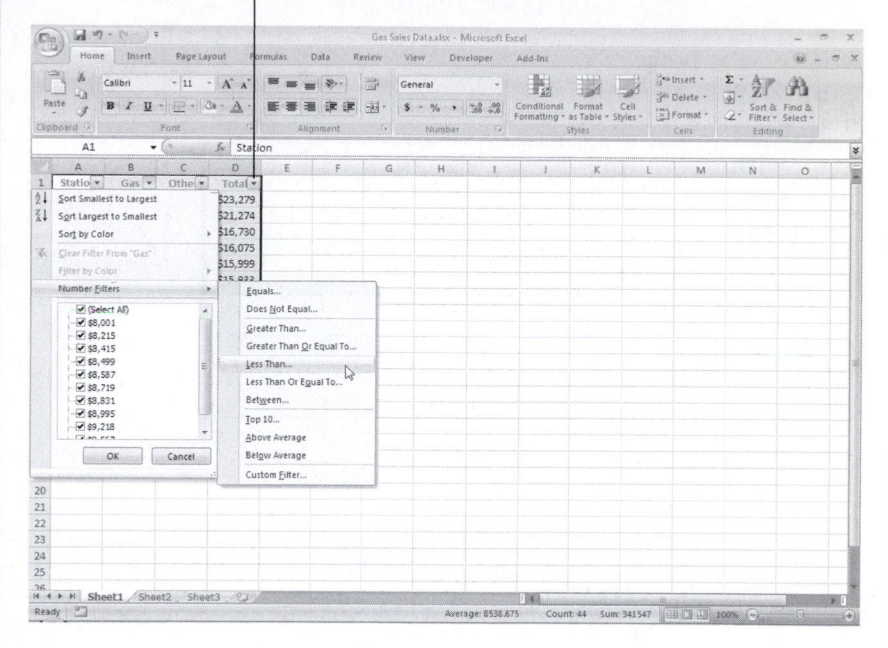

Figure 2-16 Filtering data values

Excel opens the Custom AutoFilter dialog box. From this dialog box you can specify the criteria used to filter the values in the data list.

3 Type **8500** in the input box as shown in Figure 2-17.

Figure 2-17 Creating a filter for gas sales less than $8,500

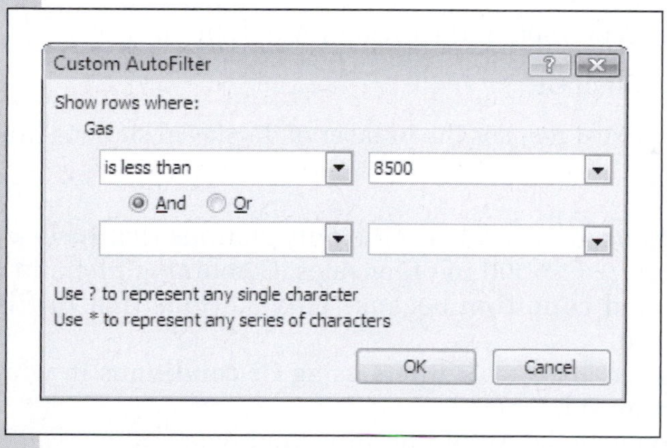

Custom AutoFilter

Show rows where:

Gas

is less than | 8500

And | Or

Use ? to represent any single character

Use * to represent any series of characters

OK | Cancel

4 Click **OK**.

Excel modifies the list of service stations to show stations 1, 2, 6, and 9. See Figure 2-18.

Figure 2-18 Stations with gas sales < $8,500

	A	B	C	D	E
1	Statio	Gas	Othe	Total	
6	2	$8,499	$7,500	$15,999	
9	9	$8,215	$7,508	$15,723	
10	1	$8,415	$7,211	$15,626	
11	6	$8,001	$6,281	$14,282	
12					
13	Total	$87,047	$83,704	$170,751	
14					

The service station data for the other stations has not been lost, merely hidden. You can retrieve the data by choosing the All option from the Gas drop-down list.

Let's say that you need to add a second filter that also filters out those service stations selling less than $7,500 worth of other products. This filter does not negate the one you just created; it adds to it.

To add a second filter:

1 Click the **Other** drop-down filter arrow, click **Number Filters** and then click **Less Than Or Equal to**.

2 Type **7500** in the Custom AutoFilter dialog box.

3 Click **OK**.

Excel reduces the number of displayed stations to Stations 1, 2, and 6.

Stations 1, 2, and 6 are the only stations that have <$8,500 in gasoline sales *and* <= $7,500 in other sales. Combining filters in this way is known as an **And condition** because only stations that fulfill both criteria are displayed.

You can also create filters using **Or conditions** in which only one of the criteria must be true.

To remove the AutoFilter from your data set, you can either stop running the AutoFilter or remove each filter individually. Try both methods now.

To remove the filters:

1 Click the **Other** drop-down filter arrow and click **Clear Filter from "Other"**.

The second filter is removed, and now only the results of the first filter are displayed.

2 Click the **Sort & Filter** button from the Editing group on the Home tab and then click **Filter** from the drop-down menu.

Excel stops running AutoFilter altogether and removes the filter drop-down arrows from the worksheet.

Using the Advanced Filter

There might be situations where you want to use more complicated criteria to filter your data. Such situations include criteria that

- Require several *And/Or* conditions
- Involve formulas and functions

Such cases are often beyond the capability of Excel's AutoFilter, but you can still do them using the Advanced Filter. To use the Advanced Filter, you must first enter your selection criteria into cells on the worksheet. Once those criteria are entered, you can use them in the Advanced Filter command.

Try this technique by recreating the pair of criteria you just entered; only now you'll use Excel's Advanced Filter.

To create a query for use with the Advanced Filter:

1 Click cell **B15**, type **Advanced Filter Criteria**, and press **Enter**.

2 Type **Gas** in cell B16 and press **Tab**. Type **Other** in cell C16 and press **Enter**.

3 Type **< 8500** in cell B17 and press **Tab**. Type **<= 7500** in cell C17 and press **Enter**.

If two criteria occupy the same row in the worksheet, Excel assumes that an And condition exists between them. In the example you just typed in, both criteria were entered into row 17, and Excel assumed that you wanted gas sales < $8,500 *and* other sales <= $7,500. Thus, these criteria match what you created earlier using the AutoFilter. Now apply these criteria to the service station data. To do this, open the Advanced Filter dialog box and specify both the range of the data you want filtered and the range containing the filter criteria.

To run the Advanced Filter command:

1. Select the cell range **A1:D11**.
2. Click the **Advanced** button from the Sort & Filter group on the Data tab. Excel opens the Advanced Filter dialog box.
3. Make sure that the Filter the list, in-place option button is selected and that $A$1:$D$11 is displayed in the List range box.
4. Enter **B16:C17** in the Criteria range box. This is the cell range containing the filter criteria you just typed in. See Figure 2-19.

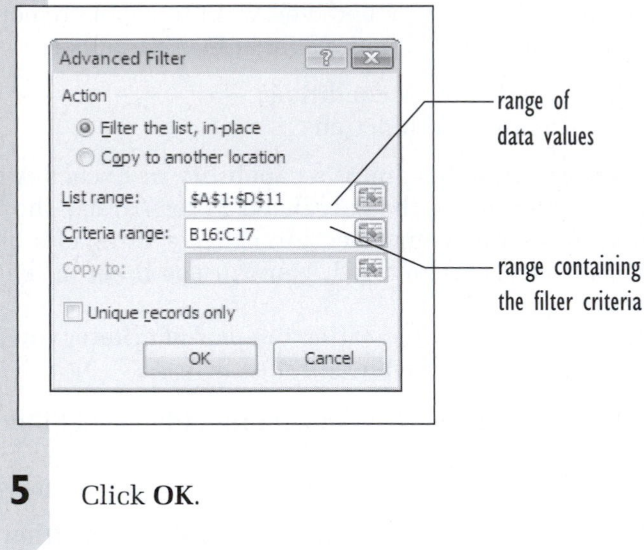

Figure 2-19 The Advanced Filter dialog box

5. Click **OK**.

As before, only Stations 1, 2, and 6 are displayed. Note that the column totals displayed in row 12 are not adjusted for the hidden values. You have to be careful when filtering data in Excel because formulas will still be based on the entire data set, including hidden values.

What if you wanted to look at only those service stations with *either* gasoline sales < $8,500 *or* other sales <= $7,500? Entering an Or condition between two different columns in your data set is not possible with the AutoFilter, but you can do it with the Advanced Filter. You do this by placing the different criteria in different rows in the worksheet.

To create an Or condition with the Advanced Filter:

1 Delete the criteria in cell C17.

2 Enter the criterion **<= 7500** in cell C18.

3 Once again, click the **Advanced** button from the Sort & Filter group on the Data tab to open the Advanced Filter dialog box.

4 Enter the cell range **A11:D11** into the List range box.

5 Change the cell reference in the Criteria range box to **B16:C18** to reflect the changes you made to the criteria.

6 Click **OK**.

Excel now displays Stations 0, 1, 2, 4, 5, 6, and 9. Each station has gas sales < $8,500 or sales of other items <= $7,500. See Figure 2-20.

Figure 2-20 Filtering data using an Or condition

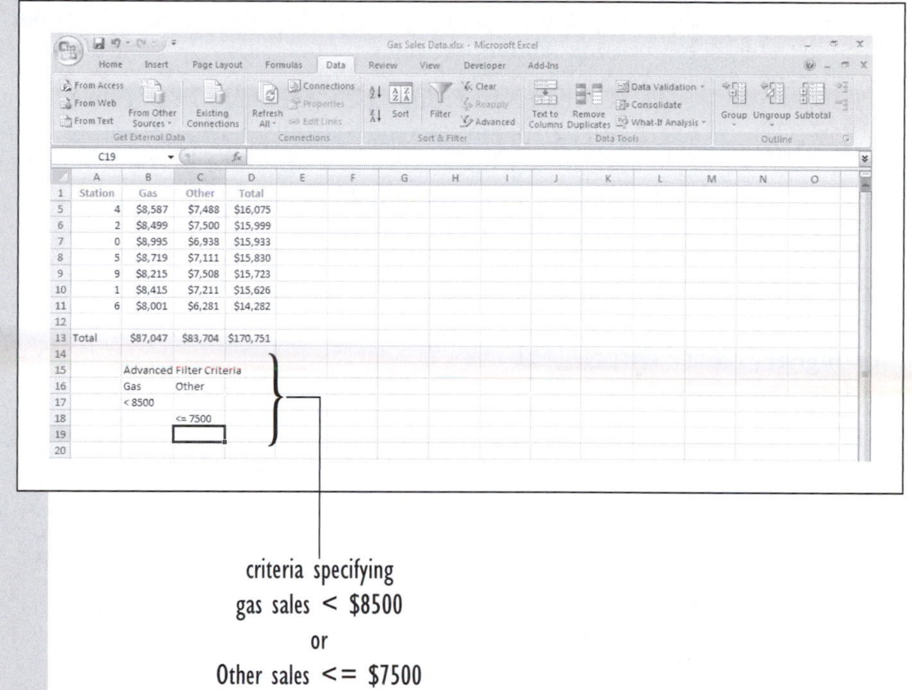

7 To view all stations again, click the **Clear** button from the Sort & Filter group on the Data tab.

Using Calculated Values

You decide to reward service station managers whose daily gasoline sales were higher than average. How would you determine which service stations qualified? You could calculate average gasoline sales and enter this number explicitly into a filter (either an AutoFilter or an Advanced Filter). One problem with this approach however, is that every time you update your service station data, you have to recalculate this number and rewrite the query.

However, Excel's AutoFilter allows you to include this information in your query automatically.

To select stations with higher-than-average gas sales:

1 Select the cell range **A11:D11** again.

2 Click the **Filter** button from the Sort & Filter group on the Data tab to display the AutoFilter drop-down arrows.

3 Click the **Gas** drop-down list arrow, click **Number Filters,** and then click **Above Average**.

As shown in Figure 2-21, the data list is filtered again, showing only the data from Service Stations 0, 3, 5, 7, and 8. Those are five service stations whose daily gas sales are higher than the average from all stations in the data list.

Figure 2-21 Filtered data for higher-than-average gas sales

	A	B	C	D	E
1	Statio	Gas	Othe	Total	
2	7	$9,567	$13,712	$23,279	
3	8	$9,218	$12,056	$21,274	
4	3	$8,831	$7,899	$16,730	
7	0	$8,995	$6,938	$15,933	
8	5	$8,719	$7,111	$15,830	
12					
13	Total	$87,047	$83,704	$170,751	
14					
15		Advanced Filter Criteria			
16		Gas	Other		
17		< 8500			
18			<= 7500		

4 Click the **Filter** button again from the Sort & Filter group on the Data tab to turn off the filter of the service station data.

For more complicated formulas, you can enter the expressions using an Advanced Filter.

You've completed your analysis of the service station data. Save and close the workbook now.

To finish your work:

1. Click the **Office** button and then click **Save**.
2. Click the **Office** button again and then click **Close**.

Importing Data from Text Files

Often your data will be created using applications other than Excel. In that case, you'll want to go through a process of bringing that data into Excel called **importing**. Excel provides many tools for importing data. In this chapter you'll explore two of the more common sources of external data: text files and databases.

A **text file** contains only text and numbers, without any of the formulas, graphics, special fonts, or formatted text that you would find in a workbook. Text files are one of the simplest and most widely used methods of storing data, and most software programs can both save and retrieve data in a text file format. Thus, although text files contain only raw, unformatted data, they are very useful in situations where you want to share data with others.

Because a text file doesn't contain formatting codes to give it structure, there must be some other way of making it understandable to a program that will read it. If a text file contains only numbers, how will the importing program know where one column of numbers ends and another begins? When you import or create a text file, you have to know how the values are organized within the file. One way to structure text files is to use a **delimiter**, which is a symbol, usually a space, a comma, or a tab, that separates one column of data from another. The delimiter tells a program that retrieves the text file where columns begin and end. Text that is separated by delimiters is called **delimited text**.

In addition to delimited text, you can also organize data with a **fixed-width file**. In a fixed-width text file, each column will start at the same location in the file. For example, the first column will start at the first space in the file, the second column will start at the tenth space, and so forth.

When Excel starts to open a text file, it automatically starts the **Text Import Wizard** to determine whether the contents are organized in a fixed-width format or a delimited format and, if it's delimited, what delimiter is used. If necessary, you can also intervene and tell it how to interpret the text file.

Having seen some of the issues involved in using a text file, you are ready to try importing data from a text file. In this example, a family-owned bagel shop has gathered data on wheat products that people eat as snacks or for breakfast. The family members intend to compare these products with the products that they sell. The data have been stored in a text file, Wheat.txt, shown in Table 2-5. The file was obtained from the nutritional information on the packages of the competing wheat products.

Table 2-5 Wheat Data

Brand	Food	Price	Total Oz.	Serving Grams	Calories	Protein	Carbo	Fiber	Sugar	Fat
SNYDER	PRETZEL	2.19	9.0	31.0	120	3	25	1	1	1.0
LENDERS	BAGEL	1.39	17.1	81.0	210	7	43	2	3	1.5
BAYS	ENG MUFFIN	2.27	12.0	57.0	140	5	27	1	2	1.5
THOMAS	ENG MUFFIN	2.94	12.0	57.0	120	4	25	1	1	1.0
QUAKER	OAT SQUARES CEREAL	5.49	24.0	57.0	210	6	44	5	10	2.5
NABISCO	GRAH CRACKER	3.17	14.4	31.0	130	2	24	1	7	3.0
WHEATIES	CEREAL	5.09	15.6	27.0	100	3	22	3	4	1.5
WONDER	BREAD	0.99	20.0	26.0	60	2	13	0	2	1.5
BROWNBERRY	BREAD	3.49	24.0	43.0	120	4	23	1	3	2.0
PEPPERIDGE	BREAD	2.89	16.0	25.5	70	2	13	1	2	1.0

To start importing Wheat.txt into an Excel workbook:

1. Click the **Office** button and then click **Open**.
2. Navigate to the Chapter02 data folder and change the file type to **Text Files (*.pm; *.txt; *.csv)**.

   Excel displays the file wheat.txt from the list of text files in the Chapter02 folder.
3. Double-click the **wheat.txt** file.

   Excel displays the Text Import Wizard to help you select the text to import. See Figure 2-22.

Figure 2-22
Text Import Wizard
Step 1 of 3

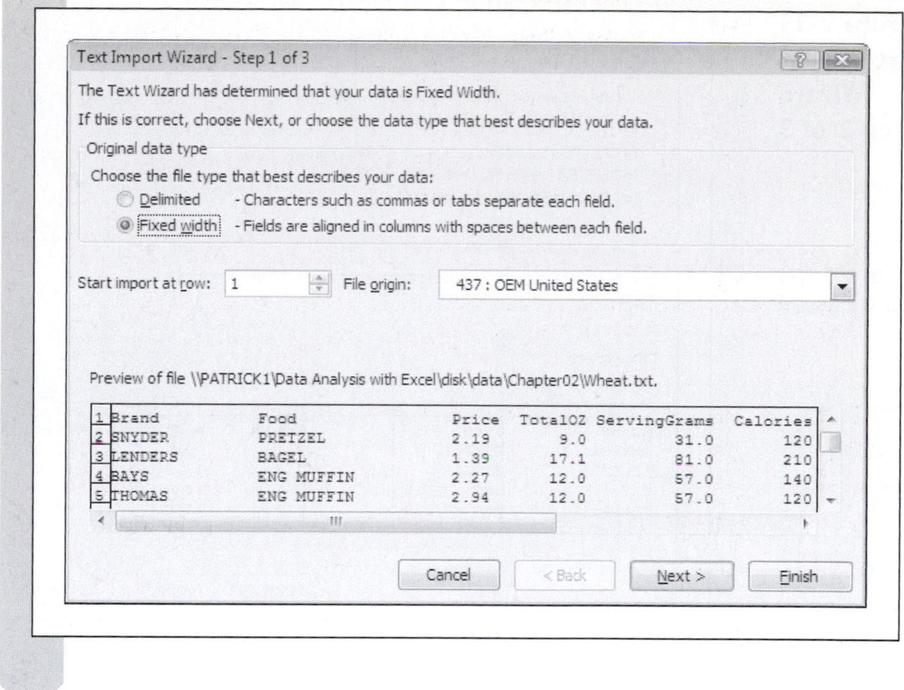

The wizard has automatically determined that the wheat.txt file is organized as a fixed-width text file. By moving the horizontal and vertical scroll bars, you can see the whole data set. Once you've started the Text Import Wizard, you can define where various data columns begin and end. You can also have the wizard skip entire columns.

To define the columns you intend to import:

1 Click the **Next** button.

The wizard has already placed borders between the various columns in the text file. You can remove a border by double-clicking it, you can add a border by clicking a blank space in the Data Preview window, or you can move a border by dragging it to a new location. Try moving a border now.

2 Click and drag the right border for TotalOZ further to the right so that it aligns with the left edge of the ServingGrams column. See Figure 2-23.

Figure 2-23 Text Import Wizard Step 2 of 3

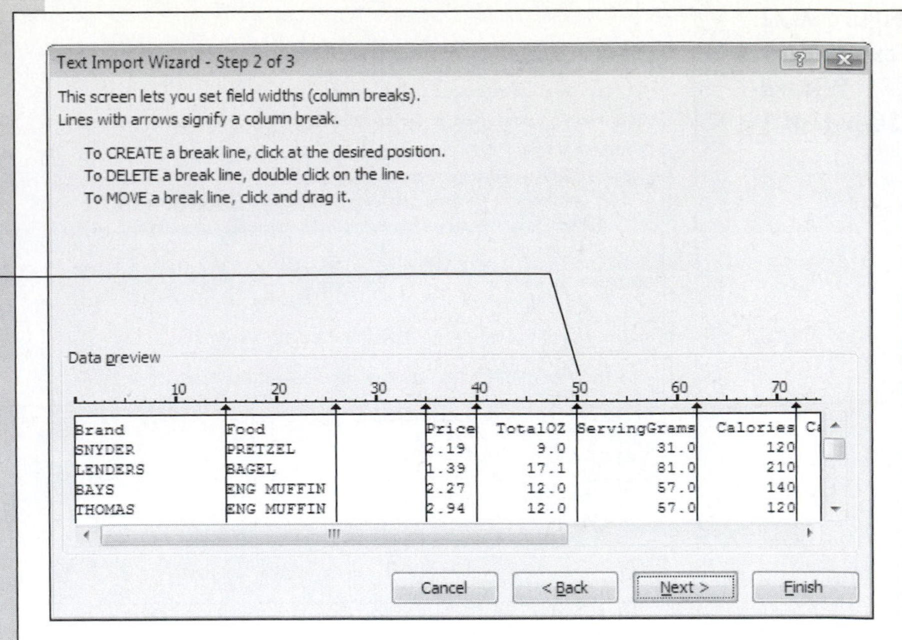

3 Click the **Next** button.

The third step of the wizard allows you to define column formats and to exclude specific columns from your import. By default, the wizard applies the General format to your data, which will work in most cases.

4 Click the **Finish** button to close the wizard.

Excel imports the wheat data and places it into a new workbook. See Figure 2-24.

Figure 2-24 Wheat data imported into Excel

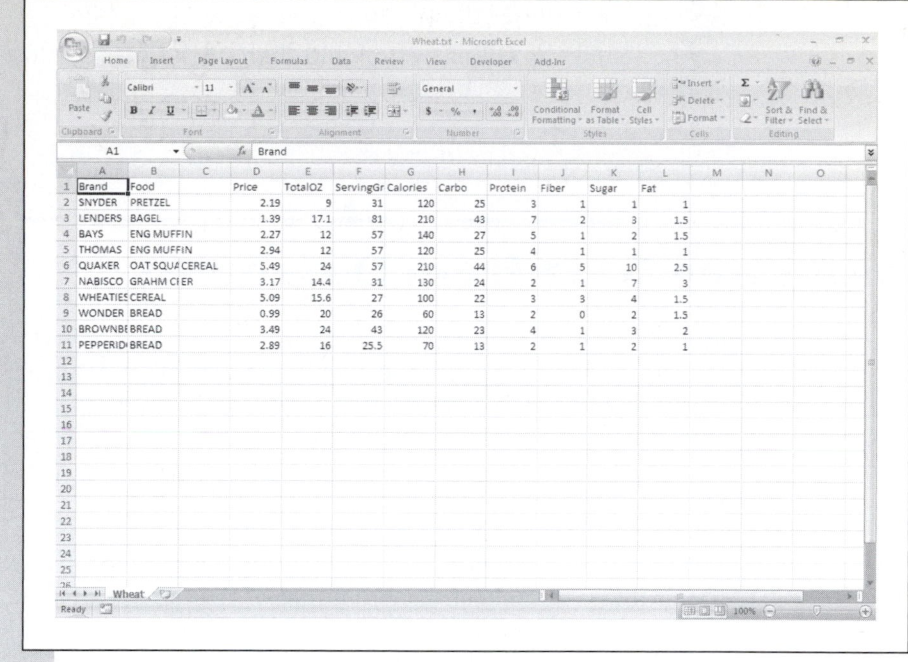

Brand	Food	Price	TotalOZ	ServingGr	Calories	Carbo	Protein	Fiber	Sugar	Fat
SNYDER	PRETZEL	2.19	9	31	120	25	3	1	1	1
LENDERS	BAGEL	1.39	17.1	81	210	43	7	2	3	1.5
BAYS	ENG MUFFIN	2.27	12	57	140	27	5	1	2	1.5
THOMAS	ENG MUFFIN	2.94	12	57	120	25	4	1	1	1
QUAKER	OAT SQUA CEREAL	5.49	24	57	210	44	6	5	10	2.5
NABISCO	GRAHM CI ER	3.17	14.4	31	130	24	2	1	7	3
WHEATIES	CEREAL	5.09	15.6	27	100	22	3	3	4	1.5
WONDER	BREAD	0.99	20	26	60	13	2	0	2	1.5
BROWNBE	BREAD	3.49	24	43	120	23	4	1	3	2
PEPPERIDI	BREAD	2.89	16	25.5	70	13	2	1	2	1

Notice that the data for the first two columns appear to be cut off, but don't worry. When Excel imports a file, it formats the new workbook with a standard column width of about nine characters, regardless of column content. The data are still there but are hidden.

To format the column widths to show all the data:

1 Press **CTRL+a** twice to select all cells in the worksheet.

2 Move the mouse pointer to the border of one of the column headers until the pointer changes to a ✛ and double-click the column border.

Excel changes the column widths to match the width of the longest cell in each column.

3 Click cell **A1** to remove the selection.

Save and close the workbook.

4 Click the **Office** button and then click **Save As**.

5 Type **Wheat Data** in the File Name box and select **Excel Workbook (*.xlsx)** from the Save As Type list box.

6. Click the **Save** button.
7. Click the **Office** button and then click **Close** to close the workbook.

Importing Data from Databases

Excel allows the user to create connections to a variety of data sources. You've already seen how to create a connection to a text file; now you'll learn how to create a connection to a database file.

A **database** is a program that stores and retrieves large amounts of data and creates reports describing that data. Excel can retrieve data stored in most database programs, including Microsoft® Access, Borland dBASE®, Borland Paradox®, and Microsoft FoxPro®.

Databases store information in **tables**, organized in rows and columns, much like a worksheet. Each column of the table, called a **field**, stores information about a specific characteristic of a person, place, or thing. Each row, called a **record**, displays the collection of characteristics of a particular person, place, or thing. A database can contain several such tables; therefore, you need some way of relating information in one table to information in another. You relate tables to one another by using **common fields**, which are the fields that are the same in each table. When you want to retrieve information from two tables linked by a common field, Excel matches the value of the field in one table with the same value of the field in the second table. Because the field values match, a new table is created containing records from both tables.

A large database can have many tables, and each table can have several fields and thousands of records, so you need a way to choose only the information that you most want to see. When you want to look only at specific information from a database, you create a database query. A **database query** is a question you ask about the data in the database. In response to your query, the database finds the records and fields that meet the requirements of your question and then extracts only that data. When you query a database, you might want to extract only selected records. In this case, your query would contain criteria similar to the criteria you used earlier in selecting data from an Excel workbook.

Using Excel's Database Query Wizard

You can import data from a database file directly, as you did with the wheat text file. You can also write a query to retrieve only portions of data from selected tables within the database file.

To see how this works, you'll import another data set containing nutritional data located in an Access database file named wheat.mdb. The database contains two tables: Product, a table containing descriptive information about each product (the name, manufacturer, serving size, price, and so on), and Nutrition, a table of nutritional information (calories, proteins, etc.). You'll import the data by creating a connection to the database file using **Microsoft Query,** a small application installed with most Office 2007 products.

To access Microsoft Query:

1 Click the **Office** button and then click **New** to open a new blank worksheet in Excel.

2 Click the **From Other Sources** button from the Get External Data group on the Data tab and then click **From Microsoft Query**.

3 Verify that the Databases dialog sheet tab is selected.

At this point, you'll choose a data source. Excel provides several choices from such possible sources as Access, dBase, FoxPro, and other Excel workbooks. You can also create your own customized data source. In this case, you'll use the Access data source because this data comes from an Access database.

4 Click **MS Access Database*** from the list of data sources in the Databases dialog sheet and click the **OK** button.

5 Navigate to the folder containing your Chapter02 data files and select the **wheat.mdb** database file. Click the **OK** button.

Excel opens the Query Wizard dialog box shown in Figure 2-25.

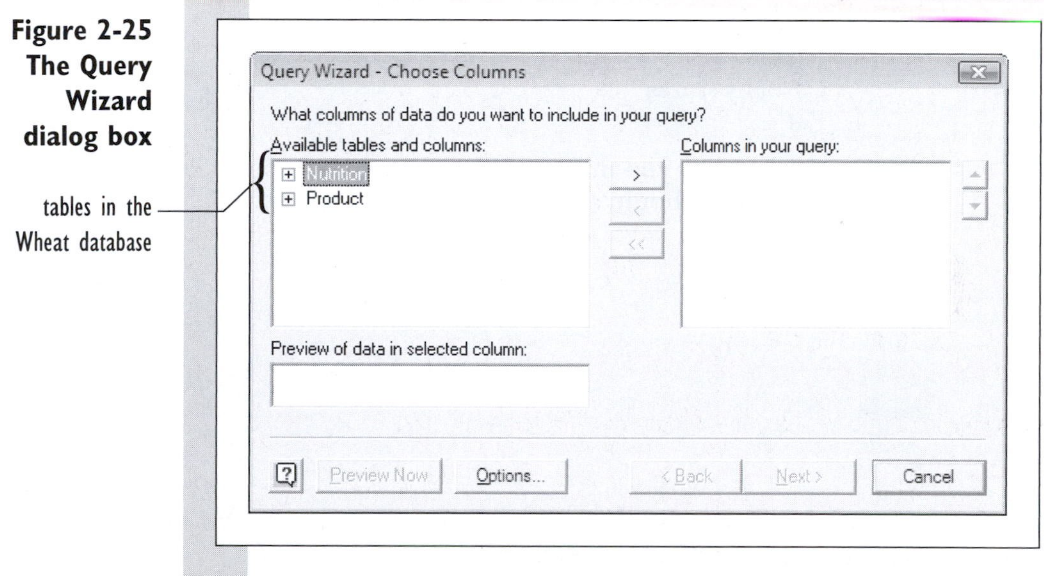

Figure 2-25 The Query Wizard dialog box

Now that you've started the Query Wizard, you are free to select the various fields that you'll import into Excel. The box on the left of the wizard shown in Figure 2-25 shows the tables in the database. As you expected, there are two: Nutrition and Product. By clicking the plus box in front of each table name, you can view and select the specific fields that you'll import into Excel. Try this now by selecting fields from both tables.

To select fields for your query:

1 Click the **plus box [+]** in front of the Product table name.

A space opens beneath the table name displaying the names of each of the fields in the table.

2 Double-click the following names in the list:

Brand
Food
Price
Package oz
Serving oz

As you double-click each field name, the name appears in the box on the right, indicating that they are part of your selection in the query. Note that you do not select the Product ID field. This field is the common field between the two tables and contains a unique id number for each wheat product. You don't have to include this in your query.

3 Click the **plus box [+]** in front of the Nutrition table name and then double-click the following field names:

Calories
Protein
Carbohydrates
Fat

Once again, you do not select the common field, Product ID. Your dialog box should appear as shown in Figure 2-26.

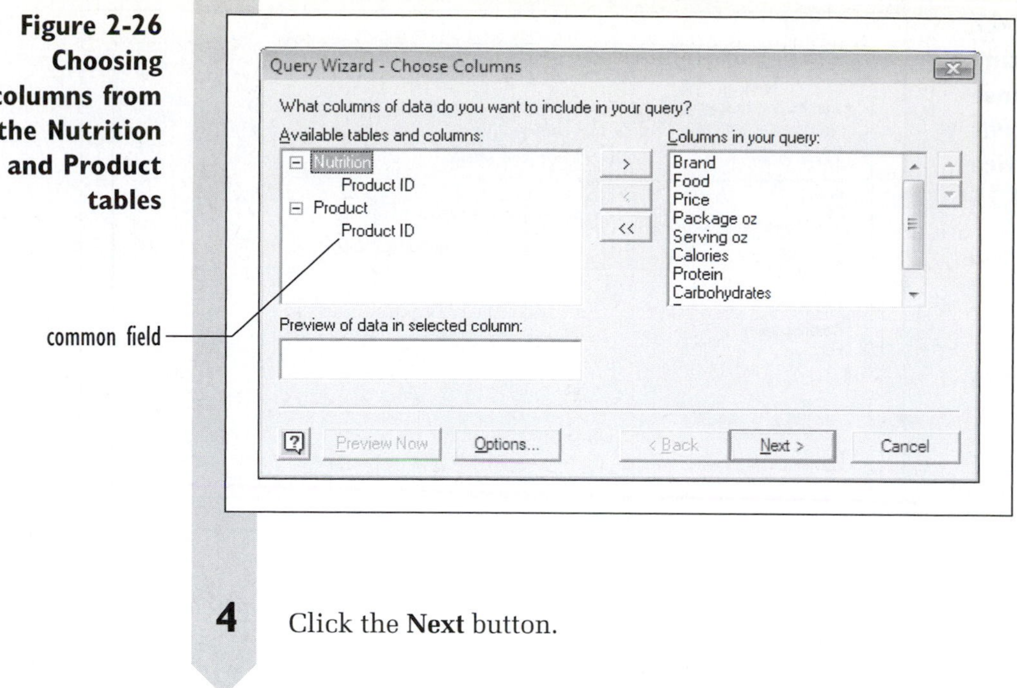

Figure 2-26 Choosing columns from the Nutrition and Product tables

4 Click the **Next** button.

After specifying which fields you'll import, you'll now have the opportunity to control which records to import and how your data will be sorted.

Specifying Criteria and Sorting Data

You can apply criteria to the data you import with the Query Wizard. At this point, your query will import all of the records from the Wheat database, but you can modify that. Say you want to import only those wheat products whose price is $1.25 or greater. You can do that at this point in the wizard. You can specify several levels of And/Or conditions for each of the many fields in your query.

To add criteria to your query:

1. Click **Price** from the list of columns to filter in the box at the left of the Query Filter dialog box.
2. In the highlighted drop-down list box at the right, click **is greater than or equal to**.
3. Type **1.25** in the drop-down list box to the immediate right. See Figure 2-27.

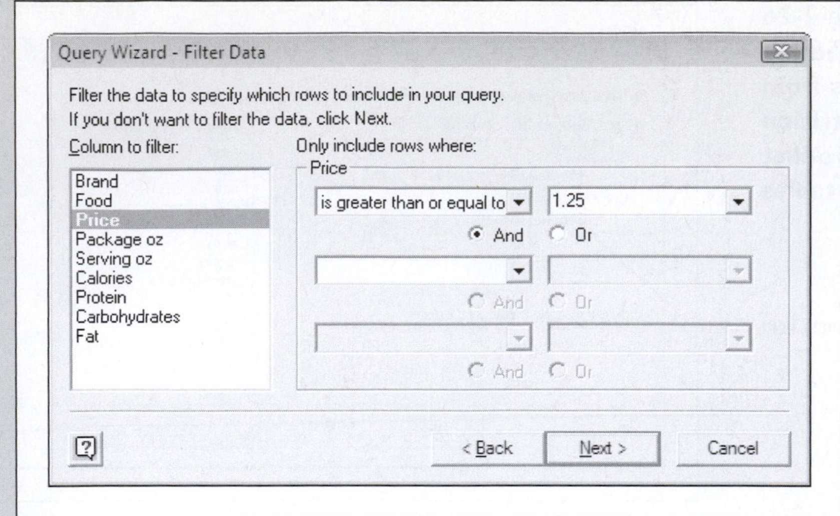

Figure 2-27 Selecting only those records whose price >= $1.25

This criterion selects only those records whose price is greater than or equal to $1.25.

4 Click the **Next** button.

The last step in defining your query is to add any sorting options. You can specify up to three different fields to sort by. In this example, you decide to sort the wheat products by the amount of calories they contain, starting with the highest-calorie product first and going down to the lowest.

To specify a sort order:

1 Select **Calories** from the Sort by list box.

2 Click the **Descending** option button. See Figure 2-28.

Figure 2-28
Sorting records in descending order of calories

Query Wizard - Sort Order

Specify how you want your data sorted.
If you don't want to sort the data, click Next.

Sort by: Calories — Ascending / Descending (selected)
Then by: — Ascending / Descending
Then by: — Ascending / Descending

< Back | Next > | Cancel

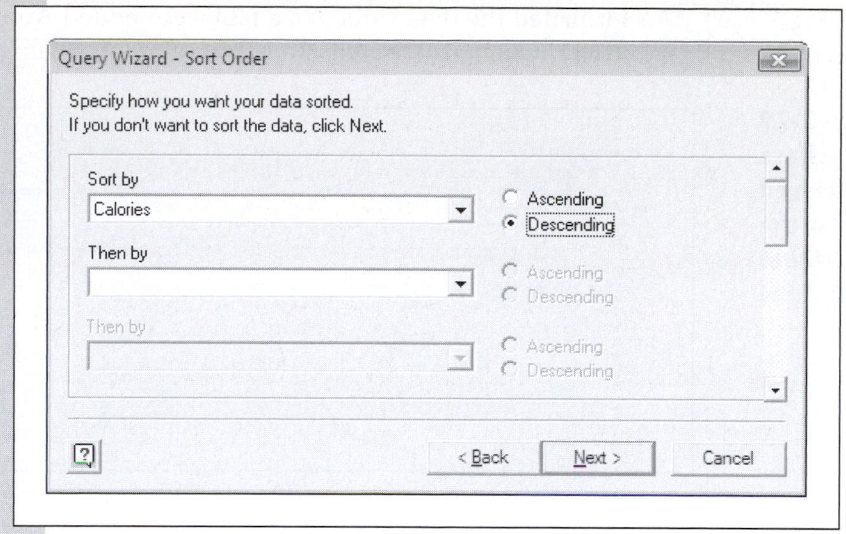

3 Click the **Next** button.

The last step in the Query Wizard is to choose where you want to send the data. You can

1. Import the data into your Excel workbook.
2. Open the results of your query in Microsoft Query. **Microsoft Query** is a program included on your installation disk with several tools that allow you to create even more complex queries.

You can learn more about these two options in Excel's online Help. In this example, you'll simply retrieve the data into your Excel workbook.

To finish retrieving the data:

1 Click the **Return Data to Microsoft Excel** option button.

2 Click the **Finish** button.

You can now specify where the data will be placed. The default will be to place the data in the active cell of the current worksheet. In this case, that is cell A1. Accept this default.

3 Click the **OK** button.

Excel connects to the Wheat database and retrieves the data shown in Figure 2-29. Note that these include only those wheat products whose price is $1.25 or greater and that the data are sorted in descending order of calories. Also note that Excel has automatically formatted the data values in a table and added AutoFilter buttons to filter the data if you so desire.

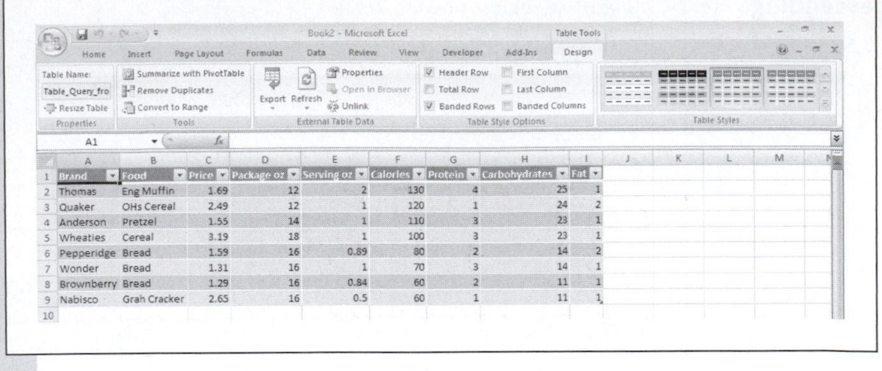

Brand	Food	Price	Package oz	Serving oz	Calories	Protein	Carbohydrates	Fat
Thomas	Eng Muffin	1.69	12	2	130	4	25	1
Quaker	OHs Cereal	2.49	12	1	120	1	24	2
Anderson	Pretzel	1.55	14	1	110	3	23	1
Wheaties	Cereal	3.19	18	1	100	3	23	1
Pepperidge	Bread	1.59	16	0.89	80	2	14	2
Wonder	Bread	1.31	16	1	70	3	14	1
Brownberry	Bread	1.29	16	0.84	60	2	11	1
Nabisco	Grah Cracker	2.65	16	0.5	60	1	11	1

Figure 2-29 Data retrieved from the Wheat database

Unlike importing from a text file, you can have Excel automatically refresh the data it imports from a database. Thus, if the source database changes at some point, you can automatically retrieve the new data without recreating the query. Commands like refreshing imported data are available from the Connections group on the Data tab. Table 2-6 describes some of these commands.

Table 2-6 Commands on the Data tab

Button	Icon	Purpose
Refresh All	Refresh All	Refresh all data queries located in the current workbook
Connections	Connections	View and modify the properties of all the data connections in the workbook
Properties	Properties	View and modify the properties of the currently selected data range

Having seen how one would import data from a database into Excel, you are ready to save and close the workbook.

To save and close the Wheat workbook:

1. Click the **Office** button and then click **Save As**
2. Type **Wheat Database** in the File name box, verify that Excel Workbook (*.xlsx) is displayed in the Save as type box, and then click **Save**.

3. Click the **Office** button and then click **Close**.
4. Click the **Office** button and then click **Exit Excel**.

Exercises

1. Air quality data had been collected from the Environmental Protection Agency (EPA) and stored in an Excel workbook. The workbook displays the number of unhealthful days (heavy levels of pollution) per year for 14 major U.S. cities in the year 1980 and then from 2000 through 2006. Open this workbook and examine the data.
   a. Open the **Pollution** workbook from the Chapter02 folder and save it as **Pollution Report**.
   b. Create a new column named **AVER00_06** that uses Excel's average() function to calculate the average pollution days for each city from 2000 through 2006.
   c. Create a new column named **DIFF06_80** that calculates the difference between the average number of pollution days from 2000 through 2006 and the number of pollution days in 1980 for each of the 14 cities.
   d. Sort the data in ascending order of the DIFF80_06 column you just created. Which cities showed an increase in the number of pollution days? Which cities showed a decrease? Which city showed the greatest improvement in terms of the decline in the number of unhealthy days?
   e. Create a new column that calculates the ratio of unhealthy days between the average from 2000 through 2006 and the value for 1980. Name the column **RATIO06_80**.
   f. Format the values in the RATIO06_80 column as a percentage to two decimal places.
   g. Sort the data in ascending order of the RATIO06_80 column. Which city showed the greatest improvement in the terms of the ratio of the 2006 average to the 1980 value?
   h. Create range names for all of the columns in your workbook.
   i. Save your workbook and write a report summarizing your observations. Does the data *prove* any conclusions you might have reached? What kind of information might be missing from this data set? Remember that you only have one year's worth of data from the 1980s versus seven years data from 2000 to 2006. In what way could the average value from those seven years not be comparable to a single year's value from 1980?

2. Data on soft drink sales shown in Table 2-7 have been saved in a text file. The file has five variables and ten cases. The first variable is the name of the soft drink brand; the next three variables are company sales in millions of 192-ounce cases for the years 2000, 2001, and 2002. (*Source: http://www.bevnet.com/news/2002/03-01-2002-softdrink.asp,* Beverage Marketing Corporation.) The final column indicates the year of origin for each brand.

Table 2-7 Soft Drink Sales Data

Brand	Cases2000	Cases2001	Cases2002	Origin
Coca-Cola	3198.0	3189.6	3288.9	1886
Pepsi	2188.0	2163.9	2156.4	1898
Mountain Dew	810.3	853.7	862.7	1946
Dr Pepper	747.4	740.0	737.4	1885
Sprite	713.9	703.3	687.9	1961
Gatorade	355.8	375.0	422.8	1965
7 Up	276.0	261.6	243.4	1929
Tropicana	301.2	307.7	292.9	1954
Minute Maid	218.0	226.5	285.3	1946
Aquafina	105.0	151.4	203.0	1994

a. Import the **Drinks.txt** file from the Chapter02 data folder into an Excel workbook (note that columns are delimited by tabs).

b. Create range names for each of the five data columns in the workbook.

c. Create two new columns displaying the change in sales from 2000 to 2002 and the ratio of the 2000 sales to the 2002 sales. Assign range names to these two new columns.
Sort the list in descending order of the difference in sales.

d. Is there any relationship between the year in which the brand was founded and the change in sales? (*Hint:* Are the older brands showing less growth than the new brands?)

e. Repeat your analysis using the ratio of sales.

f. Save the workbook in Excel format to the Chapter02 folder under the name **Soft Drinks Sales Report** and write a report summarizing your observations.

3. The NCAA requires schools to submit information on graduation rates for its student athletes. Table 2-8 shows the data for the 11 schools in the Big Ten, covering the years 1997 through 2000, indicating the graduation percentage (within six years.) The overall graduation percentage for all undergraduates is shown in the Graduated column and then is broken down by race and gender in the remaining four columns of the table for those who received athletic scholarships.

Table 2-8 Big Ten Graduation Data

University	Graduated	White Males	Black Males	White Females	Black Females
ILL	81	70	52	77	83
IND	72	61	45	76	82
IOWA	66	61	51	81	50
MICH	86	79	44	88	67
MSU	72	61	33	87	63
MINN	58	63	39	70	56
NU	93	87	79	94	100

(continued)

OSU	66	60	42	77	83
PSU	84	76	69	91	93
PU	67	66	48	84	80
WIS	77	65	50	79	64

a. Enter the data from Table 2-8 into a blank workbook and save the workbook as **Big Ten Graduation** to the Chapter02 folder.
b. Create two new columns displaying the difference between white male and white female graduation rates and the ratio between white male and white female graduation rates. How do graduation rates compare?
c. Create two more columns calculating the difference and ratio of the white female graduation rate to the overall rate from 1997 to 2000.
d. What do you observe from the data? Does one university stand out from the others?
e. Sort the data files in descending order of the ratio of white male to white female graduation rate. Create range names for all of the columns in the workbook.
f. Save your changes to the workbook and write a summary of your observations.

4. Over 4,000 television viewers were interviewed in 1984 to determine which television ads were remembered for being significant and interesting. The level of retained impressions were then compared to the advertising budgets from each firm. (*Source: Wall Street Journal*, 1984.)
   a. Open the **TV Ads** workbook from the Chapter02 folder and save it as **TV Ads Analysis**.
   b. Calculate the ratio of the retained impressions per week to the advertising budget.
   c. Create range names for the three columns in the worksheet.
   d. Sort the list in descending order of ratio values. Which firm showed the greatest bang for the buck from their advertising dollars? Print the sorted data values.
   e. Filter the data list, showing only those firms with a higher-than-average ratio of retained impressions to advertising dollars. Print the filtered values.
   f. Save your changes to your workbook and then write a report summarizing your observations.

5. The Teacher.txt file contains the average public teacher pay and spending on public schools per pupil in 1985 for 50 states and the District of Columbia as reported by the National Education Association.
   a. Open the **Teacher.txt** file from the Chapter02 data folder as a tab-delimited text file. There are four columns in the text file. The State column contains the abbreviations of the 50 states and the District of Columbia. The Pay column contains the average annual salary of public school teachers in each state and district. The Spend column contains the public school spending per pupil for each state and district. The Area column contains the area in the country for each state or district. Import all of these columns except the Area column.
   b. Create a new column calculating the ratio of the Pay column to the Spend column.

c. Create range names for all of the columns in the worksheet.
d. Sort the data in ascending order of the Ratio column.
e. Filter the data values, showing only those states or districts that have a ratio value of less than 6.
f. Save your workbook as **Teacher Salary Analysis** to the Chapter02 folder in Excel workbook format and summarize your observations.

6. Working in groups in a high school chemistry lab, students measured the mass (grams) and volume (cubic centimeters) of eight aluminum chunks. Both the mass in grams and the volume in cubic centimeters were measured for each chunk. Analyze the data from the lab.
   a. Open the **Aluminum** workbook from the Chapter02 folder and save it as **Aluminum Density Analysis**.
   b. Create a new column in the worksheet, computing the density of each chunk (the ratio of mass to volume). Apply a range name to the new column.
   c. Sort the data from the chunk with the highest density to that with the lowest.
   d. Calculate the average density for all chunks.
   e. Is there an extreme value (an observation that stands out as being different from the others)? Calculate the average density for all chunks aside from the outlier. Print your results.
   f. Which of the two averages gives the best approximation of the density of aluminum? Why?
   g. Save your changes to the workbook and summarize your results.

7. The **Economy** workbook has seven variables related to the US economy from 1947 to 1962.
   a. Open the **Economy** workbook from the Chapter02 data folder. The Deflator variable is a measure of the inflation of the dollar; arbitrarily set to 100 for 1954. The GNP column contains the Gross National Product for each year (in millions). The UnEmploy column contains the number unemployed in thousands, and the Arm Force column has the number in the armed forces in thousands. The Population column contains the population in thousands. The Total Emp contains the total employment in thousands. Save the workbook as **Economy Data**.
   b. Create range names for each column in the worksheet.
   c. Notice that values in the Population column increase each year. Use the Sort command to find out for which other columns this is true.
   d. There is an upward trend to the GNP, although it does not increase each year. Create a new column that calculates the GNP per person for each year. Name this new column GNPPOP and create a range name for the values it contains.
   e. Save your changes to the workbook and write a report summarizing your observations.

8. An analyst has collected 2007 data including salary and batting average for major league players. Examine the data that have been collected.
   a. Open the **Baseball** workbook from the Chapter02 folder and save it as **Baseball Salary Analysis**.
   b. Create range names for all of the columns in the workbook.
   c. Sort the data values in descending order of batting average.
   d. Display only those players whose career batting average is 0.310 or greater. List these players.

e. Remove the filter, displaying all values in the workbook again.
f. Add a new column to the worksheet, displaying the batting average divided by the player's salary and multiplied by 1,000,000.
g. Sort the worksheet in descending order of the new column you created. Who are the top-ten players in terms of batting average per dollar? Print your results.
h. Examine the number of years played in your sorted list. Where does it appear that most of the first-year players lie? What would account for that? (*Hint:* What are some of the other factors besides batting average that may account for a player's high salary?)
i. Save your workbook and write a report summarizing your observations.

9. An analyst has collected data on the death rates for diabetes and influenza-pneumonia for the year 2003. The data have been saved in the **Health** workbook. Your job is to examine the data values from the workbook.
   a. Open the Health workbook from the Chapter02 folder and save it as **Health Report Analysis**.
   b. Sort the data by ascending order of diabetes-related deaths. Then do a sort on influenza-pneumonia related deaths. Which states have the highest and lowest values in those categories?
   c. Use the AutoFilter to list the top-ten states in each category. Print your filtered worksheet.
   d. Turn off the filter and create a new column calculating the ratio of the diabetes-related death rate to the pneumonia-related death rate in each of the 50 states. Create a range name for the new column.
   e. Format the ratio values in the new column as percentages to two decimal places.
   f. Sort the data in ascending order of the new column. Which state or region has the highest ratio of diabetes-related deaths? African-Americans have a much higher rate of diabetes than whites. Discuss how this explains your observation of the state or region with the highest rate of diabetes-related death.
   g. Create range names for the all of the columns in your workbook.
   h. Save your changes to the workbook and write a summary of your observations.

10. The Cars workbook contains data from Consumer Reports.org®, February 1, 2008, on 275 different car models, as described in the following table:

Table 2-9 Car Data

Field	Description
Model ID	Number from 1 to 275
Model	Make and model
Type	Type of vehicle
Price	Price in dollars
HP	Horsepower
Eng size	Engine size in liters
Cyl	Number of cylinders
Eng Type	Type of engine

(continued)

MPG	Overall miles per gallon
Time0–60	0–60 time in seconds
Weight	Weight in pounds
Date	Date of Issue
Region	Original location of manufacturer
Eng type01	1 if hybrid or diesel, 0 otherwise

a. Open the **Cars** workbook from the Chapter02 folder.

b. Create range names for all of the data columns you retrieved.

Using techniques that you've learned in this chapter, answer the following questions:

c. How many cars come from the model year 2007?

d. Which car has the highest horsepower?

e. Which car has the highest horsepower relative to its weight?

f. Which car has the highest miles per gallon (MPG)?

g. Which car has the highest MPG relative to its 0–60 time? Note that this car will have fast acceleration and pretty good MPG.

h. If you sort the data by the ratio of MPG to 0–60 time, what is the **Region** of most of the cars with low values? What kind of vehicles are these? What are the Regions of most of the highest models, and what kind of vehicles do you find high on this scale? Discuss the relationship of this scale to Weight.

i. Save your workbook as Cars Performance Analysis.

Chapter 3

WORKING WITH CHARTS

Objectives

In this chapter you will learn to:

- Identify the different types of charts created by Excel
- Create a scatter plot with the Chart Wizard
- Edit the appearance of your chart
- Label points on your scatter plot
- Break a scatter plot down by categories
- Create a bubble plot
- Create a scatter plot containing several data series

In Chapter 2, you learned how to work with data in an Excel worksheet. In this chapter you'll learn how to display those data with charts. This chapter focuses primarily on two types of charts: scatter plots and bubble charts. Both are important tools in the field of statistics. You'll also learn how to use some features of StatPlus that give you additional tools in working with and interpreting your charts.

Introducing Excel Charts

A picture is worth a thousand words. Properly designed and presented, a graph can be worth a thousand words of description. Concepts difficult to describe through a recitation of numbers can be easily displayed in a chart or plot. Charts can quickly show general trends, unusual observations, and important relationships between variables. In Table 3-1, a table of monthly sales values is displayed. How do sales vary during the year? Which month in the table displays an unusual sales result? Can you easily tell?

Table 3-1 Monthly Sales Values

Date	Sales
Jan. 2010	$16,800
Feb. 2010	$19,300
Mar. 2010	$21,100
Apr. 2010	$21,200
May 2010	$20,700
Jun. 2010	$19,200
Jul. 2010	$16,100
Aug. 2010	$14,900
Sep. 2010	$12,100
Oct. 2010	$11,900
Nov. 2010	$12,500
Dec. 2010	$14,300
Jan. 2011	$17,500
Feb. 2011	$19,600
Mar. 2011	$20,900
Apr. 2011	$18,200
May 2011	$20,600
Jun. 2011	$18,800
Jul. 2011	$17,100
Aug. 2011	$14,100

It's difficult to answer those questions by examining the table. Now let's plot those values in Figure 3-1.

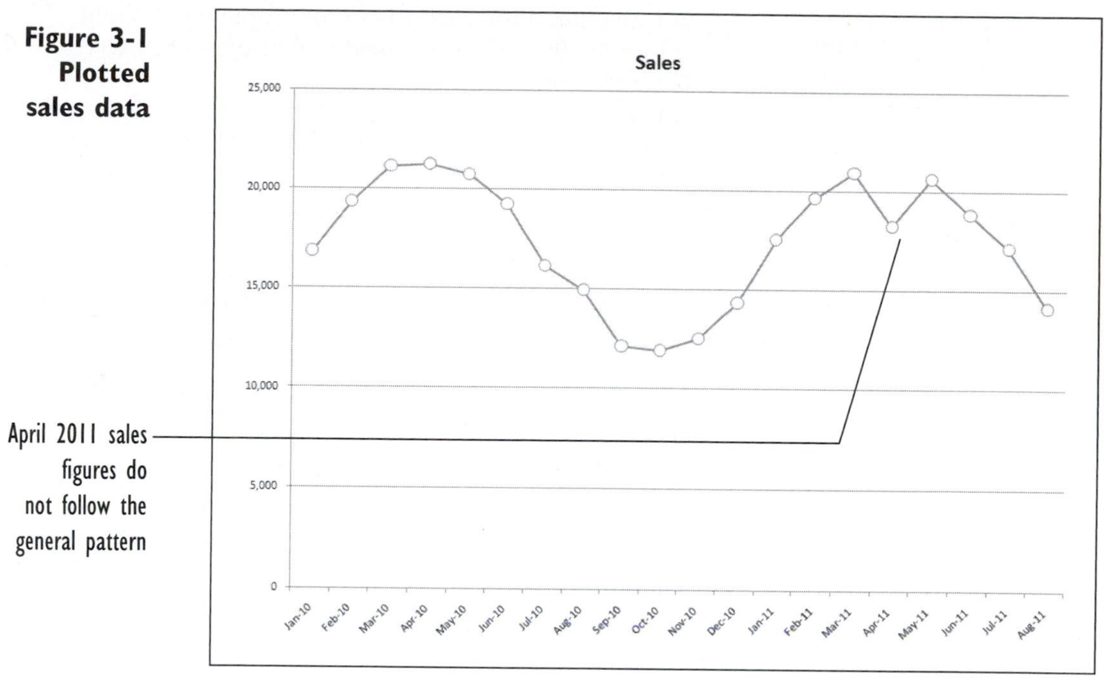

Figure 3-1 Plotted sales data

The chart clarifies things for us. We notice immediately that the sales figures seem to follow a classic seasonal curve with the highest sales occurring during the late winter—early spring months. However, the sales figures for April 2011 seem to be too low. Perhaps something occurred during this time period that should be investigated, or perhaps an erroneous value was entered. In any case, the chart has provided insights that would have been difficult to immediately grasp from a table of values alone.

Excel supports several different chart types for different situations. Table 3-2 shows a partial list of these.

Table 3-2 Excel Chart Types

Name	Icon	Description
Area		An area chart displays the magnitude of change over time or between categories. You can also display the sum of group values, showing the relationship of each part to the whole.
Column		A column chart shows how data change over time or between categories. Values are displayed vertically, categories horizontally.

(continued)

Bar		A bar chart shows how data change over time or between categories. Values are displayed horizontally, categories vertically.
Line		A line chart shows trends in data, spaced at equal intervals. It can also be used to compare values between groups.
Pie		A pie chart shows the proportional size of items that make up the whole. The chart is limited to one data series.
Doughnut		A doughnut chart, like the pie chart, shows the proportional size of items relative to the whole; it can also display more than one data series at a time.
Stock		The stock chart is used to display stock market data, including opening, closing, low, and high daily values.
XY(Scatter)		An XY(scatter) chart displays the relationship between numeric values in several data series. The chart is commonly used for scientific data and is also known as a scatter plot.
Bubble		A bubble chart is a type of scatter plot in which the size of the bubbles is proportional to the value of a third data series.
Radar		A radar chart shows values from different categories radiating from a center point. Lines connect the values within each data series.
Surface		A surface chart shows the value of a data series in relation to the combination of the values of two other data series. Surface charts are often used in topographical maps.
Cone, Cylinder, and Pyramid		Cone, cylinder, and pyramid charts are similar to bar and column charts except that they use cones, cylinders, and pyramids for the markers.

Excel includes variations of each of these chart types. For example, the column charts can display values across categories or the percentage that each value contributes to the whole across categories. Many of the charts can be displayed in 3D as well.

Most of the charts you'll create in this book will be of the XY(scatter) type. Other chart types, like stock charts, are designed for specific types of data (like stock market data), and they are not as useful for general data analysis. In addition, the StatPlus add-in included with this book gives you the capability of creating other types of charts not part of Excel's library of built-in charts. You'll learn about these charts as you read through these chapters.

Excel charts are placed in workbooks in one of two ways: either as **embedded chart objects**, which appear as objects within worksheets, or as

chart sheets, which appear as separate sheets in the workbook. Figure 3-2 shows examples of both ways of displaying a chart.

Figure 3-2 An embedded chart object and a chart sheet

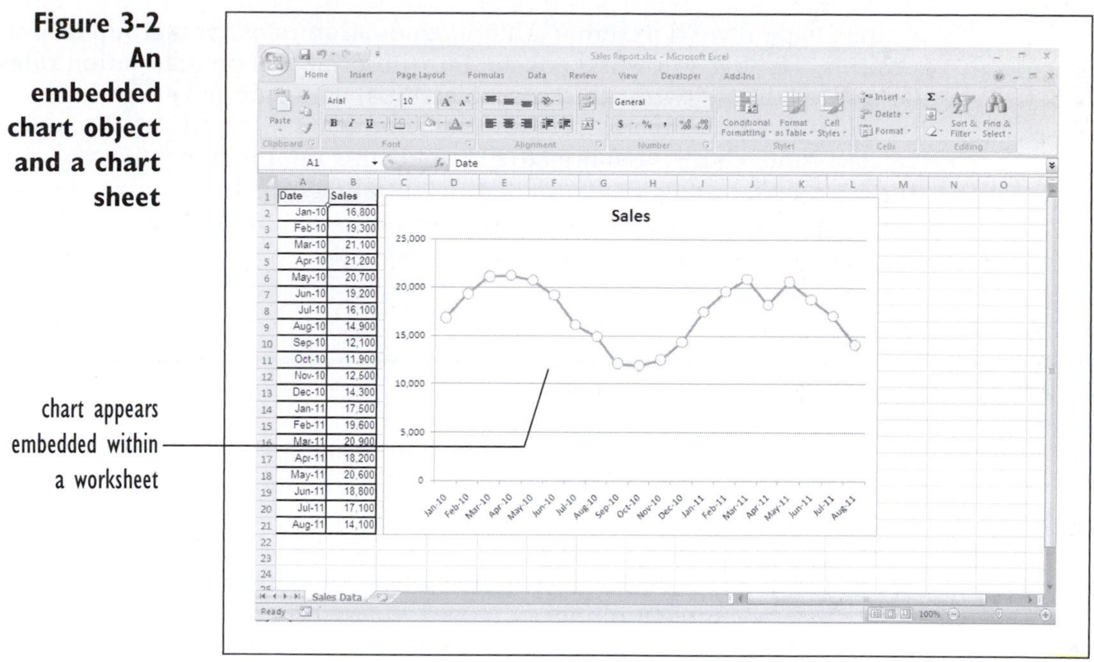

chart appears embedded within a worksheet

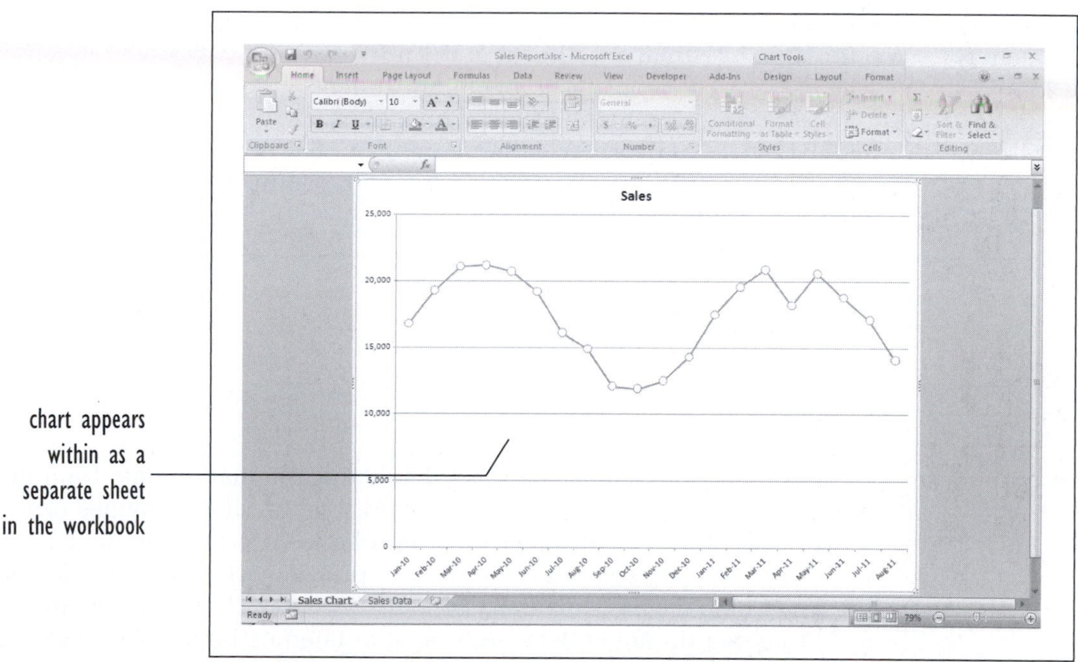

chart appears within as a separate sheet in the workbook

Introducing Scatter Plots

In this chapter, we'll examine athletic graduation rates for a group of universities. The Big Ten workbook contains information on graduation rates of student athletes (with athletic scholarships) who enrolled as freshmen at Big Ten universities in 1997, 1998, 1999, and 2000. Each NCAA Division I college or university is required to distribute this information to prospective student-athletes and parents, so that potential recruits have a way of comparing the education environment between different universities. Table 3-3 describes the range names used in the workbook.

Table 3-3 Big Ten Graduation Rates

Range Name	Range	Description
University	A2:A12	Name of the university
SAT	B2:B12	Average SAT scores of all freshmen
ACT	C2:C12	Average ACT scores of all freshmen
SAT_Calc	D2:D12	SAT calculated from ACT based on a formula from the College Board
Graduated	E2:E12	Percentage of all freshmen graduating within 6 years of enrolling.
White_Males	F2:F12	Six-year graduation rates for white male athletes
Black_Males	G2:G12	Six-year graduation rates for black male athletes
White_Females	H2:H12	Six-year graduation rates for white female athletes
Black_Females	I2:I12	Six-year graduation rates for black female athletes
Enrollment	J2:J12	Total enrollment at the university
Top_25	K2:K12	Percentage of incoming students graduating in the top 25% of their high school class
Top_25_Rate	L2:L12	Indicates whether more than 80% or less than 80% of the incoming freshmen graduated in the upper quarter of their class

To open the Big Ten workbook:

1 Start Excel and open the **Big Ten** workbook from the Chapter03 folder.

The workbook opens to a sheet displaying graduation data from the 11 universities in the Big Ten. See Figure 3-3. Range names based on the column labels of each column have already been created for you. There are some missing values in the worksheet, such as the SAT value for the University of Iowa in row 4. However for universities that do not collect SATs, a calculated estimate of the SAT is displayed in column E.

Figure 3-3
The Big Ten workbook

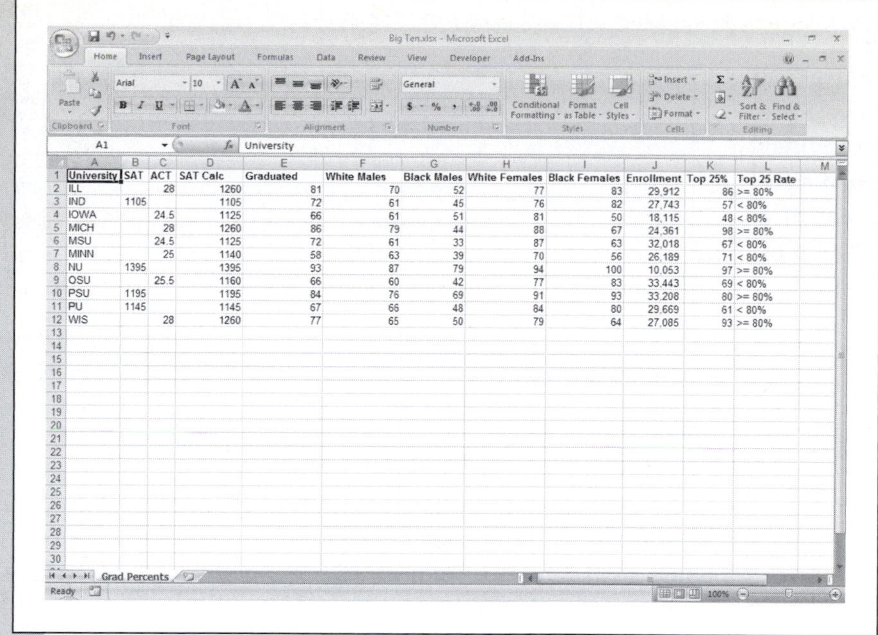

University	SAT	ACT	SAT Calc	Graduated	White Males	Black Males	White Females	Black Females	Enrollment	Top 25%	Top 25 Rate
ILL		28	1260	81	70	52	77	83	29,912	86	>= 80%
IND	1105		1105	72	61	45	76	82	27,743	57	< 80%
IOWA		24.5	1125	66	61	51	81	50	18,115	48	< 80%
MICH		28	1260	86	79	44	88	67	24,361	98	>= 80%
MSU		24.5	1125	72	61	33	87	63	32,018	67	< 80%
MINN		25	1140	58	63	39	70	56	26,189	71	< 80%
NU	1395		1395	93	87	79	94	100	10,053	97	>= 80%
OSU		25.5	1160	66	60	42	77	83	33,443	69	< 80%
PSU	1195		1195	84	76	69	91	93	33,208	80	>= 80%
PU	1145		1145	67	66	48	84	80	29,669	61	< 80%
WIS		28	1260	77	65	50	79	64	27,085	93	>= 80%

2 Save the file as **Big Ten Graduation Chart**.

A school with a low graduation rate among its student-athletes is vulnerable to investigation and possible sanctions on the part of the NCAA. We're going to explore the relationship between the average SAT score from classes of incoming first-year students and the percentage of those students in those classes who eventually graduate within six years of entering college.

One question we might ask is: Do incoming classes with high average SAT scores have higher rates of graduation? Is this true for all universities? We'll get a visual picture of this relationship by producing a scatter plot.

A **scatter plot** is a chart in which observations are represented by **points** on a rectangular coordinate system. Each observation consists of two values: One value is plotted against the vertical or ***y* axis**, and the second value is plotted against the horizontal or ***x* axis** (see Figure 3-4). In Figure 3-4 we are plotting a point with $x = 3$ and $y = 5$.

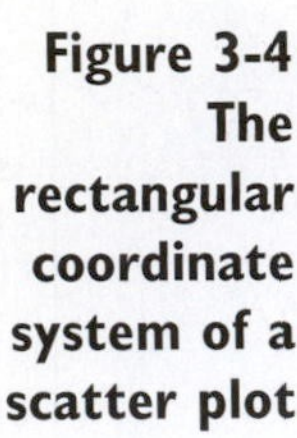

Figure 3-4 The rectangular coordinate system of a scatter plot

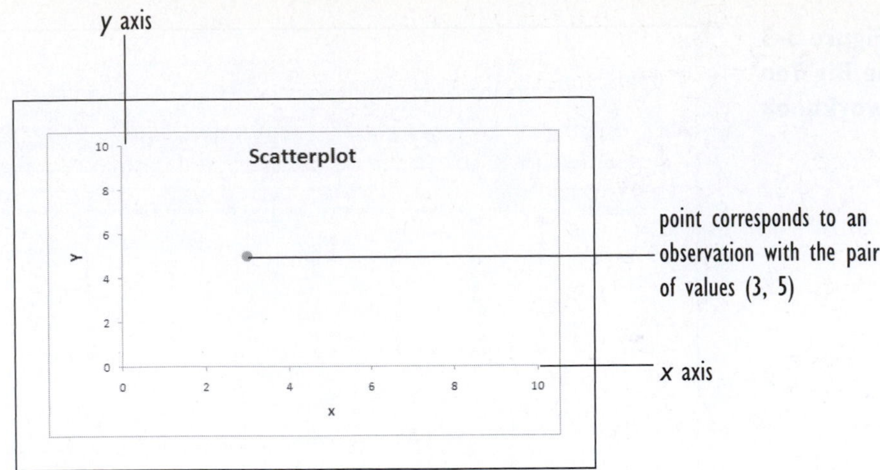

By observing the placement of the points on the scatter plot, you can get a general impression of the relationship between the two sets of values. For example, the scatter plot of Figure 3-5 shows that high values of Variable 1 are associated with low values of Variable 2. This is not a perfect association; rather, there is some scatter in the points.

Figure 3-5 Scatter plot displays a general relation between two sets of values

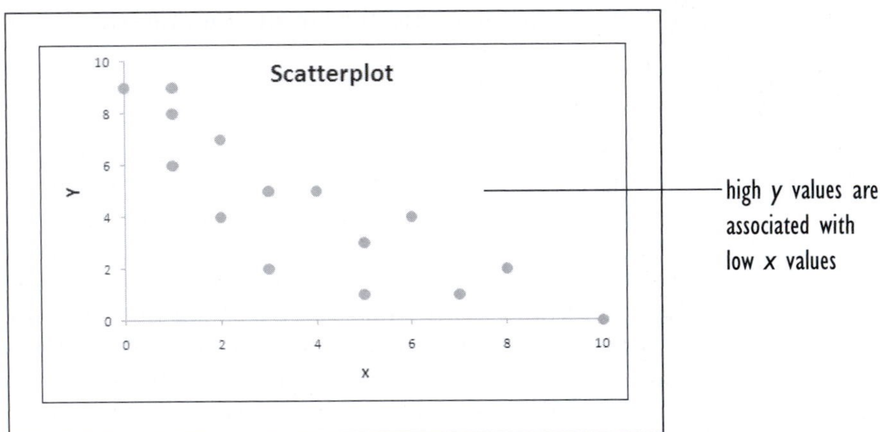

The scatter plot you'll create will have the graduation rate for each university on the *y* axis and the university's average SAT score on the *x* axis. To create a chart, you insert a chart object on the worksheet using the Excel Insert tab.

To insert a chart:

1 Select the cell range **D1:E12.**

2 Click the **Scatter** button from the Charts group on the Insert tab and then click the **Scatter with only markers** option as shown in Figure 3-6.

Figure 3-6 Inserting a scatter plot

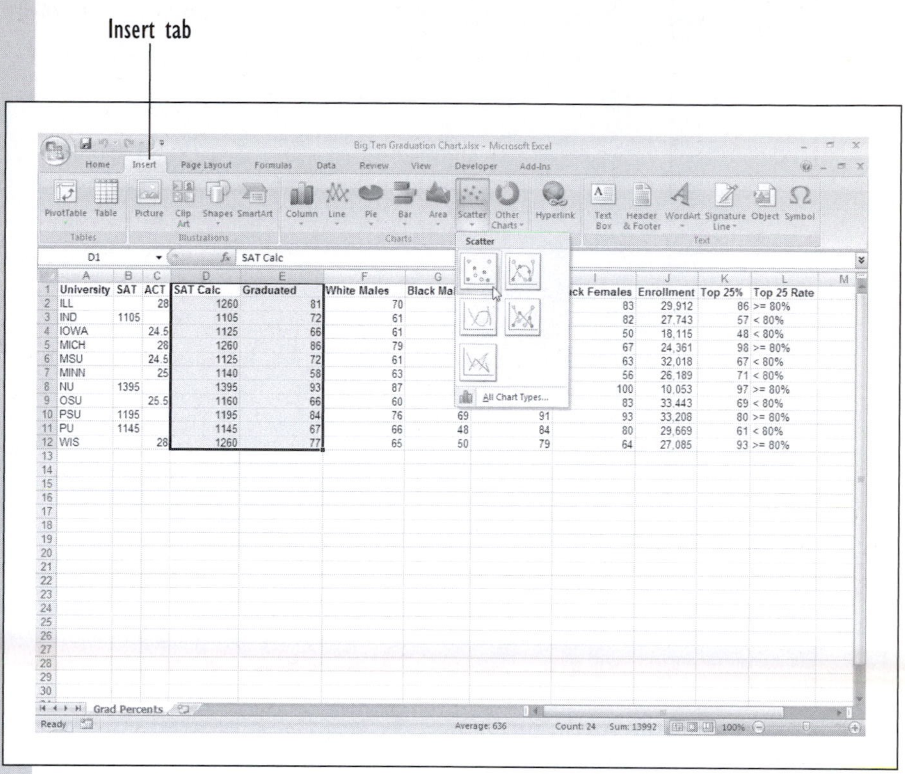

Excel inserts an embedded chart object containing the scatter plot of the Graduated values versus the SAT calculated values (see Figure 3-7).

Figure 3-7 Embedded scatter plot

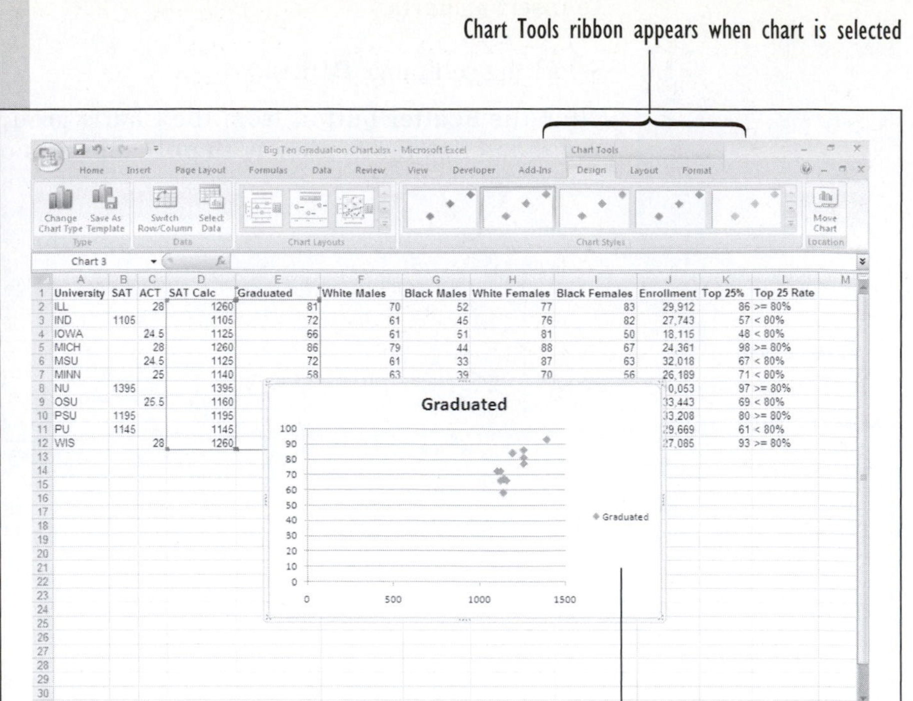

Once you've created the basic scatter plot, you can format its appearance using the commands on the Chart Tools ribbon. Note that the Chart Tools ribbon is a contextual ribbon and will appear within the Excel window whenever a chart object is selected in the new workbook. Newly created charts are selected by default.

EXCEL TIPS

- When you insert a scatter plot using Excel, the column of data values to the left will be used for the *x* values, the column on the right will be used for the *y* values. If your columns are not laid out this way, then do the following:

  a. Generate the scatter plot with the two columns as they are currently laid out.

  b. Click the Select Data button from the Data group on the Design tab of the Chart Tools ribbon to open the Select Data Source dialog box.

c. Click the Edit button within the dialog box, and then select a different data range for the Series X values and the Series Y values and the Series name.

If you don't want to manually edit your scatter plot this way, you can also create scatter plots using StatPlus. Simply click the StatPlus menu, click Single Variable Charts, and then click Fast Scatterplot. From the dialog box that appears, you can select the *x* axis and *y* axis values without regard to their column order.

- Excel supports five built-in scatter plots, allowing the user to connect the scatter plot points with straight or smoothed lines.

Editing a Chart

Using Excel's editing tools, you can modify the symbols, fonts, colors, and borders used in your chart. You can change the scale of the horizontal and vertical axes and insert additional data into the chart. To start, you'll edit the size and location of the embedded chart object you just created with Excel.

Resizing and Moving an Embedded Chart

Newly created charts are inserted as embedded objects with selection handles around the chart. When a chart is selected, you can move and resize it on the worksheet. The chart you've just created covers some of the data on the Grad Percents worksheet. Move it to a different location.

To move the embedded chart:

1 Click an empty area within the embedded chart, either above or to the right of the chart area, and hold down the mouse button. As you press the mouse button down, the pointer changes to a .

Note: If you click the title or other chart element, that element will have a selection border around it. If this happens, click elsewhere within the chart, holding down the mouse button. You don't want to select individual chart elements yet.

2 With the mouse button still pressed down, move the chart down so that the upper-left corner of the chart covers cell B14 and release the mouse button (see Figure 3-8.)

Figure 3-8 Moving an embedded chart

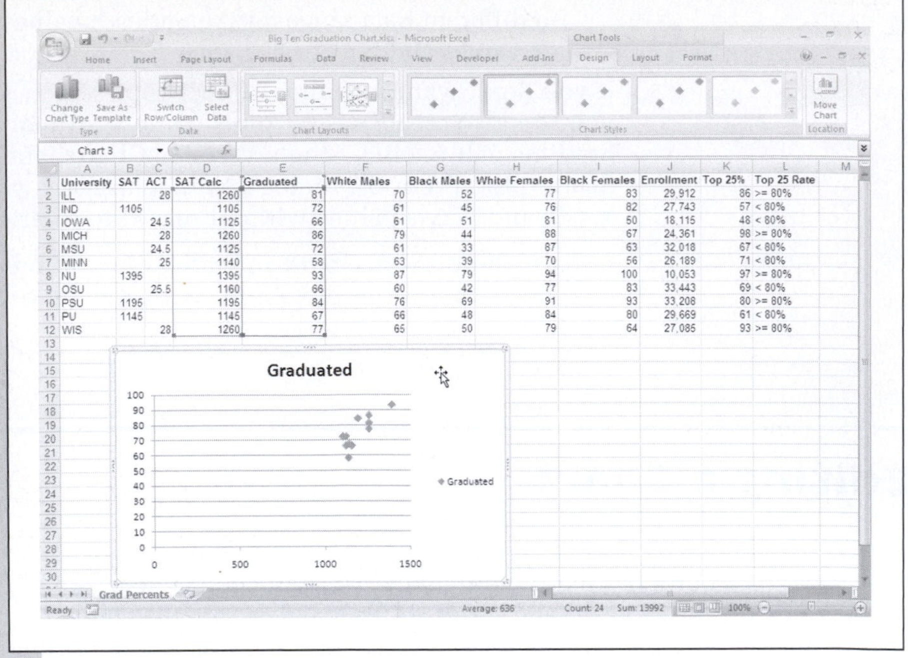

	University	SAT	ACT	SAT Calc	Graduated	White Males	Black Males	White Females	Black Females	Enrollment	Top 25%	Top 25 Rate
2	ILL		28	1260	81	70	52	77	83	29,912	86	>= 80%
3	IND	1105		1105	72	61	45	76	82	27,743	57	< 80%
4	IOWA		24.5	1125	66	61	51	81	50	18,115	48	< 80%
5	MICH		28	1260	86	79	44	88	67	24,361	98	>= 80%
6	MSU		24.5	1125	72	61	33	87	63	32,018	67	< 80%
7	MINN		25	1140	58	63	39	70	56	26,189	71	< 80%
8	NU	1395		1395	93	87	79	94	100	10,053	97	>= 80%
9	OSU		25.5	1160	66	60	42	77	83	33,443	69	< 80%
10	PSU	1195		1195	84	76	69	91	93	33,208	80	>= 80%
11	PU	1145		1145	67	66	48	84	80	29,669	61	< 80%
12	WIS		28	1260	77	65	50	79	64	27,085	93	>= 80%

Around the selected chart object at the four corners and four sides are selection handles that you can use to resize the embedded chart. As you move the pointer arrow over the handles, you'll see the pointer change to a double-headed arrow of various orientations. Each pointer allows you to resize the chart in the direction indicated. Try using the selection handles to make the chart a little larger.

To enlarge the chart:

1 Move your mouse pointer over the handle on the right edge of the chart until the pointer changes to a .

2 Drag the pointer to the right so that the embedded chart covers column I.

3 Move the pointer to bottom edge of the chart object and drag the selection handle so that the border edge covers row 35 (see Figure 3-9).

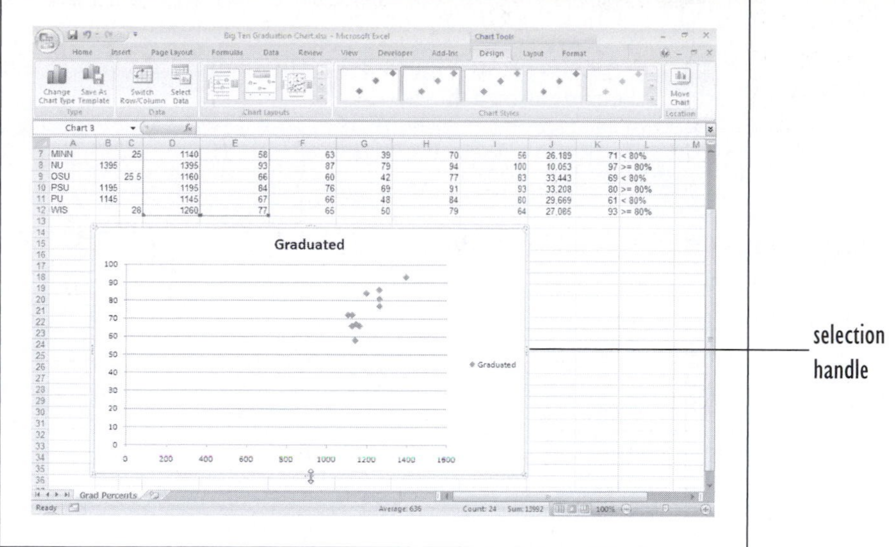

Figure 3-9 Resizing an embedded chart

Moving a Chart to a Chart Sheet

To move an embedded chart to its own chart sheet, you can use the Move Chart button from the Chart Tools ribbon. Try this now by moving the embedded chart you just created to a chart sheet.

To move an embedded chart to a chart sheet:

1. With the embedded chart still selected, click the **Move Chart** button located in the Location group of the Design tab of the Chart Tools ribbon.
2. Excel opens the Move Chart dialog box. Click the **New sheet** option button and type **Graduation Chart** in the accompanying text box as shown in Figure 3-10.

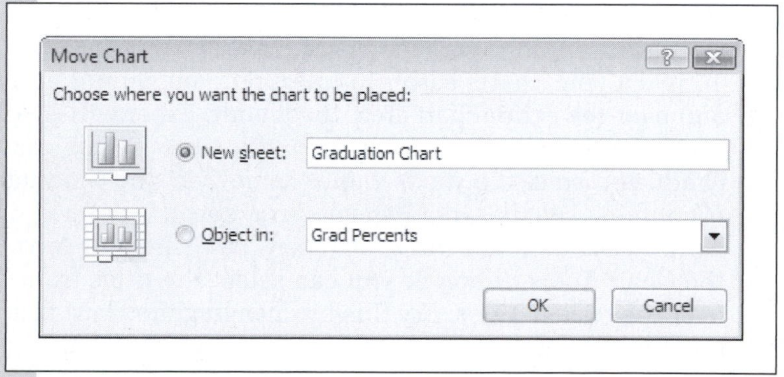

Figure 3-10 Move Chart dialog box

3 Click the **OK** button.

As shown in Figure 3-11, the chart is moved to a chart sheet named Graduation Chart.

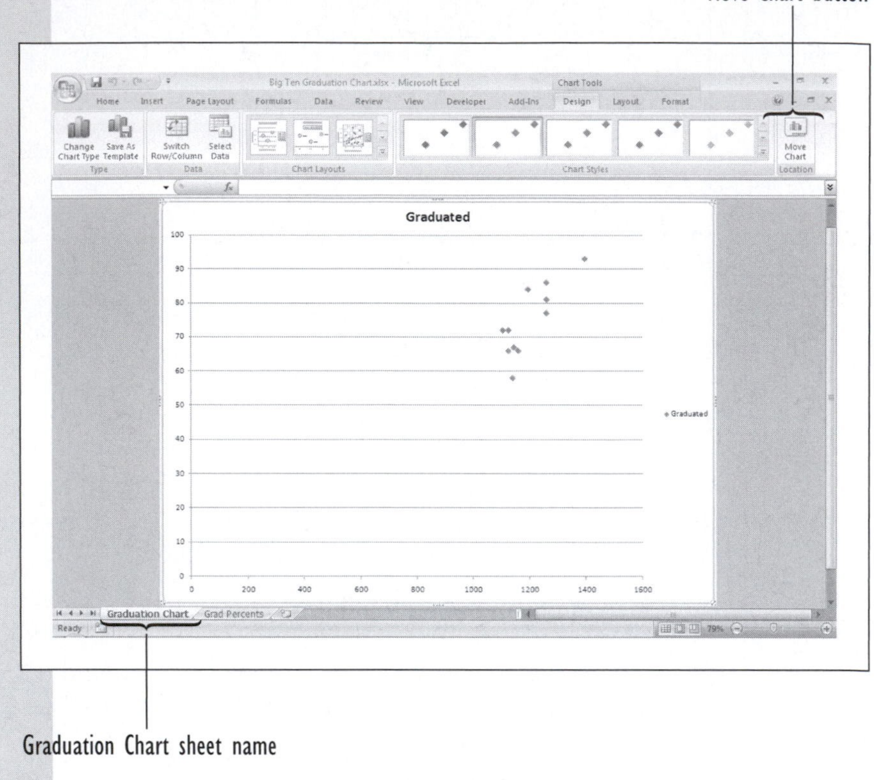

Figure 3-11 Newly inserted chart sheet

Working with Chart and Axis Titles

To make your charts easier to interpret, you should add titles to both axes and over the entire chart area. By default, Excel will display the name used for the *y* axis data values as the chart title. In this case the chart title is Graduated since the *y* axis values come from the Graduated column, which is column G on the Grad Percents worksheet.

To insert different titles, use the Chart Title and Axis Title buttons from the Chart Tools ribbon or you can select the titles from the chart and type over the current titles. Try this by changing the chart title to Big Ten Graduation Percentages.

To change the chart title:

1 Click the **Graduated** title located directly above the scatter plot.

Selection handles appear around the chart title indicating that it is selected.

2 Type **Big Ten Graduation Percentages** and press **Enter**. As shown in Figure 3-12, the title changes to reflect the new text.

Figure 3-12 Changing the chart title

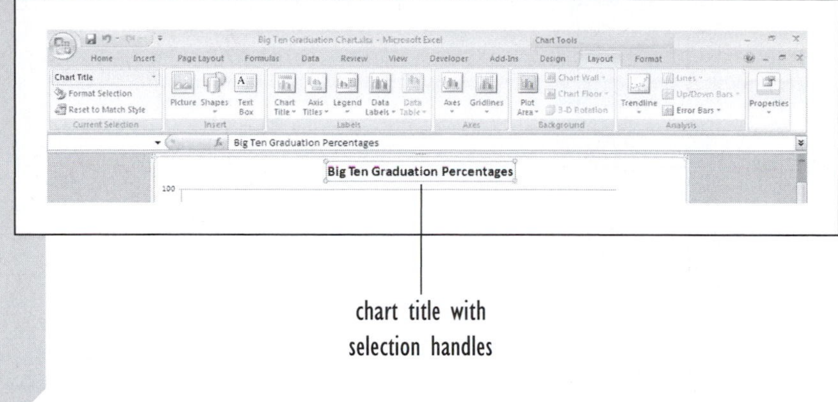

Next, add titles to both the *y* axis and *x* axis. Since these titles are not currently displayed on the chart you have to add them with the Axis Titles button.

To insert the axis titles:

1 Click the **Axis Titles** button from the Labels group on the Layout tab of the Chart Tools ribbon, click **Primary Horizontal Axis Title**, and then click **Title Below Axis**.

Excel inserts the text Axis Title below the horizontal, or *x*, axis. The Axis Title text is surrounded with selection boxes indicating that it is the currently selected object in the chart.

2 Type **Calculated SAT Values** and press **Enter**.

3 Click the **Axis Titles** button again, click **Primary Vertical Axis Title**, and then click **Rotated Title**.

4 Type **Graduation Percentage** and press **Enter**. Figure 3-13 shows the newly added axis titles.

Figure 3-13
Inserting axis titles

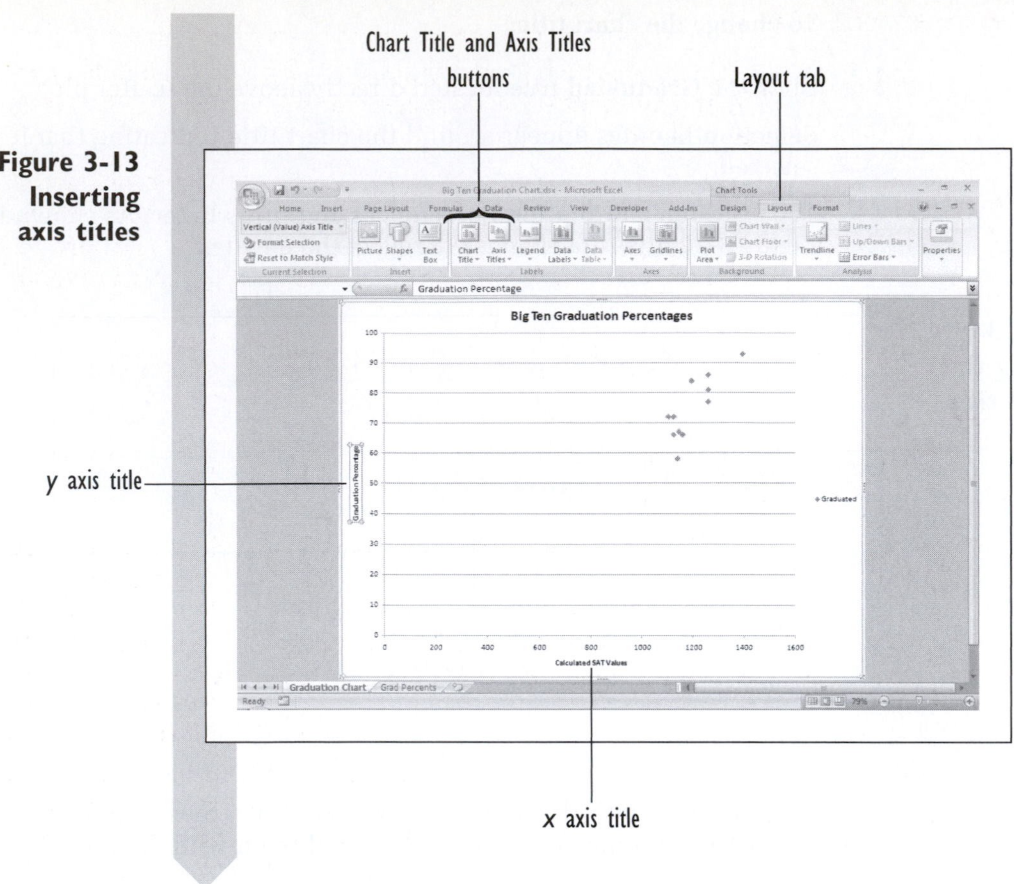

The chart and axis titles you've entered can be formatted using the same formatting buttons found on the Home tab that you use for formatting cell text. Try using these buttons now to increase the font size of the two axes titles.

To format the axis titles:

1 Click the **Home** tab with the Graduation Percentage title still selected; then click the **Font Size** button 11, changing the font size to **16**.

2 Click the **Calculated SAT Values** *x* axis title and change the font size to **14** using the same technique.

EXCEL TIPS

- You can also use the format buttons on the Home tab to change the font color, font style, alignment, and fill color of chart and axes titles.
- You can further format chart and axes titles by selecting the title and then clicking the Format Selection button from the Current Selection group on the Layout tab of the Chart Tools ribbon. The button opens a Format Axis Title dialog box containing several formatting options.

Editing the Chart Axes

Next you'll look at editing the values displayed in the chart. Even though all of the Big Ten graduation rates are 50% or greater in this chart, Excel uses a range of 0 to 100%; and even though the lowest SAT score is 1105, Excel uses a lower range of 0 in the chart. The effect of this is that all of the data are clustered in the upper right edge of the chart, leaving a large blank space to the left and below. There are some situations where you want your charts to show the complete range of possible values and other situations where you want to concentrate on the range of observed values. In that case, you rescale the axes so that the scales more closely match the range of the observed values. You can change the scale of the axes by clicking the Axes button on the Chart Tools ribbon. Start with changing the scale on the *x* axis to better match the range of calculated SAT scores for the 11 schools in the data sample.

To change the scale of the *x* axis:

1 Click the **Axes** button from the Axes group on the Layout tab of the Chart Tools ribbon, click **Primary Horizontal Axis** and then click **More Primary Horizontal Axis Options**. Excel opens the Format Axis dialog box.

2 Verify that the Axis Options tab is selected. On this tab the options for the axis scale are shown. By default, Excel automatically selects the minimum and maximum range of the axis values. You want to change the minimum from 0 to 1000, set the maximum at 1500, and set the interval between tick marks on the axis to 50 units. To make this change, do the following steps:

3 Click the **Fixed** option button for the Axis minimum and enter **1000** in the accompanying text box.

4 Click the **Fixed** option button for the Axis maximum and enter **1500** in the accompanying text box.

5 Click the **Fixed** option button for the Axis major unit and enter **50** in the accompanying text box. Figure 3-14 shows the completed Format Axis dialog box.

Figure 3-14
Setting the axis scale

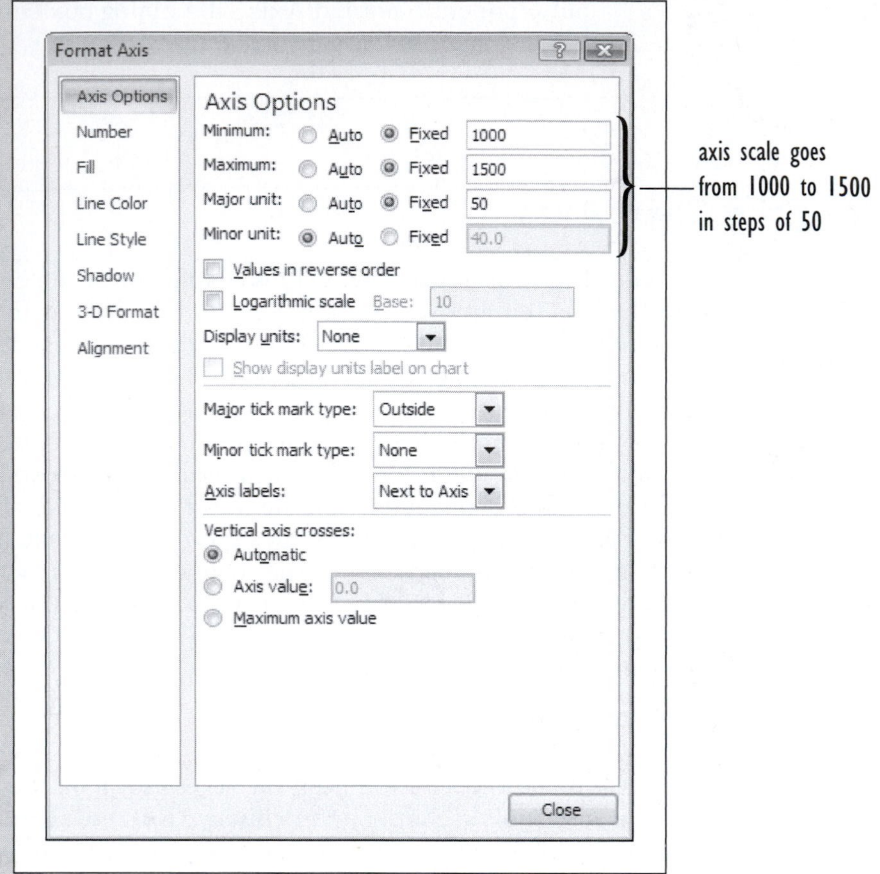

6 Click the **Close** button. Figure 3-15 shows the appearance of the chart with the revised scale for the *x* axis.

Figure 3-15 Revised scale for the *x* axis

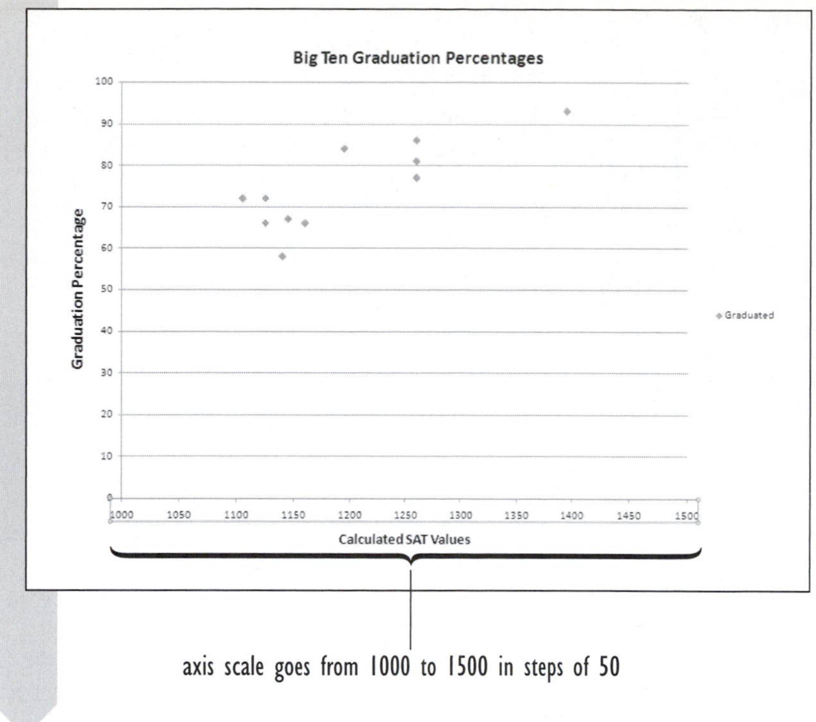

Now the data points are only compressed near the top of the chart. You can change this by setting the scale of the graduation percentages on the *y* axis to go from 50 to 100% in steps of 5%.

To change the scale of the *y* axis:

1 Click the **Axes** button from the Axes group on the Layout tab of the Chart Tools ribbon, click **Primary Vertical Axis** and then click **More Primary Vertical Axis Options**. Excel opens the Format Axis dialog box for the vertical axis.

2 Click the **Fixed** option buttons for the Axis minimum, maximum, and major units and change the minimum value to **50**, the maximum value to **100**, and the major unit to **5**.

3 Click the **Close** button. Figure 3-16 shows the revised scale for the vertical axis on the chart.

Figure 3-16
Revised scale for the y axis

Big Ten Graduation Percentages

Graduation Percentage

Calculated SAT Values

Graduated

axis scale goes from 50 to 100 in steps of 5

EXCEL TIPS

- You can display the axis scale in units of thousands, millions, and billions by selecting the appropriate options from the Axes button.
- You can display values on a log scale by selecting the Show Axis with Log Scale option on the Axes button.

Working with Gridlines and Legends

Another chart object that appears in the graduation scatter plot is gridlines. **Gridlines** are vertical or horizontal lines that match up with the major and minor units on the *x* and *y* axes. Gridlines can make it easier to line up data values within the scatter plot. By default, Excel will open a scatter plot with horizontal gridlines matching the major units on the *y* axis. You can add or remove major and minor gridlines from scatter plots using the commands on the Chart Tools ribbon. Try this by adding vertical gridlines to the graduation percentage scatter plot.

To add vertical gridlines:

1 With the chart still selected, click the **Gridlines** button from the Axes group on the Layout tab of the Chart Tools ribbon, click **Primary Vertical Gridlines** and then click **Major Gridlines**.

As shown in Figure 3-17, vertical gridlines are added to the scatter plot.

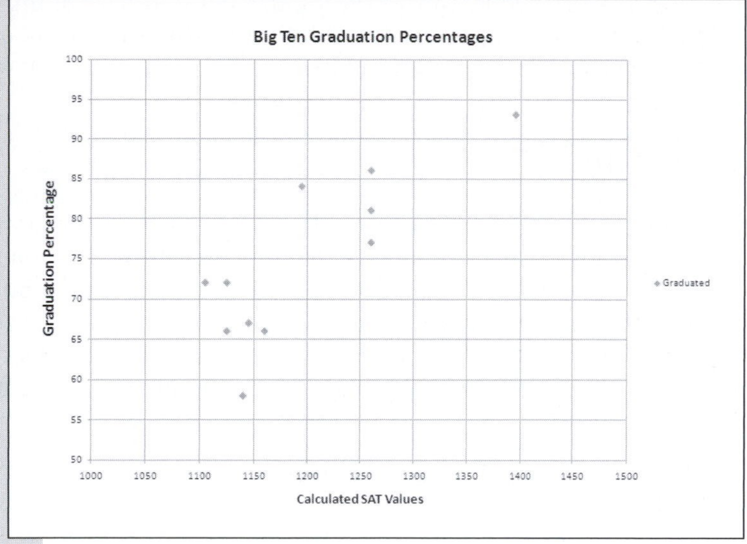

Figure 3-17 Adding vertical gridlines

You can edit the format of the gridlines by clicking More Primary Gridlines Options command found on the menu of commands for each gridline. By modifying the format, you can change the gridline's color and style as well as add drop shadows to each gridline.

The graduation percentage scatter plot also contains a legend. A **legend** is a box that identifies the patterns or colors that are assigned to the data points in a chart. When you insert a chart, Excel automatically adds a legend. In the graduation percentage chart, the legend appears on the right edge of the chart, providing the name of the *y* values (in this case values from the Graduated column). If there is only one set of data values in the chart you usually do not need a legend.

To remove the legend:

1 Click the **Legend** button from the Labels group in the Layout tab of the Chart Tools ribbon and then click **None**.

Excel removes the legend from the scatter plot

From the Legend button you can also choose commands to move the legend to different locations relative to the chart area and to format the legend's appearance including its size, fill color, and font styles.

Editing Plot Symbols

As with other parts of the chart, Excel allows the user to modify the display of the plot symbols. By default, Excel uses a blue diamond as the plot symbol. You'll change this to an empty circle. There is no button on the Chart Tools ribbon to modify the appearance of the symbols; instead you must select the symbols and format them using the Format Selection button on the Layout tab. Try this by changing the appearance of the scatter plot points to empty circles.

To change the plot symbol:

1 Click any of the plot symbols in the scatter plot. Selection handles will appear around all of the scatter plot data points.

2 Click the **Format Selection** button from the Current Selection group on the Layout tab of the Chart Tools ribbon. Excel opens the Format Data Series dialog box.

The Format Data Series dialog box is divided into different tabs that allow you to format the appearance of the plot symbols used in the selected data series.

3 Click the **Marker Options** tab and then click the **Built-in** option button. Click the **Type** drop-down list box and select the **circle** symbol.

4 Click the **Size** list box and increase the circle size to **10**. Figure 3-18 shows the selected options from the dialog box.

Figure 3-18 Selecting a plot symbol

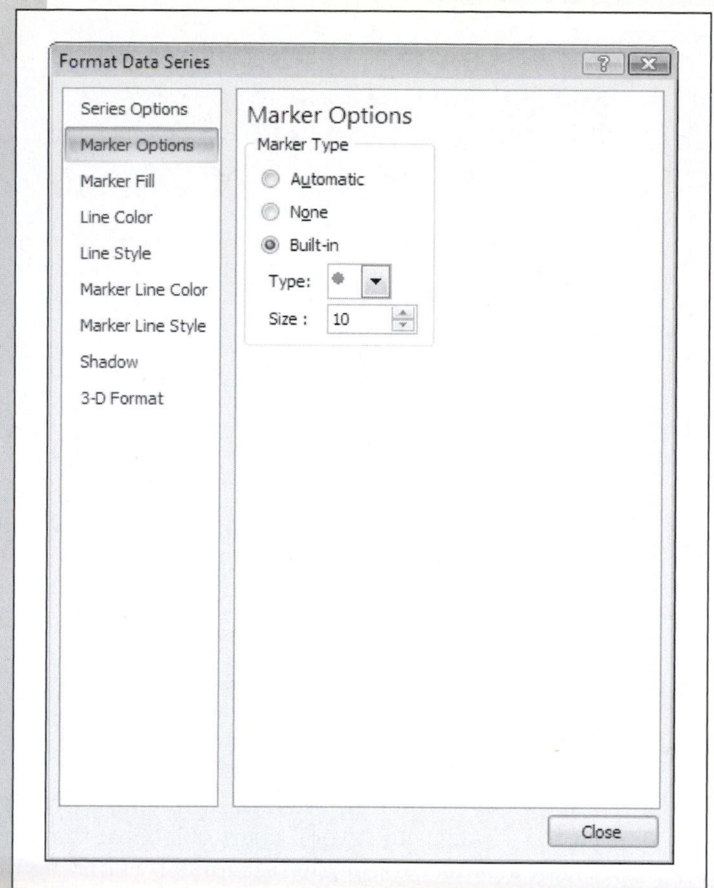

5 Click the **Marker Fill** tab and then click the **No fill** option button.

6 Click the **Close** button to accept the changes to the plot symbols.

7 Click outside of the chart sheet to deselect. Figure 3-19 shows the revised appearance of the graduation chart.

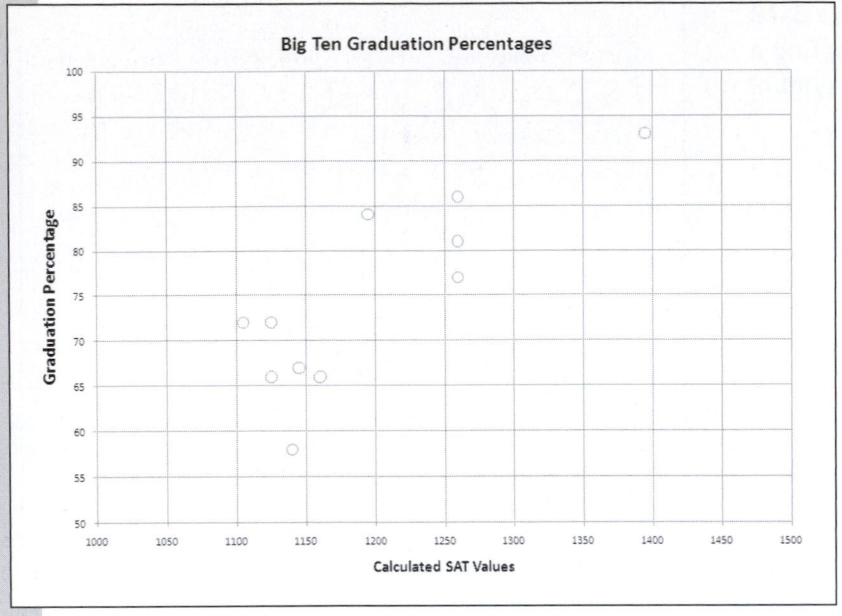

Figure 3-19 Revised symbols for the graduation chart

EXCEL TIPS

- Excel has several different built-in chart layouts for making quick changes to several chart objects at the same time. You can view a gallery of the chart layouts by clicking the Chart Layouts button located on the Design tab of the Chart Tools ribbon.
- You can change the type of chart displayed by Excel by selecting a chart and then clicking the Change Chart Type button located in the Type group of the Design tab in the Chart Tools ribbon. The Change Chart Type will then display a list of all chart types and chart templates stored on your computer.
- If you want to reuse all of the formatting choices you made for your chart in future charts, you can save your chart as a template by clicking the Save As Template button from the Type group on the Design tab of the Chart Tools ribbon.

Now that you've formatted the chart, interpret the scatter plot you've created. One of the questions you were asking was What is the relationship (if any) between the average SAT score of a freshman class and its eventual graduation rate? You can now put forward one hypothesis: Higher average SAT scores seem to be associated with higher graduation rates. That's not surprising, but there are a couple of exceptions to that relationship. For example, a freshman class of students at one university showed an average SAT score of 1140 with a graduation percentage of 58%, which might be lower than would be expected on the basis of the graduation rates for the other universities with similar average SAT scores. Which university is it?

Identifying Data Points

When you plot data, you often want to be able to identify individual points. This is particularly important for values that seem unusual. In those cases, you might want to go back to the source of the data and check to see whether there were any anomalies in how the data were collected and entered. You may have already noticed that if you pass your mouse cursor over the selected data points in the BigTen scatter plot, a screen tip appears to identify the data series name as well as the pair of values used in plotting the point (see Figure 3-20).

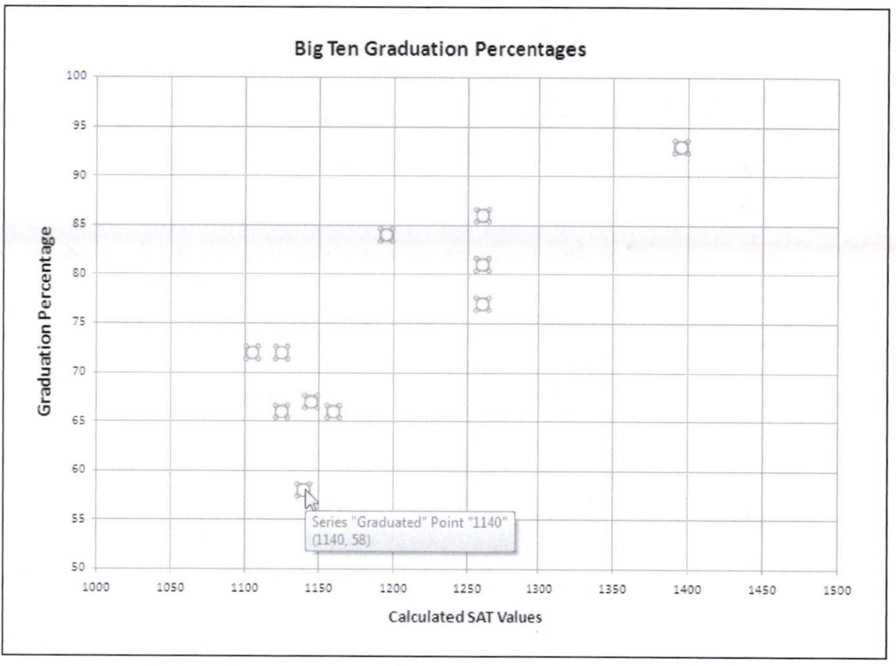

Figure 3-20 Screen tip identifying a data point

Although this information is interesting and potentially helpful, it doesn't tell you more about the source of the data point. For example, which university supplied this particular combination of SAT score and graduation percentage? One way to find out is to compare the values given in the pop-up label with the values in the worksheet. For example, you could return to the Grad Percents worksheet to see that the point identified in Figure 3-20 is from the University of Minnesota (MINN), whose freshman class had an SAT average of 1140 and an eventual graduation rate of 58%. In this fashion you could continue to compare values between the chart and the worksheet, finding out which university is associated with which data point. Of course, this is time consuming and impractical, especially for larger data sets. Excel doesn't provide any other method of identifying specific points, but the StatPlus add-in that comes with this book does provide some additional commands for this purpose (if you haven't installed StatPlus, please read the material in Chapter 1 about StatPlus and installing add-ins).

Selecting a Data Row

One of the StatPlus commands you can use to identify a particular row is the **Select Row** command. This command works only if your data values are organized into columns. To use this command, you select a single point from the chart and then click Select Row from the StatPlus menu. Try this now and identify the university that had the highest graduation percentage in the Big Ten.

To select a data row:

1. Click a data point in the scatter plot in order to select the entire data series.
2. Click the plot symbol highlighted in Figure 3-20 where the SAT value is equal to 1140 and the graduation percentage is equal to 58. Now only that plot symbol should be selected and none of the other symbols.
3. Click **StatPlus** from the Menu Commands group on the Adds-Ins tab.
4. Click **Select Row** from the StatPlus menu.

The eighth row should now be highlighted, indicating that the University of Minnesota (MINN) is the university that had the highest graduation percentage in the Big Ten (see Figure 3-21).

Figure 3-21 Selecting the data row for a plot point

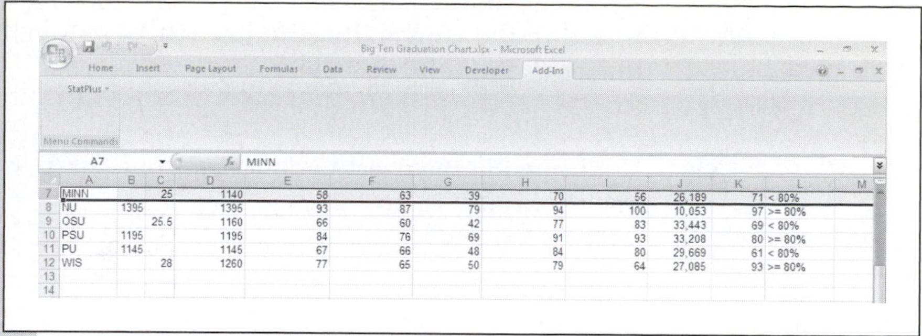

5 Click cell **A1** to remove the highlighting.

Labeling Data Points

You can also use the StatPlus add-in to attach labels to all of the points in the data series. These labels can be linked to text in the worksheet so that if the text changes, the labels are automatically updated. Use StatPlus now to add the university name to each point in the chart.

To add labels to the chart:

1. Return to the **Graduation Chart** chart sheet and then click outside of the chart to deselect it.
2. Click a plot symbol from the chart again to reselect all of the plot symbols.
3. Click **Label series points** from the StatPlus menu located in the Menu Commands group on the Adds-Ins tab. Excel opens the Label Point(s) dialog box.

   Most StatPlus commands give you the choice of entering range names or range references. Because range names have already been created for this workbook, you can select the appropriate range name from a list box. In this case, you'll use the text entered into the University column from the worksheet.

4. Click the **Labels** button to open the Input Options dialog box and then click the **Use Range Names** option button, if necessary, to select it.
5. Scroll down the list box and click **University** and then click the **OK** button.
6. Click the **Link to label cells** checkbox.

By linking to the label cells, you ensure that any changes you make to text in the University column will be automatically reflected in the labels in the scatter plot.

7 Click the **OK** button.

8 Click outside the chart to deselect. Your chart should resemble the one shown in Figure 3-22.

Figure 3-22 Identifying plot points by university

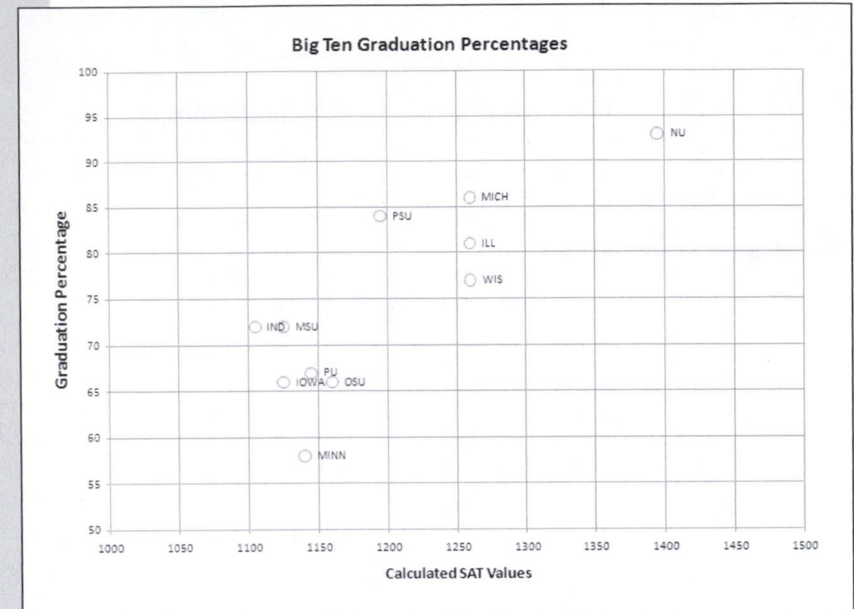

STATPLUS TIPS

- If you want to use the same text font and format in the worksheet and in the labels, click the **Copy label cell format** checkbox in the Label Point(s) dialog box.
- If you want to replace the plot symbols with labels, click the **Replace points with labels** checkbox in the Label Point(s) dialog box. Be aware, however, that once you do this, you cannot go back to displaying the plot symbols.
- To label a single point rather than all of the points in the data series, select only a single plot symbol and then apply the Label Point(s) command from the StatPlus menu.

When you label every data point, there is often a problem with overcrowding. Points that are close together tend to have their labels overlap, as is the case with the Iowa, Ohio State, and Purdue labels in Figure 3-22. This is not necessarily bad if you're interested mainly in points that lie outside the norm.

Formatting Labels

You've learned that the Big Ten university that has a low graduation rate for its students relative to the average SAT score of its freshman class is Minnesota. You might wonder why its graduation rate is so much lower than the rates for the other universities. On the basis of the values in the chart, you would expect a graduation rate between 65 and 75% for an average SAT score of around 1140, not one as low as 58%. Perhaps it is because Minneapolis–St. Paul is the largest city among Big Ten towns, and students might have more distractions there, or the composition of the student body might be different. Columbus is the next largest city, and Ohio State is next to last in graduation rate with 66%, which seems to verify this hypothesis. On the other hand, Northwestern is in Evanston, right next door to Chicago, the biggest midwestern city, so you might expect it to have a low graduation rate too. However, Northwestern is also a private school with an elite student body and high admission standards and has a graduation rate of 93%.

Minnesota's graduation rate still seems curious. You decide to mark this point for further study by changing the color of the label to boldface red.

To format a label:

1. Click any label in the chart to select all of the labels in the data series.

   Note that selection handles appear around each label. If you wanted to format all of the labels simultaneously, you could do this by applying any of Excel's formatting commands to this selected group. To format a single label, you have to select it again from the group of labels.

2. Click the **MINN** label.

   The selection label is now limited to only the Minnesota point.

3. Click the **Font Color** button from the Font group on the Home tab and change the font color to **red** (the second entry in the list of standard colors).

4. Click the **Bold** button from the Font group.

5. Click outside the chart to deselect it. The format of the MINN data label should now be boldface red.

EXCEL TIPS

- You can also select and format your data labels by clicking the Data Labels button located in the Labels group of the Chart Layout tab.

Creating Bubble Plots

Let's examine another possible impact on the graduation rate. The data set also includes the percentage of all freshmen who graduated in the top 25% of their high school class. We could create a scatter plot of the Graduated column values versus the Top 25% column values. However it may be more instructive to include the calculated SAT values in the chart. One way of observing the relationship among the three variables is through a bubble plot. A **bubble plot** is similar to a scatter plot, except that the size of each point in the plot is proportional to the size of a third value. In this case, we'll create a bubble plot of graduation rate versus SAT average, the size of each plot symbol being determined by the percentage of incoming freshmen who graduate in the top 25% of their class. Note that we won't *prove* that this value affects the graduation rate; we are merely exploring whether there is graphical evidence to suggest such a relationship. Bubble plots are another chart type supported by Excel and can be easily created using the Insert tab.

To insert a bubble plot:

1 Return to the **Grad Percents** worksheet and select the nonadjacent cell range **D1:E12;K1:K12**.

The order of the columns is important in a bubble plot. The values for the *x* axis should be listed first, then the values for the *y* axis, followed by the values that determine the size of the plot bubbles.

2 Click the **Insert** tab and then click the **Other Charts** button from the Charts group on the ribbon; then as shown in Figure 3-23, select the first **Bubble** chart option from the menu.

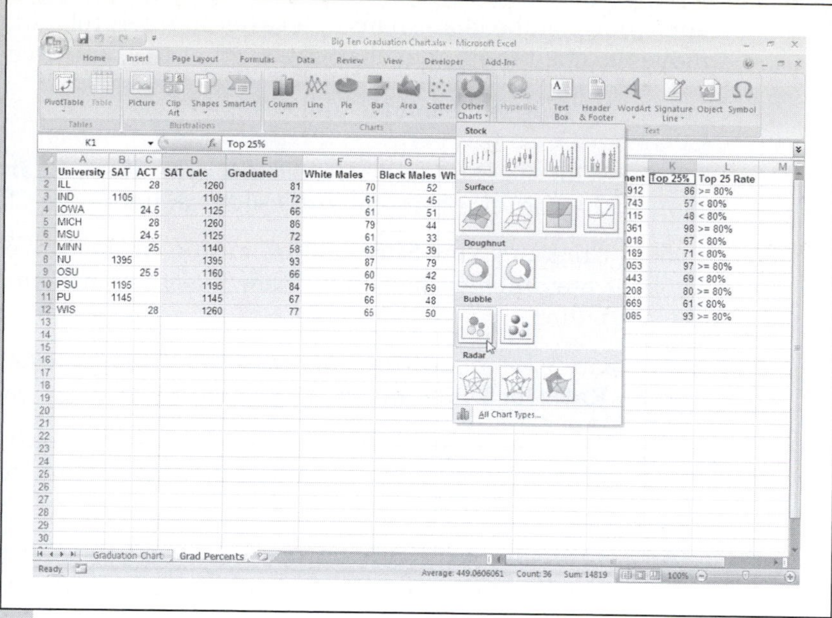

Figure 3-23 Identifying plot points by university

Excel inserts the unformatted bubble plot as an embedded chart on the Grad Percents worksheet (see Figure 3-24).

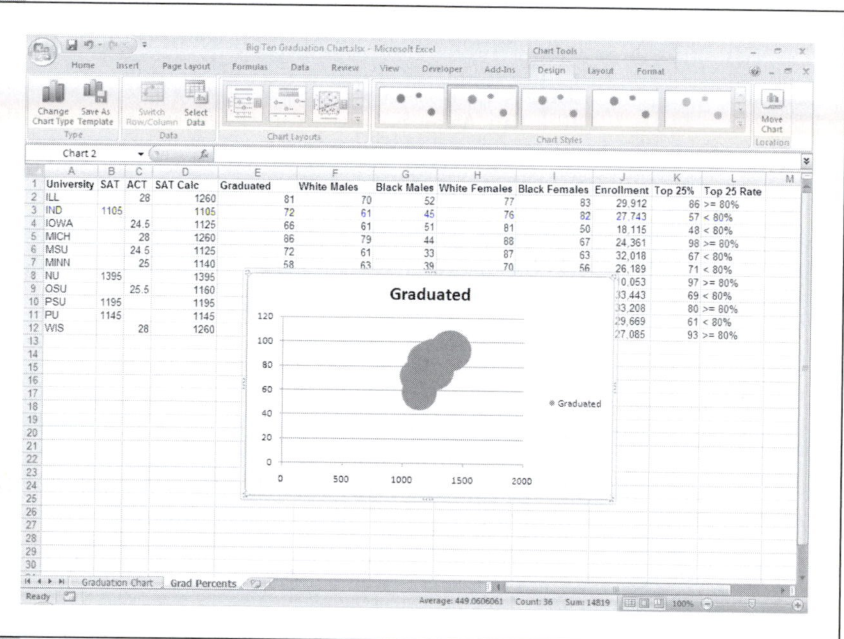

Figure 3-24 The initial bubble plot as an embedded chart object

As you did earlier in creating the scatter plot, you now have to format the appearance of the chart to make it easier to interpret and understand. You'll move the chart to its own chart sheet, add titles, and change the axis scales.

To format the bubble plot:

1 With the chart still selected click the **Move Chart** button located in the Location group of the Design tab under the Chart Tools ribbon.

2 Click the **New Sheet** option button in the Move Chart dialog box, enter **Bubble Chart** in the New Sheet text box and then click the **OK** button.

3 Click the chart title and change it from Graduated to **Graduation Rates for Different Top 25 Percent Rates**.

4 Click the **Legend** button from the Labels group on the Layout tab of the Chart Tools ribbon and then click **None** to remove the chart legend.

5 Click the **Axis Title** button from the Labels group, click **Primary Horizontal Axis** and then click **Title Below Axis**. Type **Calculated SAT Values** for the horizontal axis title. Set the font size to **14** points.

6 Click the **Axis Title** button again from the Labels group, click **Primary Vertical Axis**, and then click **Rotated Title**. Type **Graduation Percentages** for the vertical axis title. Set the font size to **16** points.

Now change the scale of the horizontal axis to go from 1000 to 1500 in intervals of 50 points.

7 Click the **Axes** button from the Axes group on the Layout tab of the Chart Tools ribbon and then click **More Primary Horizontal Axis Options**.

8 Within the Axis Options tab, set the Minimum fixed value to **1000**, the Maximum fixed value to **1500**, and the Major Unit value to **50**. Click the **Close** button.

Change the scale of the vertical axis to range from 50 to 100 in steps of 5.

9 Click the **Axes** button and then click **More Primary Vertical Axis Options**. Within the Axis Options tag of the Format Axis button set the Minimum value to **50**, the Maximum value to **100**, and the Major Unit value to **5**. Click the **Close** button.

Figure 3-25 shows the current appearance of the bubble plot.

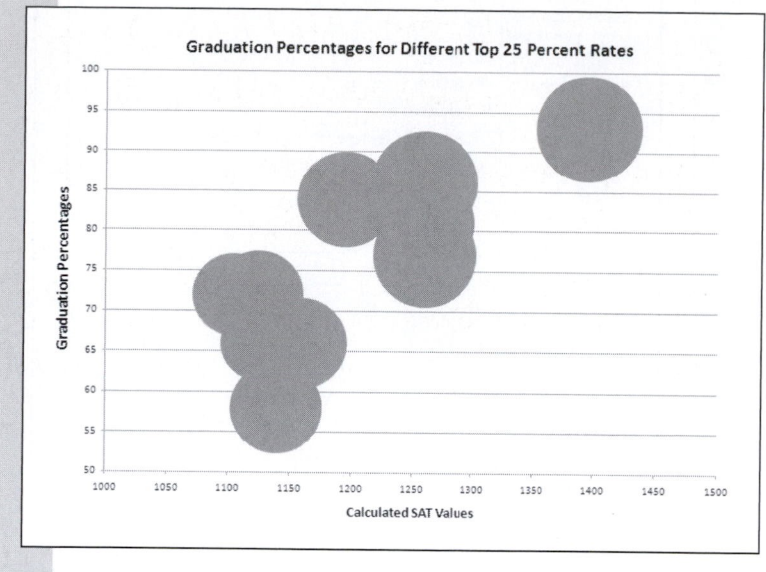

Figure 3-25 The initial bubble plot as an embedded chart object

The default appearance of the bubble symbols makes it difficult to separate one bubble from another. You can modify the plot symbols to make the chart easier to read and interpret.

To format the bubble symbols:

1. Click one of the bubbles in the chart to select all of the bubble symbols.
2. Click the **Format** tab from the Chart Tools ribbon and then click the **Format Selection** button from the Current Selection group.
3. Excel opens the Format Data series dialog box. Click the **Fill** tab and then click the **Color** button and select **Yellow** from the list of standard colors.
4. Fill colors are, by default, solid; but you can allow them to become partially transparent so that overlapping bubbles can be distinguished from one another. Drag the **Transparency** slider located below the Color button to the value **66%**. Figure 3-26 shows the modified fill color scheme for the bubbles.

Figure 3-26
Setting the fill color from the bubble symbols

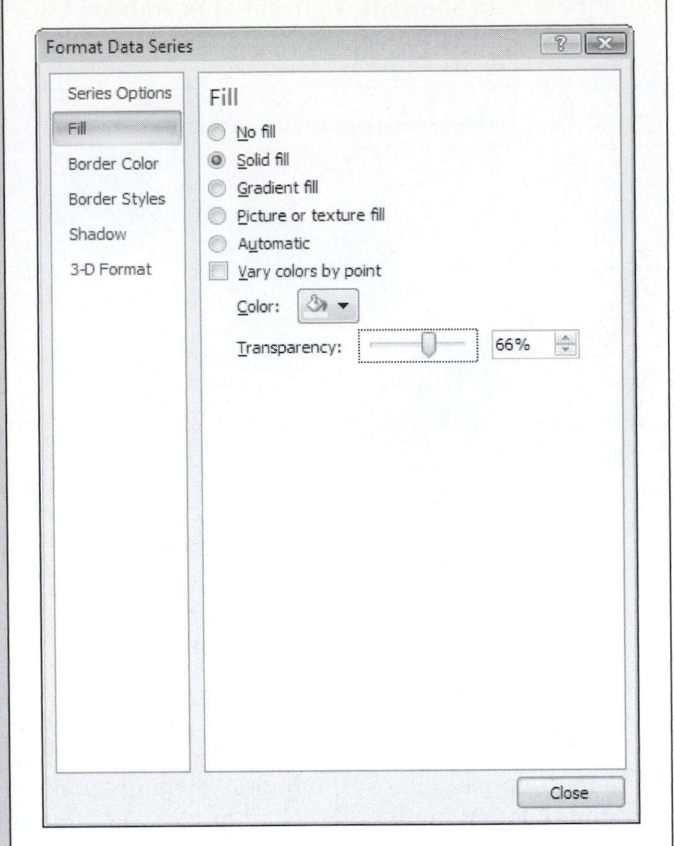

5 You still need to specify a border color for the different bubbles so that you can see where one bubble begins and the others end. Click the **Border Color** tab and then click the **Solid line** option button.

6 Click the **Close** button and then deselect the chart. The revised chart appears in Figure 3-27.

Figure 3-27 Revised bubble plot

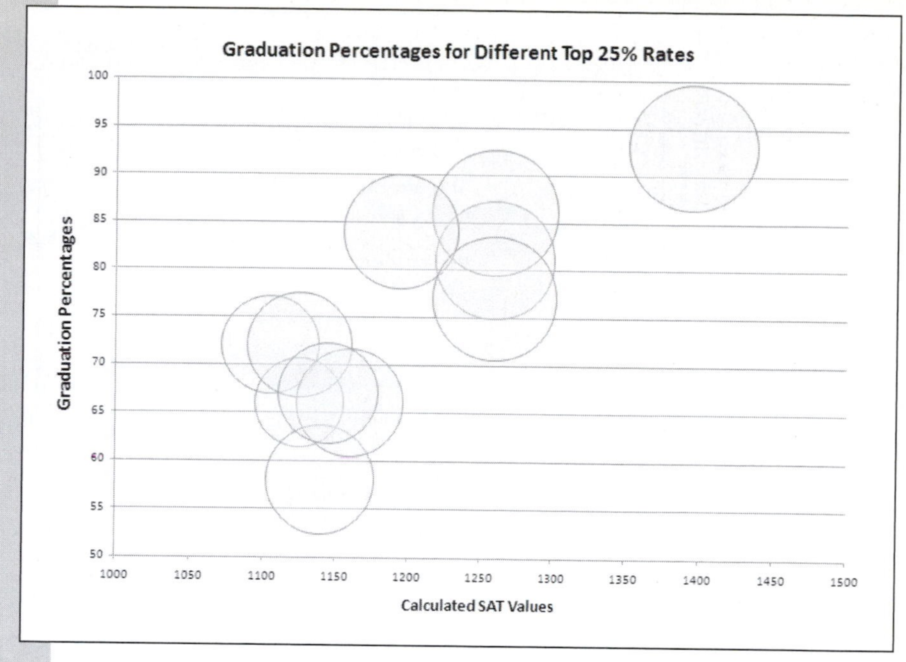

The area of each bubble in the plot is proportional to the percentage of incoming freshman-athletes who graduated in the top 25% of their class. You can change this so that the *width* of each bubble is proportional to that value. In some situations, this works better in displaying differences between one data point and another. You can also make the bubbles smaller so that there is less overlap. Investigate the effect of changing the bubble symbol size on the appearance of your chart.

To change the bubble size:

1. Click one of the bubbles in the chart to select all of the bubble symbols again.
2. Return to the Format Data series dialog box by clicking the **Format** tab from the Chart Tools ribbon and then click the **Format Selection** button from the Current Selection group.
3. If necessary click the **Series Options** tab and then click the **Width of bubbles** option button and enter **50** in the Scale bubble size to input box. This will reduce the width of the bubbles to 50% of their default size (see Figure 3-28).

Figure 3-28 Setting the width of the plot bubbles

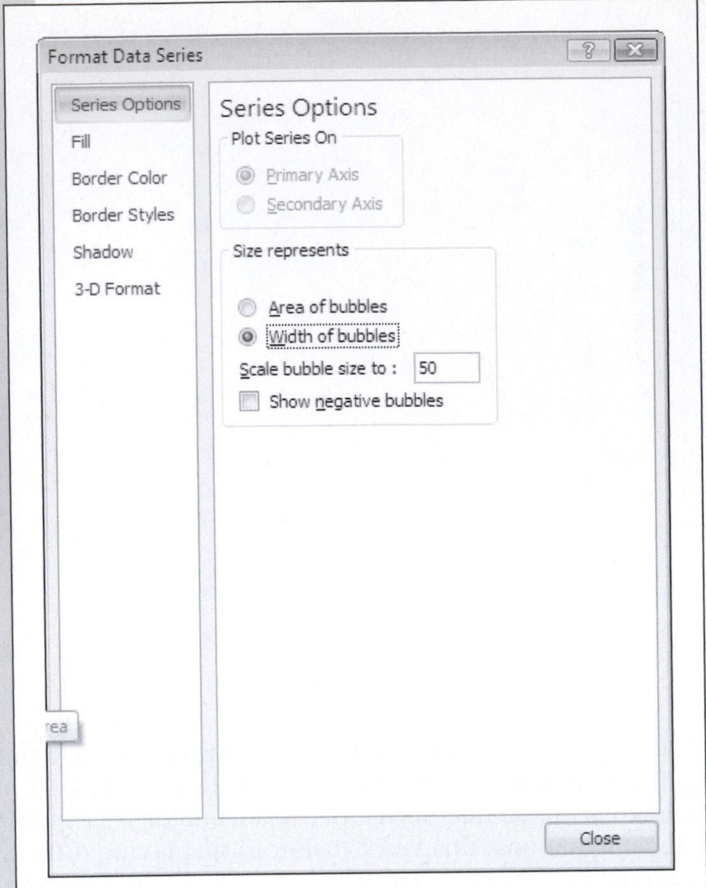

4 Click the **Close** button and then deselect the chart. The final appearance of the bubble chart is shown in Figure 3-29.

Figure 3-29 Final bubble plot

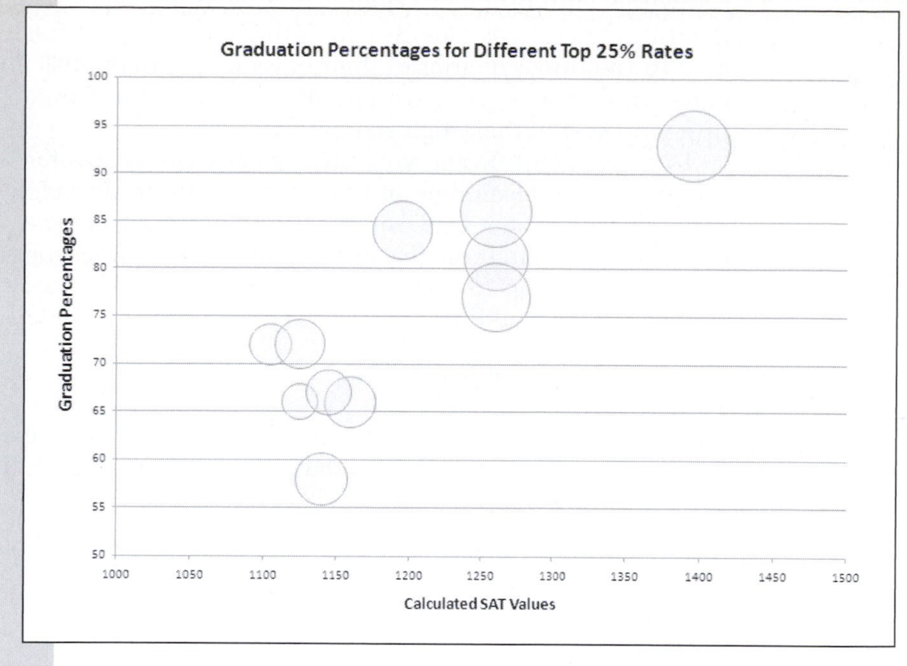

Let's evaluate what we've created. In interpreting bubble plots, the statistician looks for a pattern in the distribution of the bubbles. Are bubbles of similar size all clustered in one area on the plot? Is there a progression in the size of the bubbles? For example, do the bubbles increase in size as we proceed from left to right across the plot? Is there a bubble that is markedly different from the others? In this plot, we notice immediately that the smaller bubbles seem to be clustered more toward the left end of the plot. This would indicate that schools which have a lower percentage of incoming freshmen that graduate in the top quarter of their class also tend to have a lower ultimate graduation rate. However, it's also interesting that the bubble representing Minnesota is slightly larger than its surrounding bubbles indicating that we probably cannot argue that Minnesota's lower graduation rate is due to a lower number of incoming students who graduated in the top quarter of their class. We would probably have to do further research to discover a reason from Minnesota's slightly lower graduation rate.

Breaking a Scatter Plot into Categories

Bubble plots have the problem that it is not always easy to compare the relative sizes of different bubbles, so another approach we can take is to divide the universities into categories, plotting universities from different categories with different symbols. For example we can divide the universities into two groups: one in which the percentage of freshmen who graduated in the upper quarter of their class is less than 80% and another group consisting of schools in which 80% or more of freshmen graduated in the upper quarter of their high school class.

To do this with Excel, you have to copy the values for these universities into two separate columns and then recreate the scatter plot, plotting two data series instead of one. That can be a time-consuming process. To save time, you can use StatPlus to break the scatter plot into categories for you. You'll try this now, using the Top 25 Rate column to determine the category values (<80% or >=80%). First, you'll make a copy of the scatter plot you created earlier.

To copy the scatter plot:

1 Right-click the **Graduation Chart** sheet tab and select **Move or Copy** from the pop-up menu that appears. Excel opens the Move or Copy dialog box.

2 Click the **Create a copy** checkbox and select **Bubble Chart** from the list of chart sheets as shown in Figure 3-30.

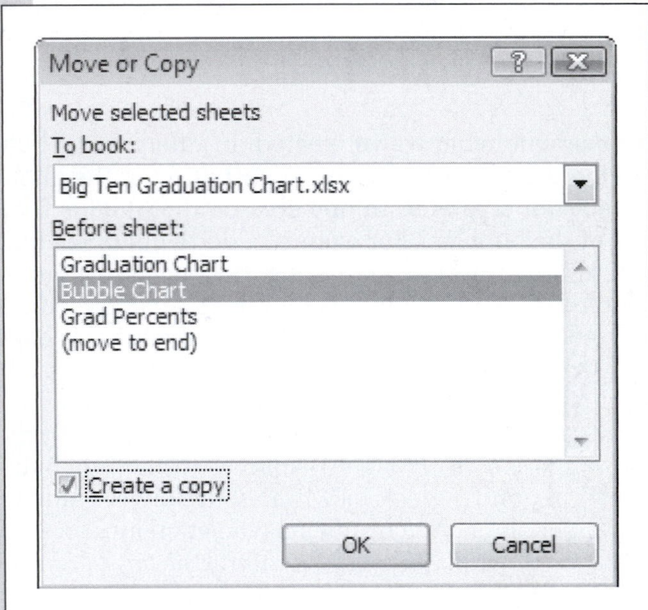

Figure 3-30 Moving or copying a chart sheet

3 Click the **OK** button.

Excel inserts a new chart sheet named Graduation Chart (2) directly before the Bubble Chart chart sheet.

4 Click any one of the plot symbol labels to select them all and then press the **Delete** key.

The plot labels disappear (you won't be using them in the plot you'll create next). Now, break the points in the scatter plot into two categories on the basis of the values in the Size column.

To break the scatter plot into categories:

1. Click **Display series by category** from the StatPlus menu located in the Menu Commands group on the Add-Ins tab.
2. Click the **Categories** button.
3. Verify that the **Use Range Names** option button is selected, click **Top_25_Rate** from the list of range names, and click **OK**.
4. Click the **Bottom** option button to display the categories' legend at the bottom of the scatter plot.
5. Click the **OK** button. Figure 3-31 displays the scatter plot broken down by categories.

Figure 3-31 Breaking the scatter plot into categories

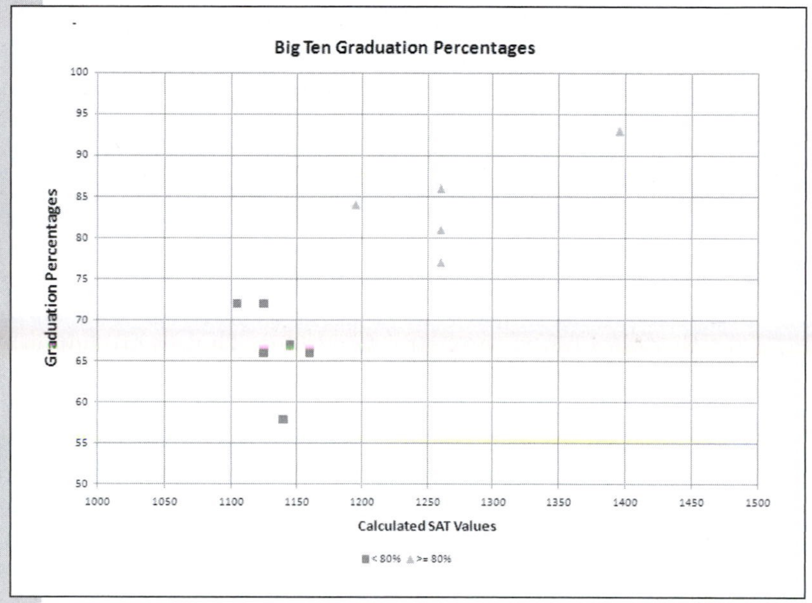

STATPLUS TIPS

- Once you break a scatter plot into categories, you cannot go back to the original scatter plot (without the categories). If this is a problem, create a copy of the scatter plot before you break it down.
- If your chart contains several data series, you can choose which series to break down into categories by selecting the series name from the Select a Series drop-down list box in the Display by Series Category dialog box.

With the data series now broken down into categories, we can compare the universities' graduation rates on the basis of the high school performance of its incoming freshman students. On the basis of the chart, we quickly see that schools in which 80% or more of the incoming freshmen graduate in the upper quarter of their high school have higher graduation percentages whereas those schools in which less than 80% of incoming freshmen graduate in the upper quarter see a lower percentage of those students graduate.

So we have two predictors of graduation percentage. One is the calculated SAT score, and the other is percentage of incoming freshmen that graduate in the upper 25% of their class. Notice, however, that within each category (<80% and >=80%), there does not seem to be a clear trend based on calculated SAT values.

Essentially both measures are telling us the same general thing: universities that attract better student athletes will also have higher graduation percentages. That's not surprising. What might be useful, however, is determining which of the calculated SAT score or the high school graduation ranking is the better predictor. That is not something we can easily determine from the chart. To answer that question, we have to perform a statistical analysis of the data, which we'll do in a future chapter.

There are other factors which we haven't investigated. Does the size or the location of the school matter? Does it make a difference if the university is a public or private institution? And we have to be aware that we are only looking at a sample of 11 universities; we don't know if any of our conclusions can apply to another sample of schools. All of these are questions for future study.

Plotting Several Variables

Before finishing, let's explore one more question. The data include graduation rates for the athletes in the freshman class broken down by gender and race. How do these graduation rates compare? A scatter plot displaying these results needs to have several data series. You can create such a scatter plot by simply selecting additional columns of data to be plotted on the y axis of the chart. In this example you'll plot the graduation rates for white male and female athletes.

To create the scatter plot:

1 Return to the Grad Percents worksheet and select the nonadjacent cell range **D1:D12;F1:F12;H1:H12**.

2 Click the **Scatter** button from the Charts group on the Insert tab and then click the first Scatter subtype, displaying a scatter plot with only markers. Excel inserts an embedded scatter plot as shown in Figure 3-32.

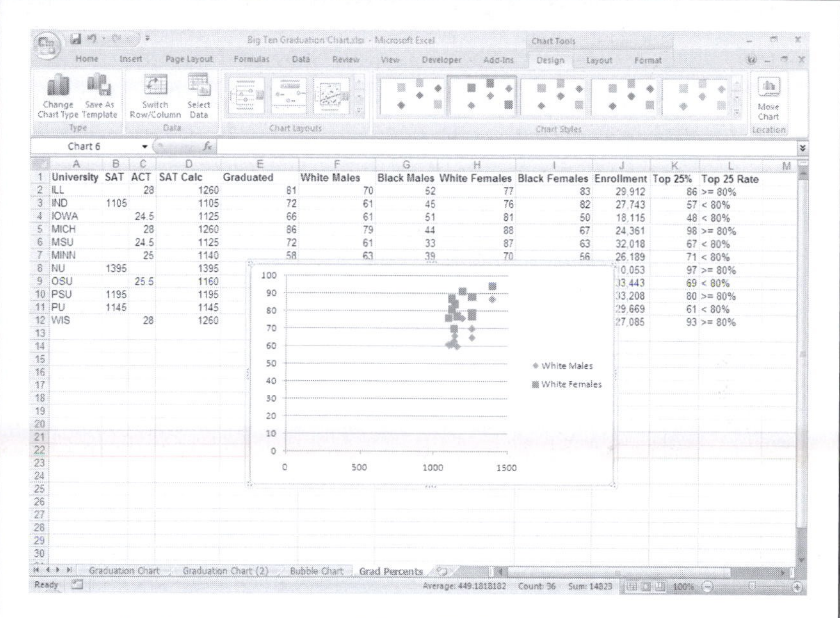

Figure 3-32 Plotting the white male and female graduation rates

3 Move the scatter plot to its own chart sheet named **Graduation Chart by Gender**.

4 Excel does not automatically add a chart title when there is more than one data series being plotted. To insert a title, click the **Chart Title** button located in the Labels group of the Layout tab on the Chart Tools ribbon and then click **Above Chart**. Enter **Graduation Percentage by Gender** for the title.

5 As you did for the other scatter plots, change the title of the x axis to **Calculated SAT Values** and the title of the y axis to **Graduation Percentages**. Set the font size of the titles to **14** points and **16** points respectively.

6 Change the scale of the x axis to range from **1000** to **1500** in intervals of 50 points. Change the scale of the y axis to range from **50** to **100** in intervals of 5 points.

7 Click the **Gridlines** button from the Axes group on the Layout tab of the Chart Tools ribbon and then click **Primary Vertical Gridlines** and **Major Gridlines** to add gridlines to the chart. Figure 3-33 shows the final formatted version of the scatter plot.

Figure 3-33 Breaking the scatter plot into categories

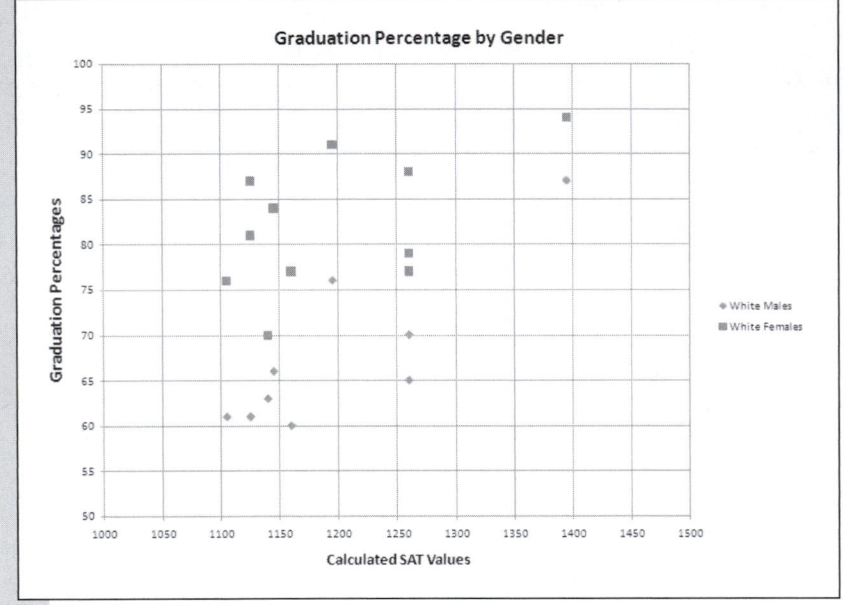

The scatter plot shows that the white female athletes generally have higher graduation rates than the white male athletes. Does this tell us something about white female and male athletes? Perhaps, but we should bear in mind that this chart plots the average graduation rates for these two groups against the average SAT score for the *entire* class of incoming freshmen. We don't have data on the average SAT score for incoming freshman male athletes or incoming freshman female athletes. It's possible that the female athletes also had higher SAT scores than their male counterparts, and thus we would expect them to have higher graduation rates. On the other hand, if their SAT scores are comparable, we might look at the college experiences of male and female athletes at these universities to see whether this would have an effect on graduation rates. Are the demands on male athletes different from those on female athletes, and does this affect the graduation rates?

STATPLUS TIPS

- You can quickly create your own scatter plots using StatPlus. To create a scatter plot with the add-in, click **Single Variable Charts** and then **Fast Scatterplot** from the StatPlus menu.
- In the same way you can quickly create your own bubble plots using StatPlus. To create a bubble plot, click **Multi-Variable Charts** and then **Fast Bubble Plot** from the StatPlus menu.

You've completed your work on the Big Ten scatter plots. You can now close the workbook, saving your changes.

Exercises

1. You decide to investigate whether there is any relationship between the graduation rate for student athletes and the race of the athlete. To do this, open the Big Ten workbook you examined in this chapter and perform the following analyses:
   a. Open the Big Ten workbook from the Chapter02 folder and save it as **Big Ten Graduation by Race**.
   b. Create a scatter plot with the calculated SAT score on the *x* axis and the white male and black male graduation rates on the *y* axis. Title the chart **Graduation Percentages by Race**. Title the *y* axis **Graduation Percentages** and the *x* axis **Calculated SAT Values**. Move your scatter plot to a chart sheet named **Male Graduation by Race**.
   c. Edit the scale of the *x* and *y* axes to reflect the range of data values.
   d. Add labels to the scatter plot, identifying each university. Compare the black male graduation rate for each university to the corresponding white male graduation rate. What do you observe?
   e. Repeat your analysis, this time comparing female graduation rates by race. Save your graph on a chart sheet named **Female Graduation by Race**.
   f. Save your workbook and write a brief report summarizing your observations.

2. In the 1980s female professors at a junior college sought help from statisticians to show that they were underpaid relative to their male counterparts. The legal action was eventually settled out of court. Investigate their claim by creating scatter plots of the salary data they acquired for their case.
   a. Open the **Junior College** workbook from the Chapter02 folder and save it as **Junior College Salary Charts**.
   b. Create a scatter plot with Salary on the *y*-axis and Years on the *x*-axis. Title the scatter plot **Employee Salaries**. Title the *y* axis **Salary** and the *x* axis **Years Employed**. Remove the legend and gridlines from the plot. Save the plot as a chart sheet named **Salary by Gender Chart**.

c. Break the scatter plot points into two categories on the basis of gender. Does the plot suggest that male salaries tend to be higher than female salaries for comparable years of employment?

d. Examine the list of other variables in the workbook. Are there other variables in that list which should be taken into account before coming to a conclusion about the relationship between gender and salary?

e. Save your workbook and write a report summarizing your observations.

3. Admission decisions to colleges are often partly based on ACT math scores and high school rank with the expectation that these scores are related to success in college. Is this always the case and is gender a factor? You've been provided with a data set to investigate this question. The data set contains columns for gender, high school rank (HS Rank), American College Testing Mathematics test score (ACT Math), and an algebra placement test score (Alg Place) from the first week of class and the final first semester calculus grades (Calc) for a group of students [see Edge and Friedberg (1984)]. Graph the data to investigate what kind of relationships appear to exist between the variables.

   a. Open the **Calculus** workbook from the Chapter03 folder and save it as **Calculus ACT Charts**.

   b. Create a scatter plot, on a separate chart sheet, plotting Calc on the *y* axis and ACT Math on the *x* axis. Label the axis appropriately. How strong does the relationship between the ACT Math score and the first semester calculus score appear to you?

   c. Break down the scatter plot by gender. Is there evidence of a difference in calculus scores based on gender?

   d. Repeat steps a through c for a scatter plot relating calculus grades to the algebra placement test.

   e. Save your workbook and then write a report summarizing your findings.

4. You've been given a data set containing the mass and volume measurements from eight chunks of aluminum as recorded in a high school chemistry lab. Graph and examine their findings.

   a. Open the **Aluminum** workbook from the Chapter03 folder and save it as **Aluminum Chart**.

   b. Create a scatter plot with mass values on the *y* axis and volume values on the *x* axis. Add major gridlines for both the *x* axis and the *y* axis.

   c. Examine your chart. There should be a data value that appears out of place. Mark this point by changing the plot symbol used for that point to a different color from the rest of the points.

   d. Do the other points seem to form a nearly straight line? The ratio of mass to volume is supposed to be a constant (the density of aluminum), so the points should fall on a line through the origin. Draw the line, and estimate the slope (the ratio of the vertical change to the horizontal change) along the line. What is your estimate for the density of aluminum?

   e. Save your workbook and write a report summarizing your findings.

5. You've been asked to investigate the relationship between protein and carbohydrates in several brands of wheat cereals and breads. Data taken from a trip to a local grocery store has been recorded and saved for you.

   a. Open the **Wheat** workbook from the Chapter03 folder and save it as **Wheat Charts**.

   b. Create a scatter plot with Protein on the *y* axis and Carbo on the *x* axis.

Label the axes appropriately. Save the chart into a chart sheet named **Protein Chart**.

c. Label each point in the scatter plot with its food name.

d. Create a bubble plot of protein versus carbohydrates with the size of the bubble determined by the amount of sugar contained in each product. Format the chart to make it easy to read and interpret. Which plot points show the largest bubbles (and thus the largest sugar content) and what food do they represent?

e. Sugar does not contain protein. Explain how this fact is reflected in the placement of the two food products with the largest sugar content in the bubble plot you just created.

f. Save your workbook and write a report summarizing your observations from the graph you created.

6. How well does a player's salary match up with his career batting average? You've been given performance and salary data of major league players from the beginning of the 2007 season (non-pitchers only). Analyze the relationship between performance and pay.

   a. Open the **Baseball** workbook from the Chapter03 folder and save it as **Baseball Salaries Chart**.

   b. Create a scatter plot with salary on the *y* axis and career batting average (AVG) on the *x* axis. Label the axis appropriately and save the plot in a chart sheet named **Salary Chart vs. AVG**.

   c. Change the lower range of the *x* axis scale to 0.15.

   d. Identify on the plot the last name of the player with the highest salary. How does this player's batting average compare to the other players in the sample?

   e. Salary values can vary in magnitude from around 200,000 up to more than 20,000,000. You can adjust for this scatter by plotting the data on a logarithmic chart. Copy your chart sheet to a new sheet named **Salary Log Chart** and then change the property of the vertical axis to show the axis with a log scale over a range from 300,000 up to 30,000,000. How does the log scale affect the vertical scatter of the data?

   f. Create a scatter plot of salary vs. home runs (HR). Save the chart on a chart sheet named **Salary Chart vs. HRs**. Compare this chart to the one you created in the Salary Chart vs. AVG sheet. Which chart shows the stronger relationship? Which do owners appear to value more: batting average or homeruns?

   g. There is a player in your chart who appears to be underpaid for the number of homeruns hit. Identify this player. Examine how many seasons the player has been in the league. Explain how this could have affected his salary value.

   h. Create a bubble plot of salary vs. batting average on a new chart sheet named **Salary vs. Avg Bubble Chart** with the values of the HR column to determine the size of each bubble (Note: Due to the order of the columns in the Player Data worksheet, you can more easily create the bubble plot using the StatPlus Fast Bubble Plot command available under Multivariable Charts menu). Set the scale of the *x*-axis to range from 0.15 to 0.35 in intervals of 0.05. Display the *y*-axis on a log scale ranging from 300,000 to 30,000,000. Make the bubble symbols partly transparent with the width of the bubble scaled down to 25. Based on your bubble plot where are the largest bubbles (and thus the players with the most home runs) concentrated?

i. There appears to be a player whose reported salary is lower than expected for his combination of batting average and homeruns. Identify that player.

j. Save your workbook and write a report summarizing your observations. Which is more important in determining the player salary: homeruns or batting average?

7. Working at the Bureau of Labor Statistics, James Longley tracked several variables related to the United States economy from 1947 to 1962. Study the data he collected.

   a. Open the **Longley** workbook from the Chapter03 folder and save the workbook as **Longley Graph**.

   b. The workbook has seven columns related to the economy. The Total column displays the total U.S. employment in thousands. The Armforce column displays the total number of people in the armed forces, listed again in thousands. Create a scatter plot of Total versus Armforce. Label and scale the chart appropriately.

   c. Add labels to each plot point indicating the year in which the datum was recorded.

   d. Four points on the lower left of the scatter plot stand out. Examine the economic and political history of the era to explain why these values are so distinct from all of the others.

   e. Aside from the four points in the lower left corner of the plot, describe the general relationship between total employment and the number of people in the armed forces.

   f. Save your workbook and write a report summarizing your observations.

8. What is the relationship between race results and reaction time (the time it takes for the runner to leave the starting block after hearing the sound of the starting gun)? You've been given a workbook containing the race results and reaction times from the first round of the 100-meter heats at the 1996 Summer Olympic games in Atlanta. Graph the data to investigate the effect that reaction time has on race results.

   a. Open the **Race** workbook from the Chapter03 folder and save it as **Race Graphs**.

   b. Create a scatter plot of race time versus reaction time. Label and scale the chart appropriately. Do you see a trend that would indicate that runners with faster reaction times have faster race times?

   c. There is a point that lies away from the others. Identify the runner corresponding to this point.

   d. Copy your chart to another chart sheet and then rescale the axes, setting the *x* axis range to 0.12 to 0.24 seconds and the *y* axis scale to 9.5 to 12.5 seconds. Is there any more indication that a relationship exists between reaction time and race time?

   e. Save your workbook and write a summary including a comment on how the scale used in plotting data can affect your perception of the results.

9. The Cars workbook contains information on car models from *Consumer Reports*®, 2003–2008. Data in the workbook include the miles per gallon (MPG) of each car as well as the time to accelerate from 0 to 60, weight, horsepower, price, etc. See Exercise 10 of Chapter 2.

   a. Open the **Cars** workbook from the Chapter03 folder and save it as **Car Graphs**.

   b. Create a scatter plot on a separate chart sheet of MPG (on the *y*-axis) versus horsepower (on the *x*-axis).

c. Label the points that are highest in MPG as measured by the height above the average MPG for those with around the same horsepower. Label each of these points with the Model (make sure that you select only these points rather than all of the points—or else you'll wait a long time for all of the labels to be added to the scatter plot). What do these cars have in common? (*Hint*: Each does not have just a traditional gasoline engine). Print your chart.
d. Copy your scatter plot to a second chart sheet. Break down the plot by company region. Do you see a relationship between Region and MPG? Which region has lowest MPG, on the average, for each given horsepower?
e. Create a bubble plot on a separate chart sheet with MPG on the y-axis, horsepower on the x-axis and the size of each bubble determined by the price of the car.
f. Rescale the bubbles to 50% of the default and relate the size of the price column to the width of the bubbles. Print the chart.
g. Is the price of the car related to the horsepower and the gas mileage? How? Describe the relationship and explain why it should be expected. Print the rescaled chart. Save your changes to the workbook.

10. Voting results for two presidential elections have been recorded for you in an Excel workbook. The workbook contains the percentage of the vote won by the Democratic candidate in 1980 and 1984, broken down by state. Graph and analyze the results.

a. Open the **Voting** workbook from the Chapter03 folder and save it as **Voting Graphs**.
b. Create a scatter plot of Dem1984 versus Dem1980 on a separate chart sheet.
c. Rescale the axes so that the minimum value for the *x*-axis and the *y*-axis is 20 and the maximum value is 60.
d. Examine the scatter plot. Does the voting percentage in 1984 generally follow the voting percentage from 1980? In other words, if the Democratic candidate received a large percentage of the vote from a particular state in 1980, did he or she do as well in 1984?
e. In one state, the candidate had a large percentage of the vote in 1980 (above 55%) but a small percentage of the vote in 1984 (about 40%). Identify this state.
e. Create a copy of the scatter plot on a separate chart sheet. Break down this new scatter plot by region.
f. Examine the location of the southern states in the scatter plot. Do they follow the general pattern shown by the other points in the plot? Interpret your answer in light of what you know of the 1980 and 1984 elections. (*Hint:* Consider whether the fact that the 1980 election involved a southern Democratic candidate and the 1984 election involved a midwestern Democratic candidate caused a change in the voting percentages of the southern states.)
g. Save your workbook and write a report summarizing your conclusions.

Chapter 4

Describing Your Data

Objectives

In this chapter you will learn:

- About different types of variables
- How to create tables of frequency, cumulative frequency, percentages, and cumulative percentages
- How to create histograms and break histograms down by groups
- About creating and interpreting stem and leaf plots
- How to calculate descriptive statistics for your data
- How to create and interpret box plots

Chapter 4 introduces the different tools that statisticians use to describe and summarize the values in a data set. You'll work with frequency tables in order to see the range of values in your data. You'll use graphical tools like histograms, stem and leaf plots, and boxplots to get a visual picture of how the data values are distributed. You'll learn about descriptive statistics that reduce the contents of your data to a few values, such as the mean and standard deviation. Applying these tools is the first step in the process of evaluating and interpreting the contents of your data set.

Variables and Descriptive Statistics

In this chapter you'll learn about a branch of statistics called descriptive statistics. In **descriptive statistics** we use various mathematical tools to summarize the values of a data set. Our goal is to take data that may contain thousands of observations and reduce it to a few calculated values. For example, we might calculate the average salaries of employees at several companies in order to get a general impression about which companies pay the most, or we might calculate the range of salaries at those companies to convey the same idea.

Note that we should be very careful in drawing any general conclusions or making any predictions on the basis of our descriptive statistics. Those tasks belong to a different branch of statistics called **inferential statistics**, a topic we'll discuss in later chapters. The goal of descriptive statistics is to describe the contents of a specific data set, and we don't have the tools yet to evaluate any conclusions that might arise from examining those statistics.

When descriptive statistics involve only a single variable, as they will in this chapter, we are employing a branch of statistics called **univariate statistics**. Now we've used the term *variable* several times in this book. What is a variable?

A **variable** is a single characteristic of any object or event. In the last chapter, you looked at data sets that contained several variables describing graduation rates of the Big Ten universities. Each column in that worksheet contained information on one characteristic, such as the university's name or total enrollment, and thus was a single variable.

Variables can be classified as quantitative and qualitative. **Quantitative variables** involve values that come in meaningful (not arbitrary) numbers. Examples of quantitative variables include age, weight, and annual income—anything that can be measured in terms of a number. The number itself can be either discrete or continuous. **Discrete variables** are quantitative variables that assume values from a defined list of numbers. The numbers on a die come in discrete values (1, 2, 3, 4, 5, or 6). The number of children in a household is discrete, consisting of positive integers and zero.

Continuous variables, on the other hand, have values from a wide range of possible values. An individual's weight could be 185, 185.5, or 185.5627 pounds. To be sure, there is some blurring in the distinction between discrete and continuous variables. Is salary a discrete variable or a continuous variable? From one point of view, it's discrete: The values are limited to dollars and cents, and there is a practical upper limit to how high a specific salary could go. However, it's more natural to think of salary as continuous.

The second type of variable is the qualitative variable. **Qualitative** or **categorical** variables are variables whose values fall into some category, indicating a quality or property of an object. Gender, ethnicity, and product name are all examples of qualitative variables. Qualitative variables are generally expressed in text strings, but not always. Sometimes a qualitative variable will be coded using numerical values. A common "gotcha" for people new to statistics is to analyze these coded values as quantitative variables. Consider the qualitative data values from Table 4-1.

Table 4-1 Qualitative Variables

ID	Gender (0 = male; 1 = female)	Ethnicity (0 = Caucasian, 1 = African American; 2 = Asian; 3 = Other)
3458924065	1	0
4891029494	0	3
3489109294	0	1

Now all of these values were entered as numbers, but does it make sense to say that the average gender is 1? Or that the sum of the ethnicities is 4? Of course not, but if you're not careful, you may find yourself doing things like that in other, more subtle cases. The point is that you should always understand what type of variables your data set contains before applying any descriptive statistic.

Qualitative variables can be classified as ordinal and nominal. An **ordinal variable** is a qualitative variable whose categories can be put into some natural order. For example, users asked to fill out a survey ranking their product satisfaction may enter values from "Not satisfied" all the way up to "Extremely satisfied." These values are categorical values, but they have a clear order of ascendancy. **Nominal variables** are qualitative variables without any such natural order. Ethnicity, state of residence, and gender are all examples of nominal variables. Table 4-2 summarizes properties of the different types of variables we've been discussing.

Table 4-2 Summary of variable properties

Quantitative Variables whose values come in meaningful (not arbitrary) numbers	**Discrete** Quantitative variables whose values derive from a list of specific numbers
	Continuous Quantitative variables that can assume values from within a continuous range of possible values
Qualitative Variables whose values fall into categories	**Ordinal** Qualitative variables whose categories can be assigned some natural order
	Nominal Qualitative variables whose categories cannot be put into any natural order

In this chapter, we'll work primarily with continuous quantitative variables. You'll learn how to work with qualitative variables in Chapter 7, where you work with tables and categorical data.

Frequency Tables

You've been asked to analyze the history of housing prices in the Southwest and have been given a random sample from the records of resales of homes in Albuquerque, New Mexico, from 2/15/1993 through 4/30/1993 (*Albuquerque Board of Realtors*). The variables in this data set have been placed in an Excel workbook with the range names and descriptions shown in Table 4-3.

Table 4-3 Housing Data Set

Range Name	Range	Description
Price	A2:A118	The selling price of each home
Square_Feet	B2:B118	The square footage of the home
Age	C2:C118	The age of the home in years
Features	D2:D118	The number of features available in the home (dishwasher, refrigerator, microwave, disposal, washer, intercom, skylight(s), compactor, dryer, handicapped-accessible, cable TV access)
NE_Sector	E2:E118	Located in the northeast sector of the city (Yes or No)
Corner_Lot	F2:F118	Located on a corner lot (Yes or No)
Offer_Pending	G2:G118	Offer pending on the home (Yes or No)
Annual_Tax	H2:H118	Estimated annual tax paid on the home

To view the Housing workbook:

1 Start Excel.

2 Open the **Housing** workbook from the Chapter04 folder.

The workbook appears as shown in Figure 4-1.

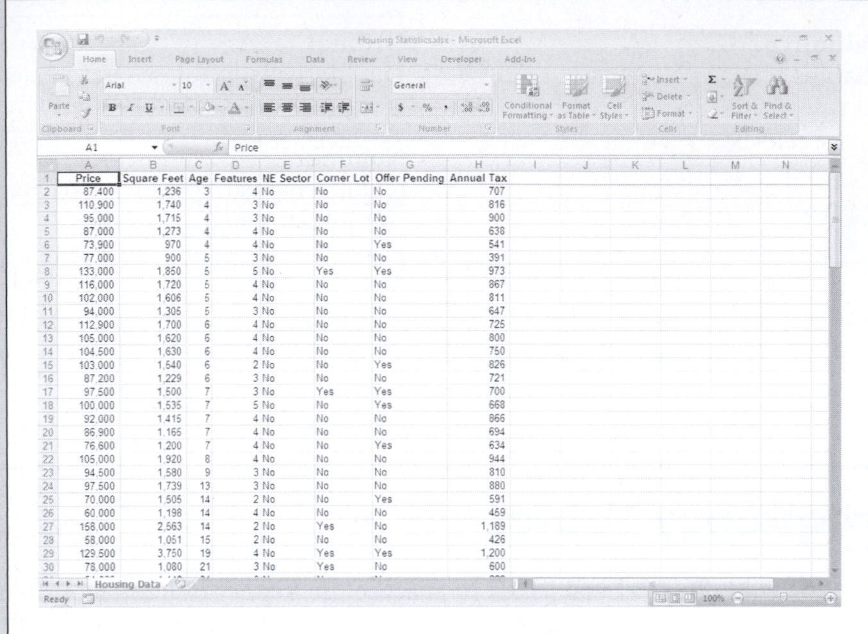

	A	B	C	D	E	F	G	H
1	Price	Square Feet	Age	Features	NE Sector	Corner Lot	Offer Pending	Annual Tax
2	87,400	1,236	3	4	No	No	No	707
3	110,900	1,740	4	3	No	No	No	816
4	95,000	1,715	4	3	No	No	No	900
5	87,000	1,273	4	4	No	No	No	638
6	73,900	970	4	4	No	No	Yes	541
7	77,000	900	5	3	No	No	No	391
8	133,000	1,850	5	5	No	Yes	Yes	973
9	116,000	1,720	5	4	No	No	No	867
10	102,000	1,606	5	4	No	No	No	811
11	94,000	1,305	5	3	No	No	No	647
12	112,900	1,700	6	4	No	No	No	725
13	105,000	1,620	6	4	No	No	No	800
14	104,500	1,630	6	4	No	No	No	750
15	103,000	1,540	6	2	No	No	Yes	826
16	87,200	1,229	6	3	No	No	No	721
17	97,500	1,500	7	3	No	Yes	Yes	700
18	100,000	1,535	7	5	No	No	Yes	668
19	92,000	1,415	7	4	No	No	No	866
20	86,900	1,165	7	4	No	No	No	694
21	76,600	1,200	7	4	No	No	Yes	634
22	105,000	1,920	8	4	No	No	No	944
23	94,500	1,580	9	3	No	No	No	810
24	97,500	1,739	13	3	No	No	No	880
25	70,000	1,505	14	2	No	No	Yes	591
26	60,000	1,198	14	4	No	No	No	459
27	158,000	2,563	14	2	No	Yes	No	1,189
28	58,000	1,051	15	2	No	No	No	426
29	129,500	3,750	19	4	No	Yes	Yes	1,200
30	78,000	1,080	21	3	No	Yes	No	600

Figure 4-1 The Housing workbook

3 Save the workbook as **Housing Statistics**.

Creating a Frequency Table

One of the first things we'll examine when studying this data set is the distribution of its values. The **distribution** is the spread of the data across a range of possible values. If you were thinking about moving to Albuquerque during the time these data were recorded, you might be interested in the distribution of home prices in the area. What is the range of housing prices in the area? What percentage of houses list for under $125,000?

As a first step in answering these types of questions, we'll create a frequency table of the home prices. A **frequency table** is a table that tabulates the number of occurrences or counts of a specific value of a given variable. Excel does not have a built-in command to create such a frequency table, but you can use the one supplied with the StatPlus add-in. Use StatPlus now to create a frequency table of the home prices in the Housing Statistics workbook.

To create a frequency table of home prices:

1. Click **Descriptive Statistics** from the StatPlus menu on the Add-Ins tab and then click **Frequency Tables**.

2. Click the **Data Values** button, click the **Use Range Names** option button, and click **Price**. Click the **OK** button.

   The Frequency Table command gives you three options for organizing your table. You can use discrete values so that the table is tabulated over individual price values, or you can organize the values into bins (you'll learn about bins shortly). For now, leave Discrete as the selected option.

3. Click the **Output** button, click the **New Worksheet** option button, and type **Price Table** in the New Worksheet name box. Click the **OK** button.

4. Click **OK** to start generating the frequency table. Figure 4-2 displays the completed table.

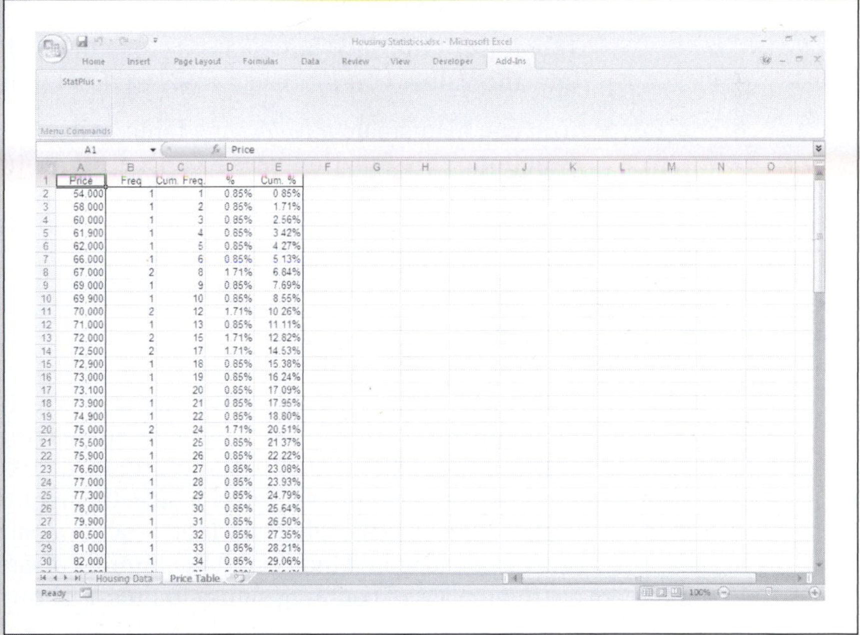

Price	Freq	Cum. Freq.	%	Cum. %
54,000	1	1	0.85%	0.85%
58,000	1	2	0.85%	1.71%
60,000	1	3	0.85%	2.56%
61,900	1	4	0.85%	3.42%
62,000	1	5	0.85%	4.27%
66,000	1	6	0.85%	5.13%
67,000	2	8	1.71%	6.84%
69,000	1	9	0.85%	7.69%
69,900	1	10	0.85%	8.55%
70,000	2	12	1.71%	10.26%
71,000	1	13	0.85%	11.11%
72,000	2	15	1.71%	12.82%
72,500	2	17	1.71%	14.53%
72,900	1	18	0.85%	15.38%
73,000	1	19	0.85%	16.24%
73,100	1	20	0.85%	17.09%
73,900	1	21	0.85%	17.95%
74,900	1	22	0.85%	18.80%
75,000	2	24	1.71%	20.51%
75,500	1	25	0.85%	21.37%
75,900	1	26	0.85%	22.22%
76,600	1	27	0.85%	23.08%
77,000	1	28	0.85%	23.93%
77,300	1	29	0.85%	24.79%
78,000	1	30	0.85%	25.64%
79,900	1	31	0.85%	26.50%
80,500	1	32	0.85%	27.35%
81,000	1	33	0.85%	28.21%
82,000	1	34	0.85%	29.06%

Figure 4-2 Frequency of housing prices

The table contains five columns. The first column, Price, lists in ascending order each home price in the sample of 117 homes. Prices in this sample range from a minimum of $54,000 to a maximum of $215,000. The second column, Freq, counts the frequency, or number of occurrences, for each value in the price column. Many prices are unique and have frequencies of 1, but other prices (such as $75,000) occur for multiple homes. The third column contains the cumulative frequency, counting the total number of homes at or less than a given price. By examining the table, you can quickly see that 24 of the homes in the sample have a price of $75,000 or less. The fourth column lists the percentage occurrence of each home price out of the total sample. For example, 1.71% of the homes are listed for exactly $75,000. Finally, the fifth and last column of the table calculates the cumulative percentage for the home prices. In this case, 24.79%—almost one-quarter of the homes—list for $77,300 or less. A table of this kind can help you in evaluating the market. For example, if you were interested in homes that list for $125,000 or less, you could quickly determine that almost 80% of the homes in this database, or 93 different listings, met that criterion.

EXCEL TIPS

- If you don't have StatPlus handy, Excel comes with an add-in called the Data Analysis ToolPak which you can use to create a frequency table. The ToolPak does not have all the frequency table options that StatPlus contains.
- If you want to count how many values in a column are equal to a specific value, you can use Excel's COUNTIF function.
- You can also create a frequency table using Excel's FREQUENCY function. This function uses Excel's array feature, which you can learn about by using the online Help.

Using Bins in a Frequency Table

By creating a frequency table, you got a clear picture of the distribution of prices in the Albuquerque area back in 1993. However, displaying individual values would be cumbersome if the sample contained 1,000 or 10,000 observations.

Rather than list individual prices, you can have the frequency table group the values by placing them in **bins**, where each bin covers a particular range of values. The frequency table would then count the number of values that fall in each bin. There are three ways of counting values in bins as shown in Figure 4-3.

1. Count those values which are ≥ the bin value and < the next bin value.
2. Count those values which are centered around the bin value (in the case of mid-point values, start counting from the lower mid-point).
3. Count those values that are ≤ the bin value but > the previous bin value.

Figure 4-3 Counting within a bin

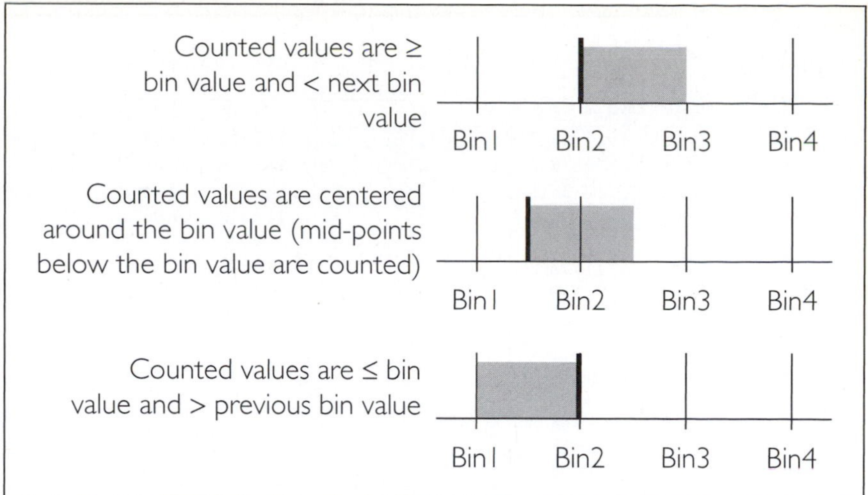
Counted values are ≥ bin value and < next bin value
Bin1 Bin2 Bin3 Bin4
Counted values are centered around the bin value (mid-points below the bin value are counted)
Bin1 Bin2 Bin3 Bin4
Counted values are ≤ bin value and > previous bin value
Bin1 Bin2 Bin3 Bin4

To interpret a frequency table that involves bins correctly, you need to know which of these methods is used in calculating the counts. We'll create another frequency table of the housing prices in the workbook, this time breaking the data down into 15 equally spaced bins.

To create a frequency table with bins:

1. Click **Descriptive Statistics** from the StatPlus menu and then click **Frequency Tables**.
2. Click the **Data Values** button and select **Price** as the data variable. Click **OK**.
3. Click the Create 15 equally spaced bins option button.

   Note that the first option button has been selected, so that counts can be calculated for values that are ≥ the bin value and < the succeeding bin value.
4. Click the **Output** button and click the **New Worksheet** option button. Type **Price Table with Bins** as the worksheet name and click the **OK** button.
5. Click the **OK** button to start generating the frequency table with bins. Figure 4-4 shows the resulting frequency table.

Figure 4-4 Frequency table with equally spaced bins

	A	B	C	D	E	F
1	Price	Freq	Cum. Freq.	%	Cum. %	
2	54,000	5	5	4.27%	4.27%	
3	64,733	19	24	16.24%	20.51%	
4	75,467	14	38	11.97%	32.48%	
5	86,200	21	59	17.95%	50.43%	
6	96,933	17	76	14.53%	64.96%	
7	107,667	11	87	9.40%	74.36%	
8	118,400	7	94	5.98%	80.34%	
9	129,133	6	100	5.13%	85.47%	
10	139,867	2	102	1.71%	87.18%	
11	150,600	4	106	3.42%	90.60%	
12	161,333	1	107	0.85%	91.45%	
13	172,067	1	108	0.85%	92.31%	
14	182,800	2	110	1.71%	94.02%	
15	193,533	1	111	0.85%	94.87%	
16	204,267	6	117	5.13%	100.00%	
17						

This frequency table gives us a little clearer picture of the distribution of housing prices back in 1993. Note that almost 80% of the prices are clustered within the first seven bins of the table (representing homes costing about $129,000 or less). Moreover, there are relatively few homes in the $160,000–$200,000 price range (only about 4% of the sample). There is, however, a small group of homes priced above $205,000.

Defining Your Own Bin Values

The bin values shown in Figure 4-4 were generated by dividing the range of prices into 15 equally spaced intervals. This resulted in cutoff values like 64,733 and 75,467. However, in an analysis of pricing we are usually more interested in even cutoff values like 60,000 and 70,000. The StatPlus Frequency Table dialog box allows you to specify your own bin values in place of automatically generated ones. Try this now, by creating a frequency table of housing prices in $10,000 increments, starting at $50,000. You will first have to enter the bin values into cells in the workbook.

To create your own bin values:

1. Click cell **G1**, type **Price**, and press **Enter**.
2. In cell G2 type **50,000** and press **Enter**. Type **60,000** in cell G3 and press **Enter**.

3. Select the range **G2:G3**, drag the fill handle down to cell **G20**, and release the mouse button.

   The values 50,000–230,000, should be now entered into the cell range G2:G20.

4. Click **Descriptive Statistics** from the StatPlus menu and then click **Frequency Tables**.

5. Click the **Data Values** button and select **Price** as the data variable.

6. Click the **Bin Values** button.

7. Click the **Use Range References** option button, select the range **G1:G20**, and then click **OK**.

8. Click the **<= bin and > previous bin** option button to control how the bin counts are determined.

9. Click the **Output** button, click the **Cells** option button, and select cell **G1**. Click the **OK** button.

10. Click the **OK** button to start generating the frequency table with your customized bin values. Figure 4-5 displays the new frequency table.

Figure 4-5 Frequency table with user-defined bin values

G	H	I	J	K	L
Price	Freq	Cum. Freq.	%	Cum. %	
50,000	2	2	1.71%	1.71%	
60,000	8	10	6.84%	8.55%	
70,000	21	31	17.95%	26.50%	
80,000	17	48	14.53%	41.03%	
90,000	17	65	14.53%	55.56%	
100,000	12	77	10.26%	65.81%	
110,000	10	87	8.55%	74.36%	
120,000	9	96	7.69%	82.05%	
130,000	4	100	3.42%	85.47%	
140,000	2	102	1.71%	87.18%	
150,000	4	106	3.42%	90.60%	
160,000	1	107	0.85%	91.45%	
170,000	0	107	0.00%	91.45%	
180,000	2	109	1.71%	93.16%	
190,000	2	111	1.71%	94.87%	
200,000	2	113	1.71%	96.58%	
210,000	4	117	3.42%	100.00%	
220,000	0	117	0.00%	100.00%	
230,000	0	117	0.00%	100.00%	

This table is a lot easier to interpret than that of Figure 4-4. Looking at the table, it's easy to discover that there were only two houses in the sample in the $140,000–$150,000 price range.

STATPLUS TIPS

- You can use the Frequency Table command to create tables that are broken down into categories on the basis of a qualitative variable. To do so, click the By button in the Frequency Table dialog box and choose the range name or range reference containing the values of the qualitative variable.
- If you forget how the bin counts are determined, place your cursor over the column title for the bin value. A pop-up comment box will appear indicating the method used.

Working with Histograms

Frequency tables are good at conveying specific information about a distribution, but they often lack visual impact. It's hard to get a good impression about how the values are clustered from the counts in the frequency table. Many statisticians prefer a visual picture of the distribution in the form of a histogram. A **histogram** is a bar chart in which each bar represents a particular bin and the height of the bar is proportional to the number of counts in that bin. Histograms can be used to display frequencies, cumulative frequencies, percentages, and cumulative percentages. Most histograms display the frequency or counts of the observations.

Creating a Histogram

Excel does not have a chart type for the histogram, but you can create one using either the Data Analysis ToolPak supplied with Excel or using the command from the StatPlus add-in. Create a histogram of the price data from the Housing workbook using the StatPlus histogram command.

To create a histogram of the home prices:

1 Click **Single Variable Charts** from the StatPlus menu and then click **Histograms**.

2 Click the **Data Values** button and select **Price** from the list of range names.

As with the Frequency Table command, you can specify the number and type of bins used to construct the bars of the histogram.

3 Click the **Chart Options** dialog tab.

4 Click the **Values** button and then click the **Use Range References** button. Select the range **G1:G20** in the Price Table with Bins worksheet. Click the **OK** button.

5 Click the **Right** option button to control how bin counts are determined. See Figure 4-6.

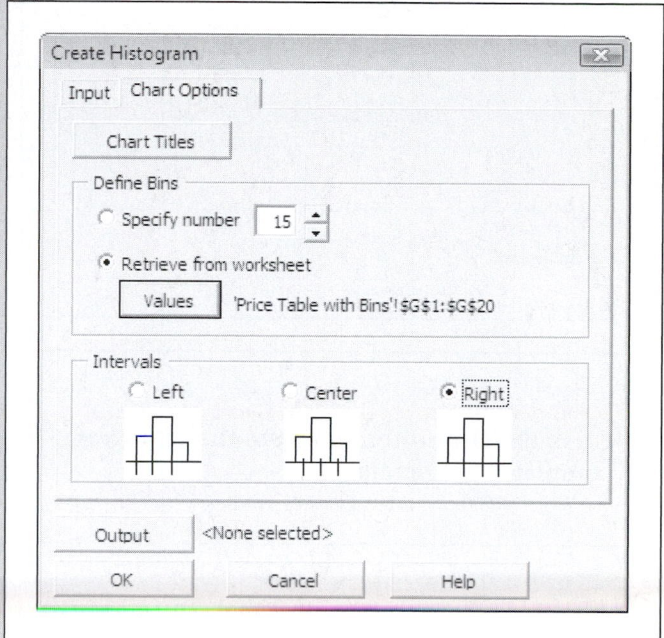

Figure 4-6 Specifying bin options for the histogram

6 Click the **Input** dialog tab to view other options for the histogram.

7 Click the **Output** button.

8 Verify that the **As a new chart sheet** option button is selected and then type **Price Histogram** in the accompanying text box.

This will send the histogram to a chart sheet named Price Histogram.

9 Click the **OK** button.

Figure 4-7 shows the completed Histogram dialog box. Note that this command allows you to create histograms of the frequency, cumulative frequency, percentage, or cumulative percentage. In most cases, histograms display the frequency of a particular variable.

Figure 4-7 Completed histogram dialog box

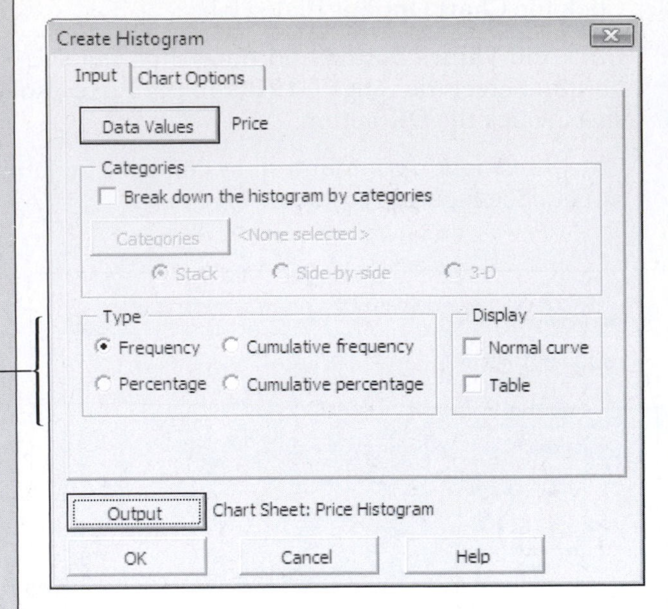

you can create four different types of histograms

10 Click the **OK** button to create the histogram. Figure 4-8 shows the completed histogram.

Figure 4-8 Histogram of housing prices

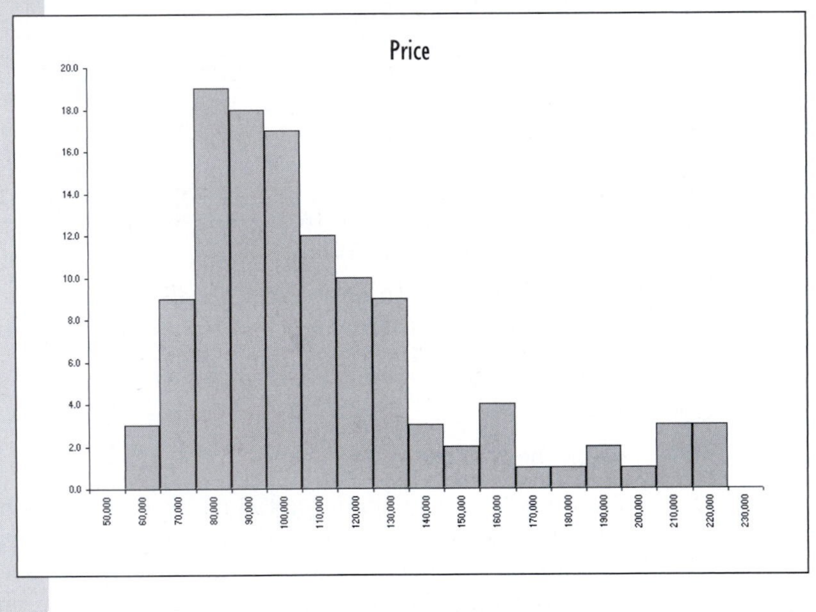

The histogram gives us the strong visual picture that most of the home prices in this 1993 sample were ≤130,000 and that most were in the \$70,000–\$100,000 range. There does not seem to be any clustering of values beyond \$130,000; rather, the data values are clustered toward the lower end of the price scale.

STATPLUS TIPS

- You can also create separate histograms for the different levels of a categorical variable (or for different variables) by using the StatPlus > Multivariable Charts > Multiple Histograms command.
- The Histogram command includes a Chart Titles button located on the Chart Options dialog sheet. By clicking this button, you can enter titles for the chart, *x* axis, and *y* axis. You can also control some of the appearance of the *x* axis and *y* axis.
- The Left option button for the bin intervals in the Histogram command is equivalent to counting observations that are ≥ bin value and < next bin value. The Center option button counts observations that are centered around the bin value (counting from the lower mid-point). The Right option button counts observations that are > bin value and ≤ next bin value.
- You can add a table to the output of the Histogram command by clicking the Table checkbox in the dialog box. This table contains count values, similar to what you would see in the corresponding frequency table.

Shapes of Distributions

The visual picture presented by the histogram is often referred to as the distribution's shape. Statisticians classify various distributions on the basis of their shape. These classifications will become important later on as we look for an appropriate statistic to summarize the distribution and its values. Some statistics are appropriate for one distribution shape but not for another.

A distribution is **skewed** if most of the values are clustered toward either the left or the right edge of the histogram. If the values are clustered toward the left edge of the histogram, this shows **positive skewness**; clustering toward the right edge of the histogram shows **negative skewness**. Skewed distributions often occur where the variable is constrained to have positive values. In those cases, values may cluster near zero, but because the variable cannot have a negative value, the distribution is positively skewed. A distribution is **symmetric** if the values are clustered in the middle with no skewness toward either the positive or the negative side. See Figure 4-9 for examples of these three types of shapes.

Figure 4-9
Distribution shapes

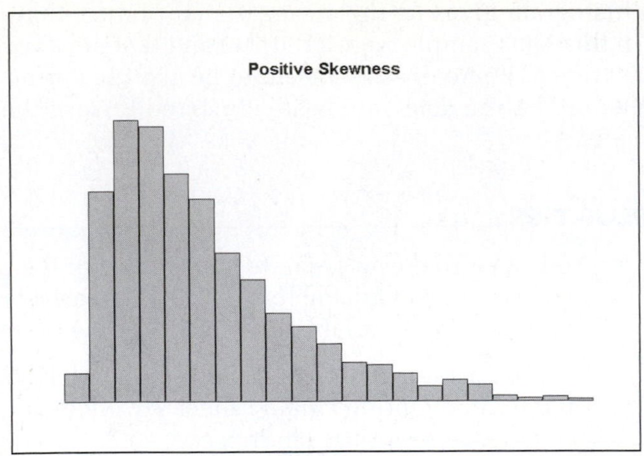

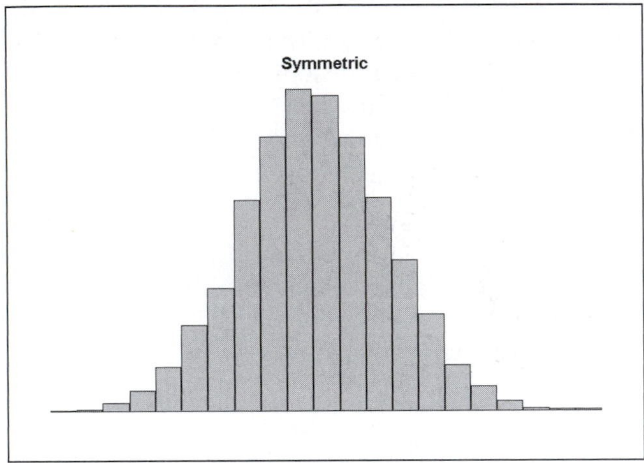

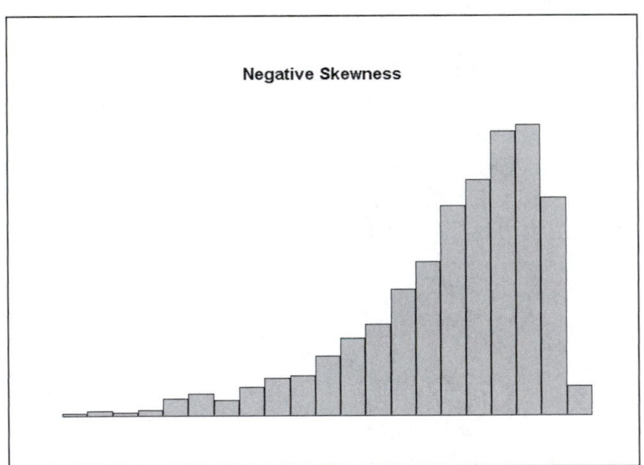

Another important component of a distribution's shape is the distribution's **tails**—the values located to the extreme left or right edge. A distribution with very extreme observations is said to be a **heavy-tailed distribution**.

The historic sample of home prices we've examined appears to be positively skewed with a heavy tail (because there are a number of houses located at the high end of the price scale). This is not surprising, because there is a practical lower limit for housing prices (around $50,000 in this sample) and an exceedingly large upper limit.

Breaking a Histogram into Categories

You can gain a great deal of insight by breaking your histogram into categories. In the current example, we may be interested in knowing how the 1993 Albuquerque prices compared when broken down by location: Were certain locations more expensive than others? One of the more desirable locations in Albuquerque at the time was the northeast sector. Was this reflected in a histogram of the sample home prices? Let's find out.

To create a histogram broken down by categories:

1 Click the **Housing Data** worksheet tab to return to the price data.

2 Click **Single Variable Charts** from the StatPlus menu and then click **Histograms**. Click the **Data Values** button, select **Price** from the list of range names as the source for the histogram, and click **OK**.

3 Click the Break down the histogram by categories checkbox.

The various categories can be displayed in a histogram as stacked on top of each other, side by side, or in three dimensions. You'll see the effect of these choices on the histogram's appearance in a moment. For now, accept the default, Stack.

4 Click the **Categories** button, click the **Use Range Names** option button, select **NE Sector**, and then click the **OK** button.

The NE Sector variable is a qualitative variable that is equal to Yes if the home is located in the northeast sector and is equal to No otherwise. Now, define the options for the histogram's bins.

6 Click the **Chart Options** dialog tab.

7 Click the **Values** button, click the **Use Range References** option button, and then select the range **G1:G20** on the Price Table with Bins worksheet. Click the **OK** button.

8 Click the **Right** option button to set how bin values will be counted in the histogram.

9 Click the **Output** button and type **Price Histogram by NE Sector** in the Chart Sheet name box. Click the **OK** button.

10 Click the **OK** button to start creating the histogram. The completed chart appears in Figure 4-10.

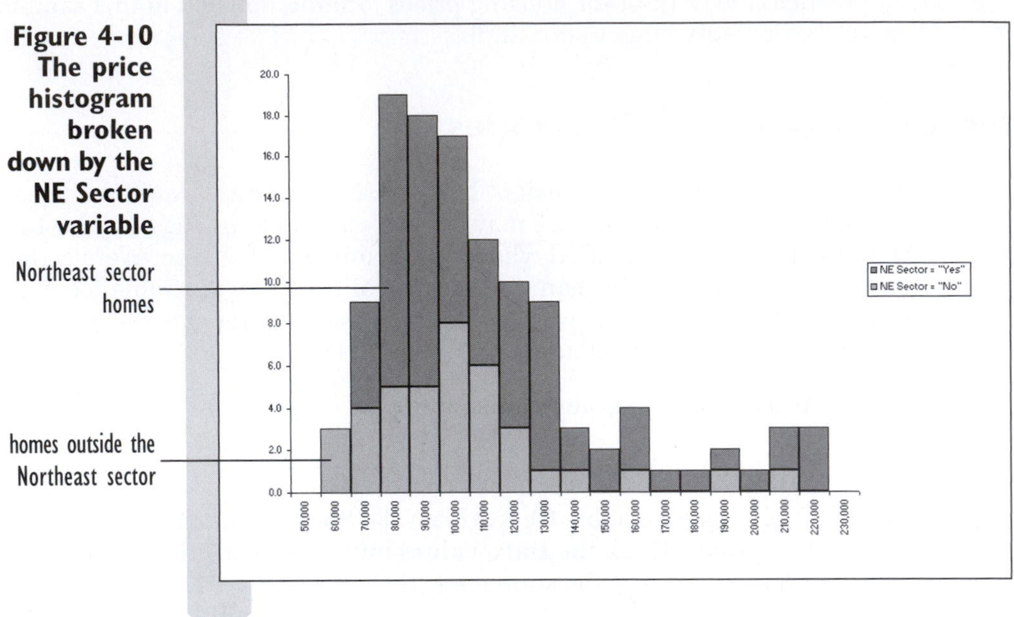

Figure 4-10 The price histogram broken down by the NE Sector variable

In this histogram, the height of each bar is still equal to the total count of values within that bin, but each bar is further broken down by the counts for the various levels of the categorical variable. The counts are stacked on top of each other. The chart makes it clear that many higher-priced homes are located in the northeast sector, though there are still plenty of northeast sector homes in the $70,000–$100,000 range.

How do the shapes of the distributions compare for the two types of homes? We can't tell from this chart, because the northeast sector homes are all stacked at uneven levels. To compare the distribution shapes, we can compare histograms side by side. We can change the orientation of the histogram by modifying the chart type employed by Excel.

To compare histograms side by side:

1 Click anywhere within the chart to select it.

2 Click the **Design** tab from the Chart Tools ribbon.

3 Click the **Change Chart Type** button on the Type group on the Design tab.

4 Excel opens the Change Chart Type dialog box. From within the dialog box click the **Column** chart type and then click the first subtype, **Clustered Column**, in the dialog box (see Figure 4-11).

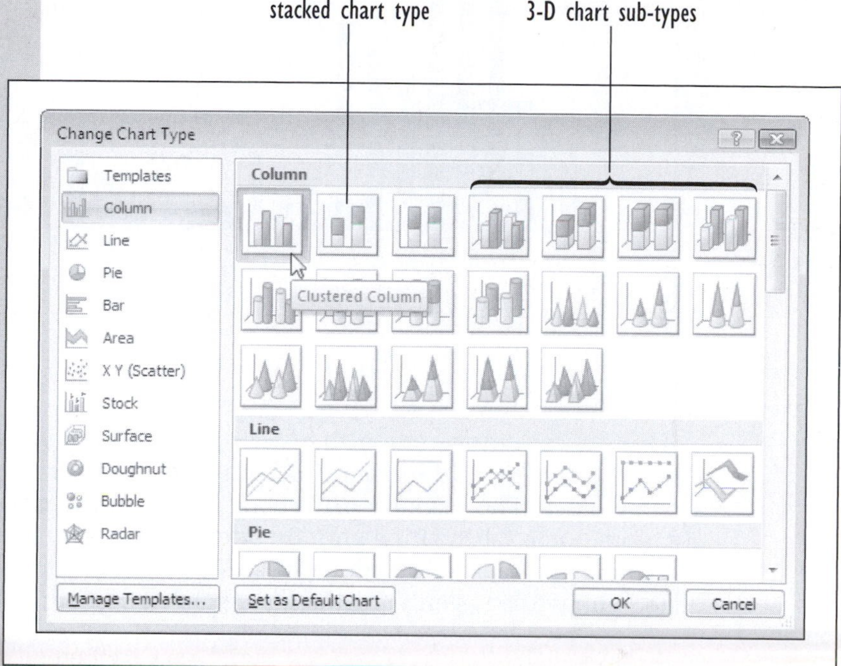

Figure 4-11
Changing the chart type

5 Click the **OK** button. As shown in Figure 4-12, Excel changes the chart type, displaying the histogram bars side by side rather than stacked.

Figure 4-12 Histogram bars displayed side by side

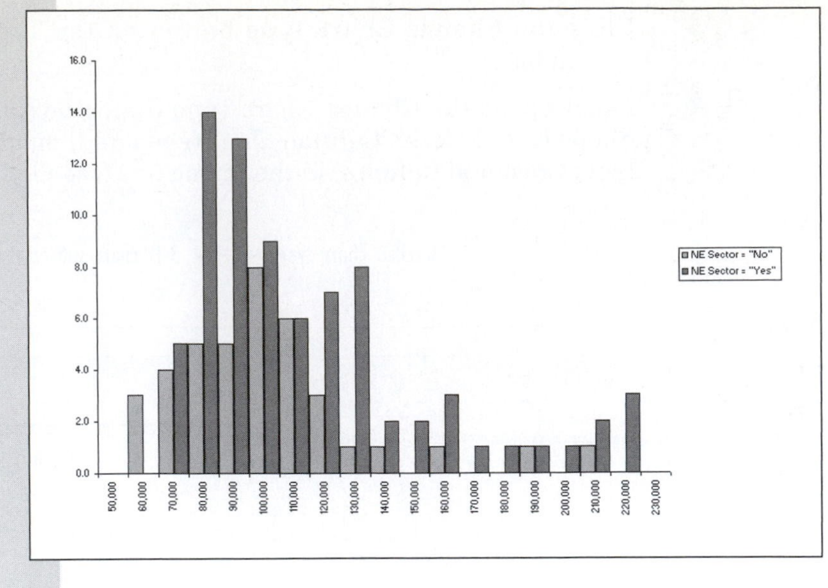

This chart shows us that the distribution of home prices is positively skewed (as we would expect) for both northeast sector and nonnortheast sector homes. The primary difference is that in the northeast sector there are more homes at the high end. By selecting the appropriate chart subtype, we can switch back and forth between side by side and stacked views of the histogram; we can even view the histogram in three dimensions.

Working with Stem and Leaf Plots

Stem and leaf plots are another way of displaying a distribution, while at the same time retaining some information about individual values from the data sample. The stem and leaf plot originally was used by statisticians as a quick way of generating a plot of a distribution using only pen and paper, but still has application even when graphical plots are so readily available.

To create a stem and leaf plot, follow these steps:

1. Sort the data values in ascending order.
2. Truncate all but the first two digits from the values (i.e., change 64,828 to 64,000, change 14,048 to 14,000, and so forth). The first of the two digits is the **stem** and the second the **leaf**. In the case of a number like 64,000, the stem is 6 and the leaf is 4.
3. List the stems in ascending order vertically on a sheet and place a vertical dividing line to the right of the stems.

4. Match each leaf to its stem, placing the leaf values in ascending order horizontally to the right of the vertical dividing line.

For example, take the following numbers:

125, 189, 232, 241, 248, 275, 291, 311, 324, 351, 411, 412, 558, 713

Truncating all but the first two digits from the list, leaves us with

120, 180, 230, 240, 240, 270, 290, 310, 320, 350, 410, 410, 550, 710

The stem and leaf pairs are therefore

(12) (18) (23) (24) (24) (27) (29) (31) (32), (35) (41) (41) (55) and (71).

Now, we list just the stems in ascending order vertically as follows:

100×	
1	
2	
3	
4	
5	
6	
7	

At the top of the stem list, we've included a multiplier, so we know our data values go from 100 to 700. Note that we've added a stem for the value 6. We include this to preserve continuity in the stem list. Now we add a leaf to the right of each stem. The first stem and leaf pair is (12), so we add 2 to the right of the stem value 1, and so on. The final stem and leaf plot appears, as follows:

100×	
1	28
2	34479
3	125
4	11
5	5
6	
7	1

The stem and leaf plot resembles a histogram turned on its side. The plot has some advantages over the histogram. From the stem and leaf plot, you can generate the approximate values of all the observations in the data set by combining each stem with its leaves. Looking at the plot above, you can quickly see that the first two stem and leaf pairs are (1.2) and (1.8). Multiplying these values by 100 yields approximate data values of 120 and 180. An added advantage is that the stem and leaf plot can be quickly generated by hand—useful if you don't have a computer handy.

This stem and leaf plot is at a disadvantage compared to the histogram in that the size of each bin is directly determined by the data values themselves. Stem and leaf plots also don't work well for large data sets where each stem will need to display a large number of leaves. One way of modifying a stem and leaf plot is to split the stems into subgroups. For example you can split a stem into two groups: those with leaves having values from 0 to 4 and those with leaves from 5 to 9. Doing this for the above chart yields the following stem and leaf plot:

```
100× |
-------
   1 | 2
   1 | 8
   2 | 344
   2 | 79
   3 | 12
   3 | 5
   4 | 11
   4 |
   5 |
   5 | 5
   6 |
   6 |
   7 | 1
   7 |
```

Another modification to the stem and leaf plot is to truncate lower and upper values in order to reduce the range of stems in the plot. This is useful in situations where you have an extreme value whose presence would greatly elongate the plot's appearance. For example, if the value 2,420 is added to the above data set, then the resulting stem and leaf plot will have a long stem with a long list of empty leafs. In this case, removing this value from the stem and leaf plot, but noting its value elsewhere, might be the best course of action. The plot might look as follows:

```
100× |
-------
   1 | 28
   2 | 34479
   3 | 125
   4 | 11
   5 | 5
   6 |
   7 | 1
--------------
2400
```

Excel does not have a command to create stem and leaf plots, but you can create one using StatPlus. Let's create a stem and leaf plot for the home price data and compare it to the histogram we created earlier. As before, we'll break the stem and leaf plot down using the values of the NE Sector variable.

To create a stem and leaf plot:

1. Return to the data set by clicking the **Housing Data** worksheet tab.
2. Click **Single Variable Charts** from the StatPlus menu and then click **Stem and Leaf**.

   This command allows you to create plots of variables located in different columns or within a single column, broken down by category levels. You'll do the latter in this case.
3. Verify that the **Use column of category levels** option button is selected and then click the **Data Values** button and select **Price** from the list of range names. Click **OK**.
4. Click the **Categories** button and select **NE_Sector** from the list of range names. Click **OK**.
5. Click the **Apply uniform stem values** checkbox. This will apply the same stem values to home prices both in the northeast sector and elsewhere.
6. Click the **Add a summary plot** checkbox. This will create a stem and leaf plot of prices for all of the homes, regardless of location.
7. Click the **Output** button, click the **New Worksheet** option button, and type **Price StemLeaf** in the New Worksheet name box. Click the **OK** button.

   Figure 4-13 shows the completed Stem and Leaf dialog box.

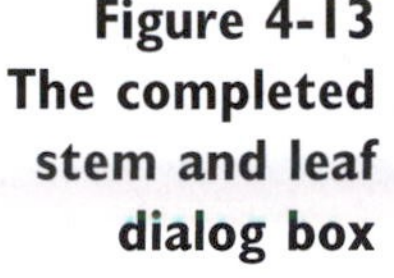

Figure 4-13 The completed stem and leaf dialog box

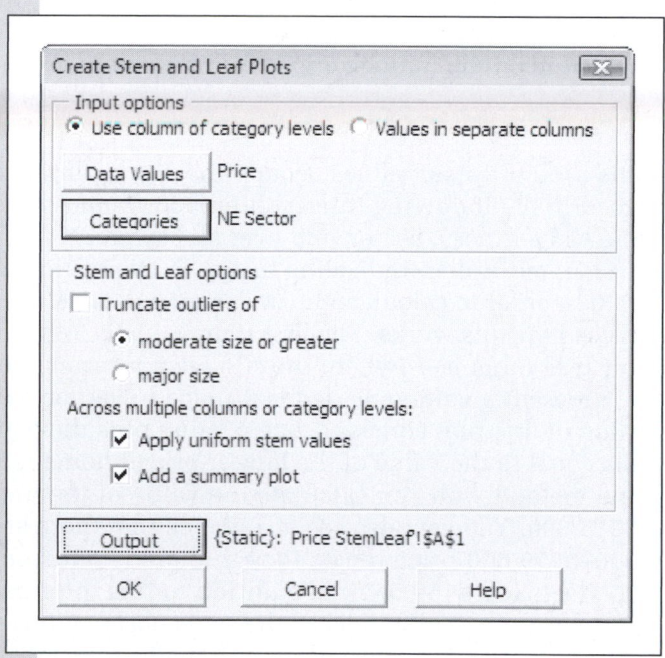

8 Click the **OK** button. Excel generates the stem and leaf plot shown in Figure 4-14.

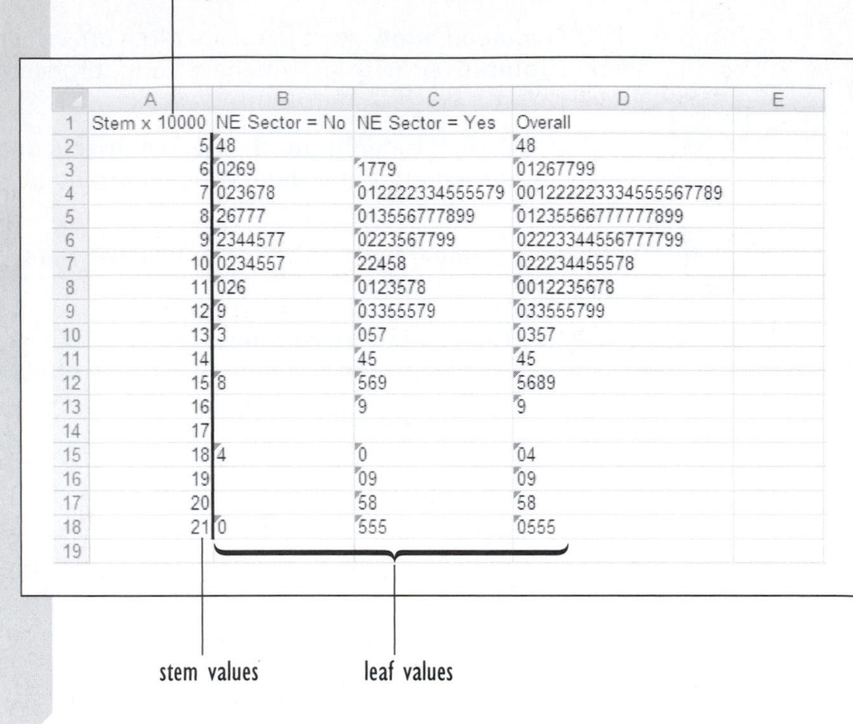

	A	B	C	D	E
1	Stem x 10000	NE Sector = No	NE Sector = Yes	Overall	
2	5	48		48	
3	6	0269	1779	01267799	
4	7	023678	012222334555579	001222223334555567789	
5	8	26777	013556777899	01235566777777899	
6	9	2344577	0223567799	02223344556777799	
7	10	0234557	22458	022234455578	
8	11	026	0123578	0012235678	
9	12	9	03355579	033555799	
10	13	3	057	0357	
11	14		45	45	
12	15	8	569	5689	
13	16		9	9	
14	17				
15	18	4	0	04	
16	19		09	09	
17	20		58	58	
18	21	0	555	0555	
19					

Figure 4-14 Stem and leaf plot of the housing data

In this plot, the stem values occupy the first column and the leaf values are placed in the following three columns for homes outside the northeast sector, in the northeast sector, and over all sectors. Note that cell A1 identifies the stem multiplier, indicating that each stem value must be multiplied by 10,000 in order to calculate the underlying data values.

Let's see how this works. The first stem value is 5; this represents 50,000. The first leaf value is 4 (where the NE_Sector variable equals No), which would represent a value one decimal place lower, or 4,000. Thus the first data value in this plot equals the stem value plus the leaf value, or 54,000, which is equal to the value of the lowest-priced home in the sample. Using the same method, you can calculate the value of the highest-priced home to be $215,000. You can also see at a glance that there are no homes in the $170,000–$179,000 price range, though there is one home priced at about $169,000 (actually $169,500). In addition to this information, you can also use the stem and leaf plot to make the same observations about the shape of the distribution that you did earlier with the histogram.

Distribution Statistics

You should always create a chart of the distribution when analyzing a data set, but once you've done that, you'll probably look for statistics that summarize key elements of the distribution. These values are sometimes called **landmark summaries** because they are used as landmarks, comparing individual values to whole populations, or whole populations to each other.

Percentiles and Quartiles

One of these landmark summaries is the ***p*th percentile**, which is a value such that roughly *p*% of the data are smaller than that value. You may have seen percentiles used in growth statistics, where the progress of a newborn child will place him or her in the 75th percentile or 90th percentile, meaning that the child's weight is equal to or above 75 or 90% of the population. In the Albuquerque data, percentiles could be used as a benchmark to compare one community of that era to another. If you knew the 10th and 90th percentiles for home price, you would have a basis for comparison between the two communities.

Perhaps the most important percentiles are the **quartiles**, which are the values located at the 25th, 50th, and 75th percentiles (the quarters). These are commonly referred to as the first, second, and third quartiles. Statisticians are also interested in the **interquartile range**, which is the difference between the first and third quartiles. Because the central 50% of the data lie within the interquartile range, the size of this value gives statisticians an idea of the width of the distribution.

One way of calculating the percentiles and quartiles of a given distribution is to create a frequency table like the one shown earlier in Figure 4-2. From the column of cumulative percents, you can determine which values correspond to the 10th, 25th, 50th, 75th, and 90th, and so on, percentiles. However, if your data set is large, this can be a cumbersome and time-consuming process. To save time, Excel has several functions that will calculate these values for you. A list of these functions is shown in Table 4-4.

Table 4-4 Excel functions to calculate percentiles and quartiles

Function	Description
PERCENTILE(*array*, *k*)	Returns the *k*th percentile of an array of values or range reference, where *k* is a value between 0 and 1.
PERCENTRANK(*array*, *x*, *significance*)	Returns the percentile of a value taken from an array of values or range reference. The number of digits is determined by the *significance* parameter.

(continued)

QUARTILE(*array, quart*)	Returns the quartile of an array of values or range reference, where *quart* is either 1, 2, or 3 for the first, second, or third quartile.
IQR(*array*)	Calculates the interquartile range of the values in an array or range reference. *StatPlus required.*

Excel allows you to work with percentiles in two different ways. You can use the PERCENTILE function to take a percentile and determine the corresponding data value, or, given the data value, you can use the PERCENTRANK function to determine its percentile.

You can create a table of percentile and quartile values by typing in the above Excel formulas, or you can have StatPlus do it for you with the Univariate Statistics command. The Univariate Statistics command also allows you to break down the variable into different levels of a categorical variable.

In this example you'll limit yourself to percentiles and quartiles. Create such a table now of the housing prices broken down by location.

To create a table of percentile and quartile values:

1. Click **Descriptive Statistics** from the StatPlus menu and then click **Univariate Statistics**.
2. Click the **Input** button and select **Price** from the list of range names.
3. Click the **Output** button, click the **New Worksheet** option button, and type **Percentiles** in the New Worksheet box. Click the **OK** button.
4. Click the **By** button and select **NE_Sector** from the list of range names.
5. Click the **Distribution** dialog tab.
6. Click each of the checkboxes for the different percentiles. See Figure 4-15.

Figure 4-15 The Univariate Statistics dialog box

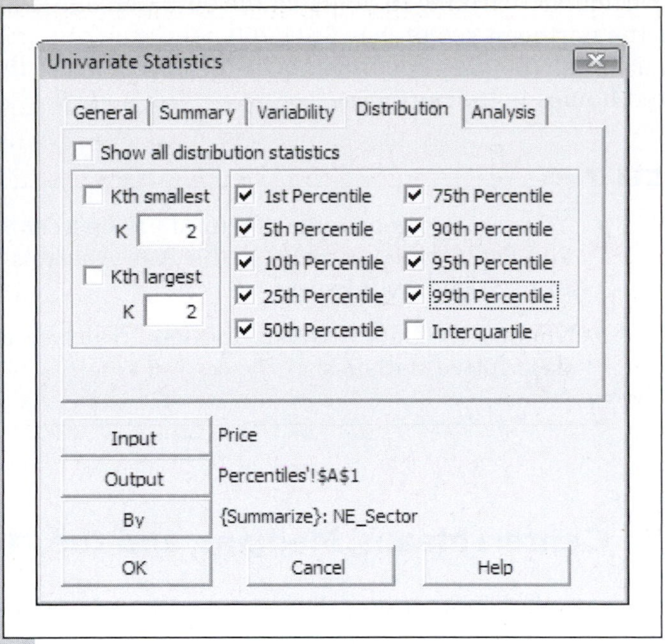

Univariate Statistics

General | Summary | Variability | Distribution | Analysis

☐ Show all distribution statistics

☐ Kth smallest K 2

☐ Kth largest K 2

☑ 1st Percentile ☑ 75th Percentile

☑ 5th Percentile ☑ 90th Percentile

☑ 10th Percentile ☑ 95th Percentile

☑ 25th Percentile ☑ 99th Percentile

☑ 50th Percentile ☐ Interquartile

Input: Price

Output: Percentiles'!$A$1

By: {Summarize}: NE_Sector

OK | Cancel | Help

7 Click the **OK** button to create the table of percentiles. Figure 4-16 shows the completed table.

Figure 4-16 Table of percentiles for the price variable

	A	B	C	D	E
1		Univariate Statistics			
2		Price			
3		NE Sector = "No"	NE Sector = "Yes"	Overall	
4	1st Percentile	55,520	65,827	58,320	
5	5th Percentile	59,800	69,850	66,800	
6	10th Percentile	65,200	72,350	70,600	
7	25th Percentile	76,800.0	80,625.0	78,000.0	
8	50th Percentile	94,000.0	98,500.0	96,000.0	
9	75th Percentile	105,000.0	125,000.0	120,000.0	
10	90th Percentile	130,200	172,650	158,760	
11	95th Percentile	160,640	205,450	200,920	
12	99th Percentile	200,272	215,000	215,000	
13					

The table of percentiles gives us some additional information about the housing prices. The values are pretty close between the two locations up to the 50th percentile, after which large differences begin to appear.

It's particularly striking to note that the 90th percentile for home prices outside the northeast sector was $130,200, whereas for northeast sector homes it was $172,650—$40,000 more. We noted earlier that there are more high-priced homes in the northeast sector.

EXCEL TIPS

- You can also get a table of cumulative percents using the Rank and Percentile command in the Data Analysis ToolPak, an add-in packaged with Excel.
- You use the Data Analysis ToolPak to also create a table of descriptive statistics.

Measures of the Center: Means, Medians, and the Mode

Another way to summarize a data set would be to calculate a statistic that summarized the contents into a single value that we would think of as the typical or most representative value. The table of percentiles suggests one such value: the 50th percentile, or **median**. Because the median is located at the 50th percentile, it represents the middle of the distribution: Half of the values are less than the median, and half are greater than the median. Based on the results from Figure 4-16, the median house price in the Albuquerque sample from 1993 was $94,000 for nonnortheast sector homes, $98,000 for northeast sector homes, and $96,000 overall.

The exact calculation of the median depends on the number of observations in the data set. If there is an odd number of values, the median is the middle value, but if there is an even number of values, the median is equal to the sum of the two central values divided by 2.

Another commonly used summary measure is the average. The **average**, or **mean**, is equal to the sum of the values divided by the number of observations. This value is usually represented by the symbol $\bar{\mathbf{x}}$ (pronounced "x-bar"), a convention we'll repeat throughout the course of this book. Expressed as a formula, this is

$$\bar{x} = \frac{\textit{Sum of values}}{\textit{Number of observations}}$$

$$= \frac{x_1 + x_2 + \cdots + x_n}{n}$$

$$= \frac{\sum_{i=1}^{n} x_i}{n}$$

The total number of observations in the sample is represented by the symbol n, and each individual value is represented by x followed by a subscript. The first value is x_1, the second value is x_2, and so forth, up to the last value (the nth value), which is represented by x_n. The formula calls for us to sum all of these values, an operation represented by the Greek symbol Σ (pronounced "sigma"), a summation symbol. In this case, we're instructed to sum the values of x_i, where i changes in value from 1 up to n; in other words, the formula tells us to calculate the value of $x_1 + x_2 + \cdots + x_n$. The average, or mean, is equal to this expression divided by the total number of observations.

How do these two measures, the median and the mean, compare? One weakness of the mean is that it can be influenced by extreme values. Figure 4-17 shows a distribution of professional baseball salaries. Note that most of the salaries are less than \$1 million per year, but there are a couple of players who make more than \$20 million per year. What, then, is a typical salary? The median value for this distribution is about \$3,500,000, but the mean salary is almost \$4,700,000. The median seems more representative of what the typical player makes, whereas the mean salary is higher as a result of the influence of a couple of much larger salaries. If you were a union representative negotiating a new contract, which figure would you quote? If you represented management, which value better reflects your expenses in salaries?

Figure 4-17 Distribution of baseball salaries

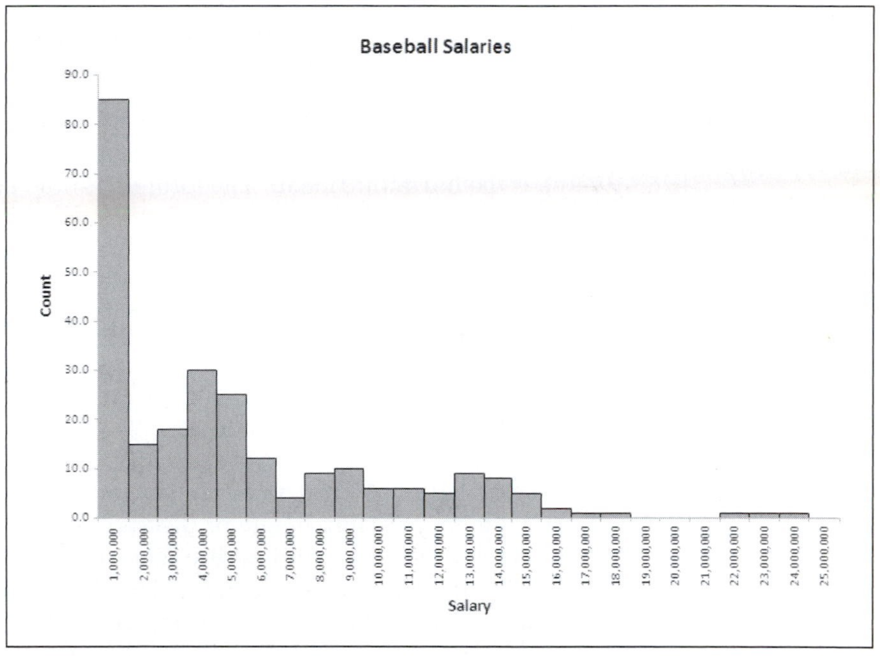

The lesson from this example is that you should not blindly accept any single summary measure. The mean is sensitive to extreme values; the median overcomes this problem by ignoring the magnitude of the upper and lower values. Both approaches have their limitations, and the best approach is to examine the data, create a histogram or stem and leaf plot of the distribution, and thoroughly understand your data before attempting to summarize it. Even then, it may be best to include several summary measures to compare.

The mean and median are the most common summary statistics, but there are others. Let's examine those now.

One method of reducing the effect of extreme values on the mean is to calculate the trimmed mean. The **trimmed mean** is the mean of the data values calculated after excluding a percentage of the values from the lower and upper tails of the distribution. For example, the 10% trimmed mean would be equal to the average of the middle 90% of the data after exclusion of values from the lower and upper 5% of the range. The trimmed mean can be thought of as a compromise between the mean and the median.

Another commonly used measure of the center is the geometric mean. The **geometric mean** is the *n*th root of the product of the data values.

$$\text{Geometric mean} = \sqrt[n]{(x_1) \cdot (x_2) \cdot \ldots (x_n)}$$

Once again, the symbols x_1 to x_n represent the individual data values from a data set with *n* observations. The geometric mean is most often used when the data come in the form of ratios or percentages. Certain drug experiments are recorded as percentage changes in chemical levels relative to a baseline value, and those values are best summarized by the geometric mean. The geometric mean can also be used in situations where the distribution of the values is highly skewed in the positive or negative direction. The geometric mean cannot be used if any of the data values are negative or zero.

Another measure, not widely used today (though the ancient Greeks used it extensively), is the **harmonic mean**. The formula for the harmonic mean *H* is

$$\frac{1}{H} = \frac{1}{n} \sum_{i=1}^{n} \frac{1}{x_i}$$

The harmonic mean can be used to calculate the mean values of rates. For example, a car traveling at a rate of *S* miles per hour to a destination and then at a rate of *T* miles per hour on the return trip, travels at an average rate equal to the harmonic mean of *S* and *T*.

Our final measure of the center is the mode. The **mode** is the most frequently occurring value in a distribution. The mode is most often used when we are working with qualitative data or discrete quantitative data, basically any data in which there are a limited number of possible values. The mode is not as useful in continuous quantitative data, because if the data are truly continuous, we would expect few, if any, repeat values.

Table 4-5 displays the Excel functions used to calculate the various measures of the distribution's center.

Table 4-5 Excel functions to calculate the distribution's center

Function	Description
AVERAGE(*array*)	Returns the average or mean of the values in an array or data range.
GEOMEAN(*array*)	Returns the geometric mean of the values in an array or data range.
HARMEAN(*array*)	Returns the harmonic mean of the values in an array or data range.
MEDIAN(*array*)	Returns the median of the values in an array or data range.
MODE(*array*)	Returns the most frequently occurring value in an array or data range.
TRIMMEAN(*array, percent*)	Returns the trimmed mean of the values in array or data range, excluding the lower and upper values where *percent* is the fractional number of data points to exclude. The function rounds the number of excluded data points down to the nearest multiple of 2. If *percent* = 0.3 and *array* contains 30 data points, 30 percent of 30 equals 9 and thus 8 points are excluded: four from the upper range and four from the lower range.

Now that you've learned a little about these functions, use the Univariate Statistics command from the StatPlus add-in to generate a table of their values.

To create a table of mean and median values:

1. Click **Descriptive Statistics** from the StatPlus menu and then click **Univariate Statistics**.
2. Click the **Input** button and select **Price** from the list of range names.
3. Click the **Output** button, click the **New Worksheet** option button, and type **Means** in the New Worksheet box. Click the **OK** button.
4. Click the **By** button and select **NE_Sector** from the list of range names.
5. Click the **Summary** dialog tab.
6. Click the **Show all summary statistics** checkbox. Figure 4-18 shows the completed dialog box.

Figure 4-18 Selecting summary statistics for the Price variable

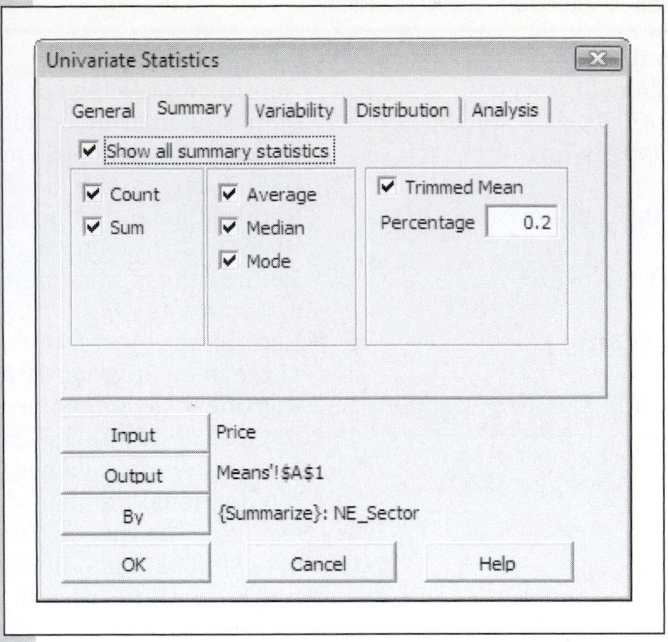

7 Click the **OK** button to create the table of values (see Figure 4-19).

Figure 4-19 Summary statistics for the Price variable

	A	B	C	D	E
1		Univariate Statistics			
2		Price			
3		NE Sector = "No"	NE Sector = "Yes"	Overall	
4	Count	39	78	117	
5	Sum	3,794,000	8,640,000	12,434,000	
6	Average	97,282.05	110,769.23	106,273.50	
7	Median	94,000.0	98,500.0	96,000.0	
8	Mode	(105000 ; 97500)	(215000 ; 125000)	97,500	
9	Trimmed Mean (0.2)	93,018.18	104,909.38	100,393.68	
10					

As we would expect, the average price of a home in Albuquerque from our historic sample is higher than the median value. This effect is more noticeable in the northeast sector homes because of the group of high-priced homes in that location. The mean home price was almost $12,000 greater than the median due to the positive skewness in the data.

Measures of Variability

The mean and median do not tell the whole story about a distribution. It's also important to take into account the variability of the data. **Variability** is a measure of how much data values differ from one another, or equivalently, how widely the data values are spread out around the center. Consider the pair of histograms shown in Figure 4-20. The mean and median are the same for both distributions, but the variability of the data is much greater for the second figure.

Figure 4-20 Distributions with low and high variability

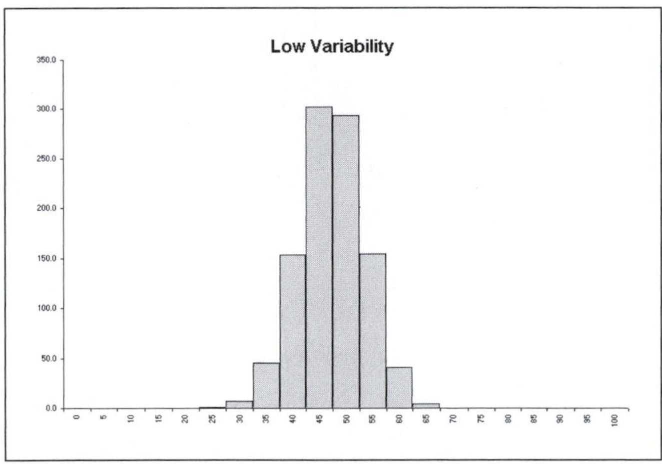

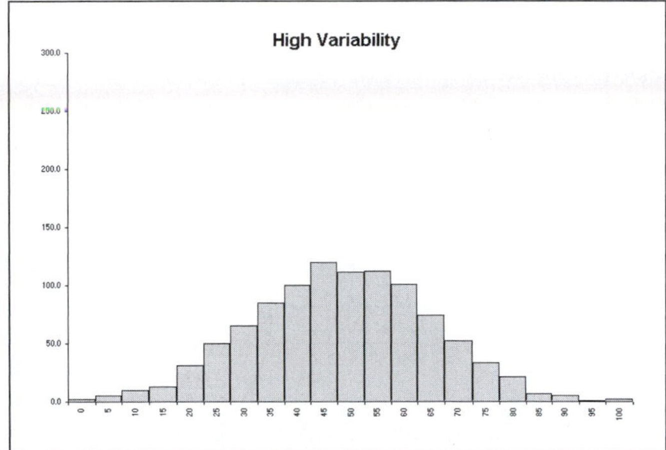

The simplest measure of variability is the **range**, which is the difference between the maximum value in the distribution and the minimum value. A large variability usually results in a large range of values. However, the range can be a poor and misleading measure of variability. As shown in Figure 4-21, two distributions can have the same range but be very different in the variability of their data.

Figure 4-21 Distributions with different variability but the same range

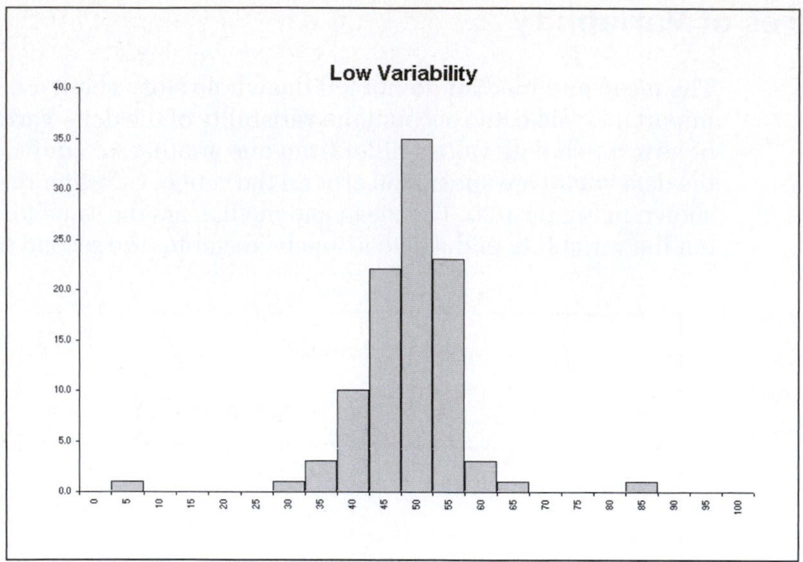

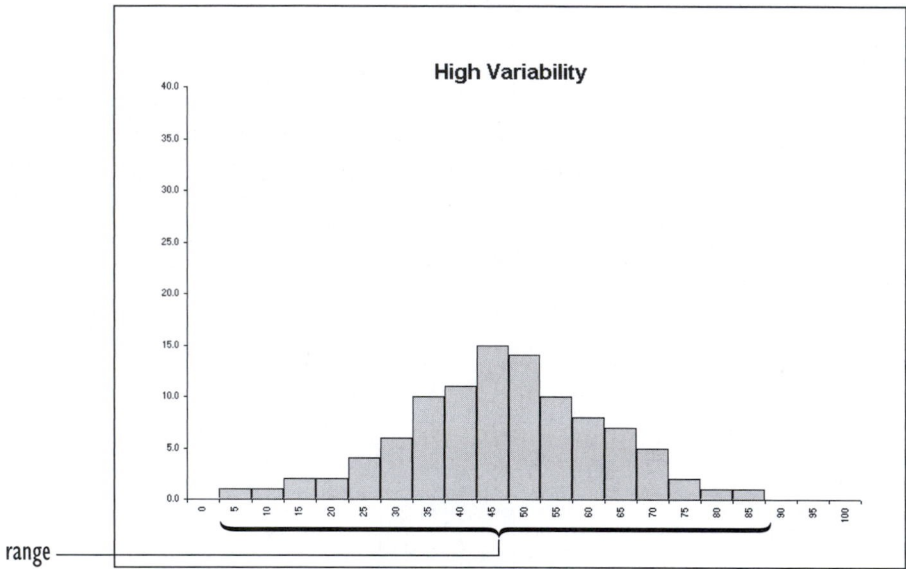

The most common measure of variability depends on the deviation of each data value from the sample average. For each data value x_i, calculate the deviation d_i, which is the difference between sample value and the sample average, or

$$d_i = x_i - \overline{x}$$

Some of these deviations will be negative (where the data value is less than the mean), and some will be positive, so we cannot simply take the average of the deviations because the positive and negative values would cancel each other out. In fact, the sum of the deviations between each sample value and the sample mean equals zero, so the average deviation is also zero.

Instead of averaging the deviations, we'll square each deviation (to make it positive) and then sum those values and divide by the number of observations minus 1. This value, known as the **variance**, is represented by s^2. The formula for calculating s^2 is

$$s^2 = \frac{\text{Sum of squared deviations}}{\text{Number of observations} - 1}$$

$$= \frac{1}{n-1}\sum_{i=1}^{n}(x_i - \bar{x})^2$$

One measure of variability, the **standard deviation** (represented by the symbol s), is calculated by taking the square root of the variance. The complete formula for the standard deviation s is

$$s = \sqrt{\frac{\sum_{i=1}^{n}(x_i - \bar{x})^2}{n-1}}$$

Why do we divide the total of the squared deviations by $n - 1$, rather than n? Recall that the sum of the deviations is known to be zero, so given the first $n - 1$ deviations, we can always calculate the remaining deviation. This means only $n - 1$ of the deviations can vary freely; the last value is constrained by the values of the preceding deviations. This figure, $n - 1$, is known as the degrees of freedom and is a value that will become more important in the chapters that follow.

The standard deviation represents the typical deviation of values from the average. A large value of s indicates a high degree of variability in the data. *High* is a relative term, and we usually speak about high degrees of variability only when comparing one distribution with another. Table 4-6 summarizes the different functions supported by Excel to describe the variability of data.

Table 4-6 Formulas to calculate variability of values in data sets

Function	Description
AVEDEV(*array*)	Returns the average of the absolute value of the deviations in an array or data range.
DEVSQ(*array*)	Returns the sum of the squared deviations in an array or data range.
MAX(*array*)	Returns the maximum value in an array or data range.
MIN(*array*)	Returns the minimum value in an array or data range.
STDEV(*array*)	Returns the standard deviation of the values in an array or data range.
VAR(*array*)	Returns the variance of the values in an array or data range.
RANGEVALUE (*array*)	Returns the range of the values in an array or range reference. *StatPlus required.*

Measures of Shape: Skewness and Kurtosis

You've seen that different distributions can be characterized by their shape. For example, a distribution may be skewed positively or negatively or may be symmetric about its midpoint. These visual judgments we make of a distribution's shape also can be quantified with a statistic. One of these is the skewness statistic. **Skewness** is a measure of the lack of symmetry in the distribution of the data values.

A positive skewness value indicates a distribution with values clustered toward the lower range of values with a long tail extending toward the upper values' range. A negative skewness indicates just the opposite, with the long tail extending toward the values lower in the data range. A skewness of zero indicates a symmetric distribution.

Another statistic, **kurtosis**, measures the heaviness of the tails in the distribution. A positive kurtosis indicates more extreme values than expected in the distribution. A negative kurtosis indicates fewer extreme values than expected. Table 4-7 shows the Excel functions used to calculate skewness and kurtosis.

Table 4-7 Excel functions to calculate skewness and kurtosis

Function	Description
KURT(*array*)	Returns the kurtosis of the values in an array or data range.
SKEW(*array*)	Returns the skewness of the values in an array or data range.

Use the Univariate Statistics command from the StatPlus menu to calculate the variability and shape statistics for the prices of homes in the Albuquerque sample.

To create a table of variability and shape statistics:

1. Click **Descriptive Statistics** from the StatPlus menu and then click **Univariate Statistics**.
2. Click the **Input** button and select **Price** from the list of range names.
3. Click the **Output** button, click the **New Worksheet** option button, and type **Price Variances** in the New Worksheet box. Click the **OK** button.
4. Click the **By** button and select **NE_Sector** from the list of range names.
5. Click the **Variabilit**y dialog tab.
6. Click the **Show all variability statistics** checkbox. See Figure 4-22.

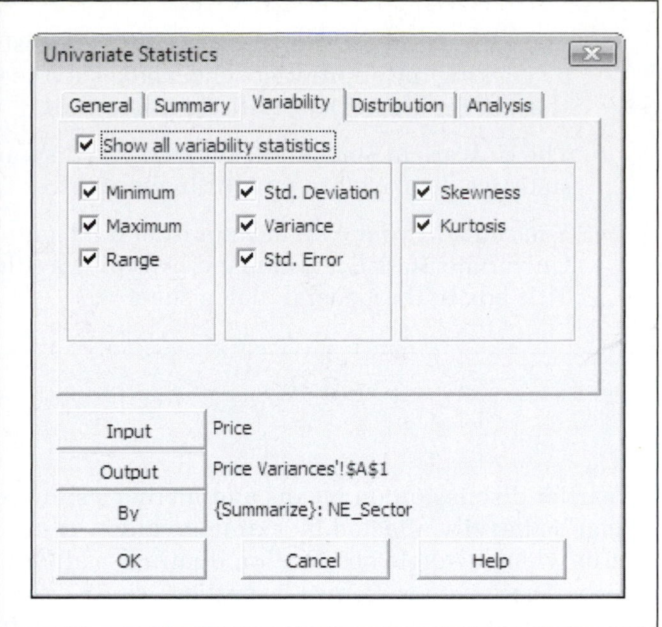

Figure 4-22 Selecting variability statistics from the Univariate Statistics dialog box

7 Click the **OK** button to create the table. Figure 4-23 shows the variability statistics generated from the Univariate Statistics command.

Figure 4-23 Variability statistics

	A	B	C	D	E
1			Univariate Statistics		
2		Price			
3		NE Sector = "No"	NE Sector = "Yes"	Overall	
4	Minimum	54,000	61,900	54,000	
5	Maximum	210,000	215,000	215,000	
6	Range	156,000	153,100	161,000	
7	Standard Deviation	32,039.522	40,154.213	38,043.699	
8	Variance	1,026,530,985.155	1,612,360,859.141	1,447,322,998.821	
9	Standard Error	5,130.430	4,546.569	3,517.141	
10	Skewness	1.722	1.233	1.375	
11	Kurtosis	4.009	0.809	1.447	
12					

On the basis of the output from Figure 4-23, we note that the variability of the 1993 Albuquerque home prices was higher in the northeast sector than outside of it (though it's interesting to note that the range of home prices was higher for nonnortheast sector homes).

STATPLUS TIPS

- You can select all summary, variability, or distribution statistics by clicking the appropriate checkboxes in the General dialog sheet of the Univariate Statistics dialog box.
- The Univariate Statistics command can display the table with statistics displayed in rows or in columns.
- You can add your own custom title to the output from the Univariate Statistics command by typing a title in the Table Title box in the General dialog sheet.

Outliers

As the earlier discussion on means and medians showed, distribution statistics can be heavily affected by extreme values. It's difficult to analyze a data set in which a single observation dominates all of the others, skewing the results. These values, known as **outliers**, don't seem to belong with the others because they're too small, too large, or don't match the properties one would expect for them. As you've seen, a large salary can affect an analysis of salary values, pushing the average salary value upward. An outlier need not be an extreme value. If you were to analyze fitness data, the records of an extremely fit 75-year-old might not be remarkable compared to all of the values in the distribution, but it might be unusual compared to the values of others in his or her age group.

Outliers are caused by either mistakes in data entry or an unusual or unique situation. A mistake in data entry is easier to deal with: You discover and correct the mistake and then redo the analysis. If there is no mistake, you have a bigger problem. In that case you have to study the outlier and decide whether it really belongs with the other data values. For example, in a study of Big Ten universities, we might decide to remove the results from Northwestern because that school, unlike the other schools, is a small, private institution. In the Albuquerque data, we might remove a high-priced home from the sample if that house were a public landmark and thus uniquely expensive.

However, and this point cannot be emphasized too strongly, *merely being an extreme value is not sufficient grounds to remove an observation.* Many advances have been made by scientists studying the observations that didn't seem to fit the expected distribution. Extreme values may be a natural part of the data (as with some salary structures). By removing those values, you are removing an important aspect of the distribution.

One possible solution to the problem of outliers is to perform two analyses: one with the outliers and one without. If your conclusions are the same, you can be confident that the outlier had no effect. If the results are extremely different, you can report both answers with an explanation of the differences involved. In any case, you should not remove an observation without good cause and documentation of what you did and why.

What constitutes an outlier? How large (or small) must a value be before it can be considered an outlier? One accepted definition depends on the interquartile range (IQR; recall that the interquartile range is equal to the difference between the third and first quartiles).

1. If a value is greater than the third quartile plus 1.5 × IQR or less than the first quartile minus 1.5 × IQR, it's a **moderate outlier**.
2. If a value is greater than the third quartile plus 3 × IQR or less than the first quartile minus 3 × IQR, it's an **extreme outlier**.

A diagram displaying the boundaries for moderate and extreme outliers is shown in Figure 4-24.

Figure 4-24 The range of moderate and extreme outliers

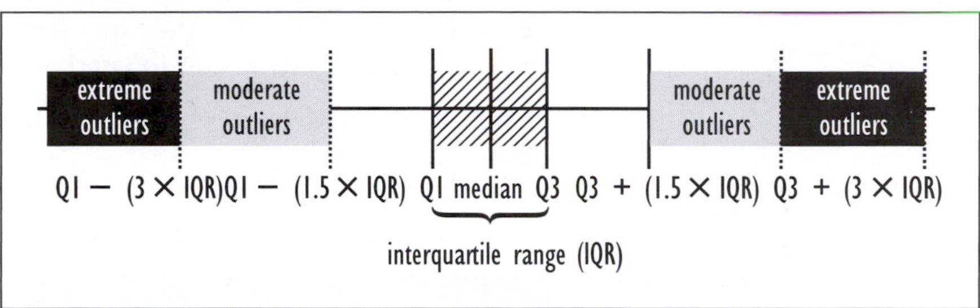

For example, if the first quartile equals 30 and the third quartile equals 80, the interquartile range is 50. Any value above 80 + (1.5 × 50), or 155, would be considered a moderate outlier. Any value above 80 + 150, or 230, would be considered an extreme outlier. The lower ranges for outliers would be calculated similarly.

This definition of the outlier plays an important role in constructing one of the most useful tools of descriptive statistics—the boxplot.

Working with Boxplots

In this section, we'll explore one of the more important tools of descriptive statistics, the boxplot. You'll learn about boxplots interactively with Excel, and then you'll apply what you've learned to the Albuquerque price data.

CONCEPT TUTORIALS: Boxplots

The files available with this book contain several instructional workbooks. The **instructional workbooks** provide interactive worksheets and macros to allow you to explore various statistical concepts on your own. The first of these workbooks that you will examine concerns the box plot. Open this workbook now.

To start the Boxplots instructional workbook:

1 Open the file **Boxplots**, located in the Explore folder.

The workbook opens to the Contents page, describing the workbook. To the left side of the display area is a column of subject titles. You can move between subject titles either by clicking an entry in the column or by clicking the arrow icon located at the top of the page.

2 Click **What is a boxplot?** from the list of subject titles. The page shown in Figure 4-25 appears.

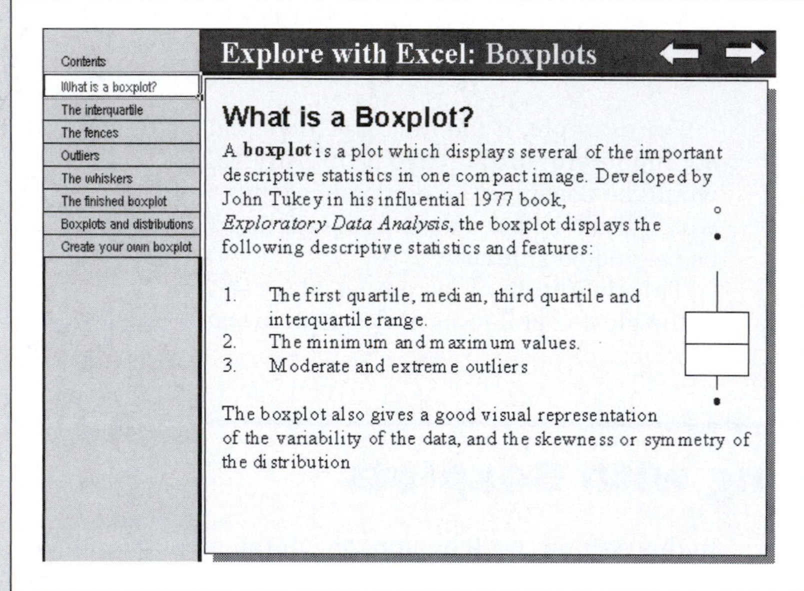

Figure 4-25 Initial worksheet from the Boxplots Explore workbook

Boxplots are designed to display in a single chart several of the important descriptive statistics, including the quartiles of the distribution as well as the minimum and the maximum. They will also identify any moderate or extreme outliers (using the definition supplied above).

3 Click **The interquartile** from the list of subject titles.

The box part of the boxplot displays the interquartile range of the distribution, ranging from the first quartile to the third. The median is shown as a horizontal line within the box. Note that the median need not be in the center of the box. The box tells you where the central 50% of the data is located. By observing the placement of the median within the box, you can also get an indication of how those values are clustered within that central 50%. A median line close to the first quartile indicates that a lot of the values are clustered in the lower range of the distribution.

Figure 4-26 The "box" from the boxplot

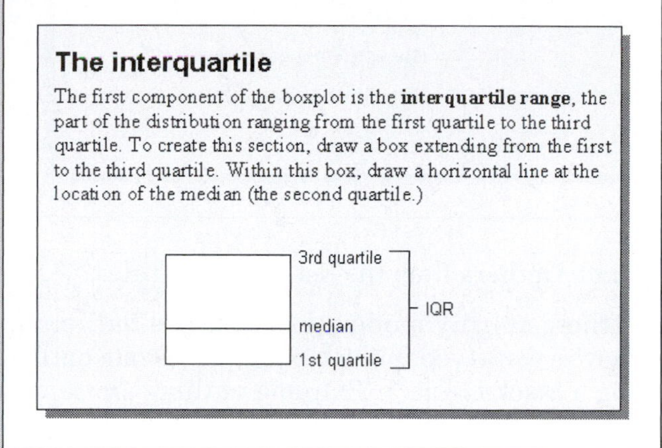

4 Click **The fences** from the list of subject titles.

The inner and outer fences of the boxplot set the boundaries between standard observations, moderate outliers, and extreme outliers. Note that the formula for the fences matches the formula for moderate and extreme outliers discussed in the previous section.

Figure 4-27
Constructing the "fences" of a boxplot

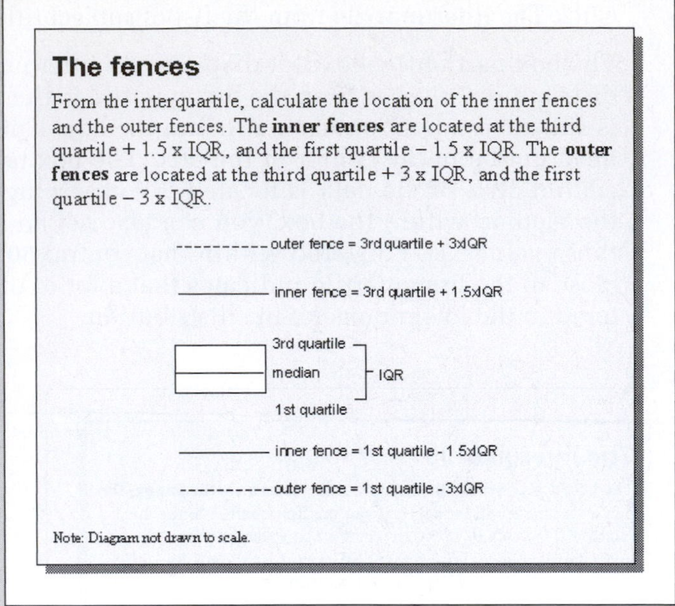

5 Click **Outliers** from the list of subject titles.

If there are any moderate or extreme outliers in the distribution, they're displayed in the boxplot. Moderate outliers are displayed using a black circle •. Extreme outliers are represented by an open circle ◦. With a boxplot you can quickly see the outliers in your distribution and their severity.

Figure 4-28
Representing outliers from the sample distribution

Outliers

Any values that lie between the inner and outer fences are **moderate outliers** and are shown with the the symbol, •. Any values that lie beyond the outer fences are **extreme outliers** and are shown with the symbol, o.

o
outer fence
•
inner fence
3rd quartile
median
1st quartile
inner fence
•
outer fence

Note: Diagram not drawn to scale.

6 Click **The whiskers** from the list of subject titles.

The final component of the boxplot are the whiskers. These are lines that extend from the boxplot to the highest and lowest points that lie inside the moderate outliers. Thus the lines indicate the smallest and largest values in the distribution that are *not* considered outliers. The length of the whisker lines also gives you a further indication of the skewness of the distribution.

Figure 4-29 Drawing the box "whiskers"

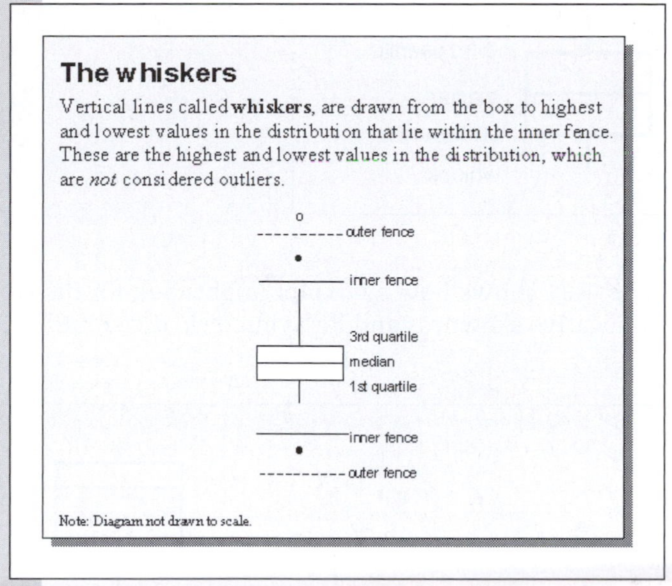
The whiskers
Vertical lines called **whiskers**, are drawn from the box to highest and lowest values in the distribution that lie within the inner fence. These are the highest and lowest values in the distribution, which are *not* considered outliers.
o
outer fence
inner fence
3rd quartile
median
1st quartile
inner fence
outer fence
Note: Diagram not drawn to scale.

In the finished boxplot, the inner and outer fences are not shown. Figure 4-30 shows a typical boxplot.

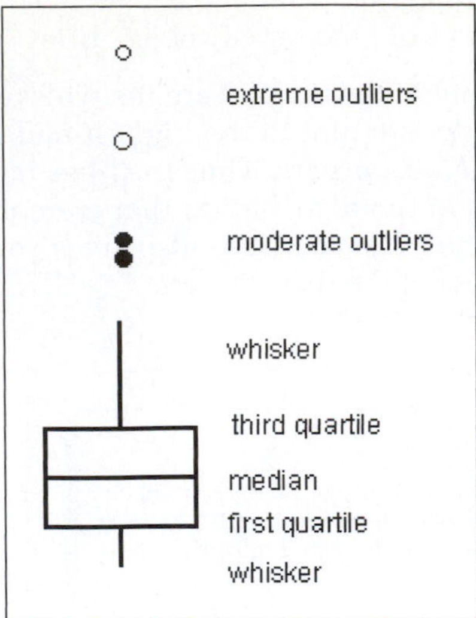

Figure 4-30 The completed boxplot

Figure 4-31 shows how a boxplot might look for distributions with positive or negative skewness and for symmetric distributions.

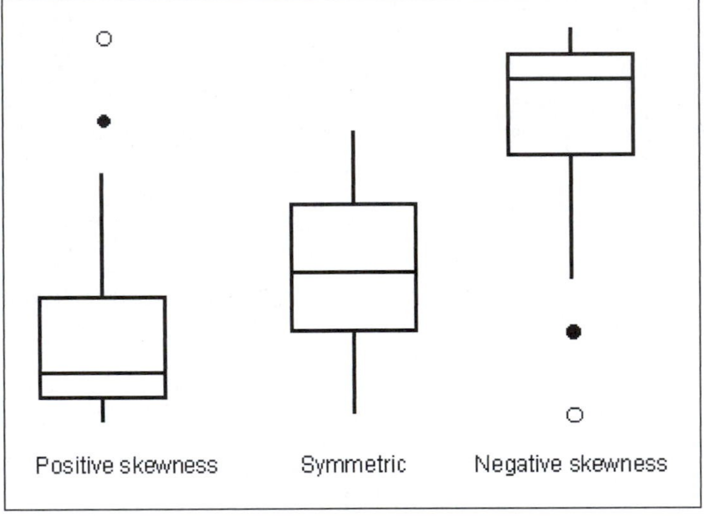

Figure 4-31 Boxplots of different distribution shapes

Now that you've learned about the structure of the boxplot, try creating a few boxplots on your own with sample data.

To create your own boxplot:

1 Click **Create your own boxplot** from the list of topics in the Boxplots workbook.

2 Enter the following numbers into the green cells located to the left of the empty chart: **0, 1, 2, 2, 3, 3, 3, 3, 4, 4, 4, 5, 5, 6, 7**.

As you enter the numbers, the chart is automatically updated to reflect the new distribution. The final form of the boxplot is shown in Figure 4-32.

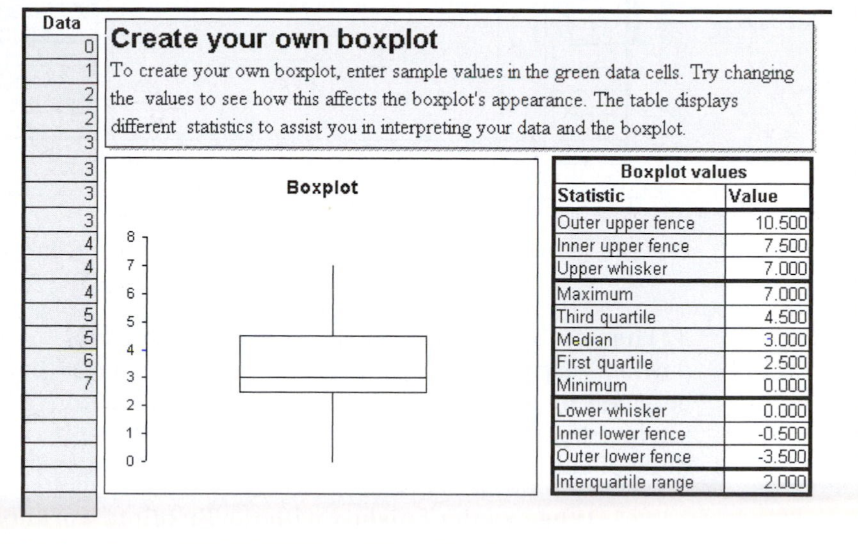
Data
0
1
2
2
3
3
3
3
4
4
4
5
5
6
7
Create your own boxplot
To create your own boxplot, enter sample values in the green data cells. Try changing the values to see how this affects the boxplot's appearance. The table displays different statistics to assist you in interpreting your data and the boxplot.
Boxplot
8
7
6
5
4
3
2
1
0

Boxplot values	
Statistic	**Value**
Outer upper fence	10.500
Inner upper fence	7.500
Upper whisker	7.000
Maximum	7.000
Third quartile	4.500
Median	3.000
First quartile	2.500
Minimum	0.000
Lower whisker	0.000
Inner lower fence	-0.500
Outer lower fence	-3.500
Interquartile range	2.000

Figure 4-32 Entering data into a sample box plot

The central 50% of the data is found in the range from 2.5 to 4.5. The median value is 3, which is not in the middle of that central 50%.

From the plot we can see that the values range from 0 to 7. There are no outliers in the distribution. Now let's see what happens if we change a few of those numbers.

3 Change the last two numbers in the sample data from 6 and 7 to **9** and **12**. Figure 4-33 shows the updated chart.

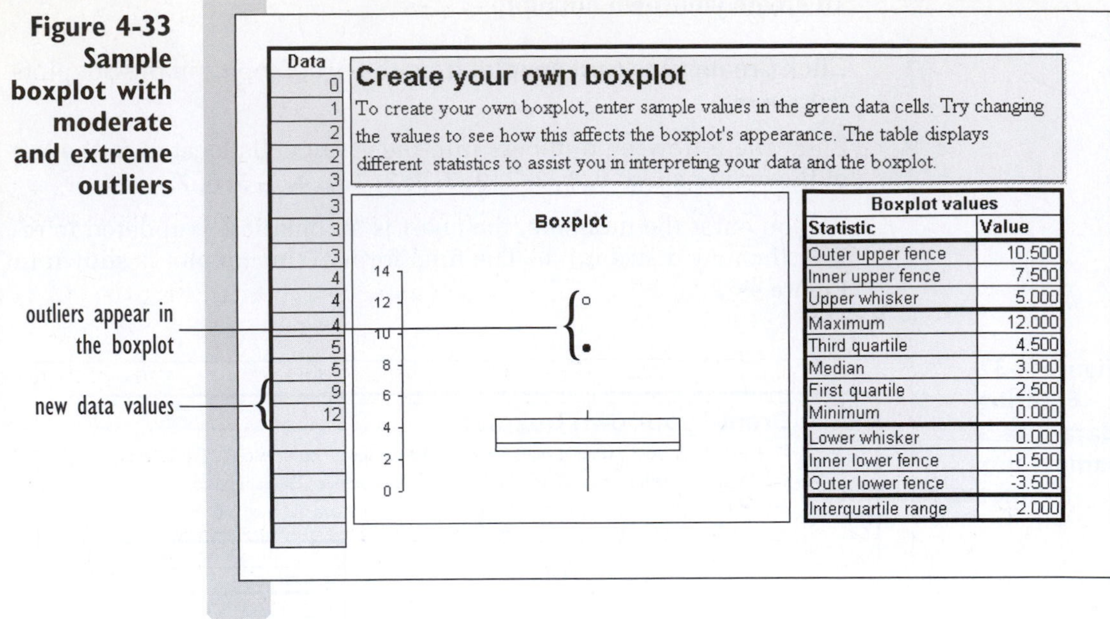
Data
0
1
2
2
3
3
3
3
4
4
4
5
5
9
12
Create your own boxplot
To create your own boxplot, enter sample values in the green data cells. Try changing the values to see how this affects the boxplot's appearance. The table displays different statistics to assist you in interpreting your data and the boxplot.
Boxplot
14
12
10
8
6
4
2
0
outliers appear in the boxplot
new data values

Boxplot values	
Statistic	**Value**
Outer upper fence	10.500
Inner upper fence	7.500
Upper whisker	5.000
Maximum	12.000
Third quartile	4.500
Median	3.000
First quartile	2.500
Minimum	0.000
Lower whisker	0.000
Inner lower fence	-0.500
Outer lower fence	-3.500
Interquartile range	2.000

Figure 4-33 Sample boxplot with moderate and extreme outliers

The new sample values appear in the boxplot as moderate and extreme outliers respectively. From the boxplot we can see that there is a large gap between the moderate outlier and the largest standard value. Continue to explore boxplots with the instructional workbook. Try different combinations of values and different types of distributions.

When you're finished with the Boxplots workbook:

1. Close the **Boxplots** workbook. You do not have to save any of your changes.
2. Return to the **Housing Statistics** workbook in Excel.

Excel does not contain any commands to create boxplots, but you can create one using StatPlus. The StatPlus command includes the added feature of displaying a dotted line representing the average value for the distribution. Try creating a boxplot comparing the NE sector from the Albuquerque data to other homes in the sample.

To create a boxplot of the price data:

1 Click **Single Variable Charts** from the StatPlus menu and then click **Boxplots.**

The Boxplots command allows you to create boxplots on the basis of values in separate columns or within a single column broken down by the levels of a categorical variable. In this case you'll use a single column, Price, and a categorical variable, NE_Sector.

2 Verify that the **Use column of category levels** option button is selected.

3 Click the **Data Values** button and choose **Price** from the list of range names.

4 Click the **Categories** button and choose **NE_Sector** from the list of range names.

5 Click the **Output** button, click the **As a new chart sheet** option button, and type **Price Boxplot** in the adjacent name box. Click the **OK** button. Figure 4-34 shows the completed dialog box.

Figure 4-34 The completed Boxplots dialog box

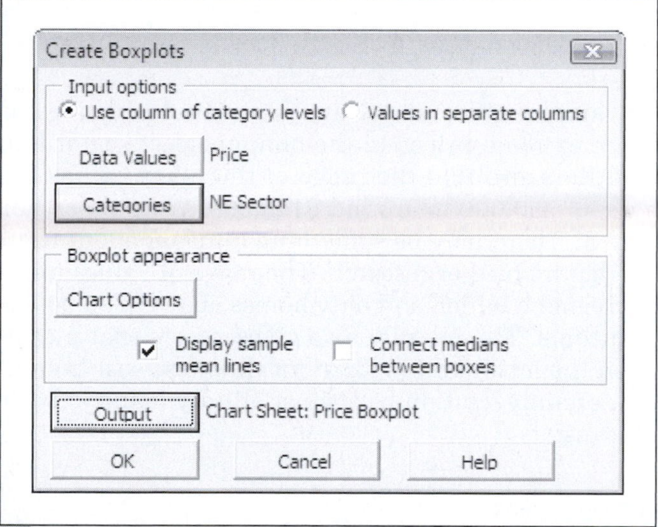

6 Click the **OK** button. The output from the Boxplot command is shown in Figure 4-35.

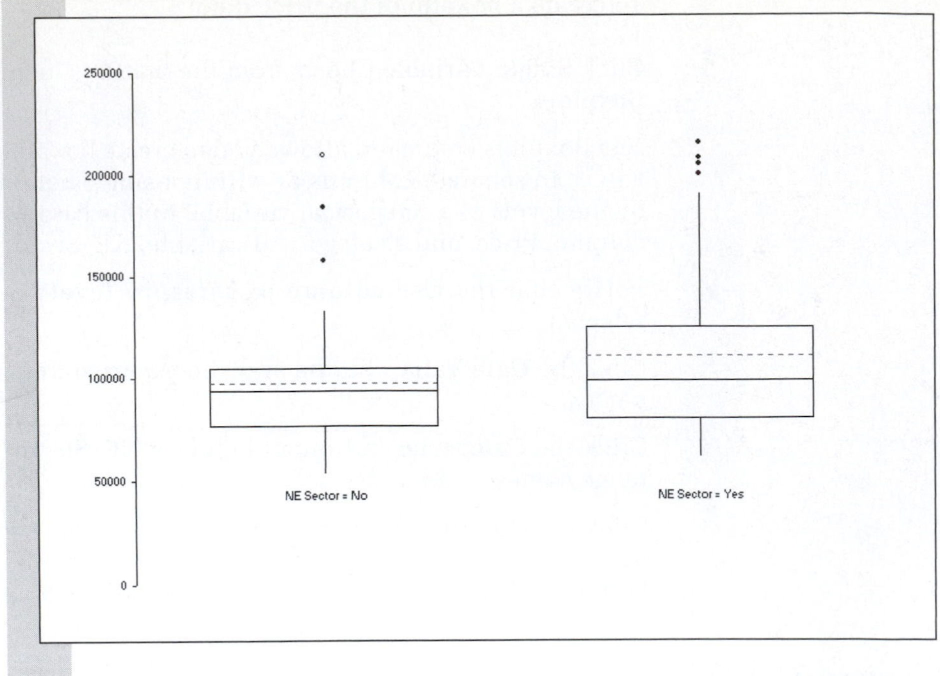

Figure 4-35 Boxplot of 1993 housing prices in the Albuquerque, New Mexico area

The boxplot gives us yet another visual picture of our data. Note that the three extreme prices in the nonnortheast sector homes are all considered outliers and that the range of the other values for that area extends from about $50,000 to around $140,000. Were those homes overpriced for their area? There may be something unusual about those three homes that would require further research. The range of values for the northeast homes is clearly much wider, and only homes above $200,000 are considered moderate outliers. The boxplot also gives us a visual picture of the difference between the mean and median for the northeast homes. This information would certainly caution us against blindly using only the mean to summarize our results.

STATPLUS TIPS

- To specify the chart's title or a title for the *x* axis or *y* axis, click the Chart Options button in the Boxplot dialog box.

You've completed your work on the Albuquerque data. You can close the workbook and exit Excel now.

Exercises

1. Define the following terms:
   - a. Quantitative variable
   - b. Qualitative variable
   - c. Continuous variable
   - d. Ordinal variable
   - e. Nominal variable

2. What is a skewed distribution? What is positive skewness? What is negative skewness?

3. What is a stem and leaf plot? What are the advantages of stem and leaf plots over histograms? What are some of the disadvantages?

4. What is the interquartile range?

5. True or false (and why): Distributions with the same range have approximately the same variability.

6. What are outliers? What is considered a moderate outlier? What is considered an extreme outlier?

7. True or false (and why): Outliers should be removed from a data set before calculating statistics on that data set.

8. What is a boxplot? What are the advantages of boxplots over histograms? What are some of the disadvantages?

9. You see the following stem and leaf plot in a technical journal:

Stem × 100	Leaf
0	336
1	01228
2	00111249
3	04
4	5
5	
6	1
7	
8	
9	0

   - a. What are the approximate values of the data set?
   - b. Is the distribution positively skewed, negatively skewed, or symmetric?
   - c. Give approximate values for the mean and the median.
   - d. What values, if any, appear to be moderate and extreme outliers in the distribution?

10. A data distribution has a median value of 22, a first-quartile value of 20, and a third-quartile value of 30. Five observations lie outside the interval from the first to the third quartile, with values of 17, 18, 40, 50, and 75.
   - a. Draw the boxplot for this distribution.
   - b. Is the skewness positive, negative, or zero?

11. You're asked to do further research on the housing market in Albuquerque, New Mexico, during the early 1990s. In this analysis you'll examine the size of the homes sold on the market and the price per square foot of each home.
   - a. Open the **Housing** workbook from the Chapter04 folder and save it as **Home Sizes**.
   - b. Create a table of univariate statistics for the size of the homes in square feet, including all distribution, variability, and summary statistics except the mode. Place the table on a worksheet named **Sq Ft. Stats**.
   - c. What are the smallest and largest houses in the sample?
   - d. If you were interested only in houses that were 2,200 square feet or higher, what percentage of the houses in the sample would meet the requirement? (*Hint*: Use the PERCENTRANK function.)

e. Create a boxplot of the size of the homes in square feet. Place the boxplot in a chart sheet named **Sq Ft. Boxplot**.
f. What value appears to be an extreme outlier in the boxplot?
g. Create a second boxplot of house size, this time breaking the boxplot down by whether the home is a corner lot. Place the plot in a chart sheet named **Sq Ft. Boxplot by Corner Lot**.
h. Interpret your boxplot in terms of the relationship between size of a house and whether it lies on a corner lot.
i. What happened to the extreme outlier you identified earlier? Discuss this in terms of the definition of *outlier* given in the text. Is this value an outlier or not?
j. Recreate the table of univariate statistics for house size, and this time break the table down by the Corner Lot variable. Place the table in a worksheet named **Sq Ft. Stats by Corner Lot**.
k. Create a new column containing the price per square foot of each house. Assign the values in the new column a range name. What type of variable is this?
l. Create a histogram with 20 evenly spaced bins of the price per square foot on a chart sheet named **PPSqFt Histogram**. Count the bins totals to the right of the cutoff points.
m. What is the shape of the distribution of price per square foot?
n. Create a boxplot of the price per square foot saved to the chart sheet **PPSqFt. Boxplot**. Are there any severe outliers? How does the median value compare to the mean value?
o. Save your workbook and then write a report summarizing your observations.

12. Data have been recorded on 50 of the largest woman-owned businesses in Wisconsin. Analyze and report the descriptive statistics on this data set.

a. Open the Woman-Owned Businesses from the Chapter04 folder and save it as Woman-Owned Business Statistics.
b. Create a table of the distribution, variability, and summary statistics except the mode for the Employees variable. Store the table in a worksheet named Employee Stats.
c. What is the average number of employees for the 50 businesses? What is the median amount? Which statistic do you think more adequately describes the size of these businesses? How does the average number of employees compare to the third quartile?
d. Create a boxplot of employees stored in a chart sheet named Employee Boxplot. How would you describe this distribution?
e. Create a new variable containing the base 10 log of the Employees variable. Assign a range name to this new column and then create a boxplot of these values in a chart sheet named Log Employee Boxplot. How does the shape of this distribution compare to the untransformed values?
f. Create a table of descriptive statistics except the mode for the log(Employees) and store the table in a worksheet named Log Employee Stats. Compare the skewness and kurtosis values between the Employees and log(Employees) variables. Explain how the difference in the distribution shapes is reflected in these two statistics.
g. Calculate the mean log(Employees) value in terms of the number of people employed (in other words, transform this value back to the original scale). How does this value compare to the geometric mean of the number of employees in each company?
h. The geometric mean is used for values that either are ratios or are best compared as ratios. Which pair of

companies is more similar in terms of size: a company totaling $50,000,000 in annual sales and a company with $10,000,000, or a company with $450,000,000 in sales and a company with $400,000,000 in sales? What are the differences in sales between the two sets of companies? What are the ratios? Does the difference or the ratio better express the similarity of the companies?

i. Save your workbook and write a report summarizing your analysis. Explain how transforming the employee values using the logarithm function affected the distribution of the data.

13. In the late 1980s, the U.S. Congress held several joint hearings on discrimination in lending practices, particularly in the mortgage industry. Refusal rates from 20 lending institutions were presented to the committee. Analyze these rates:

a. Open the **Mortgage** workbook from the Chapter04 folder and save it as **Mortgage Refusal Rates**.

b. Create a table of univariate statistics for the four data columns. Save the table in a worksheet named **Refusal Statistics**.

c. Create a boxplot of the refusal rates for the four data columns stored in a chart sheet named **Refusal Boxplots**. Label the chart appropriately.

d. Save your workbook. Including the descriptive statistics and boxplot you've created, write a report detailing your findings. What conclusion do you draw from the data? Is there any specific information that this data sample is lacking? Include a discussion of potential problems in this data set and how you would go about remedying them.

14. Average teacher salary, public school spending per pupil, and ratio of teacher salary to pupil spending for 1985 have been stored in an Excel workbook. The values are broken down by state and area. You've been asked to calculate statistics on teacher salaries on the basis of the data.

a. Open the **Teacher** workbook from the Chapter04 folder and save it as **Teacher Salaries**.

b. Create a table of univariate statistics except the mode for the teacher salaries broken down by area and overall. Save the table on the worksheet **Salary Statistics**.

c. Create a boxplot of the teacher salaries broken down by area on a chart sheet named **Salary Boxplots**.

d. Discuss the distribution of the teacher salaries for each area. There is an extreme outlier in the west area. Which state is this? Discuss why salaries for teachers in this state might be so high.

e. Create a table of univariate statistics except the mode for the ratio of teacher salary to spending per pupil broken down by area and overall. Save the table on a worksheet named **Salary Pupil Ratio Statistics**.

f. Create, on a chart sheet named **Salary Pupil Ratio Boxplots**, a boxplot of the ratio values broken down by area.

g. For the state that was an outlier in the west area in terms of teacher salary, check to see if it is also an outlier in terms of the ratio of teacher salary to public spending per pupil. Estimate the percentile of this state's salary/pupil ratio within the west area. How does that compare to its percentile for teacher's salary alone? If the cost of education per pupil is indicative of the cost of living in a state, are teachers in this particular state overpaid or underpaid relative to other states in the west area?

h. Save your changes to the workbook and write a report summarizing your observations and calculations.

15. You've been given an Excel workbook containing annual salary figures for major league baseball players (in terms of hundreds of thousands of dollars) for the 2007 season. Use the workbook to calculate statistics on the players' salaries.

  a. Open the **Baseball** workbook from the Chapter04 folder and save it as **Baseball Salary Statistics**.
  b. Create a histogram of the players' salaries with bin intervals of $1,000,000 ranging from $0 up to $25,000,000. Have the counts within each bin be ≥ the bin value and < the next bin value.
  c. Create a frequency table of the players' salaries, using the same bin intervals and options you used to create the histogram.
  d. Calculate the 10th and 90th percentiles of the salaries.
  e. Using the value for the 90th percentile, filter the player data to show only those players who were paid in the upper 10% of the salary range.
  f. What is the average player's salary? What is the median player's salary? If a player made the average salary, at what percentile would he be ranked in the data?
  g. Save your changes to the workbook and write a report summarizing your calculations.

16. You've been asked to compare the changing nature of baseball salaries. An Excel workbook has been prepared for you that contains salaries from the years 1985, 2002, and 2007. Examine these salaries and prepare a statistical report.

  a. Open the **Salary Comparison** workbook from the Chapter04 folder and save it as **Salary Comparison Statistics**.
  b. Create a histogram of the salary data broken down by year. Have the histogram display salaries in $1,000,000 intervals up to $25,000,000 and have the bars of the histogram display the percentages, not the frequencies, for each bar. Display the histogram bars side by side rather than stacked. Does the distribution of the salaries appear the same in the three years?
  c. Calculate the count, mean, median, percentiles, skewness, and kurtosis for the three years. How does the typical salary compare within the three years? What do players in the upper 10% of each year make?
  d. Has the distribution of the salaries become more skewed or less skewed or remained the same over the years? Answer this question by examining the skewness and kurtosis statistics. You've learned that the 2007 data are just for starters. How could this affect your conclusion?
  e. Save your changes to the workbook and then write a report summarizing your calculations.

17. The Cancer workbook contains data comparing the cigarette per capita for each of the 50 states and the District of Columbia to those state's rates of bladder cancer, kidney cancer, lung cancer, and leukemia per 100,000. Each state was ranked on the basis of cigarette use with 0 for low rates of cigarette use, 1 for medium, and 2 for high. Analyze the data from this workbook.

  a. Open the **Cancer** workbook from the Chapter04 folder and save it as **Cancer Statistics**.
  b. Create boxplots of the rates of bladder cancer, kidney cancer, lung cancer, and leukemia broken down by cigarette use. Label the charts and chart sheets appropriately.
  c. Create a table of univariate statistics for the rate of each illness broken down by the cigarette use category.

d. Does there appear to be any relationship between these illnesses and the level of cigarette use in the states? Defend your answer with your charts, statistics, and tables.
e. There is one state with a high level of cigarette use but a relatively low level of lung cancer. Identify this state.
f. Save your workbook and then write a report summarizing your observations and calculations.

18. The Pollution workbook contains air quality data collected by the Environmental Protection Agency (EPA). The data show the number of unhealthful days (heavy levels of pollution) per year for 14 major U.S. cities in the year 1980 and the average number of unhealthy days per year from 2000 through 2006. The workbook also contains the ratio of the 2000–2006 average to the 1980 value and the difference. A ratio value less than 1 or a difference value less than 0 indicates an improvement in the air quality. Looking at the data as a whole, is there evidence to believe that there has been improvement in the air quality? Open this workbook and examine the data.

a. Open the **Pollution** workbook from the Chapter04 folder and save it as **Pollution Boxplots**.
b. Calculate the mean and median values of the ratio and difference variables.
c. Create two boxplots. First create a boxplot of the ratio variable and then create another boxplot of the difference variable. Describe the difference between the shape of the two distributions. Is one more susceptible to extreme values than the other? Why would this be case? (*Hint:* Think about the number of unhealthy days in 1980. Which cities are most likely to show the greatest drop in absolute numbers?)
d. There is an extreme outlier in the boxplot of the difference values. Identify the city corresponding to that extreme outlier.
e. Copy the air quality data to a new worksheet without the extreme outlier you noted in part d. Redo the table of statistics and boxplots with this new set of data.
f. What are your conclusions? Have your conclusions changed without the presence of the outlier? What effect did the outlier have on the mean and median values of the ratio and difference variables? Are you justified in removing the outlier from your analysis? Why or why not?
g. Save your workbook and then write a report summarizing your calculations and observations. Which variable seems to better describe the change in air quality: the difference or the ratio?

19. The Reaction workbook contains reaction times from the first-round heats of the 100-meter race at the 1996 Summer Olympic games. Reaction time is the time elapsed between the sound of the starter's gun and the moment the runner leaves the starting block. The workbook also contains the heat number, the order of finish, and the finish group (1st through 3rd, 4th through 6th, and so forth).

a. Open the **Reaction** workbook from the Chapter04 folder and save it as **Reaction Statistics**.
b. Calculate univariate descriptive statistics for the reaction times listed. What are the average, median, minimum, and maximum reaction times?
c. Create a boxplot of the reaction times. Are there any moderate or extreme outliers in the distribution? How would you characterize the shape of the distribution?
d. Create a stem and leaf plot of the reaction times.

e. Reaction times are important for determining whether a runner has false-started. If the runner's reaction time is less than 0.1 second, a false start is declared. Where would a reaction time of 0.1 second fall on your boxplot: as a typical value, a moderate outlier, or an extreme outlier? Does this definition of a false start seem reasonable given your data?
f. Is there an association between reaction time and the order of finish? Calculate descriptive statistics for the reaction times broken down by order of finish. Pay particular attention to the mean and the median.
g. Create a boxplot of the reaction times broken down by order of finish. Is there anything in your descriptive statistics or boxplots to suggest that reaction time plays a part in how the runner finishes the race?
h. Save your changes to the workbook and then write a report summarizing your observations and calculations.

20. The Labor Force workbook shows the change in the percentage of women in the labor force from 19 cities in the United States from 1968 to 1972. You can use these data to gauge the growing presence of women in the labor force during this time period.

a. Open the **Labor Force** workbook from the Chapter04 folder and save it as **Labor Force Statistics**.
b. Calculate the difference between the 1968 and 1972 values, storing the calculations in a new column. Calculate descriptive statistics for the values in the Difference column.
c. Calculate the mean of the Difference value.
d. Create a boxplot of the Difference value. Are there any outliers present in the data? Identify which city the value comes from. What do the data tell you about the change of the presence of women in the labor force from 1968 to 1972?
e. Describe the shape of the distribution of the Difference values. Are the data positively or negatively skewed or symmetric? Can you use the mean to summarize the results from this study?
f. Save your workbook and write a report summarizing your analysis.

21. In 1970, draft numbers were determined by lottery. All 366 possible birth dates were placed in a rotating drum and selected one by one. The first birth date drawn received a draft number of 1, and men born on that date were drafted first; the second birth date received a draft number of 2; and so forth. Data from the draft number lottery can be found in the Draft workbook.

a. Open the **Draft** workbook from the Chapter04 folder and save it as **Draft Statistics**.
b. Create a box plot of the draft numbers broken down by month. Also create a table of counts, means, medians, and standard deviations. Is there any evidence of a trend in the draft numbers selected compared to the month?
c. Repeat part b, this time breaking the numbers down by quarters. Is there any evidence of a trend between draft numbers and the year's quarter?
d. Repeat part b, breaking the draft numbers by first half of the year versus second half. Is the typical draft number selected for the first half of the year close in value to the draft number for birthdays from the second half of the year?
e. Discuss your results. The draft numbers should have no relationship to the time of the year. Does this appear to be the case? What effect does breaking the numbers down into different units of time have on your conclusion?

f. Save your workbook and write a report summarizing your investigation and observations.

22. Cuckoos are known to lay their eggs in the nests of other host birds. The host birds adopt and then later hatch the eggs. The Eggs workbook contains data on the lengths of cuckoo eggs found in the nest of host birds. You've been asked to compare the length of the cuckoo eggs placed in the different nests.

   a. Open the **Eggs** workbook from the Chapter04 folder and save it as **Egg Statistics**.
   b. Create a boxplot of the egg lengths for the different birds.
   c. Calculate descriptive statistics of the egg lengths.
   d. One theory holds that cuckoos lay their eggs in the nests of a particular host species and that they mate within a defined territory. If true, this would cause a geographic subspecies of cuckoos to develop and natural selection would ensure the survival of cuckoos most fitted to lay eggs that would be adopted by a particular host. If cuckoo eggs differed in length between hosts, this would lend some weight to that hypothesis. Do the data indicate a possible difference in cuckoo egg lengths between hosts? Explain.
   e. Save your changes to the workbook and write a report summarizing your observations.

Chapter 5

PROBABILITY DISTRIBUTIONS

Objectives

In this chapter you will learn to:

- Work with random variables and probability distributions
- Generate random normal data
- Create a normal probability plot
- Explore the distribution of the sample average
- Apply the Central Limit Theorem

Up to now, you've used tools such as frequency tables, descriptive statistics, and scatter plots to describe and summarize the properties of your data. Now you'll learn about probability, which provides the foundation for understanding and interpreting these statistics. You'll also be introduced to statistical inference, which uses summary statistics to help you reach conclusions about your data.

Probability

Much of science and mathematics is concerned with prediction. Some of these predictions can be made with great precision. Drop an object, and the laws of physics can predict how long the object will take to fall. Mix two chemicals, and the laws of chemistry can predict the properties of the resulting mixture. Other predictions can be made only in a general way. Flip a coin, and you can predict that either a head or a tail will result, but you cannot predict which one. That doesn't mean that you can't say anything. If you flip the coin many times, you'll notice that roughly half the flips result in heads and half result in tails.

Flipping a coin is an example of a **random phenomenon**, in which individual outcomes are uncertain but follow a general pattern of occurrences.

When we study random phenomena, our goal is to quantify that general pattern of occurrences in order to make general predictions. How do we do this? One way is through theory. We imagine an ideal coin with two sides: a head and a tail. Because this is an ideal coin, we assume that each side is equally likely to occur during our coin flip. From this, we can define the **theoretical probability** for equally likely events:

$$\textit{Theoretical probability} = \frac{\textit{Number of possible ways of obtaining the event}}{\textit{Total number of possible outcomes}}$$

In the coin-tossing example, there is one way to obtain a head and there are two possible outcomes, so the theoretical probability of obtaining a head is 1/2, or .5.

Another way of quantifying random phenomena is through observation. For example to determine the probability of obtaining a head, we repeatedly toss the coin. From our observations, we calculate the **relative frequency** of tosses that result in heads, where

$$\textit{Relative frequency} = \frac{\textit{Number of times an event occurs}}{\textit{Number of replications}}$$

Figure 5-1 shows a chart of the results of tossing such a coin 5,000 times.

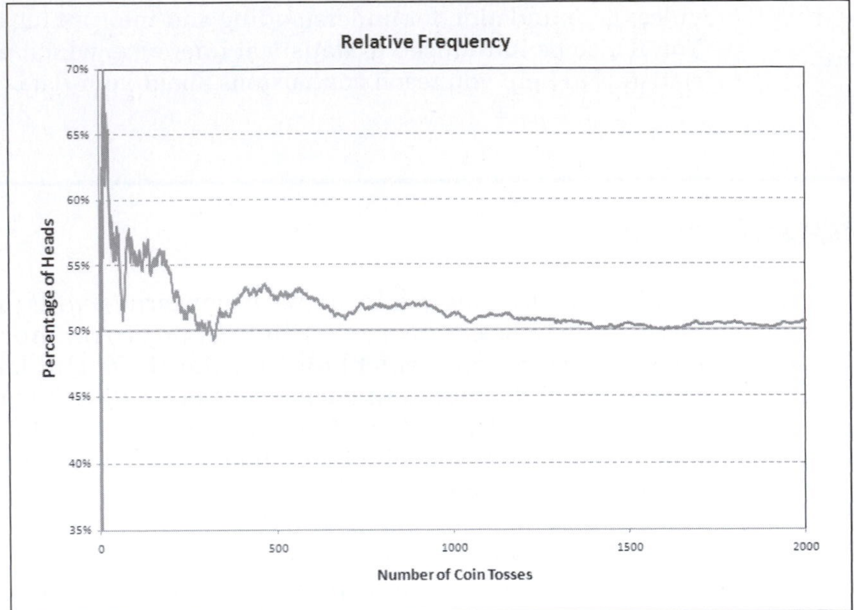

Figure 5-1 The relative frequency of tossing a head

Early in the experiment, the relative frequency of heads jumps around quite a bit, hovering above .5. As the number of tosses increases, the relative frequency narrows to around the .5 level. The **law of large numbers** states that as the number of replications increases, the relative frequency will approach the **probability** of the event, or, to put it another way, we can define the probability of an event as the value approached by the relative frequency after an indefinitely long series of trials.

Probability Distributions

The pattern of probabilities for a set of events is called a **probability distribution**. Probability distributions contain two important elements.

1. The probability of each event or combination of events must range from 0 to 1.
2. The total probability is 1.

In the coin-tossing example, there are two outcomes (head or tail), each with a probability of .5. The sum of both events is 1, so this is an example of a probability distribution. Probability distributions can be classified into two general categories: discrete and continuous.

Discrete Probability Distributions

In a **discrete probability distribution**, the probabilities are associated with a series of discrete outcomes. The probabilities associated with tossing a coin form a discrete distribution since there are two separate and distinct outcomes. If you toss a 6-sided die, the probabilities associated with that outcome also form a discrete distribution, where each side has a $\frac{1}{6}$ probability of turning up. We can write this as

$$p(y) = \frac{1}{6}; y = 1, 2, 3, 4, 5, 6$$

where $p(y)$ means the "probability of y," for integer values of y ranging from 1 to 6.

Note that *discrete* does not mean "finite." There are discrete probability distributions that cover an infinite number of possible outcomes. One of these is the **Poisson** distribution, used when the outcome event involves counts within a specified period of time. The equation for the Poisson distribution is

Poisson Distribution

$$p(y) = \frac{\lambda^y}{y!} e^{-\lambda} \quad y = 0,1, 2, \ldots$$

where λ (pronounced "lambda") is the average number of events in the specified time period and $y!$ stands for "y factorial," which is equal to the product $y(y-1)(y-2) \cdots (3)(2)(1)$. For example, $4! = 4 \times 3 \times 2 \times 1 = 24$. Lambda is an example of a **parameter**, a term in the formula for a probability distribution that defines its shape and values. Let's see what probabilities are generated for a specific value of λ.

Suppose we want to determine the number of car accidents at an intersection in a given year and we know that the average number of accidents is 3. What is the probability of exactly two accidents occurring that year? The Poisson distribution usually applies to this situation. In this case, the value of λ is 3, $y = 2$, and the probability is

$$\frac{3^2}{2!}e^{-3} = \frac{9 \cdot 0.0498}{2 \cdot 1} = 0.224$$

or the probability of exactly two accidents occurring at the intersection is about 22%. Note that the probabilities extend across an infinite number of possible integer values.

Discrete distributions can be displayed with a bar chart in which the height of each bar is proportional to the probability of the event. Figure 5-2 displays the probability distribution from $y = 0$ to $y = 10$ accidents per year.

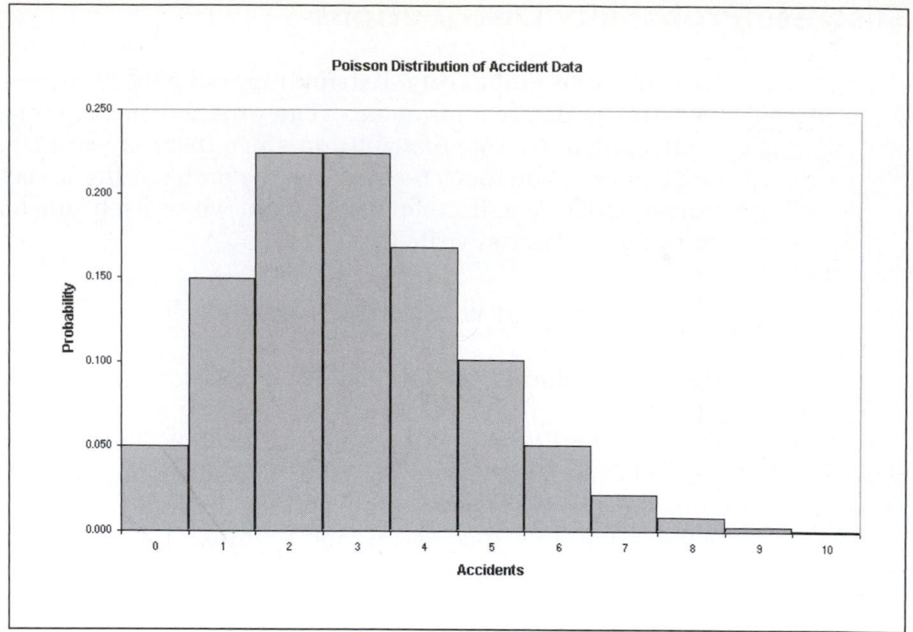

Figure 5-2 Poisson probability distribution for car accident data

To find the probability of a set of discrete events, we simply add up the individual probabilities of each event in the set. So to find the probability of two or fewer accidents occurring at the intersection we add up the probabilities of no accidents (.050), 1 accident (.149), and 2 accidents (.224) to arrive at an overall probability of .423, or about 42%. Because the total probability is 1, the probability of more than two accidents occurring at the intersection would thus be 58%.

Continuous Probability Distributions

In **continuous probability distributions**, probabilities are assigned to a range of continuous values rather than to distinct individual values. For example, consider a person shooting at a target. The distribution of shots around the bull's eye follows a continuous distribution. If the shooter is good, the probability that the shots will cluster closely around the bull's eye is very high and it is unlikely that a shot will miss the target entirely.

Continuous probability distributions are calculated using a **probability density function (PDF)**. When we plot a PDF against the range of possible values, we get a curve in which the curve's height indicates the position of the most likely values. Figure 5-3 shows a sample PDF curve.

Figure 5-3
A sample probability density function

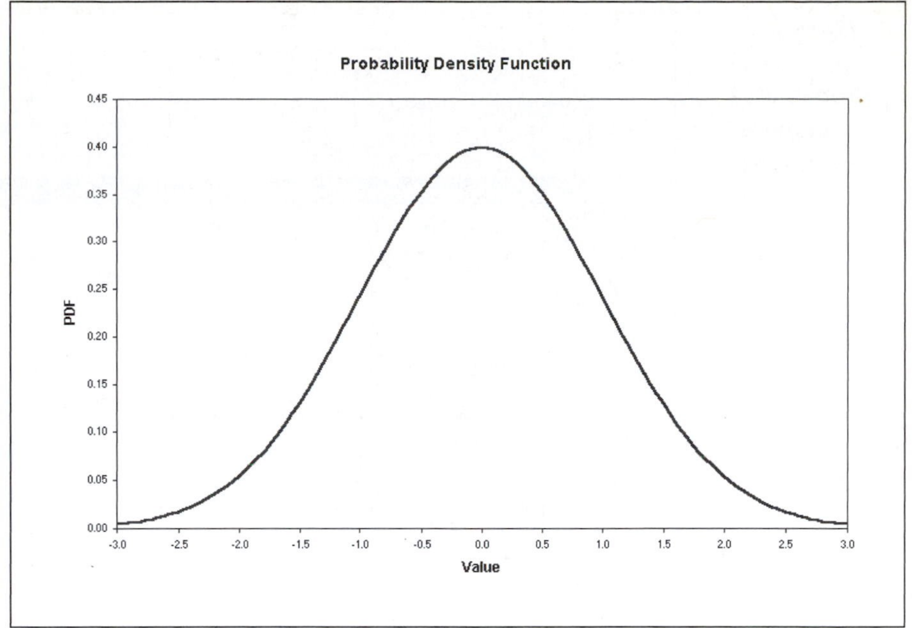

The probability associated with a range of values is equal to the area under the PDF curve. Note that unlike discrete distributions, we can't assign a probability to a specific value in a continuous distribution. The probability of any specific value is zero because its area under the curve is zero (it has a positive height, but zero width). The total area under any PDF curve must be equal to 1 because the total probability must be 1.

CONCEPT TUTORIALS
PDFs

To see the relationship between probability and the area under the PDF curve, open the instructional workbook named Probability.

To open the Probability workbook:

1. Open the file **Probability**, located in the Explore folder.

   The workbook opens to the Contents page, describing the nature of the workbook.

2. You can move through the sheets in the workbook, reviewing the material on probability discussed so far in this chapter.

3. Click **Explore a PDF** from the Table of Contents column.

   The sheet displays the curve shown in Figure 5-4.

Figure 5-4 The Probability Explore workbook

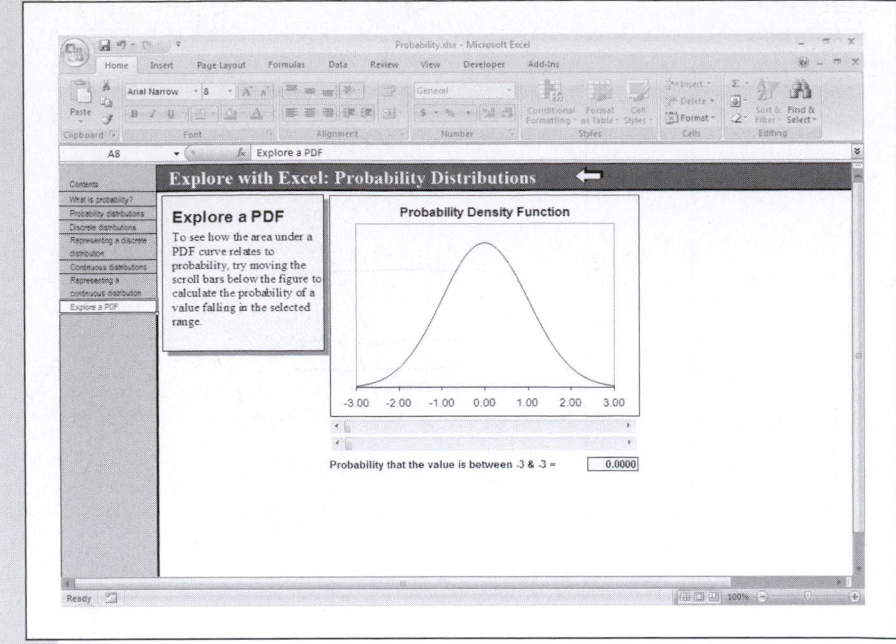

Notice the two horizontal scroll bars below the chart. You'll use these to set the range boundaries on the PDF curve. Use them now to select the range from −1 to 1 on the curve.

To set the range boundaries on the PDF curve:

1 Click or drag the **scroll button** of the bottom scrollbar until the right boundary equals 1.

2 Click or drag the **scroll button** arrow of the top scrollbar until the left boundary equals −1.

Figure 5-5 shows the PDF with the range from −1 to 1 selected. The area of this range is equal to 0.6827.

Figure 5-5 Selecting the range from -1 to 1

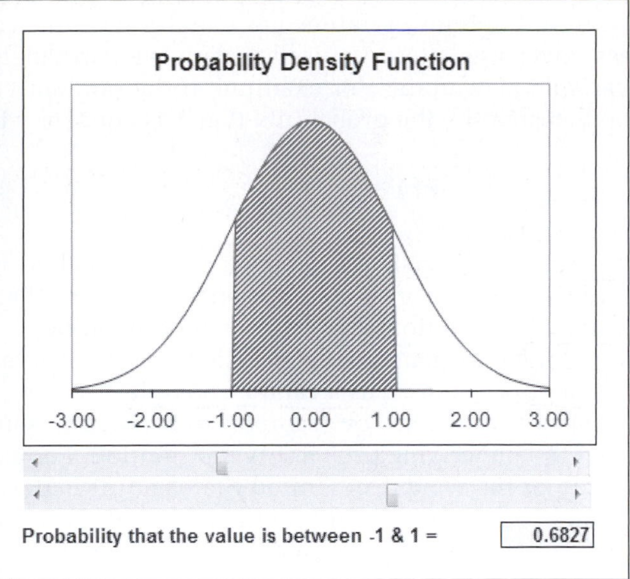

Because the area under the curve is equal to 0.6827, the probability of a value falling between −1 and 1 is equal to 68.27%. Try experimenting with other range boundaries until you get a good feel for the relationship between probabilities and areas under the curve.

3 Click the **left scroll arrow** until the left boundary equals −3.

4 Click the **right scroll arrow** until the right boundary equals 3.

Note that when you move the lower boundary to −3 and the upper boundary to 3, the probability is 99.73%, not 100%. That is because this particular probability distribution extends from minus infinity to plus infinity; the area under the curve (and hence the probability) is never 1 for any finite range.

5 Close the Probability workbook. You do not have to save any changes.

Random Variables and Random Samples

Values from a discrete or continuous probability distribution function are manifested in a random variable where a **random variable** is a variable whose values occur at random, following a probability distribution. A **discrete random variable** comes from a discrete probability distribution, and a **continuous random variable** comes from a continuous probability distribution. Random variables are usually written with a capital letter, whereas lowercase letters are used to denote a particular value that the random variable may attain. For example, if the random variable Y follows a Poisson distribution, the probability that Y is equal to y is written as

$$P(Y = y) = \frac{\lambda^y}{y!}e^{-\lambda}$$

When a random variable is assigned a value (such as when we flip a coin or record the number of traffic accidents in a year), that value is called an **observation**. A collection of several such observations is called a **sample**. If the observations are generated or selected in a random fashion with no bias, then the sample is known as a **random sample**.

In most cases, we want our samples to be random samples to give a true picture of the underlying probability distribution. For example, say we create a study of the weight of United States adult males. We want to know what type of values we would be likely to get if we picked a man at random and weighed him (here weight is our random variable). However, if our sample is biased by selecting only men in their twenties, it would not be a true random sample of all United States adult males. Part of the challenge of statistics is to remove all bias from sampling.

This is difficult to do, and subtle biases can creep into even the most carefully designed studies. By observing the distribution of values in a random sample, we can draw some conclusions about the underlying probability distribution. As the sample size increases, the distribution of the values should more closely approximate the probability distribution. To return to our example of the shooter, by observing the spread of shots around the bull's eye, we can estimate the probability distribution related to the shooter's ability to hit the target.

CONCEPT TUTORIALS
Random Samples

You can use the instructional workbook Random Samples to explore the relationship between a probability distribution and a random sample.

To use the Random Samples workbook:

1. Open the file **Random Samples** from the Explore folder. Enable any macros in the workbook.
2. Move through the sheets in this workbook, viewing the material on random variables and random samples.
3. Click **Explore a Random Sample** from the Table of Contents column.

In this worksheet, you can click the Shoot button to generate a random sample of shots at the target. You can select the underlying probability distribution and the number of shots the shooter takes. Try this now with the accuracy of the shooter set to moderate to create a sample of 50 random shots.

To generate a random sample of shots:

1. Click the **Shoot** button.
2. Click the **Moderate** button and click the spin arrow to reduce the number of shots to 50.

   See Figure 5-6.

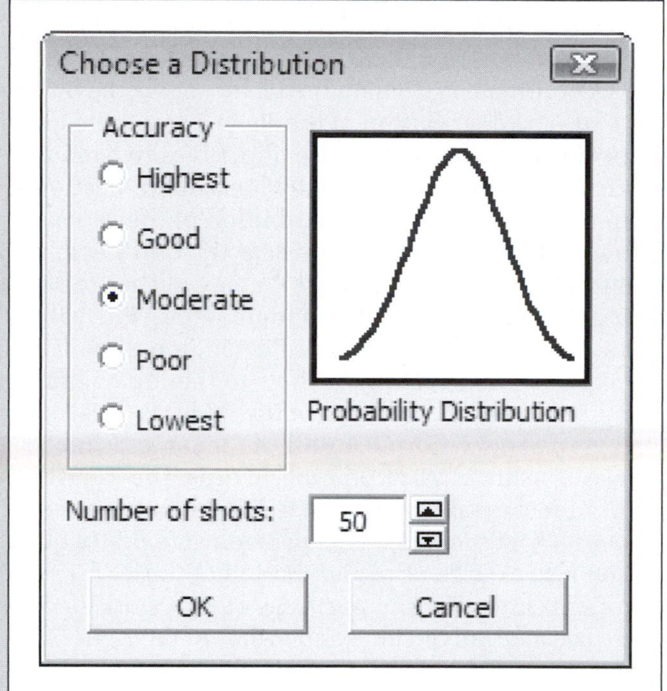

Figure 5-6 Accuracy dialog box

3. Click the **OK** button twice.

   Excel generates 50 random shots as shown in Figure 5-7 (your random sample will be different).

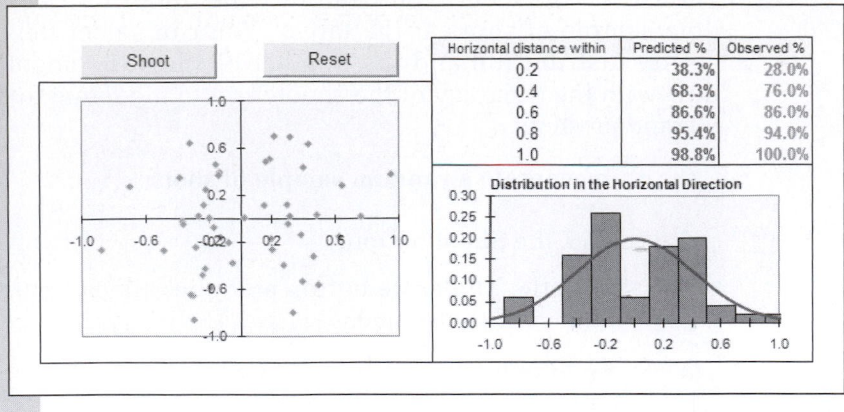

Horizontal distance within	Predicted %	Observed %
0.2	38.3%	28.0%
0.4	68.3%	76.0%
0.6	86.6%	86.0%
0.8	95.4%	94.0%
1.0	98.8%	100.0%

Figure 5-7 Randomly generated sample of shots

The *xy* coordinate system on the target shows the bull's eye, located at the origin (0, 0). The distribution of the shots around the target is described by a **bivariate** density function because it involves two random variables (one for the vertical location and one for the horizontal location of each shot). We'll concentrate on the horizontal distribution of the shots.

Although many of the shots are near the bull's eye, about a third of them are farther than 0.4 horizontal unit away, either to the left or to the right of the target. Because these are random data, your values may be different. Based on the accuracy level you selected, a probability distribution showing the expected distribution of shots to the left or right of the target is also generated in the second column of the table. In this example, the predicted proportion of shots within 0.4 unit of the target is 68.3%, which is close to the observed value of 70%. In other words, the distribution predicts that a person of moderate ability is able to hit the bull's eye within 0.4 horizontal unit about 68% of the time. This person came pretty close.

You can also examine the distribution of these shots by looking at the histogram of the shots. For the purposes of this worksheet, a shot to the left of the target has a negative value and a shot to the right of the target has a positive value. The solid curve is the probability density function of shots to the left or right of the target. After 50 shots, the histogram does not follow the probability density function particularly closely. As you increase the number of shots taken, the distribution of the observed shots should approach the predicted distribution.

To increase the number of shots taken:

1. Click the **Shoot** button again.
2. Click the **Moderate** button and click the spin arrow to increase the number of shots to 500.

Figure 5-8 compares the distribution of the random sample after 50 shots and after 500 shots. Note that the larger sample size more closely follows the underlying probability distribution.

Figure 5-8 Distribution after 50 and 500 observations

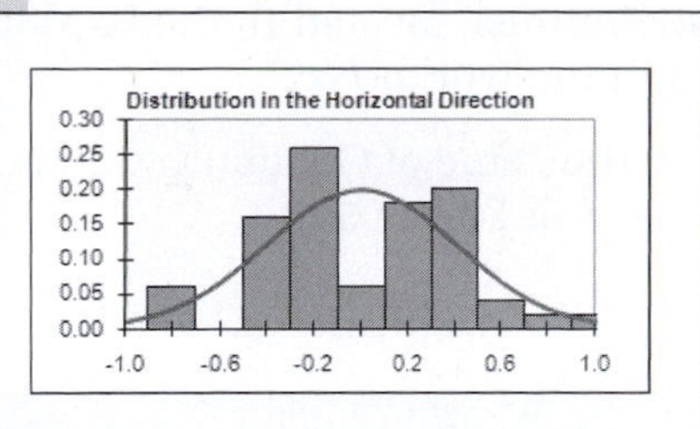

sample size = 50

sample size = 500

3 Try generating some more random samples of various sample sizes. When you are finished, close the Random Samples workbook.

The Normal Distribution

In the Exploring Random Samples workbook, you worked with a distribution in the form of a bell-shaped curve, called the **normal distribution**. This common probability distribution is probably the most important distribution in statistics. There are many real-world examples of normally distributed data, and normally distributed data are assumed in many statistical tests (for reasons you'll understand shortly). The probability density function for the normal distribution is

Normal Probability Density Function

$$f(y) = \frac{1}{\sigma\sqrt{2\pi}} e^{-(y-\mu)^2/2\sigma^2} \qquad \sigma > 0, \quad -\infty < \mu < \infty, \quad -\infty < y < \infty$$

The normal distribution has two parameters, μ (pronounced "mu") and σ (pronounced "sigma"). The μ parameter indicates the center, or mean, of the distribution. The σ parameter measures the standard deviation, or spread, of the distribution. To see how these parameters affect the distribution's location and shape, you can work with the instructional workbook named Distributions.

CONCEPT TUTORIALS
The Normal Distribution

To explore the normal distribution:

1 Open the file **Distributions**, located in the Explore folder. Enable the macros contained in the workbook.

2 Click **Normal** from the Table of Contents column. The Normal worksheet opens as shown in Figure 5-9.

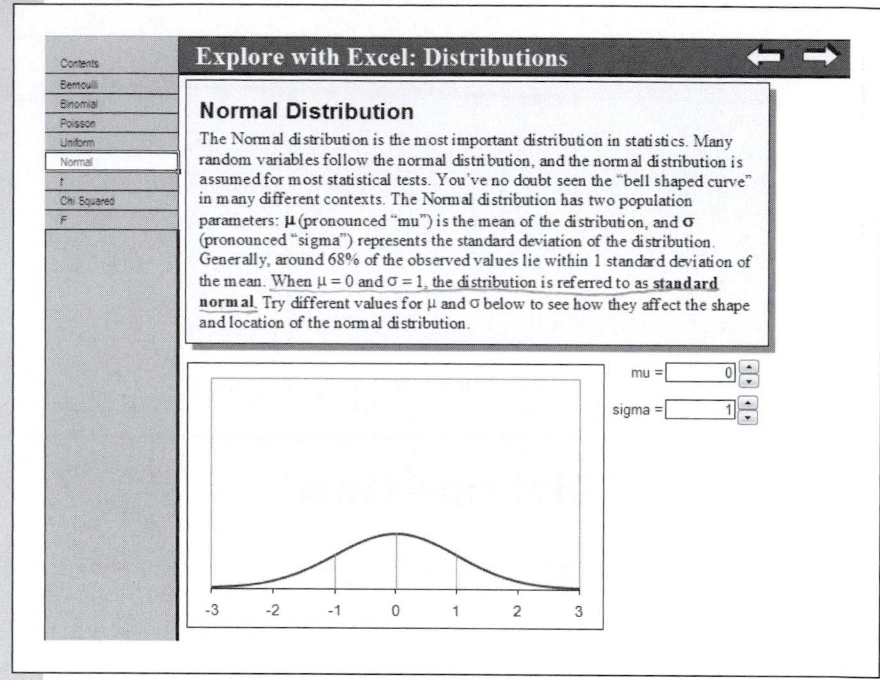

Figure 5-9 The Normal Distribution worksheet

The Normal workbook opens with μ set to a value of 0 and σ set to a value of 1. A normal distribution with these parameter values is referred to as **standard normal**. Now observe what happens as you alter the values of μ and σ.

To change the values of μ and σ:

1 Click the **up spin button** next to the mu box to change the value of μ to **2**.

Note that the distribution shifts to the right as the center of the distribution now lies over the value 2.

2. Click the **down spin button** next to the mu box to change the value of μ back to **0**.
3. Click the **down spin button** next to the sigma box to reduce the value of σ to **0.3**. The distribution tightens around the center.
4. Click the **up spin button** next to the sigma box to increase the value of σ to **1.5**. The distribution spreads out, indicating a wider range of probable values.

   Figure 5-10 shows the normal curve for a variety of μ and σ values.

Figure 5-10 The normal distribution for varying values of σ

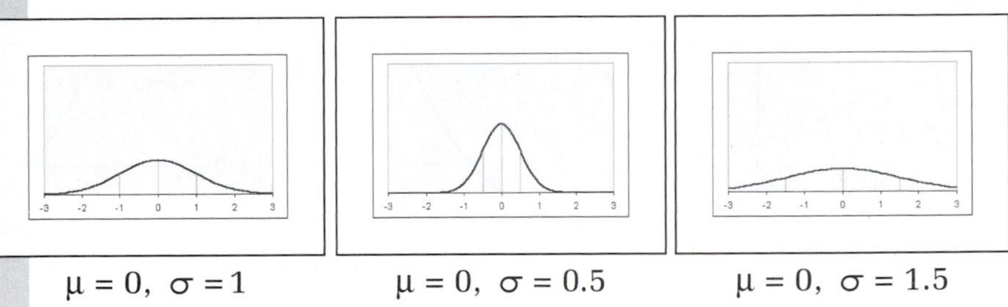

5. Examine other values of μ and σ to continue to explore how those changes affect the distribution. Close the Distributions workbook without saving any changes when you're finished.

In the normal distribution, about 68.3% of the values lie within 1 σ, or 1 standard deviation, of the mean μ. About 95.4% of the values lie within 2 standard deviations of the mean, and more than 99% of the values lie within 3 standard deviations of the mean. See Figure 5-11.

Figure 5-11 Probabilities under the normal curve

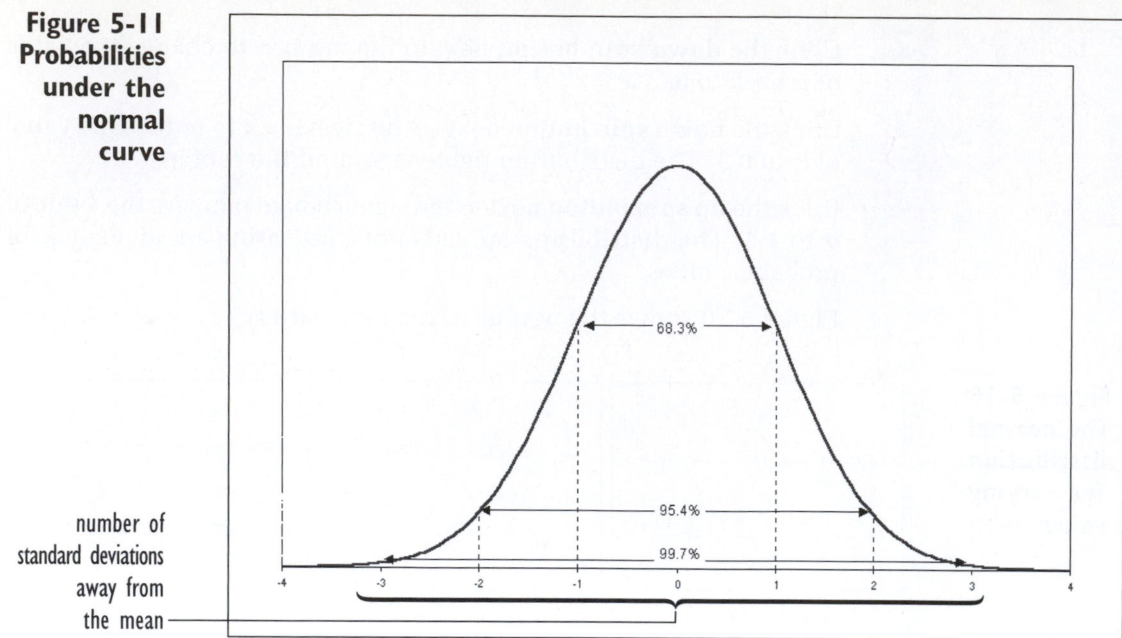

Because normally distributed data appear so often in statistical studies, these benchmarks are an important rule of thumb. For example, if you are trying to calculate a range that will incorporate most of the data, taking the mean ±2 standard deviations is a fast way of estimating that range.

Excel Worksheet Functions

Excel includes several functions to work with the normal distribution described in Table 5-1.

Table 5-1 Functions with the Normal Distribution

Function	Description
NORMDIST(*y*, *mean*, *std_dev*, *type*)	Uses the normal distribution with μ = mean and σ = *std_dev*. Setting *type* = TRUE calculates the probability of $Y \leq y$. Setting *type* = FALSE calculates the value of the probability density function at *y*.
NORMINV(*p*, *mean*, *std_dev*)	Returns the value *y* from the normal distribution for μ = mean and σ = *std_dev*, such that $p(Y \leq y) = p$.

(continued)

NORMSDIST(*y*)	Returns the probability of $Y \leq y$ for the standard normal distribution.
NORMSINV(*p*)	Returns the value y from the standard normal distribution such that $p(Y \leq y) = p$.
NORMBETW(*lower, upper, mean, std_dev*)	Calculates the probability from the normal distribution with μ = mean and σ = *std_dev* for the range lower $\leq y \leq$ upper. *StatPlus required.*

For example, if you want to calculate the probability of a random variable from a normal distribution with $\mu = 50$ and $\sigma = 4$ having a value ≤ 40, apply the Excel formula

=NORMDIST(40, 50, 4, TRUE)

and Excel returns the value .00621, indicating that there is a 0.621% probability of a value less than or equal to 40 from such a distribution. The value of the PDF at that point returned by the formula

=NORMDIST(40, 50, 4, FALSE)

is .004382, which is the height of the probability distribution function at that point in the PDF curve. On the other hand, if you want to calculate a value on the PDF for a particular probability, you use the NORMINV() function. The formula

=NORMINV(0.90, 50, 4)

returns the value 55.12621, indicating that in a normal distribution with $\mu = 50$ and $\sigma = 4$, there is a 90% probability that a random variable will have a value of 55.12621 or less.

Using Excel to Generate Random Normal Data

Now that you've learned a little about the normal distribution, you can use Excel to randomly generate observations from a normal distribution. You'll start by creating a single sample of 100 observations coming from a normal distribution with $\mu = 100$ and $\sigma = 25$. To do this, you need to have the StatPlus add-in installed on Excel.

To create 100 random normal values:

1 Open a new blank workbook in Excel, click cell **A1**, and type **Normal Data** in the cell.

2 Click **Create Data** from the StatPlus menu and then click **Random Numbers**.

The Random Numbers command presents a dialog box from which you can create random samples from a large variety of distributions. In this case you'll choose the normal distribution.

3 Click **Normal** from the Type of Distribution list box.

4 Type **1** in the Number of Samples to Generate box.

5 Type **100** in the Size of Each Sample box.

6 Type **100** in the Mean box.

7 Type **25** in the Standard Deviation box.

8 Click the **Output** button, click the **Cell** option button, and select cell **A2** as your output destination. Click the **OK** button to close the Output Options dialog box.

Figure 5-12 shows the completed Create Random Numbers dialog box.

Figure 5-12 The Create Random Numbers dialog box

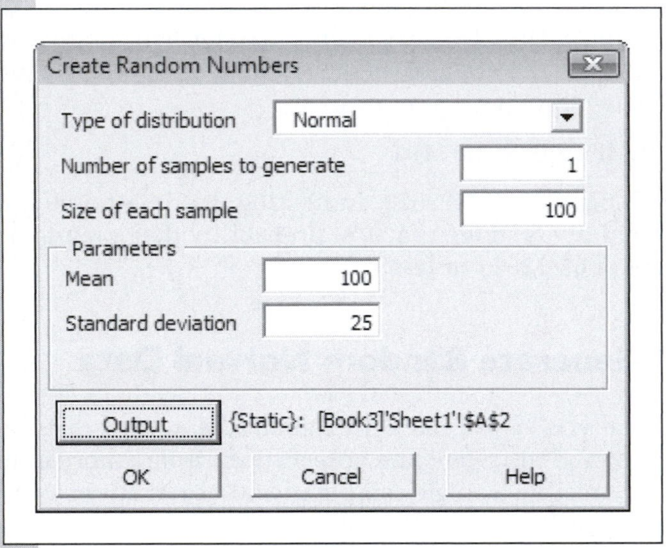

9 Click the **OK** button.

Excel generates a random sample of 100 observations following a normal distribution with mean 100 and standard deviation 25. See Figure 5-13.

Figure 5-13
One hundred random normal observations

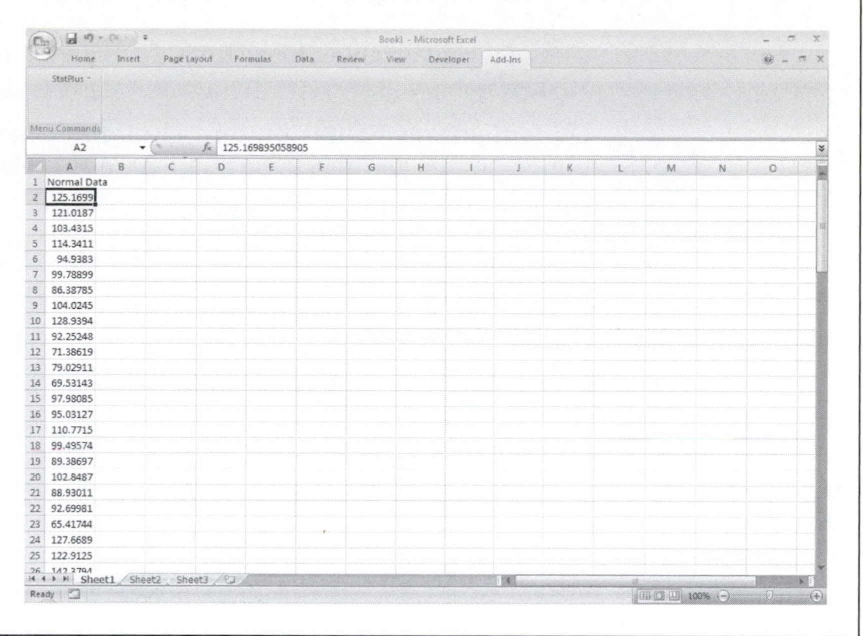

Because these are randomly generated values, your numbers will look different.

EXCEL TIPS

- You can also create random samples using the Analysis ToolPak add-in available with Excel. To create a random sample, load the Analysis ToolPak, click the Data Analysis button from the Analysis group on the Data tab and select Random Number Generation from the Data Analysis dialog box.
- StatPlus adds several new functions to Excel to generate random numbers, including the RANDNORM command to create a random number from a normal distribution.

Charting Random Normal Data

Now that you've created a random sample of normal data, your next task is to create a histogram of the distribution. The StatPlus Histogram command also includes an option to overlay a normal curve on your histogram to compare the distribution of your data with the normal distribution.

To create a histogram of the random sample:

1 Click **Single Variable Charts** from the StatPlus menu and then click **Histograms**.

2 Click the **Data Values** button, click the **Use Range References** option button, and select the range **A1:A101**. Click the **OK** button.

3 Click the **Normal curve** checkbox.

4 Click the **Output** button and type **Normal Histogram** in the As a New Chart Sheet box to send the chart to a new chart sheet. Click the **OK** button.

Figure 5-14 shows the histogram of the randomly generated values from the normal distribution

Figure 5-14 Histogram of the 100 random normal values

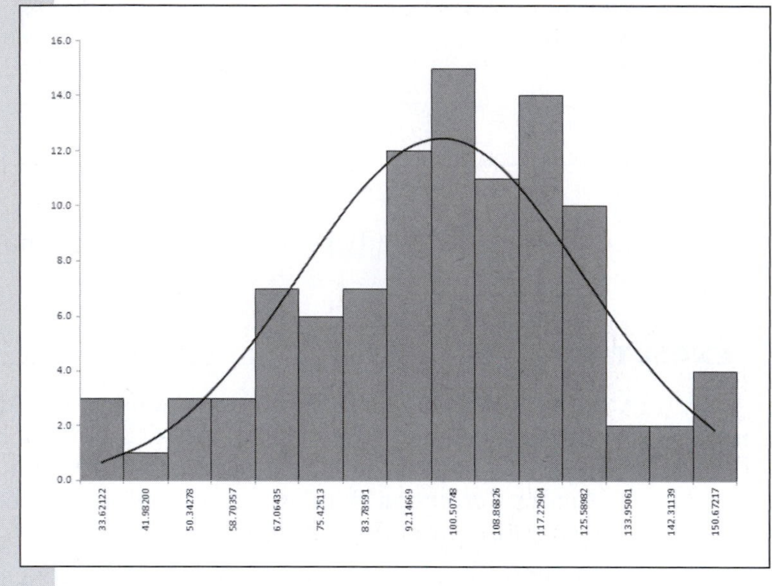

The histogram does not follow the normal curve exactly, but as you saw earlier, if you increase the size of the random sample, the distribution of the sample values approaches the underlying probability distribution. A sample size of 100 is perhaps still too small. Because you generated these values, you already know that the data are normally distributed, but suppose you observed these values in an experiment or study. Would the chart shown in Figure 5-14 convince you that you're working with normal data? It's not always easy to tell from a histogram whether your data are normal, so statisticians have developed some procedures to check for normality.

The Normal Probability Plot

To check for normality, statisticians compute normal scores for their data. A **normal score** is the value you would expect if your sample came from a standard normal distribution. As an example, for a sample size of 5, here are the five normal scores:

$$-1.163, -0.495, 0, 0.495, 1.163$$

To interpret these numbers, think of generating sample after sample of standard normal data, each sample consisting of five observations. Now, take the average of the smallest value in each sample, the second smallest value, and so forth up to the average of the largest value in each sample. Those averages are the normal scores. Here, we would expect the largest value from a random sample of five standard normal values to be 1.163 and the smallest to be −1.163.

Once you've generated the appropriate normal scores, plot the largest value in your data set against the largest normal score, the second largest value against the second largest normal score, and so forth. This is called a **normal probability plot**. If your data are normally distributed, the points should fall close to a straight line.

StatPlus includes a command to calculate normal scores and create a normal probability plot. Use it now to plot your random sample of normal data.

To create a normal probability plot:

1. Click **Single Variable Charts** from the StatPlus menu and then click **Normal P-plots**.
2. Click the **Data Values** button, click the **Use Range References** option button, and select the range **A1:A101** on your worksheet. Click the **OK** button.
3. Click the **Output** button and type **Normal P-plot** in the As a New Chart Sheet box to send the chart to a new chart sheet. Click the **OK** button.
4. Click the **OK** button to start creating the normal probability plot.

Figure 5-15 shows the resulting plot (yours will look slightly different because you've generated a different set of random values).

Figure 5-15 Normal probability plot

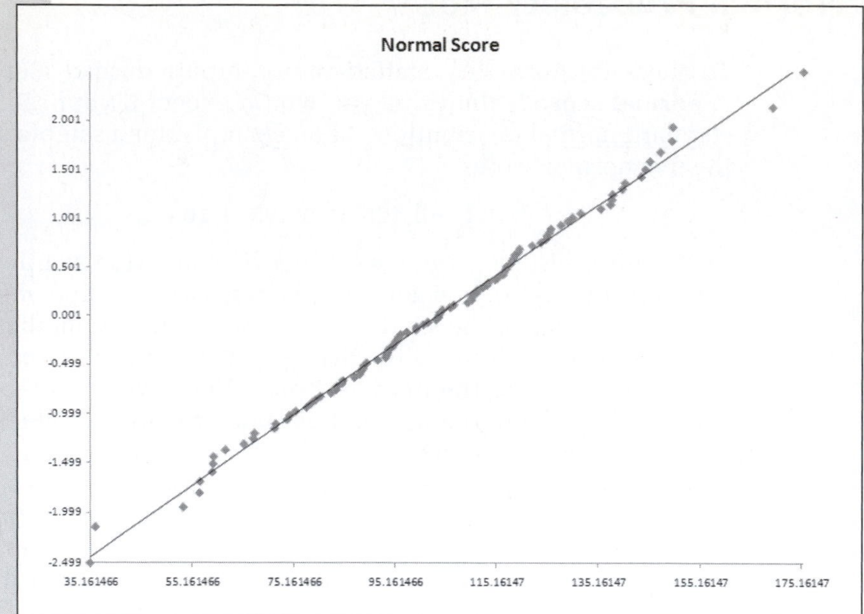

The points follow a general straight line trend fairly well, but with some departures at the left end of the scale. If the sample size were larger it might follow the line even more closely.

5 Close your workbook. You do not have to save any of the random data or plots you created.

Let's apply this technique to some real data. The Baseball workbook from the Chapter05 data folder contains information about baseball player salaries and batting averages. Do the batting averages follow a normal distribution? Let's find out. We'll start by creating a histogram of the batting average data.

To create a histogram of the batting average data:

1 Open the **Baseball** workbook located on the companion website.

2 Save the workbook as Baseball Batting Averages

3 Click **Single Variable Charts** from the StatPlus menu and then click **Histograms**.

4 Click the **Data Values** button, select **AVG** from the list of range names, and click **OK**.

5 Click the **Normal Curve** checkbox.

6 Click the **Output** button and send the histogram to a chart sheet named **Batting Average**.

7 Click the **OK** button to start creating the histogram and normal curve.

Figure 5-16 shows the resulting chart.

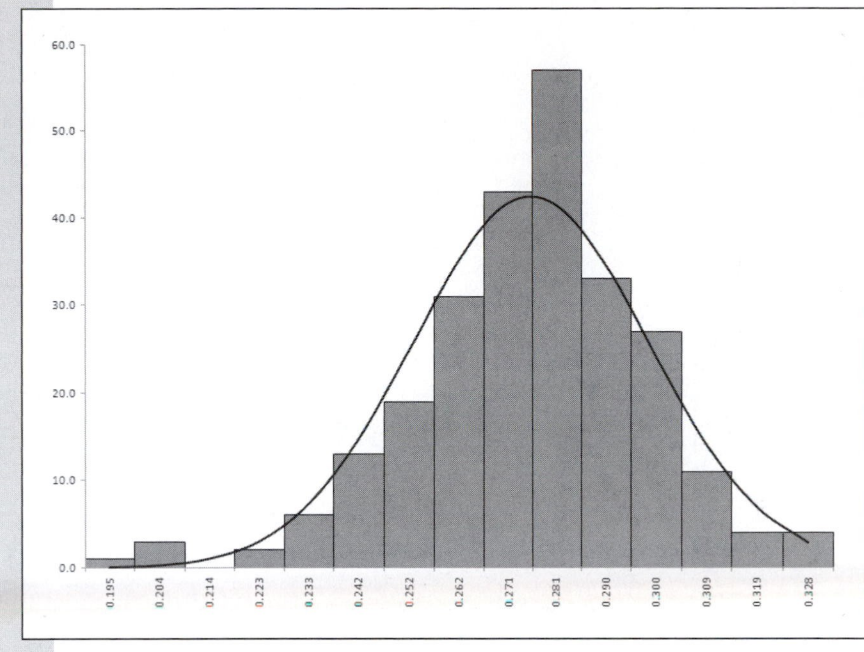

Figure 5-16 Distribution of the batting average data

The distribution of the batting average values appears to follow the superimposed normal curve pretty well (certainly no worse than the sample of random numbers generated earlier). There is no indication that the batting averages do not follow the normal distribution. Let's further check this assumption with a normal probability plot.

To create a normal probability plot of the batting average data:

1 Return to the **Baseball Salaries** worksheet.

2 Click **Single Variable Charts** from the StatPlus menu and then click **Normal P-plots**.

3 Click the **Data Values** button and select **AVG** from the list of range names. Click the **OK** button.

4 Click the **Output** button and send the probability plot to a chart sheet named **Batting Average P-plot**.

5 Click the **OK** button to start creating the normal probability plot. See Figure 5-17.

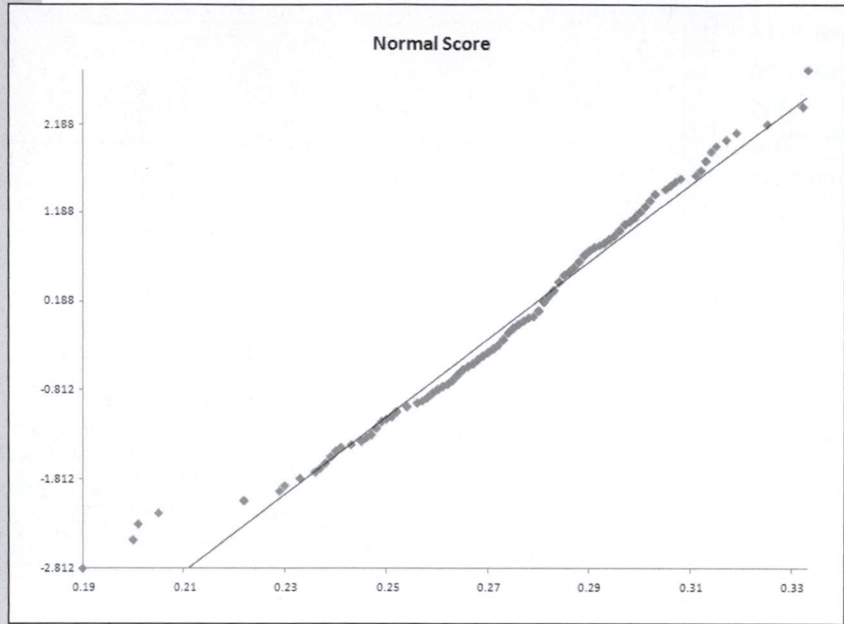

Figure 5-17 Normal probability plot of the batting average data

6 Save your changes and close the workbook.

The batting average data follow a general straight line trend on the normal probability plot pretty well. The only serious departures from the line occur at either end of the distribution of normal scores. You can examine the normal scores to determine what the batting averages at the end of the distribution would be if the sample data more closely followed the normal distribution.

To convert a normal score back to the scale of the original data, you multiply the normal scores by the standard deviation of the observed values and then add the sample average. In this case, the average batting average is 0.275528, and the standard deviation is 0.022529. If the largest normal score is

2.812, this translates into an expected batting average of 2.812 × 0.022529 + 0.275528 = 0.3388—a value slightly higher than the observed maximum batting average of 0.333. On the other end of the scale the lowest normal is -2.812, which corresponds to batting average of 0.2122, greater than the observed value of 0.19. So although the batting average appears to generally follow the normal distribution, the values at either end of the sample are less than would be expected from normal data.

One of the advantages of the normal probability plot is that if your data are skewed in either the positive or the negative direction, this will be clearly displayed in the plot. Positively skewed data fall below the straight line on both ends of the plot, whereas negatively skewed data rise above the straight line at both ends of the plot. Figure 5-18 shows a histogram and normal probability plot of the salaries of the baseball players in the workbook. The data are clearly not normal as the distribution is heavily weighted toward lower salaries. The salaries are below the line at both ends because of positive skewness.

Figure 5-18 Distribution of the baseball player salary data

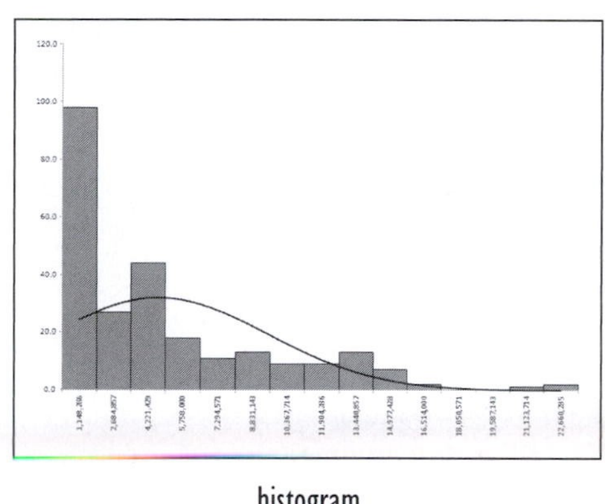

histogram

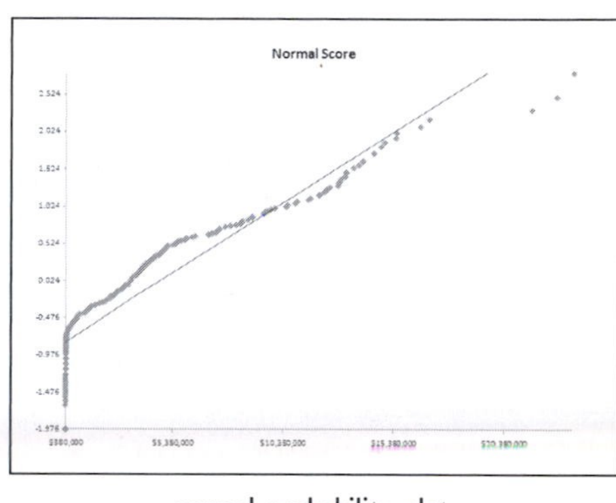

normal probability plot

Parameters and Estimators

When investigating the properties of a probability density function the parameter values of the function were known; however, most of the time we don't know the values of these parameters, so we have to use the data to estimate them using statistics. For example in the normal distribution, we have two parameters μ and σ. We can estimate the value of μ by calculating the sample average $\overline{x}$ and the value of σ by calculating the sample standard deviation s (see Chapter 4 for a description of these statistics).

The values $\overline{x}$ and s have a special and important property: They are not only estimators of μ and σ but are also **consistent estimators**, which means that as the size of the random sample is increased the values of $\overline{x}$ and s come closer and closer to the true parameter values. With a large enough sample size, $\overline{x}$ and s will estimate the true values of μ and σ to whatever degree of precision is required. Figure 5-19 shows how the value of $\overline{x}$ approaches the value of μ as the sample size increases.

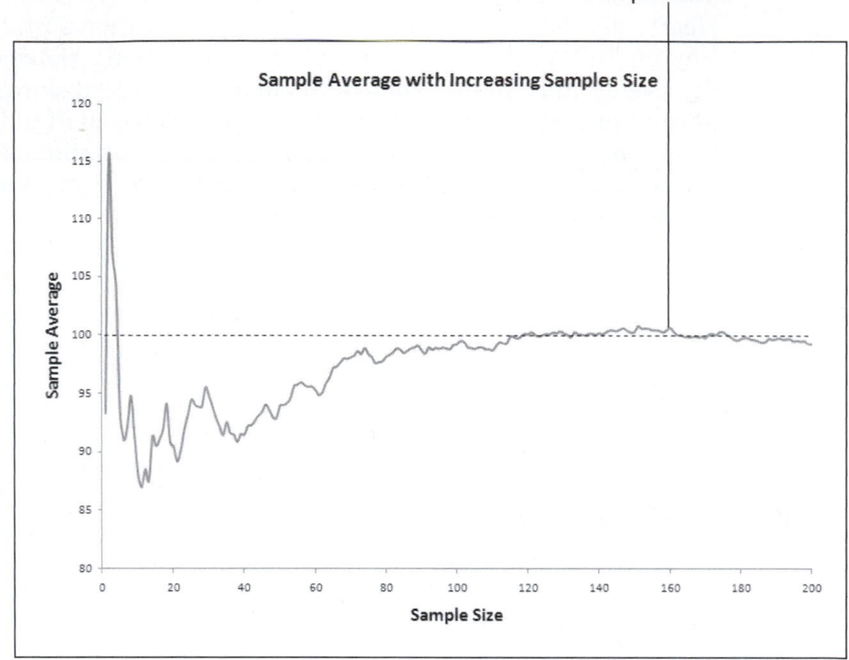

Figure 5-19 Sample averages as the sample size increases

The key question is How large must the sample be to estimate accurately the value of μ? To answer this question, we have to examine the properties of the sample average.

The Sampling Distribution

Because the sample average is calculated by taking the average of random variables, it is also a random variable following its own probability distribution called the **sampling distribution**. If we know the form of the sampling distribution. we can make inferences about the sample average.

We'll start our exploration of the sample average by creating nine random samples of standard normal data, each sample containing 100 observations.

From those random values, we'll create a new column of data containing the average of each of the samples. The distribution of those sample averages will approximate the underlying sampling distribution.

To create 100 samples with nine observations each:

1. Open a new blank workbook in Excel.
2. Click **Create Data** from the StatPlus menu and then click **Random Numbers.**
3. Select **Normal** from the Type of Distribution list box.
4. Enter **16** in the Number of Samples to Generate box.
5. Enter **100** in the Size of Each Sample box.
6. Enter **0** in the Mean box and **1** in the Standard Deviation box.
7. Click the **Output** button and select cell **A1** as the output cell on the current worksheet. Click the **OK** button.

   The completed dialog box is shown in Figure 5-20.

Figure 5-20 The completed Create Random Numbers dialog box

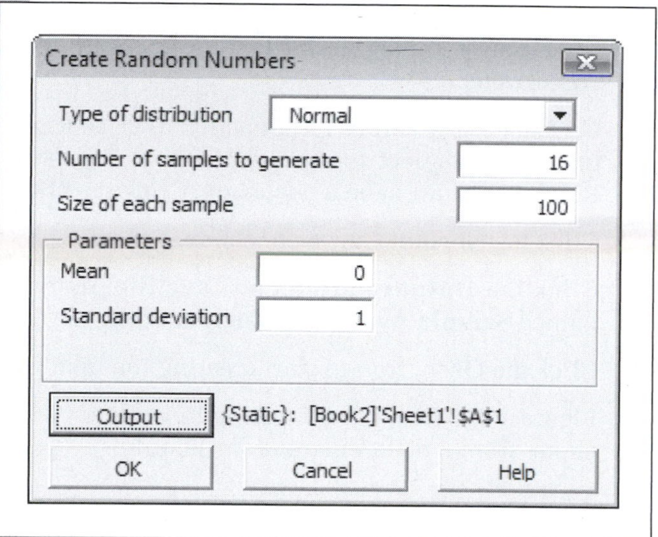

8. Click the **OK** button to start generating the random samples.

Excel creates 16 columns each containing 100 rows of random values from a standard normal distribution. Now take the average of each row; this provides you with 100 rows of sample averages with each average drawn from a sample of 16 random normal values.

To calculate the averages of the 1,000 random samples:

1. Click cell **Q1**, type the formula **=average(A1:P1)**, and press **Enter**.
2. Click cell **Q1** again and drag the fill handle down to cover the range **Q1:Q100**. Column Q now contains the average of each of the 100 samples on the worksheet.

The column of averages you just created should be much less variable than each of the individual samples, because the average smoothes out the highs and the lows of the values found within each sample. What kind of distribution does it have? Let's investigate by creating a histogram of the sample averages.

To create a histogram of the sample averages:

1. Click **Single Variable Charts** from the StatPlus menu and then click **Histograms**.
2. Click the **Data Values** button, click the **Use Range References** option button, and select the range **P1:P100**. *Deselect* the Range Includes Row of Column Labels checkbox. Click the **OK** button.
3. Click the **Normal Curve** checkbox.
4. Click the **Output** button and save the histogram to a chart sheet named **Sample Average Histogram**.
5. Click the **OK** button to start creating the histogram.

Figure 5-21 shows the resulting histogram (yours will differ since it comes from a different random sample).

Figure 5-21 Histogram of sample averages

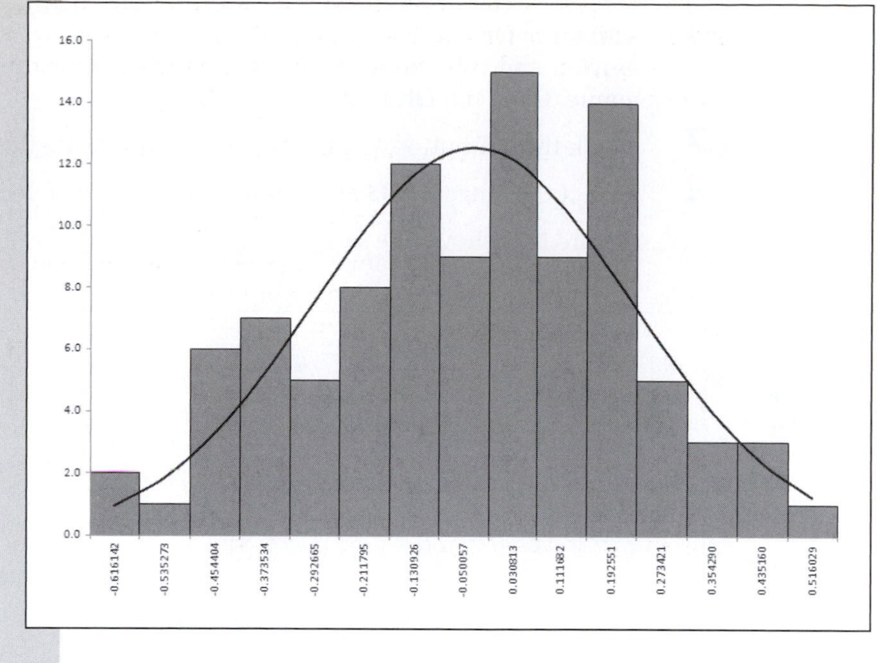

The distribution of the sample averages looks normal and, in fact, is normal. The distribution is centered at 0, as you would expect for averages based on samples taken from the standard normal distribution. The distribution differs from the standard normal in one respect: The sampling distribution is much narrower around the mean. Most of the standard normal values lie between −2 and 2, but here most of the values lie between −0.5 and 0.5. Apparently the σ for the sampling distribution of the sample average is smaller than the value of σ for a standard normal. To verify this, calculate descriptive statistics for the sample averages.

To calculate descriptive statistics for the sample averages:

1. Return to the worksheet containing the sample average values.
2. Click Descriptive Statistics from the StatPlus menu and then click Univariate Statistics.
3. Click the **Input** button, click the **Use Range References** option button, and select the range **P1:P100**. *Deselect* the Range Includes a Row of Column Labels checkbox. Click the **OK** button.
4. Click the **Summary** tab and click the **Count** and **Average** checkboxes.
5. Click the **Variability** tab and click the **Std. Deviation** checkbox.

6. Click the **Output** button, and then click the **New Worksheet** option button and type **Sample Average Statistics** for the new worksheet name. Click the **OK** button.

7. Click the **OK** button to generate the sample statistics.

8. Select the range **B4:B5** and reduce the number of decimal places to 3.

   Figure 5-22 shows the sample output (yours will be slightly different).

Figure 5-22 Sample average statistics

	A	B	C
1		**Univariate Statistics**	
2		Variable 1	
3	Count	100	
4	Average	-0.03379233	
5	Standard Deviation	0.255759294	
6			

9. Close the workbook with the random samples. You don't have to save your changes.

As you would expect, the average of the sample averages is near zero since the averages come from standard normal values in which μ is equal to zero. The standard deviation of the sample averages is 0.256. Thus the sampling distribution of the average values taken from samples with sample sizes of 16 appears to follow a normal distribution with μ of 0 and a σ of about 0.25. Is there a relationship between sample size and the value for the standard deviation?

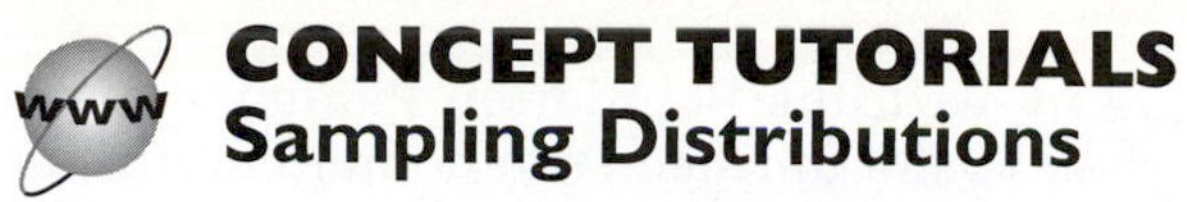

CONCEPT TUTORIALS
Sampling Distributions

You can use the exploration workbook Population Parameters to explore how sample size affects the distribution of the sample average.

To explore sampling distributions:

1. Open the file **Population Parameters**, from the Explore folder, enabling the macros the workbook contains.
2. The workbook contains information on parameters and sampling distributions. Review the material.
3. Click Exploring Sampling Distributions from the Table of Contents.
4. The worksheet displays a histogram of 150 sample averages. A scroll bar allows you to change the size of each sample from 1 up to 16.
5. Move the scroll bar and observe how the shape of the distribution changes on the basis of the different sample sizes. Also note how the value of the standard deviation changes.

Figure 5-23 shows the sampling distribution for different sample sizes. Do you see a pattern in the values of the standard deviation compared to the sample size?

Figure 5-23 Variability Statistics

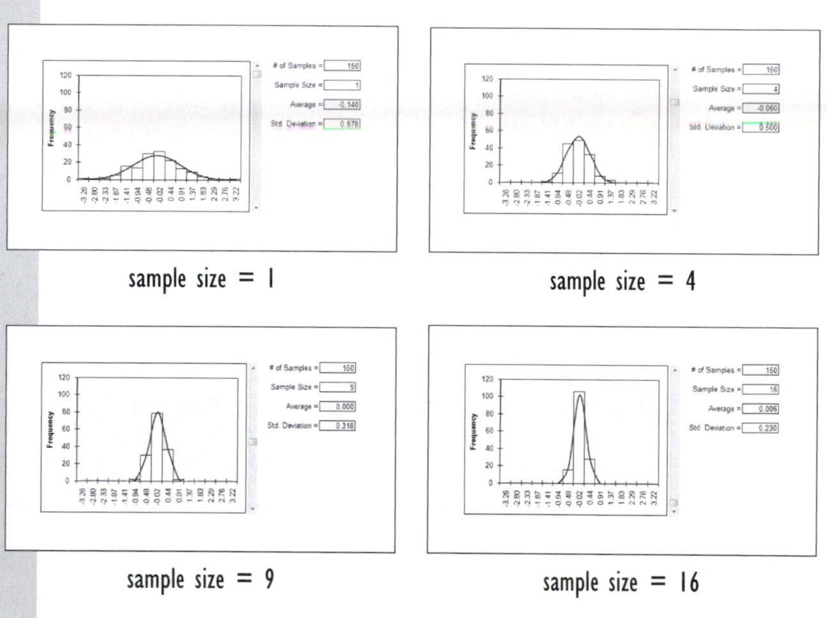

6 Continue working with the Population Parameters workbook and close it when you're finished. You do not have to save your changes.

These values illustrate an important statistical fact. If a sample is composed of n random variables coming from a normal distribution with mean μ and standard deviation σ, then the distribution of the sample average will also be a normal distribution with mean μ but with standard deviation $\sigma/\sqrt{n}$. For example, the distribution of a sample average of 16 standard normal values is normal with a mean of 0 and a standard deviation of $= 1/\sqrt{16} = \frac{1}{4}$, or .25.

The Standard Error

The standard deviation of $\overline{x}$ is also referred to as the **standard error** of $\overline{x}$. The value of the standard error gives us the information we need to determine the precision of $\overline{x}$ in estimating the value of μ. For example, suppose you have a sample of 100 observations that comes from a standard normal distribution, so that the value of μ is 0 and of σ is 1. You've just learned that $\overline{x}$ is distributed normally with a mean of 0 and a standard deviation of 0.1 (because $0.1 = 1\sqrt{100}$).

Let's apply this to what you already know about the normal distribution, namely that about 95% of the values fall within 2 standard deviations of the mean. This means that we can be 95% confident that the value of $\overline{x}$ will be within 0.2 units of the mean. For example, if $\overline{x} = 5.3$, we can be 95% confident that the value of μ lies somewhere between 5.1 and 5.5. To be even more precise, we can increase the sample size. If we want $\overline{x}$ to fall within 0.02 of the value of μ 95% of the time, we need a sample of size 10,000, because if $\overline{x}$ is 5.3 with a sample size of 10,000, we can be 95% confident that μ is between 5.28 and 5.32. Note that we can never discover the *exact* value of μ, but we can with some high degree of confidence narrow the band of possible values to whatever degree of precision we wish.

The Central Limit Theorem

The preceding discussion applied only to the normal distribution. What happens if our data come from some other probability distribution? Can we say anything about the sampling distribution of the average in that case? We can, by means of the Central Limit Theorem. The **Central Limit Theorem** states that if you have a sample taken from a probability distribution with mean μ and standard deviation σ, the sampling distribution of $\bar{x}$ is approximately normal with a mean of μ and a standard deviation of $\sigma/\sqrt{n}$. The remarkable thing about the Central Limit Theorem is that the sampling distribution of $\bar{x}$ is approximately normal, no matter what the probability distribution of the individual values is. As the sample size increases, the approximation to the normal distribution becomes closer and closer. Now you see why the normal distribution is so important in the field of statistics.

To see the effect of the Central Limit Theorem, you can use the instructional workbook named The Central Limit Theorem.

CONCEPT TUTORIALS
The Central Limit Theorem

To use the Central Limit Theorem workbook:

1. Open the **Central Limit Theorem** file from the Explore folder. Enable the macros in the workbook.
2. Review, in the workbook, the concepts behind the Central Limit Theorem.
3. Click Explore the Central Limit Theorem from the Table of Contents.

The Central Limit Theorem worksheet opens. See Figure 5-24.

Figure 5-24 The Central Limit Theorem workbook

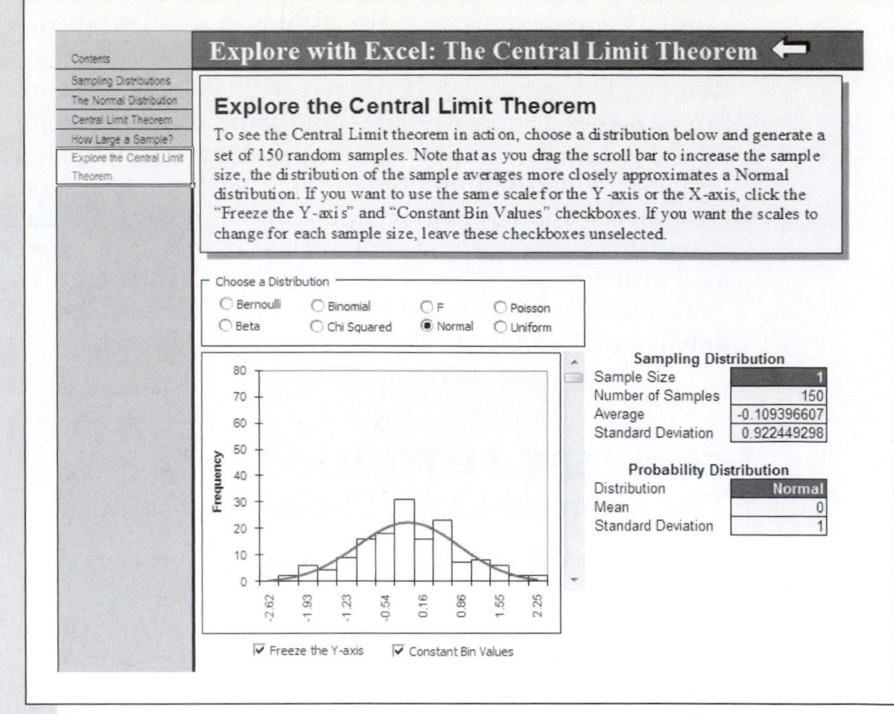

The worksheet lets you generate 150 random samples from one of eight common probability distributions with up to 16 observations per sample.

The worksheet also calculates and displays the distribution of the sample averages. You can change the sample size by dragging the scroll bar up or down. The worksheet opens, displaying the distribution of the sample average for the standard normal distribution. To see how the worksheet works, move the scroll bar, increasing the sample size from 1 to 16.

To change the sample size:

1 Drag the scroll bar down. The sample size increases from 1 to 16.

As you drag the scroll bar down, the histogram displays the change in the sample size, and the standard deviation decreases from 1 to about 0.25.

2 Drag the scroll bar back up to return the sample size to 1.

Note that if you want to view the histogram with different bin values under different sample sizes, you can deselect the Constant Bin Values checkbox. When the bin values are the same, you can compare the spread of the data from one histogram to another because the same bin values are used for all charts. Deselecting the Constant Bin Values checkbox fits the bin values to the data and gives you more detail, but it's more difficult to compare histograms from different sample sizes. You can also "freeze" the *y* axis to retain the *y* axis scale from one sample size to another, making it easier to compare one chart with another. To scale the *y* axis to the data, unselect the Freeze the Y-Axis checkbox.

Now that you've viewed the sampling distribution for the standard normal, you'll choose a different distribution from the list. Another commonly used probability distribution is the uniform distribution. In the **uniform distribution**, probability values are uniform across a continuous range of values. The probability density function for the uniform distribution is

Uniform Probability Density Function

$$f(y) = \frac{1}{b - a} \quad -\infty < b < \infty, \quad -\infty < a < b$$

where b is the upper boundary and a is the lower boundary of the distribution. Figure 5-25 displays the Uniform distribution where $a = -2$ and $b = 2$.

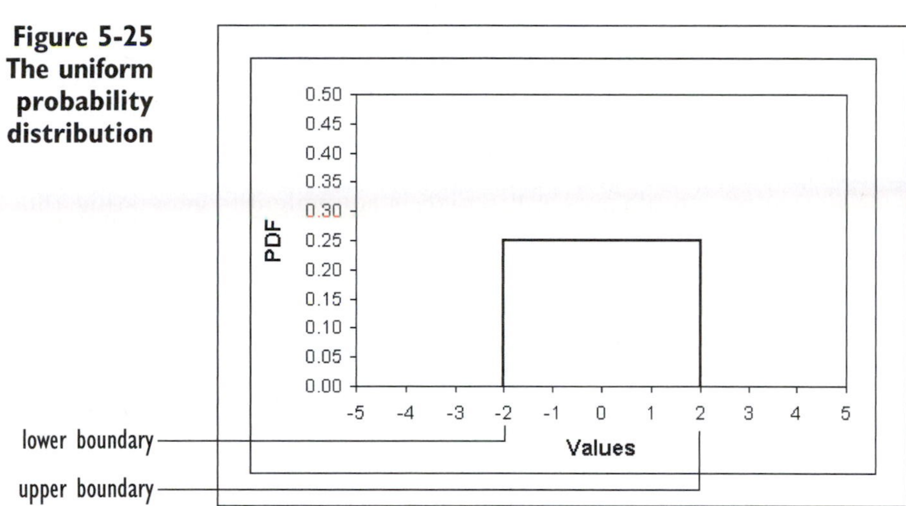

Figure 5-25 The uniform probability distribution

The mean μ and standard deviation σ for the uniform distribution are

$$\mu = \frac{b + a}{2} \qquad \sigma = \frac{b - a}{\sqrt{12}}$$

Thus if $a = -2$ and $b = 2$, $\mu = 0$ and $\sigma = 4/\sqrt{12} = 1.1547$.

Having learned something about the uniform distribution, let's observe the sampling distribution of the sample average.

To generate the sampling distribution of the uniform distribution:

1 Click the **Uniform** option button on the Central Limit Theorem worksheet.

2 Enter **−2** in the Minimum box and **2** in the Maximum box. See Figure 5-26.

Figure 5-26 Setting the parameters for the uniform distribution

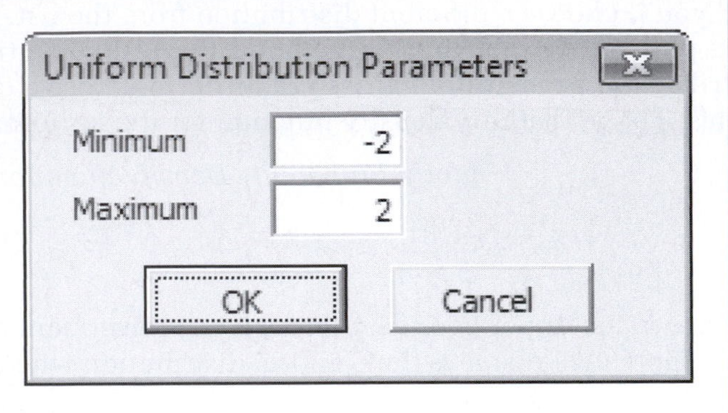

3 Click the **OK** button.

Excel generates 150 random samples for the uniform distribution.

The initial sample size is 1, which is equivalent to generating 150 different observations from the uniform distribution. The initial average value should be close to 0 and the initial standard deviation value should be close to 1.15.

4 Drag the scrollbar down to increase the sample size from 1 to 16.

Figure 5-27 shows the sampling distribution for the average under different sample sizes (your charts and values will be slightly different).

You may want to unfreeze and freeze the *y* axis in order to display the histograms in a more detailed scale.

Figure 5-27 Sampling distribution for average from the uniform distribution

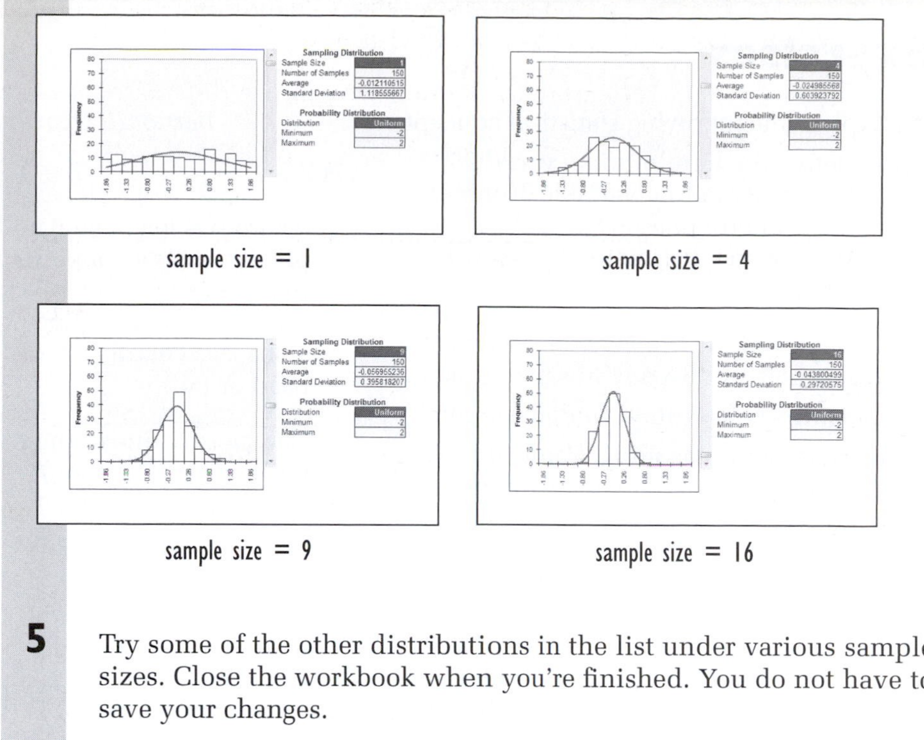

5 Try some of the other distributions in the list under various sample sizes. Close the workbook when you're finished. You do not have to save your changes.

A few final points about the Central Limit Theorem should be considered. First, the theorem applies only to probability distributions that have a finite mean and standard deviation. Second, the sample size and the properties of the original distribution govern the degree to which the sampling distribution approximates the normal distribution. For large sample sizes, the approximation can be very good, whereas for smaller samples, the approximation might not be good at all. If the probability distribution is extremely skewed, a larger sample size will be necessary. If the distribution is symmetric, the sample size usually need not be very large. How large is large? If the original distribution is symmetric and already close in shape to a normal distribution, a sample size of 15 or 20 should be large enough. For a highly skewed distribution, a sample size of 40 or 50 might be required. Usually the *Central* Limit Theorem can be safely applied if the sample size is 30 or more.

The Central Limit Theorem is probably the most important theorem in statistics. With this theorem, statisticians can make reasonable inferences about the sample mean without having to know about the underlying probability distribution. You'll learn how to make some of these inferences in the next chapter.

Exercises

1. Explore the following statistical concepts:

a. Define the term *random variable.*

b. How is a random variable different from an observation?

c. What is the distinction between $\overline{x}$ and μ?

2. A sample of the top 50 women-owned businesses in Wisconsin is undertaken. Does this constitute a random sample? Explain your reasoning. Can you make any inferences about women-owned businesses on the basis of this sample?

3. The administration counts the number of low-birth-weight babies born each week in a particular hospital. Assume, for the sake of simplicity, that the rate of low-birth-weight births is constant from week to week.

a. Of the distributions that we have studied, which one is applicable here?

b. If the average number of low-birth-weight babies is 5, what is the probability that no low-birth-weight babies will be born in a single week?

c. The administration counts the low-birth- weight babies every week and then calculates the average count for the entire year. What is the approximate distribution of the average?

4. The results of flipping a coin follow a probability distribution called the **Bernoulli distribution.** A Bernoulli distribution has two possible outcomes, which we'll designate with the numeric values 0 and 1. The probability function for the Bernoulli distribution is

Bernoulli Distribution

$$P(Y = 1) = p, \quad P(Y = 0) = 1 - p \qquad 0 < p < 1$$

where p is between 0 and 1. For example, if we tossed an unbiased coin and indicated the value of a head with 1 and a tail with 0, the value of p would be .5 since it is equally likely to have either a head or tail.

The mean value of the Bernoulli distribution is p. The standard deviation is $\sqrt{p(1-p)}$. In the flipping coin example, the mean value is equal to 0.5 and the standard deviation is $\sqrt{0.5(1 - 0.5)} = 0.5$.

a. You toss a die, recording a 1 for the values 1 through 3, and a 0 for values 4 through 6. What is the mean value? What is the standard deviation?

b. You toss a die, recording a 1 for a value of 1 or 2, and a 0 for the values 3, 4, 5, and 6. What is the mean value? What is the standard deviation?

c. You toss a die, recording a 1 for a value of 1, and a 0 for all other values. What is the mean value? What is the standard deviation?

5. If you flip 10 coins, what is the probability of getting exactly 5 heads? To answer this question, you have to refer to the **Binomial distribution**, which is the distribution of repeated trials of a Bernoulli random variable. The probability function for the Binomial distribution is

Binomial Distribution

$$P(Y = y) = \binom{n}{y} p^y (1 - p)^{n-y} \qquad y = 0,1,2, \ldots, n$$

$$\text{where } \binom{n}{y} = \frac{n!}{y!(n - y)!}$$

where p is the probability of the event (such as getting a head) and n is the number of trials. For example, to calculate the probability of getting exactly 5 heads in 10 tosses, the formula is

$$P(Y = 5) = \binom{10}{5}(0.5)^5(1 - 0.5)^{(10-5)} = 0.2461$$

or 24.6%. In other words, there is about a 1 in 4 chance of getting exactly 5 heads out of 10 tosses. To calculate the probability of getting 5 or fewer heads, we add the probabilities for the individual numbers: $p(Y = 0)$, $p(Y = 1)$, $p(Y = 2)$, $p(Y = 3)$, $p(Y = 4)$, and $p(Y = 5)$.

The mean of the Binomial distribution is np. The standard deviation is $\sqrt{np(1 - p)}$. For example, if we flip 100 coins with $p = .5$, the expected number of heads is $100 \times .5 = 50$ and the standard deviation is $\sqrt{0.5 \cdot 0.5 \cdot 100} = 5$.

a. If you toss 20 coins, what is probability of getting exactly 10 heads?

b. If you toss 50 coins, what is the expected number of heads? What is the standard deviation?

c. You toss 10 dice, recording a 1 for a 1 or a 2, and a 0 for a 3, 4, 5, or 6, what is the expected total? What is the standard deviation?

d. You toss 10 dice, recording a 1 for a 1 and a 0 for the other numbers, what is the expected total? What is the standard deviation?

6. Excel includes the BINOMDIST function to calculate values of the binomial distribution. The syntax of the function is

BINOMDIST(*number, trials, prob, cumulative*)

where *number* is the number of successes, *trials* is the number of trials, *prob* is the probability of success, and *cumulative* is TRUE to calculate the cumulative value of the probability function and FALSE to calculate the value for the specified number value. For example, the formula

BINOMDIST(5, 10, 0.5, FALSE)

returns the value 0.246. To calculate the cumulative value (in this case the probability of getting 5 or fewer heads out of 10), use the formula

BINOMDIST(5, 10, 0.5, TRUE)

which returns the value .623. Thus, there is a 62.3% probability of getting 5 or fewer heads out of 10. Use Excel to answer the following questions:

a. What is the probability of getting exactly 10 heads out of 20 coin tosses? What is the probability of getting 10 or less?

b. What is the probability of getting 10 heads out of 15 coin tosses? What is the probability of getting more than 10?

c. You toss 10 dice, recording a 1 for a 1 or 2, and a 0 for a 3, 4, 5, or 6. If you total up your scores, what is the probability of scoring exactly 3 points? What is the probability of scoring 3 or fewer?

d. You toss 100 dice, recording a 1 for a 1 and a 0 for the other numbers, what is the probability of recording a score of exactly 20? What is the probability of scoring 20 or less?

e. If you toss a coin 10 times, what is the probability of recording 4, 5, or 6 heads?

7. The mean of the Poisson distribution is λ and the standard deviation is $\sqrt{\lambda}$, where λ is the expected count per interval.

a. The number of accidents at a factory in a year follows a Poisson distribution with an expected value of 10 accidents per year. What is the value of λ? What is the standard deviation?

b. If you collect 25 years of accident information at this factory, how could the number of accidents per year be used to estimate $\overline{x}$? What would be the standard error of this estimate?

8. Excel includes a function to calculate values of the Poisson distribution. The syntax of the function is

POISSON(*x, mean, cumulative*)

where *x* is the number of counts, *mean* is the expected number of counts, and *cumulative* is TRUE to calculate the cumulative value of the probability function and FALSE to calculate the density for the specified *x* value. Use this function to calculate the cumulative and specific probabilities of the following values:

a. $\lambda = 2, x = 2$
b. $\lambda = 2, x = 3$
c. $\lambda = 2, x = 4$
d. $\lambda = 2, x = 5$

9. Excel includes a function to calculate the probabilities associated with the standard normal distribution. The syntax of the function is

NORMSDIST(*z*)

The function returns the probability of a standard normal random variable having a value $\leq z$. Use this function to calculate the following probabilities:

a. $z = .5$
b. $z = 1$
c. $z = 1.65$
d. $z = 1.96$
e. What is the probability of a standard normal random variable having a value of exactly 2.0?

10. The Excel function

NORMSINV(*prob*)

returns the inverse of the standard normal distribution. For a given cumulative probability *prob*, the function returns the value of *z*. Use this function to calculate *z* values for the following probabilities of the standard normal distribution:

a. .05
b. .10
c. .50
d. .90
e. .95
f. .975
g. .99

11. Excel includes a function to calculate the probability for a random variable coming from any normal distribution. The syntax of the function is

NORMDIST(*x, mean, std_dev, cumulative*)

where *x* is the value of the random variable, *mean* is the mean of the normal distribution, *std_dev* is the standard deviation of the distribution, and *cumulative* is TRUE to calculate the cumulative value of the probability function and FALSE to calculate the pdf for the specified *x* value. Use this function to calculate the cumulative probabilities for the following values:

a. *x* = 1.96, *mean* = 0, *std_dev* = 1
b. *x* = 1.96, *mean* = 0, *std_dev* = 0.5
c. *x* = 1.96, *mean* = 0, *std_dev* = 0.25
d. *x* = −1.96, *mean* = 0, *std_dev* = 1
e. *x* = 5, *mean* = 5, *std_dev* = 2

12. The Excel function

NORMINV(*prob, mean, std_dev*)

calculates the inverse of the normal distribution. For a cumulative probability of *prob*, a mean value of *mean*, and a standard deviation of *std_dev*, the function returns the value of x. Use this function to calculate the x values for the following:

a. *mean* = 5, *std_dev* = 2, *prob* = .10
b. *mean* = 5, *std_dev* = 2, *prob* = .20
c. *mean* = 5, *std_dev* = 2, *prob* = .50
d. *mean* = 5, *std_dev* = 2, *prob* = .90
e. *mean* = 5, *std_dev* = 2, *prob* = .95
f. *mean* = 5, *std_dev* = 2, *prob* = .99

13. Open the Baseball workbook from the Chapter05 folder. You want to analyze the batting average statistics from the workbook. The mean career batting average is .263 and the standard deviation is .02338.

a. Open the workbook and save it as **Batting Average Analysis**.
b. Assuming that the batting averages are normally distributed, use Excel's NORMDIST function to find the probability that a player will bat .300 or better. (*Hint:* Calculate 1 − probability that a player will bat less than .300.)
c. How many players batted .300 or better? Compare this to the expected number.
d. Save your workbook and summarize your findings.

14. The Housing workbook contains a sample of 117 housing prices for Albuquerque, New Mexico, during the early 1990s. You've been asked to analyze this historic data set.

a. Open the **Housing** workbook from the Chapter05 folder and save it as **Housing Price Analysis**.
b. Create a histogram (with a normal curve) and a normal probability plot of the housing prices. Do the data appear to follow a normal distribution?
c. Calculate the average housing price and the standard deviation of the housing price. Because this is a sample of all of the house prices in Albuquerque, the average serves as an estimate of the mean house price.
d. Create a new column containing the log of the home price values. Create a histogram with a normal curve and a normal probability plot. Modify the x axis label Number format to display the log values to three decimal places. Do the transformed values appear more normally distributed than the untransformed values?
e. Save your changes to the workbook. Write a report summarizing your observations and calculations.

15. The dispersion of shots used in shooting at the target in the Random Samples workbook follows a bivariate normal distribution (a combination of two normal distributions, one in the vertical direction and one in the horizontal direction). The value of σ for each level of accuracy is

Accuracy	**Standard Deviation**
Highest	0.1
Good	0.2
Moderate	0.4
Poor	0.6
Lowest	1.0

a. Open the **Random Samples** workbook from the Explore folder and create a distribution of shots around the target with good accuracy.
b. Explain why the predicted percentages have the values they have.
c. For a shooter with the lowest accuracy, how many shots would the person have to take before she or he could assume with 95% confidence

that the average horizontal location of her or his shots was within 0.2 unit of the bull's eye?

d. How many shots would a shooter with the highest accuracy have to take before achieving similar confidence in the average placement of his or her shots?

16. Study the properties of the Poisson distribution by doing the following:

a. Open a blank workbook and, using the Create Random Data command from StatPlus, create 16 columns of 100 rows of Poisson random values with $\lambda = 0.25$.

b. Create another column in your workbook containing the average values from those columns.

c. Create a histogram of the first column of random Poisson values with a superimposed normal curve. Create a second histogram of the column averages with a superimposed normal curve. Compare the two curves.

d. Calculate the average and standard deviation of the two columns.

e. How do these calculated values compare to the value of λ? See Exercise 7 for more information on the Poisson distribution.

f. Save your workbook as **Poisson Distribution Analysis** to the Chapter05 folder.

17. Repeat questions a. through d. of Exercise 16 with 16 columns of 100 rows of binomial random values where the number of trials is 16 and the value of p is .25. How do the averages in part d compare with the value $p = .25$? What about the standard deviations computed in part d? Save your workbook as **Binomial Distribution Analysis** to the Chapter05 folder.

18. Repeat questions a. through d. of Exercise 16 with 16 columns of 100 rows of Bernoulli random values where $p = .25$. How do the averages in part d compare with the value $p = .25$? What about the standard deviations computed in part d? Save your workbook as **Bernoulli Distribution Analysis** to the Chapter05 folder.

19. Repeat questions a. through d. of Exercise 16 with 16 columns of 100 rows of uniform random values where the lower boundary is 0 and the upper boundary is 100. Save the workbook as **Uniform Distribution Analysis** to the Chapter05 folder.

20. *True or false*: According to the Central Limit Theorem, as the size of the sample increases, the distribution of the observations approximates a normal distribution. Defend your answer.

21. You want to collect a sample of values from a uniform distribution where μ is unknown but $\sigma = 10$.

a. How large a sample would you need to estimate the value of μ within 2 units with a confidence level of 95%?

b. How large a sample would you need to estimate the value of μ within 2 units with a confidence level of 99%?

c. If the sample size is 25 and μ is 50, what is the probability that the sample average will have a value of 48 or less?

22. At the 1996 Summer Olympic games in Atlanta, Linford Christie of Great Britain was disqualified from the finals of the men's 100-meter race because of a false start. Christie did not react before the starting gun sounded, but he did react in less than 0.1 second.

According to the rules, anyone who reacts in less than a tenth of a second *must* have false-started by anticipating the race's start. Christie bitterly protested the ruling, claiming that he had just reacted very quickly. Using the reaction times from the first heat of the men's 100-meter race, try to weigh the merits of Christie's claim versus the argument of the race officials that no one can react as fast as Christie did without anticipating the starting gun.

a. Open the **Reaction** workbook from the Chapter05 folder and save it as **Reaction Time Analysis**.

b. Create a histogram of the reaction times. Where would a value of 0.1 second fall on the chart?

c. Calculate the mean and standard deviation of the first heat's reaction times. Use these values in Excel's NORMDIST function and calculate the probability that an individual would record a reaction time of 0.1 or less.

d. Create a normal probability plot of the reaction times. Do the data appear to follow the normal distribution?

e. Save your workbook and write a report summarizing your conclusions. Include in your summary a discussion of the difficulties in determining whether Christie anticipated the starter's gun. Are the data appropriate for this type of analysis? What are some limitations of the data? What kind of data would give you better information regarding a runner's reaction times to the starter's gun (specifically, runners taking part in the finals of an Olympic event)?

Chapter 6

STATISTICAL INFERENCE

Objectives

In this chapter you will learn to:

- Create confidence intervals
- Apply a hypothesis test
- Use the *t* distribution in a hypothesis test
- Perform a one-sample and a two-sample *t* test
- Analyze data using nonparametric approaches

The concepts you learned in Chapter 5 provide the basis for the subject of this chapter, statistical inference. Two of the main tools of statistical inference are confidence intervals and hypothesis tests. In this chapter, you'll apply these tools to reach conclusions about your data. You'll be introduced to a new distribution, the *t* distribution, and you'll see how to use it in performing statistical inference. You'll also learn about nonparametric tests that make fewer assumptions about the distribution of your data.

Confidence Intervals

In the previous chapter, you learned two very important facts about distributions and samples.

1. A sample average will approximately follow a normal distribution with mean μ and standard deviation $\sigma/\sqrt{n}$, where μ is the mean of the probability distribution the sample is drawn from, σ is the standard deviation of the probability distribution, and n is the size of the sample. Another way of writing this is

$$\bar{x} \sim N\left(\mu, \sigma/\sqrt{n}\right)$$

2. In a normal distribution, about 95% of the time, the values fall within 2 standard deviations of the mean.

From these two facts, we can calculate how precisely the sample average estimates the value of μ. For example, if $\sigma = 10$ and our sample size is 25, the sample average will approximately follow a normal distribution with mean μ and standard deviation 2, so 95% of the time, the sample average will fall within 4 units of μ. This indicates that if the sample average is 20, we could construct a **confidence interval** from about 16 to 24 that should, with 95% confidence, "capture" the value of μ. If we want this confidence interval to be smaller, we simply increase the sample size. A sample of 100 observations would result in a 95% confidence interval for μ ranging from about 18 to 22.

The use of the 2 standard deviations rule is an approximation. What if we wanted a more exact estimate of the 95% confidence interval, or what if we wanted to construct other confidence intervals, such as a 99% confidence interval? How would we go about doing that?

z Test Statistic and *z* Values

In order to derive a more general expression of the confidence interval, we first have to express the sample average in terms of a standard normal distribution. We can do this by subtracting the value of μ and dividing by the standard error. The calculated value will then follow a standard normal distribution; that is,

$$\frac{\bar{x} - \mu}{\sigma / \sqrt{n}} \sim N(0, 1)$$

This value is called a **z test statistic**. We can then compare the z test statistic to z values taken from a standard normal distribution. A **z value**, usually written as z_p, is the point z on a standard normal curve such that for random variable Z, $P(Z \leq z_p) = p$. For example, $z_{0.95} = 1.645$ because 95% of the area under the curve is to the left of 1.645. See Figure 6-1.

Figure 6-1
The z value

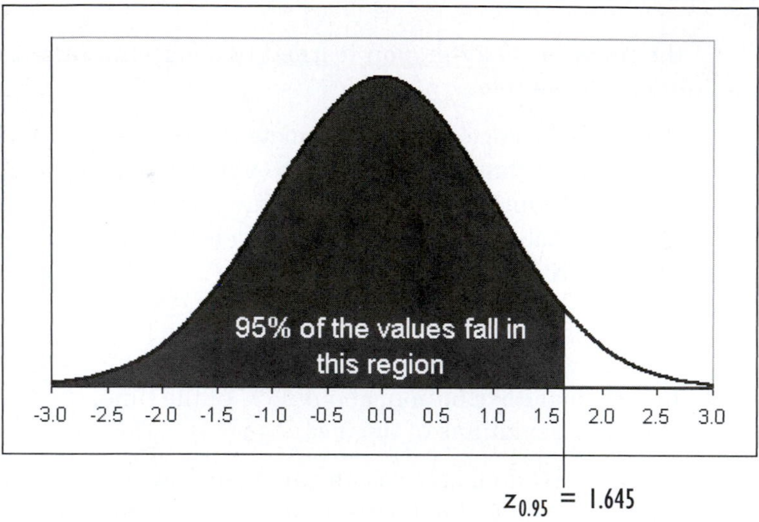

Figure 6-1 shows a one-sided z value, but for confidence intervals, we're more interested in a two-sided z value, where p is the probability of the value falling in the center of the distribution and α (which equals $1 - p$) is the probability of its falling in one of the two tails. For a two-sided range of size p, these z values are $-z_{1-\alpha/2}$ and $z_{1-\alpha/2}$. In other words, for a random variable Z, $P(-z_{1-\alpha/2} < Z < z_{1-\alpha/2}) = 1 - \alpha = p$. If we want to find the central 95% of the standard normal curve, $p = 0.95$, $\alpha = 0.05$, and $z_{1-0.05/2} = z_{0.975} = 1.96$. This means that 95% of the values on a standard normal curve lie between -1.96 and 1.96. See Figure 6-2.

Figure 6-2
Two-sided z values

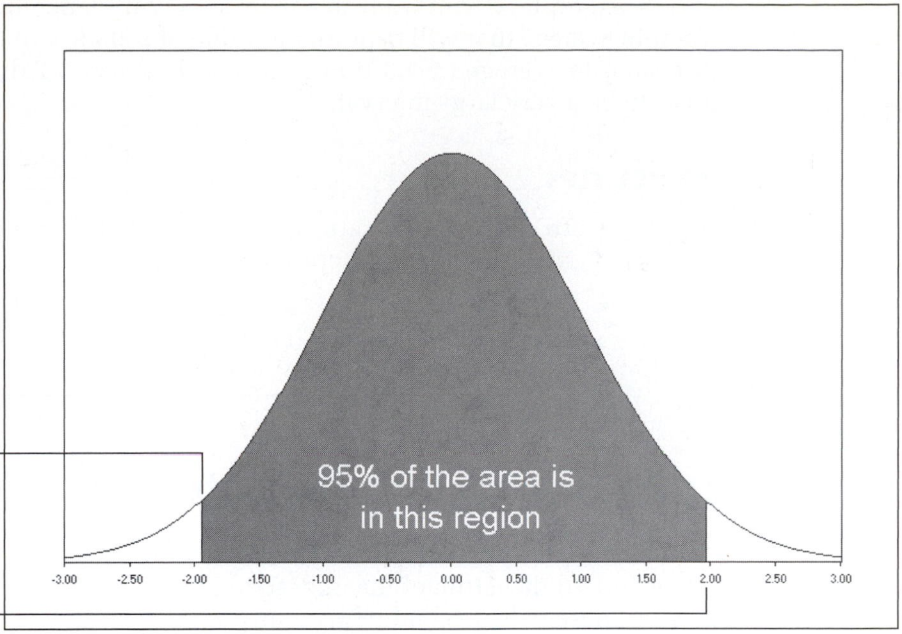

$-z_{0.975} = -1.96$

$z_{0.975} = 1.96$

We can use a two-sided z value to construct a general expression for the confidence interval. The more general expression is

$$P\left(\overline{x} - z_{1-a/2}\frac{\sigma}{\sqrt{n}} < \mu < \overline{x} + z_{1-a/2}\frac{\sigma}{\sqrt{n}}\right) = 1 - a$$

Thus the upper and lower confidence limits for μ are $\overline{x} \pm z_{1-a/2}\sigma/\sqrt{n}$. For example, if $\alpha = 0.05$, then $z_{1-0.05/2} = 1.96$ and the 95% confidence limits are $\overline{x} \pm 1.96 \times \sigma/\sqrt{n}$, which is pretty close to our rule-of-thumb estimate of $\pm$ 2 standard errors from the sample average. Table 6-1 shows confidence intervals of various sizes using this approach.

Table 6-1 Confidence Intervals

$1 - \alpha$	$z_{1-\alpha/2}$	**Confidence Band**
0.800	1.282	$\overline{x} \pm 1.282 \times \sigma/\sqrt{n}$
0.900	1.645	$\overline{x} \pm 1.645 \times \sigma/\sqrt{n}$
0.950	1.960	$\overline{x} \pm 1.960 \times \sigma/\sqrt{n}$
0.990	2.576	$\overline{x} \pm 2.576 \times \sigma/\sqrt{n}$
0.999	3.290	$\overline{x} \pm 3.290 \times \sigma/\sqrt{n}$

For example, if you want to construct a confidence interval around the sample average that will capture the value of μ 99.9% of the time, calculate the sample average $\pm$ 3.3 times the standard error. Admittedly, this will tend to be a very large interval.

EXCEL TIPS

- To calculate the value of $z_{1-\alpha/2}$ with Excel, use the function NORMSINV(x), where $x = 1 - \alpha/2$.
- To find the probability associated with a z test statistic, use the function NORMSDIST(z), where z is the z test statistic.

Calculating the Confidence Interval with Excel

You can use Excel's functions to calculate a confidence interval if you know the standard deviation of the underlying probability distribution. For example, suppose you are conducting a survey on the cost of a medical procedure as part of research on health care reform. The cost of the procedure follows the normal distribution, where $\sigma = 1{,}000$. After sampling 50 different hospitals at random, you calculate the average cost to be $5,500. What is the 90% confidence interval for the value of μ—the mean cost of all hospitals? (That is, how far above and below $5,500 must you go to say, "I'm 90% confident that the mean cost of this procedure lies in this range"?)

To calculate the 90% confidence interval:

1. Start Excel and open a blank workbook.
2. Type **Average** in cell A1, **Std. Error** in cell B1, **Alpha** in cell C1, **Lower** in cell D1, and **Upper** in E1.
3. Click cell **A2** and type **5500** (the observed sample average).
4. Type **=1000/sqrt(50)** in cell B2. This is the standard error of the sample average.
5. Type **10%** in cell C2. This is the alpha value for your confidence interval.
6. Type **=A2−B2*NORMSINV(1-C2/2)** in cell D2. Note that we use the NORMSINV(0.95) function to return the z value from the standard normal distribution.
7. Type **=A2+B2*NORMSINV(1-C2/2)** in cell E2. Figure 6-3 shows the resulting 90% confidence interval.

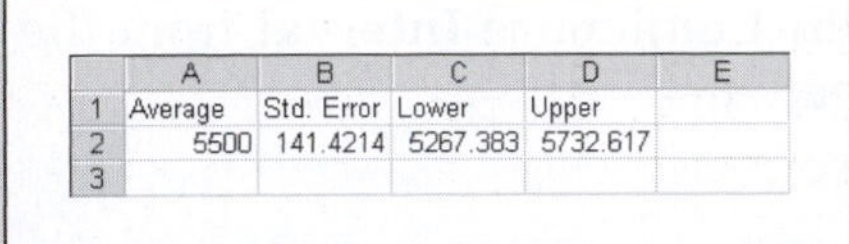

	A	B	C	D	E
1	Average	Std. Error	Lower	Upper	
2	5500	141.4214	5267.383	5732.617	
3					

Figure 6-3
Using Excel to calculate a confidence interval

8 Close your workbook. You do not have to save the changes.

Excel returns a 90% confidence band ranging from $5,267.38 to $5,732.62. If you were trying to estimate the mean cost of this procedure for all hospitals, you could state that you were 90% confident that the cost was not less than $5,267.38 or more than $5,732.62.

Interpreting the Confidence Interval

It's important that you understand what is meant by statistical confidence. When you calculated the confidence interval for the cost of the hospital procedure, you were *not* stating that the probability of the mean cost falling between $5,267.38 and $5,732.62 was .90. That would incorrectly imply that the range you calculated and the mean cost are random variables. They're not. After drawing a specific sample and from that sample calculating the confidence interval, we're no longer working with random variables but with actual observations. The mean cost is also some fixed (but unknown) number and is not random. The term *confidence* refers to our confidence in the procedure we used to calculate the range. The term *90% confident* means that we are confident that our procedure will capture the value of μ 90% of the times it is used.

CONCEPT TUTORIALS
The Confidence Interval

To get a visual picture of the confidence interval in action, you can use the Explore workbook named Confidence Intervals to read about, and work with, a confidence interval.

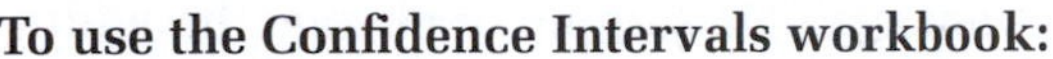

To use the Confidence Intervals workbook:

1 Open the **Confidence Intervals** workbook located in the Explore folder. Enable any macros in the workbook.

2 Move through the sheets in this workbook, viewing the material on z values and confidence intervals.

3 Click **Explore the Confidence Interval** from the Table of Contents column. See Figure 6-4.

Figure 6-4 The Confidence Intervals workbook

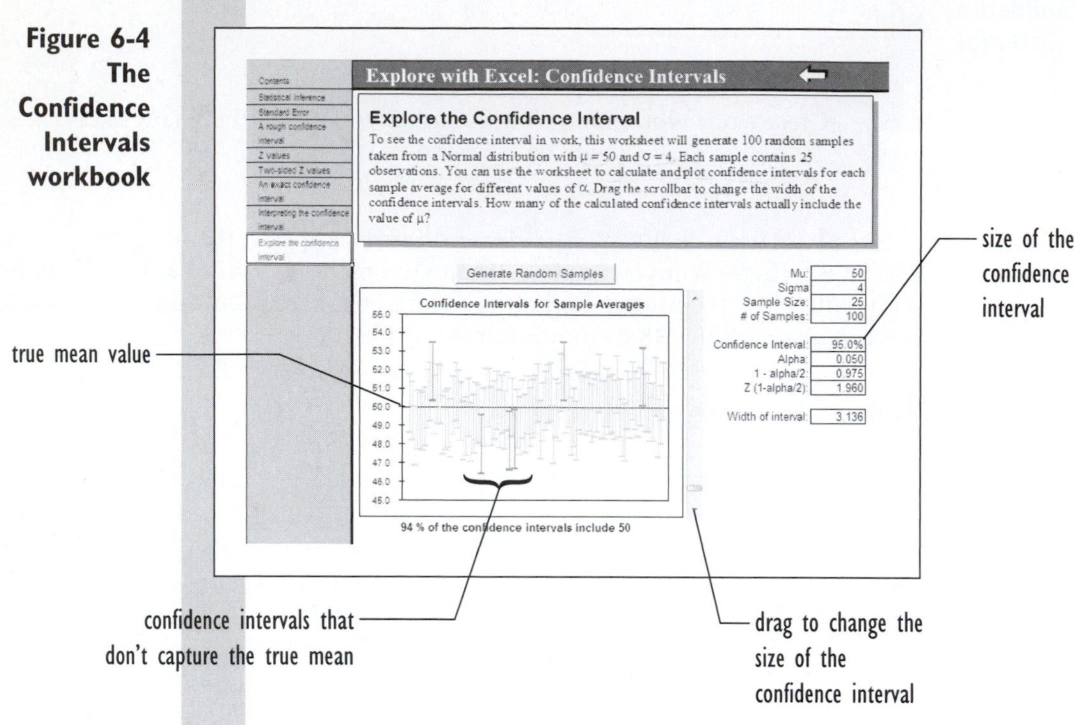

The worksheet in Figure 6-4 shows 100 simulated confidence intervals taken from a normal distribution with $\mu = 50$ and $\sigma = 4$. Each of the 100 samples contains 25 observations, so the standard error of each sample average is 0.8.

If a confidence interval captures the true value of μ, it shows up on the chart as a vertical green line. If a confidence interval does not include μ, it shows up as a vertical red line. You would expect that 95 of the 100 samples would include the value of μ in their confidence intervals. Because there is some random variation, Figure 6-4 shows that only 94% of the sample confidence intervals include the value of μ. Using this worksheet, you can generate a new random sample or change the width of the confidence band. Try this now by reducing the confidence interval from 95 to 75%.

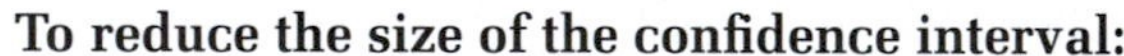

To reduce the size of the confidence interval:

1 Drag the vertical scroll bar up until the value **75.0%** appears in the highlighted Confidence Interval box. See Figure 6-5.

Figure 6-5
75% confidence intervals

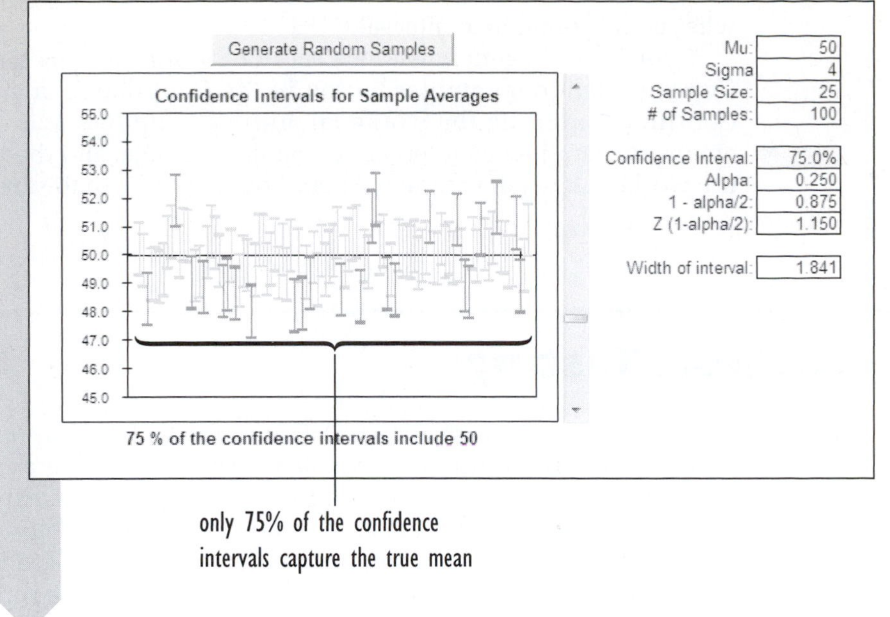

By reducing the confidence interval to 75%, you've reduced the width of the confidence band, but at the cost of many more red lines appearing on your chart. If you were relying upon this confidence interval to capture the value of μ, you would run a great risk of making an error. Let's go the other way and increase the confidence interval.

To increase the size of the confidence interval:

1. Drag the vertical scroll bar down until the value **99.0%** appears in the Confidence Interval box.

All of the confidence intervals now capture the value of μ, but the size of the confidence bands has greatly increased. As you can see, there is a trade-off in using the confidence interval. Selecting too small a value could result in missing the value of μ. Selecting a larger value will almost certainly capture μ, but at the expense of having a range of values too broad to be useful. Statisticians have generally favored the 95% confidence interval as a compromise between these two positions.

An important lesson to learn from this simulation is to not take the sample average at face value. Confidence intervals help you quantify how precisely the sample average estimates the value of μ. The next time you hear in the news that a study has shown that a drug causes a mean decrease in blood pressure or that polls predict a certain election result, you should ask, "And what is the confidence interval?"

If you want to generate a new set of random samples, you can click the Generate Random Samples button in the Confidence Intervals workbook. Continue exploring the workbook until you understand the relationship among the confidence interval, the sample average, and the value of μ. Close the workbook when you're finished. You do not have to save your changes.

Hypothesis Testing

Confidence intervals are one way of performing statistical inference; another way is hypothesis testing. In a **hypothesis test**, you formulate a theory about the phenomenon you're studying and examine whether that explanation is supported by the statistical evidence. In statistics, we formulate a hypothesis first, then collect data, and then perform a statistical test. The order is important. If we formulate our hypothesis *after* collecting the data, we run the risk of having a biased test, because our hypothesis might be designed to fit the data. To guard against a biased test, the hypothesis should be tested on a new set of data. Figure 6-6 displays a classical approach to developing and testing a theory.

Figure 6-6 The steps in developing and testing a hypothesis

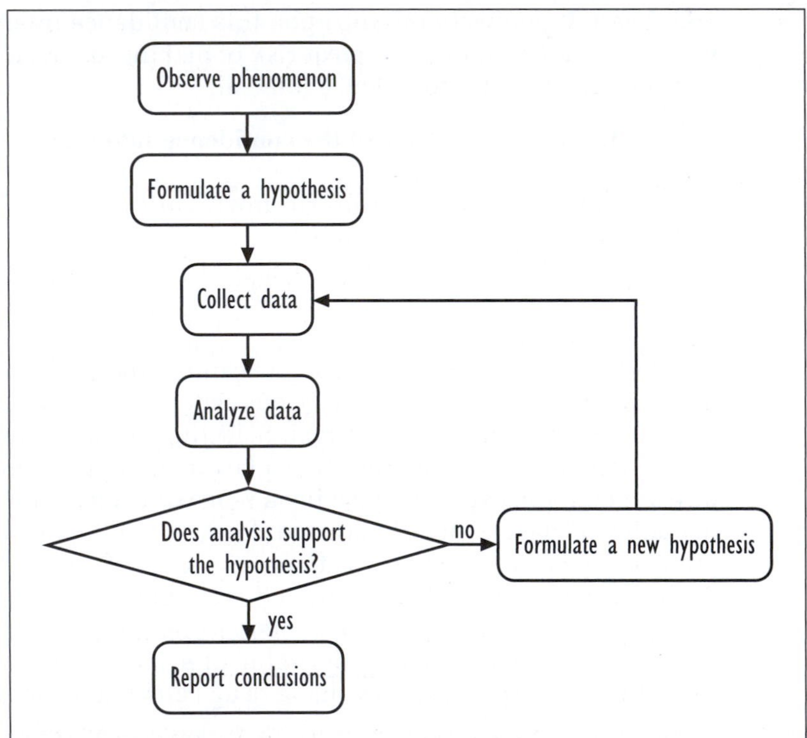

There are four elements in a hypothesis test:

1. A null hypothesis, H_0
2. An alternative hypothesis, H_a
3. A test statistic
4. A rejection region

The **null hypothesis**, usually labeled $\mathbf{H_0}$, represents the default or status quo theory about the phenomenon that you're studying. You accept the null hypothesis as true unless you have convincing evidence to the contrary. The **alternative hypothesis**, or $\mathbf{H_a}$, represents an alternative theory that is automatically accepted as true if the null hypothesis is rejected. Often the alternative hypothesis is the hypothesis you want to accept. For example, a new medication is being studied that claims to reduce blood pressure. The null hypothesis is that the medication *does not* affect the patient's blood pressure. The alternative hypothesis is that the medication *does* affect the patient's blood pressure (in either a positive or a negative direction).

The **test statistic** is a statistic calculated from the data that you use to decide whether to reject or to accept the null hypothesis. The **rejection region** specifies the set of values of the test statistic under which you'll reject the null hypothesis (and accept the alternative).

Types of Error

We can never be sure that our conclusions are free from error, but we can try to reduce the probability of error. In hypothesis testing, we can make two types of errors:

1. **Type I error**: Rejecting the null hypothesis when the null hypothesis is true
2. **Type II error**: Failing to reject the null hypothesis when the alternative hypothesis is true

The probability of Type I error is denoted by the Greek letter α, and the probability of Type II error is identified by the Greek letter β.

Generally, statisticians are more concerned with the probability of Type I error, because rejecting the null hypothesis often results in some fundamental change in the status quo. In the blood pressure medication example, incorrectly accepting the alternative hypothesis could result in prescribing an ineffective drug to thousands of people. Statisticians will set a limit, called the **significance level**, that is the highest probability of Type I error allowed. An accepted value for the significance level is 0.05. This means we set up a region that has probability .05 if the null hypothesis is true, and we reject H_0 if the data fall in this region.

Reducing Type II error becomes important in the design of experiments, where the statistician wants to ensure that the study will detect an effect if a true difference exists. An analysis of the probability of Type II error can aid the statistician in determining how many subjects to have in the study.

An Example of Hypothesis Testing

Let's put these abstract ideas into a concrete example. You work at a plant that manufactures resistors. Previous studies have shown that the number of defective resistors in a batch follows a normal distribution with a mean of 50 and a standard deviation of 15. A new process has been proposed that will reduce the number of defective resistors, saving the plant money. You put the process in place and create a sample of 25 batches. The average number of defects in a batch is 45. Does this prove that the new process reduces the number of defective resistors, or is it possible that the process makes no difference at all, and the 45 is simply a random aberration?

Here are our hypotheses.

H_0: There is no change in the mean number of defective resistors under the new process.

H_a: The mean number of defective resistors *has* changed.

Or, equivalently,

H_0: The mean number of defective resistors in the new process is 50.

H_a: The mean number of defective resistors is *not* 50.

Acceptance and Rejection Regions

To decide between these two hypotheses, we assume that the null hypothesis is true. Let μ_0 be the mean under the null hypothesis. This means that under the null hypothesis,

$$P\left(-z_{1-\alpha/2} < \frac{\overline{x} - \mu_0}{\sigma/\sqrt{n}} < z_{1-\alpha/2}\right) = 1 - \alpha$$

Multiplying by the standard error and adding μ_0 to each term in the inequality, we get

$$P\left(\mu_0 - z_{1-\alpha/2}\frac{\sigma}{\sqrt{n}} < \overline{x} < \mu_0 + z_{1-\alpha/2}\frac{\sigma}{\sqrt{n}}\right) = 1 - \alpha$$

This means that the sample average should be in the range $\mu_0 \pm z_{1-\alpha/2}\,\sigma/\sqrt{n}$ with probability $1 - \alpha$, *if the null hypothesis is true.* Now let α be our significance level, so that if the sample average lies outside this range, we'll reject the null hypothesis and accept the alternative. These outside values would constitute the rejection region, mentioned earlier. The values within the range constitute the **acceptance region**, under which we'll accept the null hypothesis. The upper and lower boundaries of the acceptance region are known as **critical values**, because they are critical in deciding whether to accept or to reject the null hypothesis.

Let's apply this formula to our example: $\mu_0 = 50$, $\sigma = 15$, $n = 25$, and we set $\alpha = 0.05$ so that the probability of Type I error is 5%. The acceptance region is therefore

$$\begin{aligned} &= 50 \pm 1.96 \times 15/\sqrt{25} \\ &= 50 \pm 5.88 \\ &= (44.12, 55.88) \end{aligned}$$

Any value that is less than 44.12 or greater than 55.88 will cause us to reject the null hypothesis. Because 45 falls in the acceptance region, we accept the null hypothesis and do not conclude that the new process decreases the number of defective resistors in a batch.

p Values

The *p value* is the probability of a value as extreme as the observed value. We can calculate that by examining the *z* test statistic. For the manufacturing example, the *z* test statistic is

$$\begin{aligned} z &= \frac{\bar{x} - \mu_0}{\sigma/\sqrt{n}} \\ &= \frac{45 - 50}{15/\sqrt{25}} \\ &= -1.67 \end{aligned}$$

The probability of a standard normal value of less than -1.67 is 0.0478. To calculate the *p* value, we need to take into account the terms of the alternate hypothesis. In this case, the alternative hypothesis was that the new process made no difference (either positive or negative) in the number of defects. Thus, we need to calculate the probability of an extreme value 1.67 units from 0 in *either* direction. Because the standard normal distribution is symmetric, the probability of a value being < -1.67 is equal to the probability of a value being > 1.67, so we can simply double the probability, resulting in a *p* value of $2 \times 0.0478 = 0.0956$.

This was an example of a **two-tailed test**, in which we assume that extreme values can occur in either direction. We can also construct a **one-tailed test**, in which we consider differences in only one direction. A one-tailed test could have these hypotheses.

H_0: The mean number of defective resistors in the new process is 50.

H_a: The mean number of defective resistors is < 50.

We use a one-tailed test if something in the new process would *absolutely* rule out the possibility of an increase in the number of defective resistors.

If that were the case, we would not need to double the probability, and the p value would be 0.0478, so the sample average lies outside the acceptance region if $\alpha = 0.05$. We would call this result statistically significant and would reject the null hypothesis, accepting the hypothesis that the new process reduces the number of defective resistors.

It sounds like we've got something for nothing, but we haven't. We've attained significant results at the cost of assuming something that we hadn't assumed before. Because it's easier to achieve "significant" results in one-tailed tests, they should be used with extreme caution and only when warranted by the situation. You should always state your alternative hypothesis *before* doing your analysis (rather than deciding on a one-tailed test after seeing the results with the two-tailed test).

EXCEL TIPS

- To calculate the p value with Excel, first calculate the value of the z test statistic using the Excel function z=(AVERAGE(*data range*)$-\mu_0$)/(σ/SQRT(n)), where *data range* is the range of cells in your worksheet containing the sample values, μ_0 is the mean under the null hypothesis, σ is the standard deviation of the probability distribution, and n is the sample size.
- For a one-tailed test where z is negative, the p value = NORMSDIST(z).
- For a one-tailed test where z is positive, the p value = 1 − NORMSDIST(z).
- For a two-tailed test where z is negative, the p value = 2 × NORMSDIST(z).
- For a two-tailed test where z is positive, the p value = 2 × (1 − NORMSDIST(z)).

CONCEPT TUTORIALS
Hypothesis Testing

You can get a visual picture of the principles of hypothesis testing by opening the Hypothesis Testing workbook.

To use the Hypothesis Testing workbook:

1 Open the **Hypothesis Testing** workbook, located in the Explore folder. Enable any macros in the workbook.

2 Move through the workbook, reviewing the material on hypothesis testing.

3 Click **Explore Hypothesis Testing** from the Table of Contents column. See Figure 6-7.

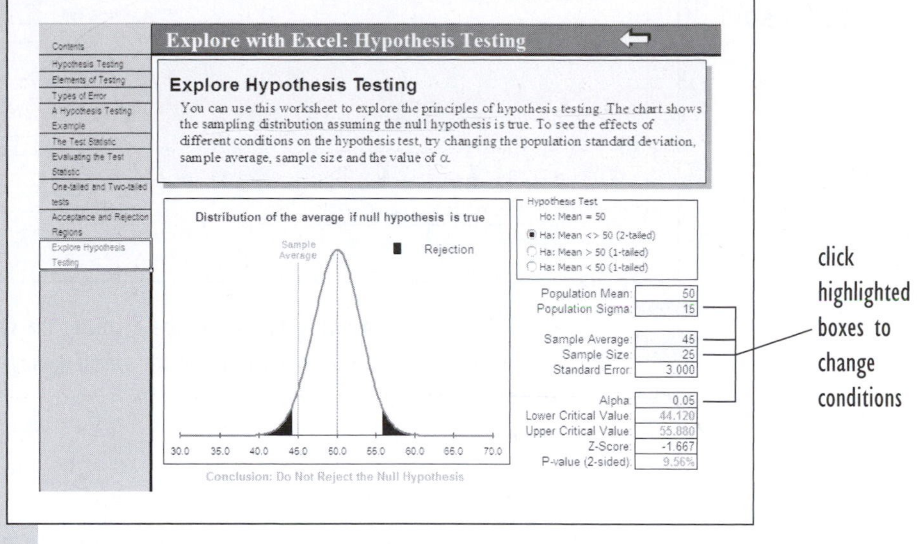

Figure 6-7 The Hypothesis Testing workbook

The workbook shows the sampling distribution under the null hypothesis, where it is assumed that $\mu = 50$. The rejection region is displayed on the chart in black. This workbook allows you to vary four values—σ, $\overline{x}$, the sample size, and α—to see their effect on the hypothesis test. You can also choose whether to perform a one-tailed or a two-tailed test. The results of your choices are automatically displayed on the workbook and in the chart. By working with different values of these factors, you can get a clearer picture of how hypothesis testing works.

For example, what impact would doubling the sample size have on the hypothesis test, assuming all other factors remained the same? Let's find out.

To increase the sample size:

1 Click the **Sample Size** box, and change the sample size from 25 to **50**. See Figure 6-8.

Figure 6-8 Changing the sample size from 25 to 50

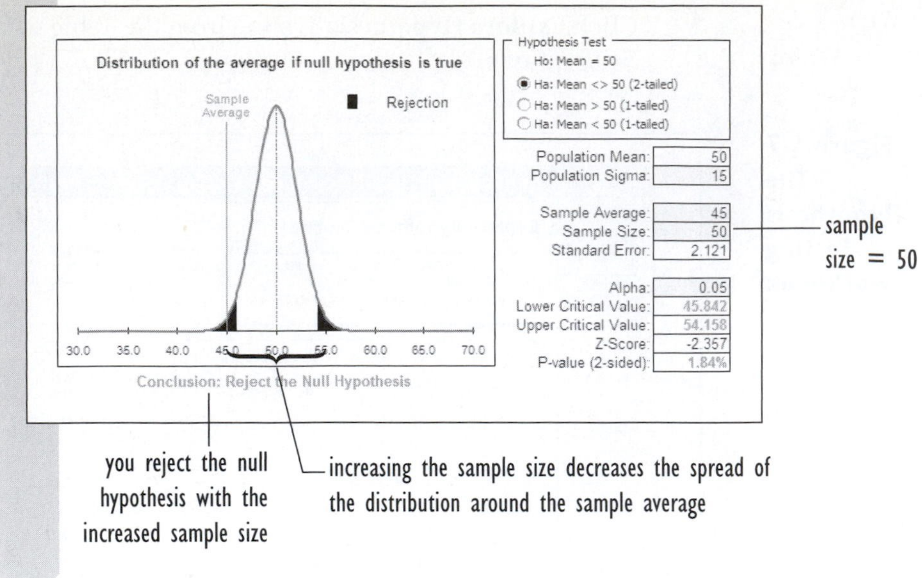

The width of the acceptance region shrinks from 11.76 (44.12 to 55.88) with a sample size of 25 to 8.32 (45.84 to 54.16) with a sample size of 50. The observed sample average lies within the rejection region, so you reject the null hypothesis with a p value of 1.84%.

Now let's see what happens when you increase the value of σ from 15 to 20.

To increase the value of σ:

1 Click the **Population Sigma** box, and change the value from 15 to **20**.

Because the value of σ has increased, the value of the standard error has increased too, from 2.121 to 2.828. The lower critical value has fallen to 44.456 and the p value has increased to 7.71%, so you do not reject the null hypothesis. Variability is one of the most important factors in hypothesis testing; much of statistical analysis is concerned with reducing or explaining variability.

Finally, let's find out what our conclusions would be if we used a one-tailed test, where H_a is the hypothesis that the mean < 50.

To change to a one-tailed test:

2 Click the **Ha: Mean < 50 (1-tailed)** option button.

The chart changes to a one-tailed test. See Figure 6-9.

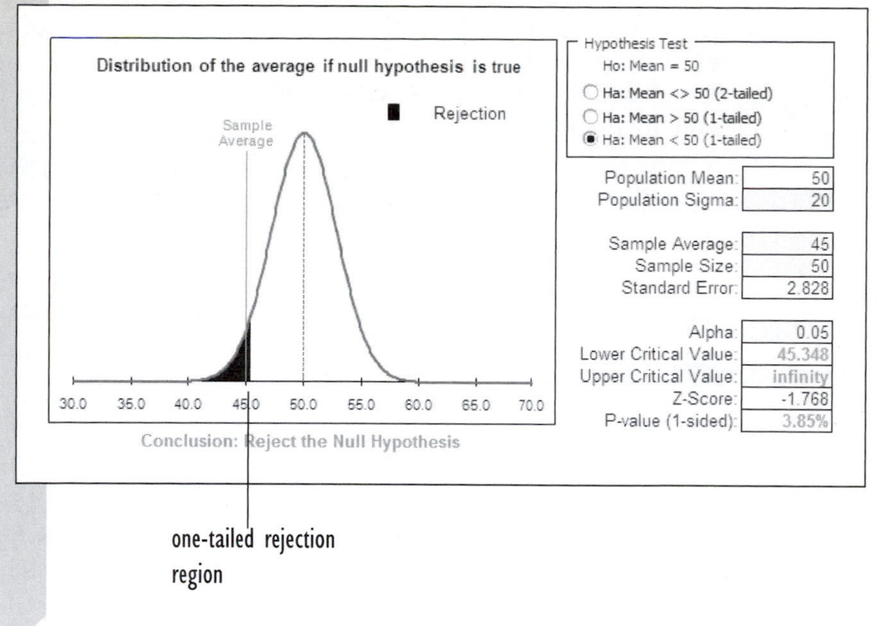

Figure 6-9 Switching from a two-tailed to a one-tailed test

Under a one-tailed test, we reject the null hypothesis. Of course, we have to be very careful with this approach, because we are changing our hypothesis *after* seeing the data. If this were an actual situation, changing the hypothesis like this would be inappropriate. The better course would be to draw another random sample of 25 batches and test the new hypothesis on that set of data (and only if we have compelling reasons for doing a one-tailed test).

Try other combinations of hypothesis and parameter values to see how they affect the hypothesis test. Close the workbook when you're finished. You do not have to save your changes.

Additional Thoughts about Hypothesis Testing

One important point you should keep in mind when hypothesis testing is that accepting the null hypothesis does *not* mean that the null hypothesis is true. Rather, you are stating that there is insufficient reason to reject it. The distinction is subtle but important. To state that accepting the null hypothesis means that $\mu = 50$ excludes the possibility that μ actually equals 49 or 49.9 or 49.99. But you didn't test any of these possibilities. What you *did* test was whether the data are incompatible with the assumption that $\mu = 50$. You found that in some cases, they are not compatible.

You've looked at two approaches to statistical inference: the confidence interval and the hypothesis test. For a particular value of α, the width of the confidence interval around the sample average is equal to the width of the two-sided acceptance region around μ_0. This means that the following two statements imply each other:

1. The value μ_0, lies outside the $(1 - \alpha)$ confidence interval around $\bar{x}$.
2. Reject the null hypothesis that $\mu = \mu_0$ at the α significance level.

The *t* Distribution

Up to now, you've been assuming that the value of σ is known. What if you didn't know the value of σ? One solution is to substitute the standard deviation of the sample values, s for σ in the hypothesis-testing equations. However, there are problems with this approach. If s underestimates σ, then you'll overestimate the significance of the results, perhaps causing you to reject the null hypothesis falsely. Or if s overestimates the value of σ, you could accept the null hypothesis when the null hypothesis isn't true.

Early in the twentieth century, William Gosset, working at the Guinness brewery in Ireland, became worried about the uncertainty caused by substituting s for σ. He believed that the resulting error could be especially bad for small sample sizes. What Gosset discovered was that when you substitute s for σ, the ratio

$$\frac{\bar{x} - \mu}{s / \sqrt{n}}$$

does *not* follow the standard normal distribution; rather, it follows a distribution called the ***t* distribution**.

The t distribution is a probability distribution centered around zero and characterized by a single parameter called the **degrees of freedom**, which is equal to the sample size, n, minus 1. For example, if the sample size is 20, the degrees of freedom equal 19. The t distribution is similar to a standard normal distribution except that it has heavier tails. As the sample size increases, the t distribution approaches the standard normal, but for smaller sample sizes there can be big differences.

CONCEPT TUTORIALS
The *t* Distribution

To see how the *t* distribution differs from the standard normal, open the Distributions workbook.

To explore the *t* distribution:

1. Open the **Distributions** workbook, located in the Explore folder.
2. Click ***t*** from the Table of Contents column. Review the material and scroll to the bottom of the worksheet.
3. Click the Spin Arrow buttons located next to the Degrees of freedom box to change the degrees of freedom for the *t* distribution. See Figure 6-10.

Figure 6-10 The *t* distribution

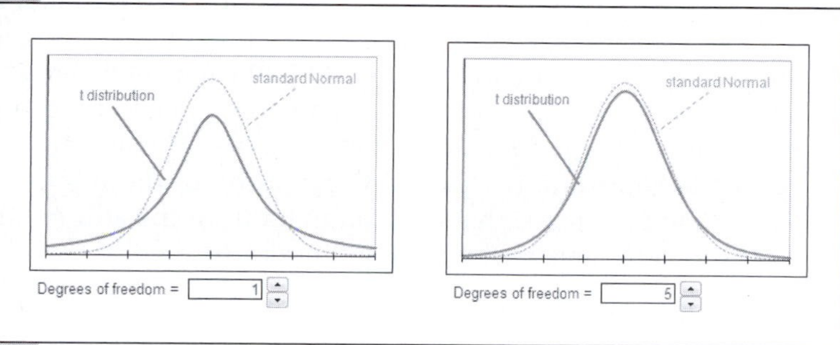

as the degrees of freedom increase, the *t* distribution more closely approximates the normal distribution

The workbook opens, displaying the *t* distribution with 1 degree of freedom. Notice that the *t* distribution has heavier tails than the superimposed standard normal distribution. As you increase the degrees of freedom, the *t* distribution more closely approximates the standard normal.

Continue exploring the *t* distribution by changing the degrees of freedom and observing the changing curve. Close the workbook when you're finished. You do not have to save your changes.

Working with the t Statistic

Excel provides several functions to work with the *t* distribution. Two of these are displayed in Table 6-2.

Table 6-2 Two *t* distribution functions in Excel

Function	Description
TDIST(*t*, *df*, *tails*)	Returns the *p* value for the *t* distribution for a given value of *t*, degrees of freedom *df*, and *tails* = 1 (one tailed) or *tails* = 2 (two tailed)
TINV(*p*, *df*)	Returns the two-tailed *t* value from the *t* distribution with degrees of freedom *df* for a *p* value = *p*. For a one-tailed *t* value, replace *p* with 2 × *p*.

Let's use Excel to apply the *t* distribution to a problem involving textbook prices. The college administration claims that students should not expect to spend more than an average of $500 each semester for books. A student associated with the school newspaper decides to investigate this claim and interviews 25 randomly selected students. The average spent by the 25 students is $520, and the standard deviation of these purchases is $50. Is this significant evidence that the statement from the administration is wrong?

First, let's construct our hypotheses.

H_0: The average cost (μ_0) of textbooks is $500.

H_a: The average cost of textbooks is not equal to $500.

Now we construct the ***t* statistic** t_{n-1}.

$$t_{n-1} = \frac{\overline{x} - \mu_0}{s/\sqrt{n}} = \frac{520 - 500}{50/\sqrt{25}} = \frac{20}{10} = 2$$

To test the null hypothesis with Excel:

1. Open a new blank workbook.
2. In cell A1, type **=TDIST(2,24,2)** and press **Enter.**

   In this example, 2 is the value of the *t* statistic, 24 is the degrees of freedom, and we enter 2 because this is a two-tailed test.

The TDIST function returns a *p* value of 0.05694, so we do not reject the null hypothesis at the 5% level. Thus we conclude that there is not sufficient evidence that the college administration is underestimating the price of textbooks. If we had used the z test statistic rather than the t statistic in this example, the p value would have been 0.0455, and we would have erroneously rejected the null hypothesis.

Constructing a *t* Confidence Interval

Still we have a sample average that doesn't completely match the administration's claim. Let's construct a 95% confidence interval for the mean value to see in what range of values the true mean might lie. Because we don't know the value of σ, we can't use the confidence interval equation discussed earlier; we'll have to use one based on the t distribution. The expression for the **t confidence interval** is

$$\left(\bar{x} - t_{1-\alpha/2,\, n-1}\frac{s}{\sqrt{n}},\ \bar{x} + t_{1-\alpha/2,\, n-1}\frac{s}{\sqrt{n}}\right)$$

Here, $t_{1-\alpha/2,\, n-1}$ is the point on the t distribution with $n - 1$ degrees of freedom, such that the probability of a t random variable being less than it is $1 - \alpha/2$. To calculate this value in Excel, you use the TINV function. However, in the TINV function, you enter the value of α, not $1 - \alpha/2$. For example, to calculate the value of $t_{1-\alpha/2,\, n-1}$, enter the function TINV(α, $n-1$). Use this information to construct a 95% confidence interval.

To construct a 95% confidence interval for the price of textbooks:

1. In cell A2, type **=220−TINV(0.05,24)*50/SQRT(25)** and press **Tab**.
2. In cell B2, type **=220+TINV(0.05,24)*50/SQRT(25)** and press **Enter**.

The 95% confidence interval is (199.36, 240.64). We do not expect the mean price of textbooks to be much less than $200, nor should it be much greater than $240. By comparison, the confidence interval based on the standard normal distribution is (200.40, 239.60), so the confidence intervals are very close in size. Notice that the t confidence interval includes 200, which means that 200 is not ruled out by the data, in agreement with our hypothesis test. This agreement is not a coincidence, as can be shown with a little algebra.

The Robustness of *t*

When you use the t distribution to analyze your data, you're assuming that the data follow a normal distribution. What are the consequences if this turns out not to be the case? The t distribution has a property called **robustness**, which means that even if the assumption of normality is moderately violated, the p values returned by the t statistic will still be fairly accurate. As long as the distribution of the data does not violate the assumption of normality in an extreme way, you can use the t distribution with confidence.

Applying the *t* Test to Paired Data

The t distribution becomes useful in analyzing **paired data**, where observations come in natural pairs and you wish to explore the difference between two pairs. For example, a doctor might measure the effect of a drug by measuring the physiological state of patients before and after administering the drug. Each patient in this study has two observations, and the observations are paired with each other. To determine the drug's effectiveness, the doctor looks at the difference between the before and after readings.

The Labor workbook contains data on the percentage of women in the workforce in 1968 and 1972 taken from a sample of 19 cities. The workbook contains the following variables:

Table 6-3 Data on percentage of women in the workforce

Range Name	Range	Description
City	A2:A20	The name of the city
Year_68	B2:B20	The percent of women in the workforce in 1968
Year_72	C2:C20	The percent of women in the workforce in 1972
Diff	D2:D20	The change in percentage from 1968 to 1972 for each city

To open the Labor workbook:

1. Open the **Labor** workbook from the Chapter06 folder.
2. Save the file as **Labor Analysis** to the same folder. The workbook appears as shown in Figure 6-11.

Figure 6-11 The Labor workbook

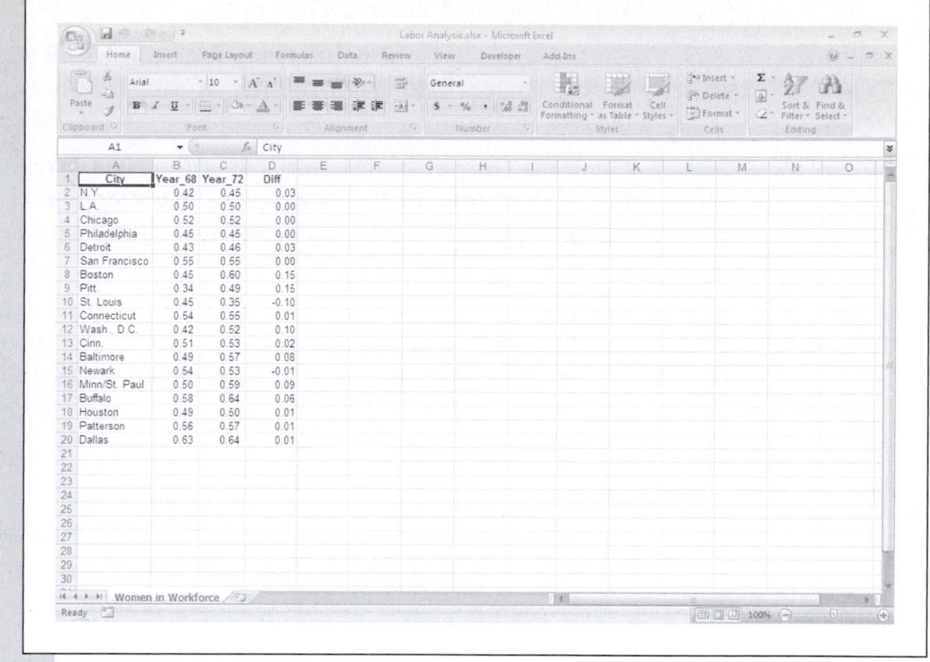

City	Year_68	Year_72	Diff
N.Y.	0.42	0.45	0.03
L.A.	0.50	0.50	0.00
Chicago	0.52	0.52	0.00
Philadelphia	0.45	0.45	0.00
Detroit	0.43	0.46	0.03
San Francisco	0.55	0.55	0.00
Boston	0.45	0.60	0.15
Pitt.	0.34	0.49	0.15
St. Louis	0.45	0.35	-0.10
Connecticut	0.54	0.55	0.01
Wash., D.C.	0.42	0.52	0.10
Cinn.	0.51	0.53	0.02
Baltimore	0.49	0.57	0.08
Newark	0.54	0.53	-0.01
Minn/St. Paul	0.50	0.59	0.09
Buffalo	0.58	0.64	0.06
Houston	0.49	0.50	0.01
Patterson	0.56	0.57	0.01
Dallas	0.63	0.64	0.01

There are two observations from each city, and the observations constitute paired data. You've been asked to determine whether this sample of 19 cities demonstrates a statistically significant increase in the percentage of women in the workforce. Let μ be the mean change in the percentage of women in the workforce. You have two hypotheses.

H_0: $\mu = 0$ (There is no change in the percentage from 1968 to 1972.)

H_a: $\mu \neq 0$ (There is some change, but we're not assuming the direction of the change.)

You can use StatPlus to test these hypotheses and create a t confidence interval for the change in percent from 1968 to 1972. Do this now by analyzing the Diff variable to test whether the average difference is significantly different from zero.

To test whether there is a significant difference:

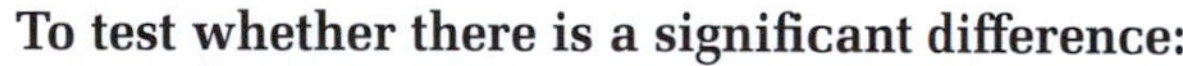

1 Click **One Sample Tests** from the StatPlus menu and then click **1 Sample *t* test.**

We use the one-sample t test because we are essentially looking at one sample of data—the sample of paired differences.

2 Verify that the **1-sample *t* test** option button is selected.

3 Click the **Input** button and then click the **Use Range Names** option button. Select **Diff** from the list of range names and click the **OK** button.

4 Click the **Output** button and select the **New Worksheet** option button. Enter ***t* test** for the new worksheet name and click the **OK** button. Figure 6-12 shows the completed dialog box.

Note that we could have also selected the Paired *t* test (two columns) and then selected the Year_68 and Year_72 columns. Note also that the dialog box will test a null hypothesis of $\mu = 0$ versus the alternative hypothesis of not equal to 0. You can change these values if you wish to test other hypotheses.

Figure 6-12 The One-Sample or Paired *t* Test dialog box

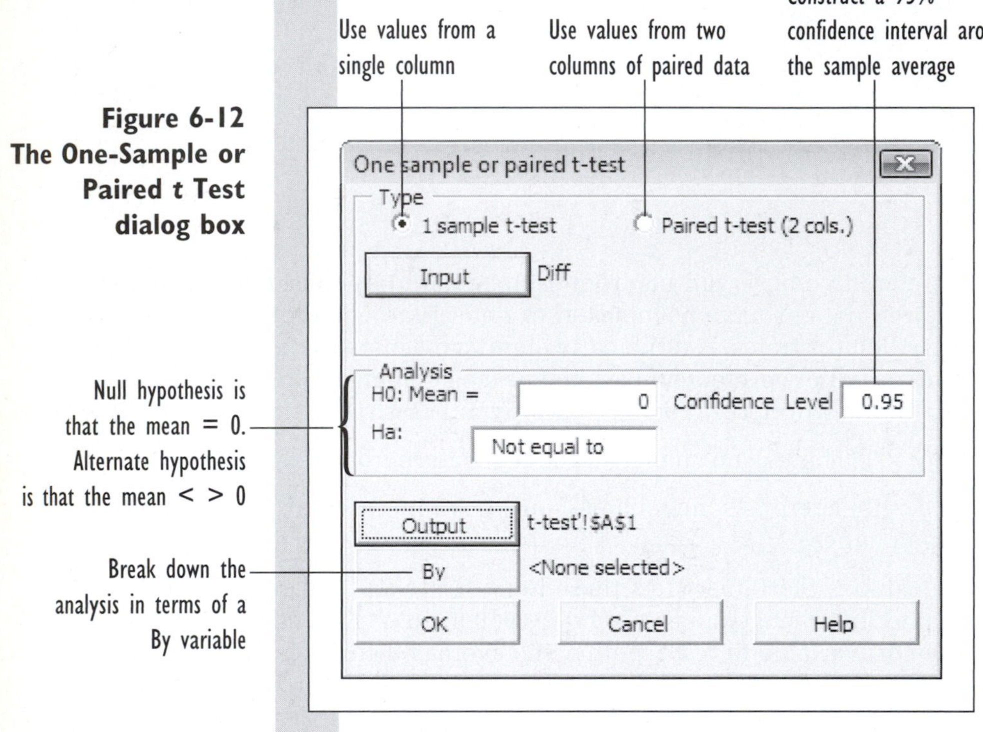

5 Click the **OK** button to generate the output from the *t* test. See Figure 6-13.

Figure 6-13 t Test analysis of the labor data

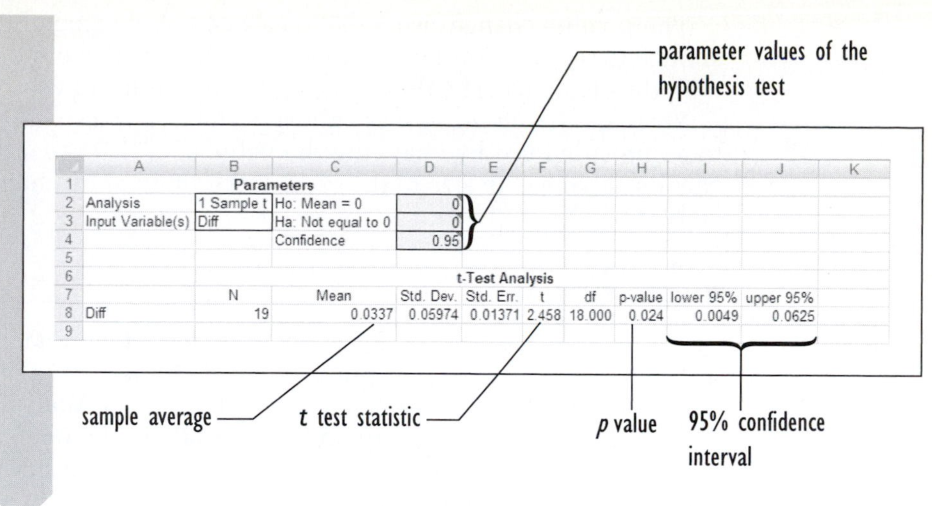

	A	B	C	D	E	F	G	H	I	J	K
1		Parameters									
2	Analysis	1 Sample t	Ho: Mean = 0	0							
3	Input Variable(s)	Diff	Ha: Not equal to 0	0							
4			Confidence	0.95							
5											
6				t-Test Analysis							
7		N	Mean	Std. Dev.	Std. Err.	t	df	p-value	lower 95%	upper 95%	
8	Diff	19	0.0337	0.05974	0.01371	2.458	18.000	0.024	0.0049	0.0625	
9											

On the basis of our analysis, there is an average increase in the percentage of women in the labor force of 3.37 percentage points between 1968 and 1972. This is statistically significant with a p value of 0.024, so we reject the null hypothesis and accept the alternative. There has been a significant change in women's participation in the workforce in those four years. The 95% confidence interval for this estimate ranges from 0.49 percentage point up to 6.25 percentage points. Notice the interval does not include 0 and the hypothesis test rejects 0, so the interval and the test agree. It is not hard to show that they will always agree.

An economist, viewing other data on this topic, claims that the percentage of women in the workforce had actually increased by 5 points from 1968 to 1972. Do your data conflict with this statement? Let's test the hypothesis.

$$H_0: \mu = 0.05$$

$$H_a: \mu \neq 0.05$$

Rather than rerunning the StatPlus command, we can simply enter the new hypothesis directly into the t test worksheet, if we have StatPlus set to create dynamic output. Otherwise, we have to rerun the command.

To test the new hypothesis:

1. Click cell **D2**, change the value from 0 to **0.05**, and press **Enter**.

The p value changes from 0.024 to 0.249. A p value of 0.249 is not small enough to reject the null hypothesis. We conclude that our data do not conflict with the claim made by this economist in a significant way.

You can also change other values in the hypothesis test. You can switch to a one-sided test by changing the value of cell D3 to either -1 or 1. You can also change the size of the confidence interval by changing the value in cell D4.

EXCEL TIPS

- You can also perform a paired t test of your data using the Analysis ToolPak, supplied by Excel. To perform a paired t test, load the Analysis ToolPak and click the Data Analysis button from the Analysis group on the Data tab. In the Data Analysis dialog box, click t Test: Paired Two Sample for Means and specify the two columns containing the paired data. This command does not calculate the confidence interval for you, so you have to calculate that using the formulas supplied in this chapter.

You should not accept your analysis at face value without further investigating the assumptions of the t test. One of these assumptions is that the data follow a normal distribution. The t test is robust, but that doesn't mean you shouldn't explore the possibility that the data are seriously nonnormal. Do this now by creating a histogram and normal probability plot of the data in the Diff column.

To create a histogram of the difference data:

1. Click **Single Variable Charts** from the StatPlus menu and then click **Histograms**.
2. Click the **Data Values** button and select **Diff** from the list of range names in the workbook.
3. Click the **Normal Curve** checkbox to add a normal curve to the histogram.
4. Click the **Output** button and save your histogram to the **Histogram** chart sheet.
5. Click the **OK** button twice to create the histogram. See Figure 6-14.

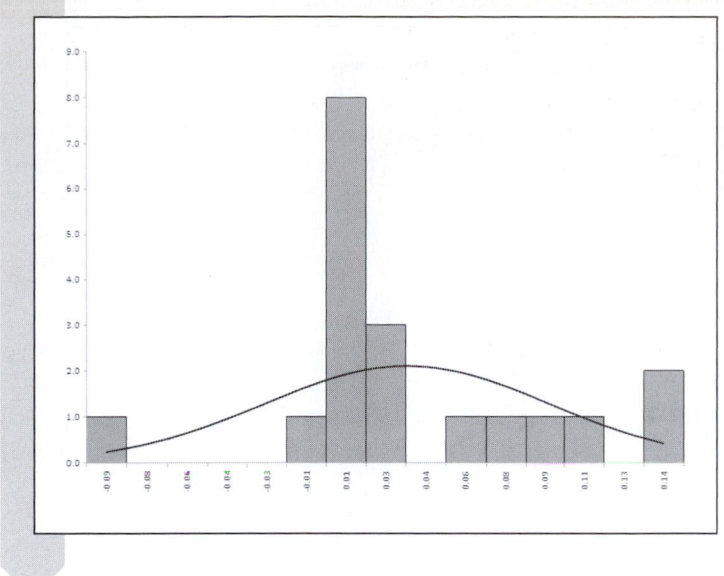

Figure 6-14
Histogram of the difference data

The observed difference values don't appear to follow the normal curve particularly well. Let's see whether the normal probability plot provides more information.

To create a normal probability plot of the difference data:

1. Click **Single Variable Charts** from the StatPlus menu and then click **Normal P-plots**.
2. Click the **Data Values** button and select **Diff** from the list of range names in the workbook.
3. Click the **Output** button and save your chart to the **P-Plot** chart sheet.
4. Click the **OK** button twice to create the normal probability plot. See Figure 6-15.

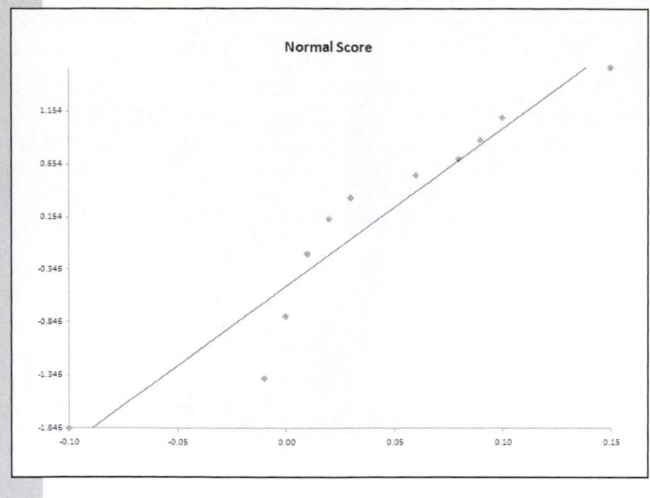

Figure 6-15 Normal probability plot of the Diff data

From the two plots, there is enough graphical evidence to make us worry that the data do not follow the normal distribution. This is a problem because now we can't feel completely comfortable about the *p* values the *t* test gave us. Because the assumption of normality may have been violated, we can't be sure that the *p* value is accurate.

Applying a Nonparametric Test to Paired Data

A **parametric test** assumes a specific distribution such as the normal distribution, and the *t* test is an example. A **nonparametric** test does not assume a particular distribution for the data. Most nonparametric tests are based on ranks and not the actual data values (this frees them from assuming a particular distribution). The study of nonparametric statistics can fill an entire textbook. We'll just cover the high points and show how to apply a nonparametric test to your data.

The Wilcoxon Signed Rank Test

The nonparametric counterpart to the *t* test is the Wilcoxon Signed Rank test. In the **Wilcoxon Signed Rank test**, we rank the absolute values of the original data from smallest to largest, and then each rank is multiplied by the sign of the original value (−1, 0, or 1). In case of a tie, we assign an average rank to the tied values. Table 6-4 shows the values of a variable, along with the values of the signed ranks.

Table 6-4 Signed Ranks

Variable Values	Signed Ranks
18	7.0
4	2.0
15	6.0
−5	−3.5
−2	−1.0
10	5.0
5	3.5

There are seven values in this data set, so the ranks go from 1 (for lowest in absolute value) up to 7 (for the highest in absolute value). The lowest in absolute value is −2, so that observation gets the rank 1 and then is multiplied by the sign of the observation to get the sign rank value −1. The value 4 gets the sign rank value 2 and so forth. Two observations, −5 and 5, are tied with the same absolute value. They should get ranks 3 and 4 in our data set, but because they're tied, they both get an average rank of 3.5 (or −3.5).

Next we calculate the sum of the signed ranks. If most of the values were positive, this would be a large positive number. If most of the values were negative, this would be a large negative number. The sum of the signed ranks in our example equals $7 + 2 + 6 - 3.5 - 1 + 5 + 3.5 = 19$.

The only assumption we make with the Wilcoxon Signed Rank test is that the distribution of the values is symmetric around the median. If under the null hypothesis we assume that the median = 0, this would imply that we should have as many negative ranks as positive ranks and that the sum of the signed ranks should be 0. Using probability theory, we can then determine how probable it is for a collection of 7 observations to have a total signed rank of 19 or more if the null hypothesis is true. Without going into the details of how to make this calculation, the p value in this particular case is 0.133, so we would not reject the null hypothesis. In addition to calculating p values, you can also calculate confidence intervals using the Wilcoxon test statistic.

One advantage in using ranks instead of the actual values is that the hypothesis test is much less sensitive to the effect of outliers. Also, nonparametric procedures can be applied to situations involving ordinal data, such as surveys in which subjects rank their preferences. The downside of nonparametric tests is that they are not as efficient as parametric tests when the data *are* normally distributed. This means that for normal data you need a larger sample size in order to detect statistically significant effects (5% larger when the Wilcoxon Signed Rank test is used in place of the t test). Of course, if the data are not normally distributed, you can often detect statistically significant effects with smaller sample sizes using nonparametric procedures. The nonparametric test can be *more* efficient in those cases.

The bottom line is that if there is some question about whether to use a parametric or a nonparametric test, do the analysis both ways.

Excel does not include any nonparametric tests, but you can use StatPlus to generate test results using the Wilcoxon Signed Rank test. Apply this test now to the work force data. Your hypotheses are

$$H_0\text{: Median population difference} = 0$$
$$H_a\text{: Median population difference} \neq 0$$

To analyze the difference data using the Wilcoxon Signed Rank test:

1. Click **One Sample Tests** from the StatPlus menu and then click **1 Sample Wilcoxon Sign Rank test.**
2. Verify that the **1-sample W-test** option button is selected.
3. Click the **Input** button, select **Diff** from the list of range names, and click the **OK** button.
4. Click the **Output** button and select the **New Worksheet** option button. Enter **W-test** for the new worksheet name and click the **OK** button.
5. Click the **OK** button to generate the output from the Wilcoxon Signed Rank test. See Figure 6-16.

	A	B	C	D	E	F	G	H	I	J
1		Parameters								
2	Analysis	1 Sample Sign Rank	Ho: Median = 0	0						
3	Input Variable(s)	Diff	Ha: Not equal to 0	0						
4			Confidence	0.95						
5										
6			Wilcoxon Sign Rank Analysis							
7		N	N < 0	N = 0	N > 0	Median	p-value	lower 95%	upper 95%	
8	Diff	19	2	4	13	0.010	0.009	0.005	0.060	
9										

Figure 6-16 Wilcoxon signed rank analysis of the Diff data

The results of our analysis using the Wilcoxon Signed Rank test are similar to the results with the *t* test. We still reject the null hypothesis, this time with a stronger *p* value of 0.009. The 95% confidence interval is pretty close to what we calculated before, (0.005, 0.060). Moreover, we learn that the number of cities whose percentage of women in the workforce increased from 1968 to 1972 was 13 and that only 2 cities showed a decrease in percentage points (4 cities were unchanged). This information strengthens our earlier conclusion that there was a significant increase in women in the workforce during those four years.

EXCEL TIPS

- Excel doesn't include a command to perform the signed rank test, but you can approximate the *p* values and confidence intervals of the signed rank test by calculating the sign ranks of your data and then performing a one-sample *t* test on those values.
- To calculate the ranks of your data, use Excel's RANK function. If your data contain ties, you should use the RANKTIED function. *StatPlus required.*
- To calculate the signed ranks of your data, use the SIGNRANK function. *StatPlus required.*

The Sign Test

Another nonparametric test that makes even fewer assumptions than the Wilcoxon Signed Rank test is the Sign test. In the **Sign test**, we ignore the values entirely and simply count the number of positive values and the number of negative values. We then test whether there are more positive (or negative) values than there should be. The test is similar to what we might use to test whether a two-sided coin is fair.

The Sign test is usually less efficient (requiring a larger sample size) than either the *t* test or the signed rank test, except in cases where the data come from a heavy tailed distribution. In those cases, the Sign test may be more effective than either the *t* test or the signed rank test.

Let's apply the Sign test to our data set. Our hypotheses are

H_0: Probability of a negative value = probability of a positive value
H_a: Probability of a negative value ≠ probability of a positive value

To analyze the difference data using the Sign test:

1. Click **One Sample Tests** from the StatPlus menu and then click **1 Sample Sign test**.
2. Verify that the **1-sample s test** option button is selected.
3. Click the **Input** button, select **Diff** from the list of range names, and click the **OK** button.
4. Click the **Output** button and select the **New Worksheet** option button. Enter ***s* test** for the new worksheet name and click the **OK** button.

5 Click the **OK** button to generate the output from the Sign test. See Figure 6-17.

Figure 6-17 Sign test analysis of the Diff data

parameter values of the hypothesis test

	A	B	C	D	E	F	G	H	I	J	K
1		Parameters									
2	Analysis	1 Sample Sign	Ho: Median = 0	0							
3	Input Variable(s)	Diff	Ha: Not equal to 0	0							
4			Confidence	0.95							
5			(Approximately)	0							
6											
7						Sign Test Analysis					
8		N	N < 0	N = 0	N > 0	Median	p-value	lower 95%	upper 95%	Achieved Confidence	
9	Diff	19	2	4	13	0.010	0.007	0.0000	0.0661	0.950	
10											

number of values <, =, and > 0

median difference

p value

95% confidence interval

Even under the Sign test, we reject the null hypothesis and accept the alternative, concluding that the percentage of women in the workforce has significantly increased. The *p* value for the Sign test is 0.007. We can also construct a 95% confidence interval with the Sign test, but because of the nature of the test, we can only approximate the confidence interval at this level. The output in Figure 6-17 shows the approximate 95% confidence interval of the change in the percentage of women in the workforce to be (0.00, 0.66). We can also find exact confidence intervals under the Sign test that are closest to 95% without going over or going under 95%.

To find the exact confidence intervals under the Sign test:

1 Click cell **D5** and type **−1**.

The output changes to give you the exact confidence interval that is *at most* 95%. In this case that is a 93.6% confidence interval, and it ranges from 0.00 to 0.060. Note that you cannot change the output if you are creating static output.

2 Click cell **D5** and type **1**.

The output changes and displays the exact confidence interval that is *at least* 95%. Excel displays the 98.1% confidence interval: (0.00, 0.80).

You've completed your research with the Labor Analysis workbook. You've found that there is sufficient evidence in this sample of 19 cities to conclude that there has been an increase in the percentage of women participating in the workforce during the four-year period from 1968 to 1972.

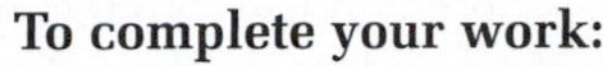

To complete your work:

Save your changes to the Labor Analysis workbook and close the file.

The Two-Sample *t* Test

In the one-sample or paired *t* test, you compared the sample average to a fixed value specified in the null hypothesis. In a **two-sample *t* test**, you compare the averages from two independent samples to determine whether a significant difference exists between the samples. For example, one sample might contain the cholesterol levels of patients taking a standard drug, while the second sample contains cholesterol data on patients taking an experimental drug. You would test to see whether there is a statistically significant difference between the two sample averages.

To compare the sample averages from normally distributed data, you have a choice of two *t* tests. One test statistic, called the **unpooled two-sample *t* statistic**, has the form

$$t = \frac{(\bar{x}_1 - \bar{x}_2) - (\mu_1 - \mu_2)}{\sqrt{\dfrac{s_1^2}{n_1} + \dfrac{s_2^2}{n_2}}}$$

where $\bar{x}_1$ and $\bar{x}_2$ are the sample averages for the first and second samples, s_1 and s_2 are the sample standard deviations, n_1 and n_2 are the sample sizes, and μ_1 and μ_2 are the means of the two distributions.

This form of the *t* statistic allows for samples that come from distributions with different standard deviations, having values of σ_1 and σ_2. On the other hand, it may be the case that both distributions share a common standard deviation σ. If that is the case, we can construct a *t* statistic by *pooling* the estimates of the standard deviation from the two samples into a single estimate, which we'll label as *s*. The value of *s* is

$$s = \sqrt{\frac{(n_1 - 1)s_1^2 + (n_2 - 1)s_2^2}{n_1 + n_2 - 2}}$$

The **pooled two-sample *t* statistic** would then be equal to

$$t = \frac{(\bar{x}_1 - \bar{x}_2) - (\mu_1 - \mu_2)}{s\sqrt{\dfrac{1}{n_1} + \dfrac{1}{n_2}}}$$

Comparing the Pooled and Unpooled Test Statistics

There are important differences between the two test statistics. The unpooled statistic, although we refer to it as a t test, *does not* strictly follow a t distribution. However, we can closely approximate the correct p values for this statistic by assuming it does and then compare the test statistic to a t distribution with degrees of freedom equal to

$$df = \frac{\left(\frac{S_1^2}{n_1} + \frac{S_2^2}{n_2}\right)^2}{\left[\frac{\left(\frac{S_1^2}{n_1}\right)^2}{n_1 - 1} + \frac{\left(\frac{S_2^2}{n_2}\right)^2}{n_2 - 1}\right]}$$

Here s_1 and s_2 are the standard deviations of the values in the first and second samples. The degrees of freedom for this statistic generally result in a fractional value. In actual practice, you'll probably never have to make this calculation yourself; your statistics package will do it for you.

For the pooled statistic, the situation is much easier. The pooled t statistic *does* follow a t distribution with degrees of freedom equal to

$$df = n_1 + n_2 - 2$$

If the standard deviations are different and you apply the pooled t statistic to the data, you run the risk of reporting an erroneous p value. To guard against this problem, it may be best to perform both a pooled and an unpooled test and then compare the results. If they agree, report the pooled t, because this test statistic is more widely known. Use the unpooled t if the two tests disagree. You should also examine the standard deviations of the two samples and determine whether they're close in value.

Working with the Two-Sample t Statistic

To see how the two-sample t test works, let's consider two groups of students: One group has learned to write using a standard teaching approach, and the other has learned using a new teaching method. There are 25 students in each group. At the end of the session, each student writes an essay that is graded on a 100-point scale. The average grade for the group 1 students is 75 with a standard deviation of 8. The average for the group 2 students is 80 with a standard deviation of 6. Could the difference in sample averages be attributed to differences between the teaching methods? We assume that the distribution of the data in both groups is normal. Our hypotheses are

$$\text{H}_0\text{: } \mu_1 - \mu_2 = 0$$
$$\text{H}_a\text{: } \mu_1 - \mu_2 \neq 0$$

where μ_1 is the assumed mean of the first distribution and μ_2 is the mean of the second distribution. Notice that we are not making any assumptions about what the actual values of μ_1 and μ_2 are; we are interested only in the difference between them. Because the standard deviations of the two samples are close in value, we'll use a pooled t test to test the null hypothesis. First, we must calculate the value of the pooled standard deviation, s:

$$s = \sqrt{\frac{(n_1 - 1)s_1^2 + (n_2 - 1)s_2^2}{n_1 + n_2 - 2}}$$

$$= \sqrt{\frac{(25 - 1)8^2 + (25 - 1)6^2}{25 + 25 - 2}}$$

$$= 7.07$$

Thus, the t statistic equals

$$t = \frac{(\bar{x}_1 - \bar{x}_2) - (\mu_1 - \mu_2)}{s\sqrt{\frac{1}{n_1} + \frac{1}{n_2}}}$$

$$= \frac{(75 - 80) - 0}{7.07\sqrt{\frac{1}{25} + \frac{1}{25}}}$$

$$= -2.5$$

which should follow a t distribution with 48 degrees of freedom. Is this a significant value?

To evaluate the *t* statistic:

1 Open a blank workbook. In cell A1 type **=TDIST(2.5,48,2)** and press **Enter**.

Note that we enter the value 2.5 rather than −2.5 because the TDIST function works with positive values. We enter the value 2 as the third parameter because this is a two-sided test.

Excel returns a p value of 0.016, and we reject the null hypothesis. We conclude that the evidence supports the conclusion that students learn better how to write using the new teaching method.

Testing for Equality of Variance

Part of the challenge in performing a two-sample *t* test is determining whether to use a pooled or unpooled estimate of the variance. Equal variances across different samples is called **homogeneity of variance**. Three ways to test for equal population variances are: the *F* test, Bartlett's test, and Levene's test.

The ***F* test** is used for comparing only two sample variances. The test statistic is

$$F = \frac{\text{larger variance}}{\text{smaller variance}}$$

where *F* follows the *F* distribution with $(n_1 - 1, n_2 - 1)$ degrees of freedom. Here n_1 is the number of observations in the sample with the larger variance, and n_2 is the size of the sample with the smaller variance. You'll learn more about the *F* distribution in Chapter 9. The *F* test requires normal data and it performs badly if the data deviate from normality.

Bartlett's test can be used to test for homogeneity of variance from more than two samples. Bartlett's test is sensitive to the effect of nonnormal data. If the sample data come from a nonnormal distribution, then Bartlett's test may simply be testing for nonnormality rather than homogeneity of variance. Because of its computational complexity, we will not discuss how to calculate the Bartlett test statistic here.

Levene's test can also be used to test for homogeneity of variance from more than two samples. The Levene test statistic is less sensitive to departures from normality than the Bartlett test statistic. On the other hand, if the sample data do follow a normal distribution, then Bartlett's test has better performance.

Generally it's a good idea to use all three tests (if possible) with your data. If none of the tests reject the hypothesis of variance homogeneity, you can probably use a pooled variance estimate in your analysis. If one or more of the tests reject this hypothesis, you should definitely consider performing an unpooled analysis.

To calculate the values of these test statistics, use the functions described in Table 6-5.

Table 6-5 Homogeneity of Variances functions in Excel

Function	Description
FDIST(*array1*, *array2*)	Returns the *p* value of the *F* test for the values in *array1* and *array2*.
BARTLETT (*column1*, *column2*, . . .)	Returns the *p* value of the Bartlett test statistic for values in *column1*, *column2*, and so forth. *StatPlus required.*

(continued)

BARTLETT2(*values, category*)	Returns the *p* value of the Bartlett test statistic for values in the *values* column. Samples are indicated in the *category* column. *StatPlus required.*
LEVENE(*column1, column2, . . .*)	Returns the *p* value of the Levene test statistic for values in *column1, column2,* and so forth. *StatPlus required.*
LEVENE2(*values, category*)	Returns the *p* value of the Levene test statistic for values in the *values* column. Samples are indicated in the *category* column. *StatPlus required.*

The values of these test statistics will also be displayed in the output of StatPlus's two-sample *t* test command.

Applying the *t* Test to Two-Sample Data

The Nursing Home workbook contains data from a random sample of nursing homes collected by the Department of Health and Social Services in New Mexico in 1988. The following variables were collected:

Table 6-6 Nursing Home data

Range Name	Range	Description
Beds	A2:A53	The number of beds in the home
Medical_Days	B2:B53	Annual medical inpatient days (hundreds)
Total_Days	C2:C53	Annual total patient days (hundreds)
Revenue	D2:D53	Annual total patient care revenue ($hundreds)
Salaries	E2:E53	Annual nursing salaries ($hundreds)
Expenses	F2:F53	Annual facility expenses ($hundreds)
Location	G2:G53	Rural and nonrural homes

To open the Nursing Home workbook:

1. Open the **Nursing Home** workbook from the Chapter06 folder.
2. Save the workbook as **Nursing Home Analysis** to the same folder. The workbook appears as shown in Figure 6-18.

Figure 6-18 The Nursing Home workbook

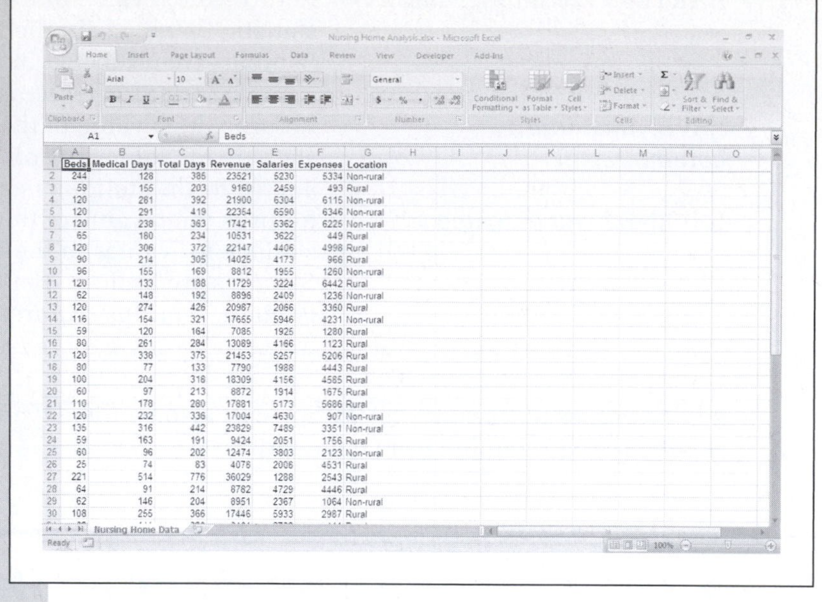

Beds	Medical Days	Total Days	Revenue	Salaries	Expenses	Location
244	128	385	23521	5230	5334	Non-rural
59	155	203	9160	2459	493	Rural
120	281	392	21900	6304	6115	Non-rural
120	291	419	22354	6590	6346	Non-rural
120	238	363	17421	5362	6225	Non-rural
65	180	234	10531	3622	449	Rural
120	306	372	22147	4406	4998	Rural
90	214	305	14025	4173	966	Rural
96	155	169	8812	1955	1260	Non-rural
120	133	188	11729	3224	6442	Rural
62	148	192	8896	2409	1236	Non-rural
120	274	426	20987	2066	3360	Rural
116	154	321	17655	5946	4231	Non-rural
59	120	164	7085	1925	1280	Rural
80	261	284	13089	4166	1123	Rural
120	338	375	21453	5257	5206	Rural
80	77	133	7790	1988	4443	Rural
100	204	318	18309	4156	4585	Rural
60	97	213	8872	1914	1675	Rural
110	178	280	17881	5173	5686	Rural
120	232	336	17004	4630	907	Non-rural
135	316	442	23829	7489	3351	Non-rural
59	163	191	9424	2051	1756	Rural
60	96	202	12474	3803	2123	Non-rural
25	74	83	4078	2006	4531	Rural
221	514	776	36029	1288	2543	Rural
64	91	214	8782	4729	4446	Rural
62	146	204	8951	2367	1064	Non-rural
108	255	366	17446	5933	2987	Rural

You've been asked to examine the data and determine whether there was a difference in revenue generated by rural and nonrural nursing homes during this time period. Here are your hypotheses.

H_0: Mean population revenue for rural nursing homes = Mean population revenue for nonrural homes

H_a: Mean population revenue for rural nursing homes ≠ Mean population revenue for nonrural homes

To test the null hypothesis you can use the StatPlus Two-Sample *t* test command. We initially assume that there is no homogeneity in variance between the two samples; however, we'll reexamine this assumption as we go along.

To perform a two-sample *t* test on the nursing home revenues:

1. Click **Two Sample Tests** from the StatPlus menu and then click **2-Sample *t* test.**
2. Verify that the **Use column of category values** option button is selected.

   Your data can be organized in one of two ways: (1) with two separate columns for each sample or (2) with one column of data values and one column of category values. The nursing home data are organized in the second way, with the Location column indicating whether a particular nursing home is rural or nonrural.

3 Click the **Data Values** button and select **Revenue** from the list of range names.

4 Click the **Categories** button and select **Location** from the range names list.

5 Click the **Use Unpooled Variance** option button.

6 Click the **Output** button and direct the output to a new worksheet named **t test**. Set the output type to **Dynamic**.

Figure 6-19 shows the completed dialog box.

Figure 6-19 The Two-Sample t-Test dialog box

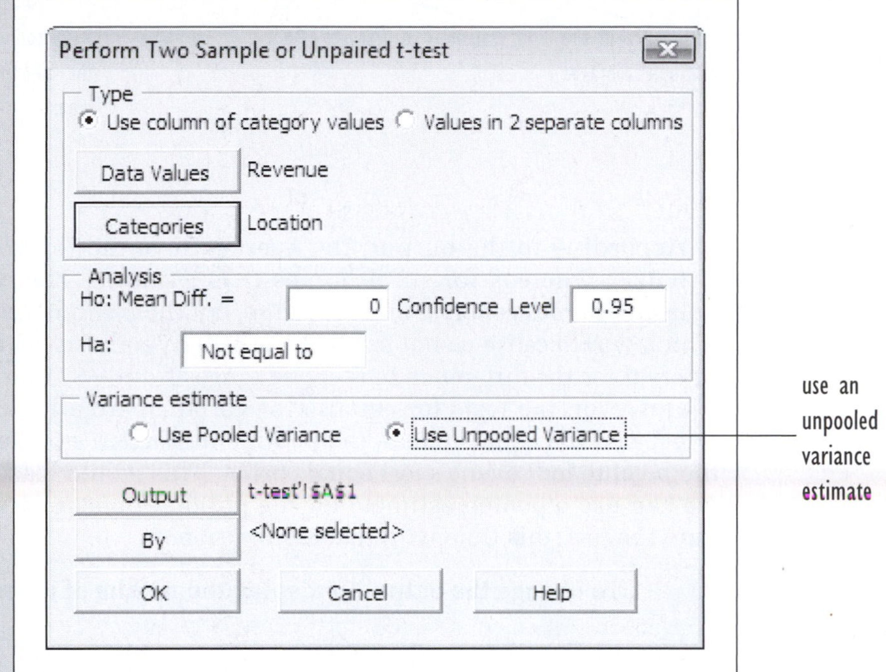

7 Click the **OK** button to generate the two-sample *t* test. Figure 6-20 shows the completed output.

Figure 6-20 Results of the two-sample t test analysis with unpooled variance

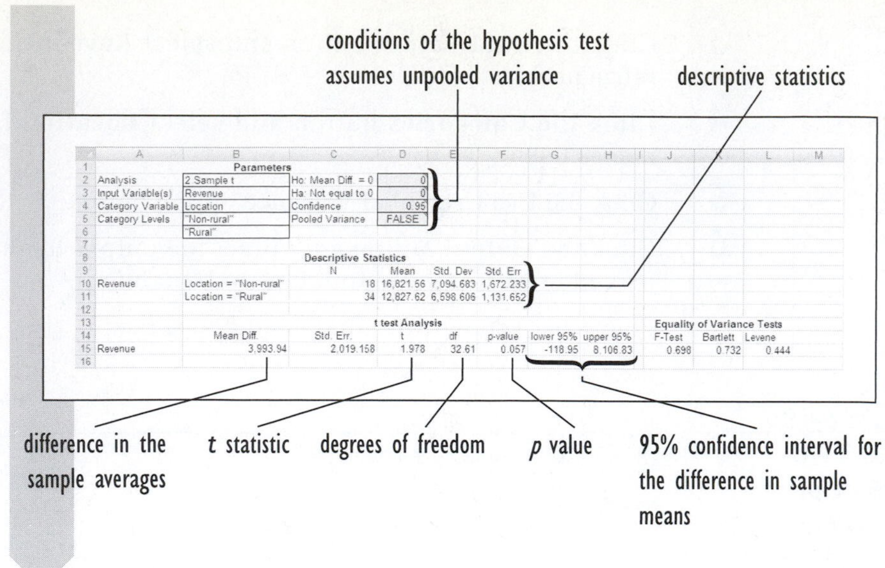

According to the output, the average revenue for nonrural homes is \$16,821, whereas for rural homes it is \$12,827. The data suggest that nonrural homes generate more revenue, but the *p* value for the *t* test is 0.057, which would cause us not to reject the null hypothesis. The 95% confidence interval for the difference in revenue ranges from −\$118 to \$8,106.

However, the tests for equality of variance are all nonsignificant. The *F* test *p* value equals 0.698, the *p* value of Bartlett's test equals 0.732, and the *p* value for Levene's test equals 0.444. This would lead us to believe that we can use a pooled estimate for the population standard deviation. Let's redo the test, this time with that assumption.

To change the output to display the results of a pooled test:

1 Click cell **D5** in the ***t* Test** worksheet.

Cell D5 contains the value TRUE if a pooled test is used and FALSE for unpooled test.

2 Replace FALSE with **TRUE** in cell D5.

The output changes to display the pooled test results. See Figure 6-21.

Figure 6-21 Results of the two-sample t test analysis with pooled variance

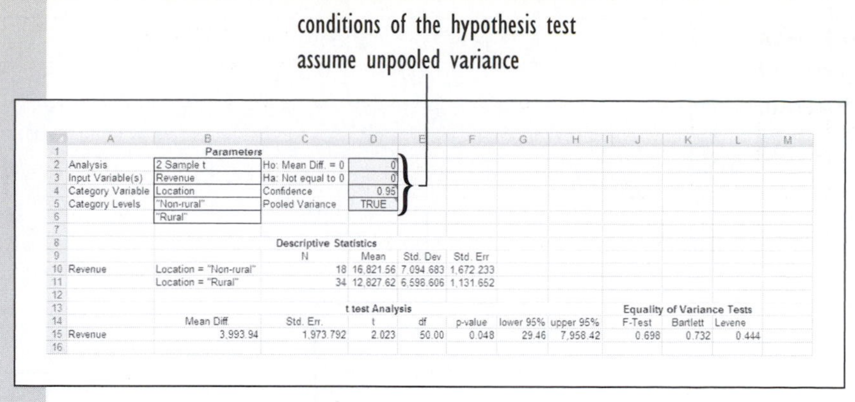

	A	B	C	D	E	F	G	H	I	J	K	L	M
1		Parameters											
2	Analysis	2 Sample t	Ho: Mean Diff. = 0	0									
3	Input Variable(s)	Revenue	Ha: Not equal to 0	0									
4	Category Variable	Location	Confidence	0.95									
5	Category Levels	"Non-rural"	Pooled Variance	TRUE									
6		"Rural"											
7													
8			Descriptive Statistics										
9			N	Mean	Std. Dev	Std. Err							
10	Revenue	Location = "Non-rural"	18	16,821.56	7,094.683	1,672.233							
11		Location = "Rural"	34	12,827.62	6,598.606	1,131.652							
12													
13				t test Analysis						Equality of Variance Tests			
14		Mean Diff.	Std. Err.	t	df	p-value	lower 95%	upper 95%		F-Test	Bartlett	Levene	
15	Revenue	3,993.94	1,973.792	2.023	50.00	0.048	29.46	7,958.42		0.698	0.732	0.444	
16													

Under the pooled test the *p* value changes to 0.048, leading us to reject the null hypothesis and accept the alternative: Nonrural homes generate more revenue than rural homes. The 95% confidence interval for the difference in revenue ranges from $29 to $7,958. Notice that the *t* test and the confidence interval agree, as they always do: the test rejects 0 as the population mean and the interval does not include 0.

We have a problem. Depending on which test we use, we reach a different conclusion. The difference between the two tests depends on the assumption we make about the population standard deviation. Should we use the pooled or the unpooled results? To get a clearer picture, it would be a good idea to create a chart of the two distributions with a boxplot.

To create a boxplot of the two samples:

1. Click **Single Variable Charts** from the StatPlus menu and then click **Boxplots.**
2. Verify that the **Use a column of category levels** option button is selected.
3. Click the **Data Values** button and select **Revenue** from the list of range names.
4. Click the **Categories** button and select the range name **Location.**
5. Click the **Output** button and send the chart output to the chart sheet **Revenue Boxplots.**
6. Click the **OK** button twice to generate the boxplots. See Figure 6-22.

Figure 6-22 Boxplots of the revenue data

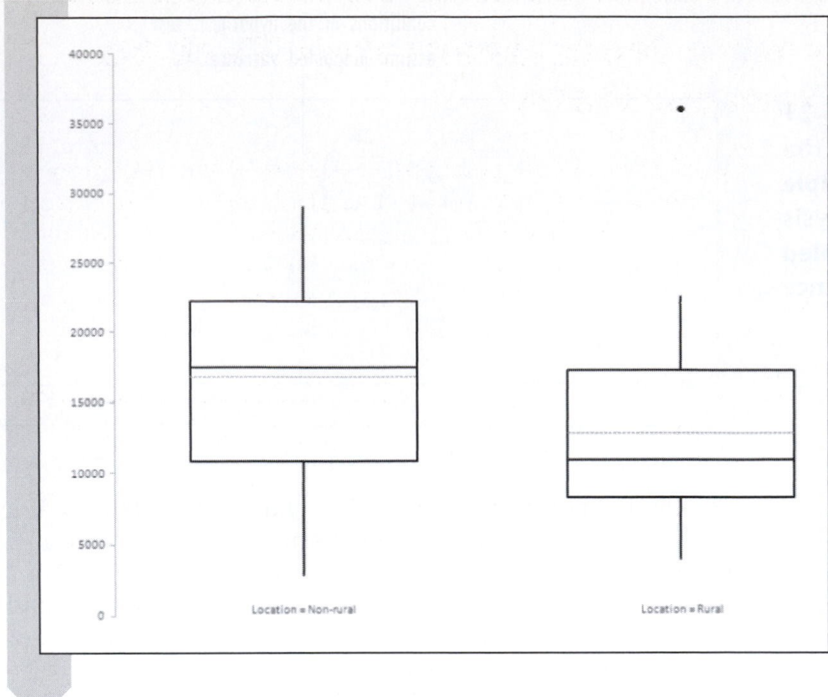

Figure 6-22 indicates that there is a moderate outlier for revenues generated by rural nursing homes. This outlier may affect the results of our *t* tests. One way of dealing with this outlier is to switch to a nonparametric test, which is less influenced by the presence of outlying values.

EXCEL TIPS

- You can also perform a two-sample *t* test of your data using the Analysis ToolPak, supplied by Excel. To perform a two-sample *t* test, first load the Analysis ToolPak and then click the Data Analysis button from the Analysis group on the Data tab.
- For a pooled test, click *t* Test: Two Sample Assuming Equal Variances and specify the two columns containing the paired data.
- For an unpooled test, click *t* Test: Two Sample Assuming Unequal Variances.
- The Analysis ToolPak does not include confidence intervals for the two-sample *t*.

Applying a Nonparametric Test to Two-Sample Data

The two-sample nonparametric test is the **Mann-Whitney test**. In the Mann-Whitney test we rank all of the values from smallest to largest and then sum the ranks in each sample. Unlike the Wilcoxon test, we do not rank the absolute data values or multiply the ranks by the sign of the original data. Table 6-7 shows an example of two sample data along with the calculated ranks.

Table 6-7 Two-Sample data

Sample 1 Values	Ranks	Sample 2 Values	Ranks
22	12.0	−3	3.0
16	11.0	−1	4.0
1	5.0	2	6.0
−4	1.5	8	9.0
7	8.0	−4	1.5
3	7.0		
9	10.0		

Note that we don't need to have equal sample sizes. Our null hypothesis is that both samples have the same median value. In this example, the sum of the Sample 1 ranks is 54.5, and the sum of the Sample 2 ranks is 23.5. We can use probability theory to determine the probability of the first sample having a rank sum of 54.5 or greater if the null hypothesis were true. In this case, that p value would be 0.176, which would not support rejecting the null hypothesis.

When using the Mann-Whitney test, we also need to calculate the median difference between the two samples. This is done by calculating the difference for each pair of observations taken from Sample 1 and Sample 2 and then determining the median of those differences. For the data in Table 6-6, there are 35 pairs, starting with the difference between 22 and −3 (the first observations in the samples) and going down to the difference between 9 and −4 (the last observations). The median of these 35 differences is 7. By comparison, the difference of the sample averages is 7.31, so the median difference is pretty close. When the sample sizes get large, these calculations cannot be easily done by hand.

The Mann-Whitney test makes only four assumptions.

1. Both samples are random samples taken from their respective probability distributions.
2. The samples are independent of each other.
3. The measurement scale is at least ordinal.
4. The two distributions have the same shape.

The Mann-Whitney test relies on ranks; this lessens the effect of outliers. The downside is that under certain situations, the Mann-Whitney test will not be as efficient as the two-sample *t* test in detecting differences between samples. Also, some researchers may not be familiar or comfortable with this nonparametric approach.

Let's apply the Mann-Whitney test to the nursing home data. Our hypotheses are

H_0: Median population revenue for rural nursing homes = Median population revenue for nonrural homes

H_a: Median population revenue for rural nursing homes ≠ Median population revenue for nonrural homes

To apply the Mann-Whitney test to the nursing home data:

Return to the Nursing Home Data worksheet and click **Two Sample Tests** from the StatPlus menu and then click **2 Sample Mann-Whitney Rank test**.

Verify that the **Use column of category values** option button is selected.

Click the **Data Values** button and select **Revenue** from the list of range names.

Click the **Categories** button and select the range name **Location**.

Click the **Output** button and send your output to a new worksheet named **MW-test**.

Click **OK** to generate the Mann-Whitney analysis for the nursing home revenue. See Figure 6-23.

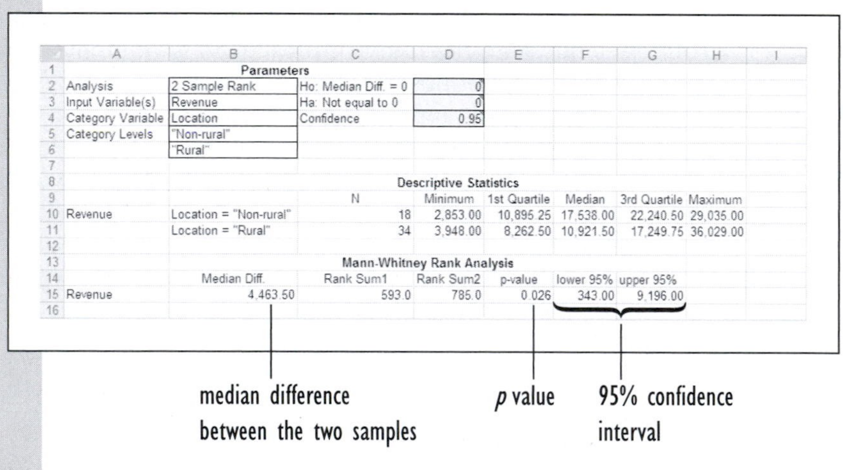

	A	B	C	D	E	F	G	H	I
1		Parameters							
2	Analysis	2 Sample Rank	Ho: Median Diff. = 0	0					
3	Input Variable(s)	Revenue	Ha: Not equal to 0	0					
4	Category Variable	Location	Confidence	0.95					
5	Category Levels	"Non-rural"							
6		"Rural"							
7									
8				Descriptive Statistics					
9			N	Minimum	1st Quartile	Median	3rd Quartile	Maximum	
10	Revenue	Location = "Non-rural"	18	2,853.00	10,895.25	17,538.00	22,240.50	29,035.00	
11		Location = "Rural"	34	3,948.00	8,262.50	10,921.50	17,249.75	36,029.00	
12									
13				Mann-Whitney Rank Analysis					
14		Median Diff.	Rank Sum1	Rank Sum2	p-value	lower 95%	upper 95%		
15	Revenue	4,463.50	593.0	785.0	0.026	343.00	9,196.00		
16									

Figure 6-23 Mann-Whitney analysis of the revenue data

From the output in Figure 6-23, we note that the median revenue from nonrural nursing homes is $17,538, whereas the median revenue of rural hospitals is $10,921. The median difference is $4,463 (remember that the median difference is *not* the difference between the two medians). This difference is statistically significant with a *p* value of 0.026. Note that by using a test that reduces the influence of outliers, we achieve a more significant *p* value.

Our final decision: We reject the null hypothesis that revenue generated by rural homes was equal to revenue generated by nonrural homes and accept the hypothesis that they were not equal. The 95% confidence interval gives a range of values for this difference. We conclude that the median difference is not less than $343 and not more than $9,196.

To complete your work:

Save your changes to the workbook and close the file.

Final Thoughts about Statistical Inference

The previous example displays some of the challenges and dangers in doing statistical inference. It is tempting to see a *p* value or a confidence interval as the authoritative answer to your research. However, to use the tools of statistical inference properly, you should always be aware of the limitations of your statistical tests. Here are some general rules you should follow when performing statistical inference.

1. State your hypotheses clearly and, if possible, *before* collecting and analyzing your data.
2. Understand the nature and limitations of the statistical tests you use. Be aware of any assumptions that the test makes about the nature of your data. Try to verify that these assumptions are met (or at least that there is no evidence that they are being violated).
3. Graph your data; it will help you more easily detect any departures from the assumptions of your statistical test. Calculate descriptive statistics of your data for the same reason.
4. If appropriate, perform more than one kind of statistical test. A different test, such as a nonparametric one, may provide important insight into your data.
5. Your goal is *not* to reject the null hypothesis. A study that fails to reject the null hypothesis is not a failure, nor is a low *p* value a sign of success (especially if you're rejecting the null hypothesis in error). Your goal should be to determine what, if any, conclusions you can reach about your data in a fair and impartial way and then to ascertain how reliable those conclusions are.

Exercises

1. *True or false (and why):* A 95% confidence interval that covers the range $(-5, 5)$ tells you that the probability is 95% that μ will have a value between -5 and 5.

2. *True or false (and why):* Accepting the null hypothesis means that the null hypothesis is true.

3. *True or false (and why):* Rejecting the null hypothesis means that the null hypothesis is false.

4. Consider a sample of 25 normally distributed observations with a sample average of 50.

   a. Calculate the 95% confidence interval if $\sigma = 20$.

   b. Calculate the 95% confidence interval if σ is unknown but if the sample standard deviation = 20.

5. The nationwide mean price for a three-year-old Honda Civic DX is $11,500 with a known standard deviation of $600. You check the newspaper and find 9 three-year-old Civics in San Francisco selling for an average price of $12,000. You wonder whether the cost of Civics in San Francisco is higher than for the rest of the nation.

   a. State your question about the price of Civics in terms of a null and an alternative hypothesis. What are you assuming about the distribution of Civic prices?

   b. Will the alternative hypothesis be one or two sided? Defend your answer.

   c. Test your null hypothesis. Do you accept or reject it and at what *p* value? Construct a 95% confidence interval for Civic prices in San Francisco.

   d. Redo your analysis, but this time assume that the sample size is 10 with a sample average of $12,000 and a sample standard deviation of $600. Assume that you don't know the value of the nationwide standard deviation.

6. In tests of stereo speakers, ten American-made speakers had an average performance rating of 90 with a standard deviation of 5. Five imported speakers had an average rating of 85 with a standard deviation of 4.

   a. Write a null and an alternative hypothesis comparing the two types of speakers.

   b. Test the null hypothesis. What is the *p* value?

   c. If you decide to change the significance level to 10%, does your conclusion change?

7. Derive the formula for the *t* confidence interval based on the definition of the *t* statistic shown earlier in this chapter.

8. You want to continue to study the nursing home data discussed in this chapter. Explore whether there is a significant difference between rural and nonrural homes in terms of size (as expressed by the number of beds in the homes).

   a. Open the **Nursing Home** workbook from the Chapter06 folder. Save the workbook as **Nursing Home Beds** to the same folder.

   b. Write down a set of hypotheses for exploring the question of whether the numbers of beds in rural and nonrural homes differ.

   c. Apply a two-sample *t* test to the data. Report your results assuming a pooled estimate of the standard deviation and assuming an unpooled estimate. What are the *p* value and confidence interval under each assumption?

d. Create a boxplot of the Beds variable for the different locations. Do the plot and the descriptive statistics give you any help in determining whether to use a pooled or a nonpooled test? Explain your answer.

e. Apply the Mann-Whitney test to the data. What are your hypotheses? What is your conclusion? Include information on the *p* value and confidence interval in your discussion.

f. Save your changes to the workbook and then write a report summarizing your results. Do your conclusions differ on the basis of which test you apply? Which statistical test would you report and why?

9. Return to the nursing home data from this chapter. This time you've been asked to explore whether rural homes are used at a lower rate than nonrural homes after adjusting for the differing size of the homes.

   a. Open the **Nursing Home** workbook from the Chapter06 folder and save it as **Nursing Home Usage Rates**.

   b. Create a new variable named Days_Beds equal to the ratio of the total number of patient days to the number of beds in the home. Format the data to display three decimal places.

   c. Compare the average value of the Days_Beds variable for rural and nonrural homes. What are your hypotheses? What test or tests will you use to evaluate your null hypothesis?

   d. Create a boxplot of the Days_Beds variable for the two locations.

   e. Save your changes to the workbook and then write a report summarizing your results, including any descriptive statistics, *p* values, and confidence intervals you created during your analysis. Is there evidence to suggest that rural homes are being utilized at a lower rate?

10. Draft numbers from the Vietnam War have been recorded for you. (See the Chapter 4 exercises for a discussion of the draft lottery.) It's been claimed that people whose birthday fell in the second half of the year had lower draft numbers and therefore were more likely to be drafted. Explore this claim.

   a. Open the **Draft** workbook from the Chapter06 folder and save it as **Draft Number Analysis.**

   b. Write down the null and alternative hypotheses for your study.

   c. Create a two-sample *t* test to analyze your hypotheses. Do you use a pooled or an unpooled test? Which type of test does the distribution of the data support?

   d. Create a histogram of the distribution of draft numbers broken down by whether the number was assigned in the first half of the year or the second. What probability distribution does the data resemble? What property of the *t* statistic allows you still to apply the *t* test to your data?

   e. Calculate a 95% confidence interval for the average draft number for people born in the first half of the year and then for people born in the second half of the year. (*Hint*: You can do this using StatPlus's 1-sample *t* test command, specifying the Half variable as the BY variable.)

   f. Save your workbook and write a report summarizing your results. What is the mean difference in draft number between people born in the first half of the year and those born in the second half? Is this a significant difference? What are the practical ramifications of your conclusions?

11. The Junior College workbook contains salary information for faculty at a college. The female faculty members claim that they are underpaid relative to their male counterparts. Investigate their claim.

    a. Open the **Junior College** workbook from the Chapter06 folder and save it as **Junior College Salary Analysis**.
    b. Write down your null and alternative hypotheses. What is the significance level for this test?
    c. Perform a two-sample *t* test on the salary data broken down by gender. Does it make any difference whether you perform a pooled or an unpooled test? Do the data suggest that there is a salary difference between male and female faculty members? Create histograms of the distribution of salary data for male and female instructors.
    d. Redo the two-sample *t* test, this time breaking down the analysis of salary versus gender by the Rank_Hired variable. Are there significant gender differences in terms of salary for the various employee ranks? (*Note*: Some combinations of gender and rank hired will have sample sizes of 0. This will result in Excel displaying a #VALUE! result in the workbook. You can ignore these employee ranks because there are no values to investigate.)
    e. Save your workbook and summarize your conclusions. Is there evidence that the college has underpaid its female faculty? If so, does this difference exist for all teaching ranks? Why does this study not prove sexual discrimination? What factors have been ignored?

12. The Big Ten workbook has graduation information on Big Ten schools. (See Chapter 3 for a discussion of this data set.) Explore whether there is a difference in the graduation rates between white male athletes and white female athletes.

    a. Open the **Big Ten** workbook from the Chapter06 folder and save it as **Big Ten Graduation Analysis**.
    b. State your null and alternative hypotheses.
    c. Perform a paired *t* test of the white male and white female athlete graduation rates. Is there statistically significant evidence of a difference in the graduation rates? What is a 95% confidence interval for the difference? What is the 90% confidence interval?
    d. Redo the analysis using the Wilcoxon Signed Rank test and the Sign test.
    e. Why is this an example of paired data?
    f. Save your workbook and write a report summarizing your conclusions. Can you apply your results to universities in general? Defend your answer.

13. The Mortgage workbook contains information on refusal rates from 20 lending institutions broken down by race and income status from the late 1980s. It was claimed in reports to Congress that lending institutions had significantly higher refusal rates for minorities. Examine the statistical basis for that claim.

    a. Open the **Mortgage** workbook from the Chapter06 folder and save it as **Mortgage Refusal Analysis**.
    b. State your null and alternative hypotheses.
    c. Apply a paired *t* test to the refusal rates for minority and white applicants. What is the 95% confidence interval for the difference in refusal rates? What is the *p* value for the test?
    d. Create a histogram and normal probability plot of the difference in refusal rate. Do the data appear normal?

e. Redo your analysis using the Wilcoxon Signed Rank test. How do your results compare to the paired *t* test?
f. Redo questions b through e using the refusal rates for high-income whites and minorities. How do the results of the two analyses compare, especially in terms of the confidence interval for the difference in refusal rate? Is there evidence to suggest that there is no refusal rate gap for higher-income minorities?
g. Save your changes to the workbook and write a report summarizing your conclusions.

14. The Teacher.xls workbook stores average teacher salary, public school spending per pupil, and the ratio of teacher to pupil spending for 1985, broken down by state and region. Analyze the results stored in this workbook.

a. Open the **Teacher** workbook from the Chapter06 folder and save it as **Teacher Analysis**.
b. Construct a 95% *t* confidence interval for each of the numeric variables, broken down by area.
c. Construct a 95% Wilcoxon Signed Rank confidence interval for the numeric variables, by area.
d. Save the changes to your workbook and write a report summarizing the results of your analysis.

15. The Pollution workbook contains data on the number of unhealthful pollution days for 14 U.S. cities comparing 1980 values to average values from 2000 to 2006. Analyze what impact environmental regulations have had on pollution.

a. Open the **Pollution** workbook from the Chapter06 folder and save it as **Pollution Analysis**.
b. State your null and alternative hypotheses for this analysis.
c. Create histograms of the ratio and difference between the 1980 and 2006 values. Does the distribution of those two variables appear to follow the normal distribution?
d. Analyze the difference and ratio values using a one-sample *t* test, *s* test, and a Wilcoxon Signed Rank test.
e. Save your changes to the workbook. Write a report summarizing your observations. Where do you see significant changes in the number of pollution days? Is this true for all statistical tests? Given the distribution of the data, which test appears to be the most appropriate for these values?

16. In a NASA-funded study, seven men and eight women spent 24 days in seclusion to study the effects of gravity on circulation. Without gravity, there is a loss of blood from the legs to the upper part of the body. The study started with a nine-day control period in which the subjects were allowed to walk around. Then followed a ten-day bed-rest period in which the subjects' feet were somewhat elevated to simulate weightlessness in space. The study ended with a five-day recovery period in which the subjects again were allowed to walk around. Every few days, the researchers measured the electrical resistance at the calf, which increases when there is a blood loss. The electrical resistance gives an indirect measure of the blood loss and indicates how the subject's body responds to the conditions. You've been asked to examine whether the male subjects and the female subjects differed in how they responded to the study. You're to perform your analysis for each of the days in the study.

a. Open the **Space** workbook from the Chapter06 folder and save it as **Space Biology Analysis**.

b. State your null and alternative hypotheses.
c. Perform a two-sample t test comparing the value of the Resistance variable between the male and female subjects, broken down by day. You do not have to summarize your results across days.
d. On what day or days is there a significant difference between the two groups? Do your results change if you use an unpooled rather than a pooled estimate of the standard deviation?
e. Create a scatterplot of Resistance versus Days. Break the scatterplot down by gender using the StatPlus command shown in Chapter 3. Describe the effect displayed in the scatter plot (you may want to change the scatter plot scales to view the data better).
f. Redo your analysis using the Mann-Whitney test. Do your conclusions from part b change any with the nonparametric procedure?
g. Save your workbook and write a report summarizing your findings. Explain how (if at all) the male and female subjects differed in their response to the study. Include in your discussion the various parts of the study (control period, bed-rest period, etc.) and how the patients responded during those specific intervals. Include any pertinent statistics.

17. The Math workbook contains data from a study analyzing two methods of teaching mathematics. Students were randomly assigned to two groups: a control group that was taught in the usual way with a relaxed homework and quiz schedule, and an experimental group that was regularly assigned homework and given frequent quizzes. Students in the experimental group were allowed to retake their exams to raise their grades (though a different exam was given for the retake). The final exam scores of the two groups were recorded. Investigate whether there is compelling evidence that students in the experimental group had higher scores than those in the control group.
   a. Open the **Math** workbook from the Chapter06 folder and save it as **Math Scores Analysis**.
   b. State your null and alternative hypotheses. Is this a one-sided test or a two-sided test? Why?
   c. Perform a two-sample t test on the final exam score. Use a pooled estimate of the standard deviation. What is the 95% confidence interval for the difference in scores? What is the p value? Do you accept or reject the null hypothesis? Do your conclusions change if you use an unpooled test?
   d. Chart the distribution of the final exam scores for the two groups. What do the charts tell about the distributions? Do the charts cast any doubt on your conclusions in part c? Why?
   e. Do a second analysis of the data using the Mann-Whitney Rank test. How do these results compare to the two-sample t? Are your conclusions the same?
   f. Save your changes to the workbook and write a report summarizing your findings and reporting your conclusion. Is there a significant change in the exam scores under the experimental approach?

18. The Voting workbook contains the percentage of the presidential vote that the Democratic candidate received in 1980 and 1984, broken down by state and region. You've been asked to investigate

the difference between the 1980 and 1984 voting patterns.

a. Open the **Voting** workbook from the Chapter06 folder and save it as **Voting Analysis**.
b. Do a paired *t* test for the voting percentage, broken down by region. Summarize your findings for all regions as well.
c. For which regions was there a significant change in the voting percentage? For which regions was there no significant change? What was the overall change in the voting percentage across all regions?
d. Save your changes to the workbook and write a report summarizing your findings, including your descriptive statistics, *p* values, and confidence intervals.

19. The Calculus workbook shows the first semester calculus scores for male and female students. Analyze the data set to determine whether there is a significant difference between the two groups.

a. Open the **Calculus** workbook from the Chapter06 folder and save it as **Calculus Scores Analysis**.
b. State your null and alternative hypotheses.
c. Perform a two-sample *t* test on the calc values using a pooled estimate of the variance. What is the 95% confidence interval of the difference between the two groups? Do you reject or accept the null hypothesis? At what *p* value?
d. Chart the distribution of the calc values for the two groups. Do the distributions appear normal? What property of exam scores makes it unlikely that these exam scores follow the normal distribution? (*Hint*: Test scores are usually constrained to fall between 0 and 100.) What property of the *t* distribution might allow you to use the *t* test anyway?
e. Repeat your analysis using the Mann-Whitney non-parametric test.
f. Save your changes to the workbook and write a report summarizing your findings and stating your conclusions.

20. The Reaction workbook contains information on reaction times and race times, recorded by sprinters running in the first three rounds of the 100-meter dash at the 1996 Summer Olympics. You're asked to determine whether there is evidence that the sprinter's reaction time (the time it takes for the sprinter to leave the starting block at the sound of the gun) changes as he advances in the competition.

a. Open the **Reaction** workbook from the Chapter06 folder and save it as **Reaction Time Analysis**.
b. Use the paired *t* test and analyze the differences between the following variables: React 1 vs. React 2, React 1 vs. React 3, and React 2 vs. React 3. Calculate the 95% confidence interval for each difference pair, and test for statistical significance at the 5% level. Are there any pairs of rounds in which there is a significant difference in the average reaction time?
c. Create three new columns in the Reaction Times worksheet displaying the three paired differences, and then create three normal probability plots of those differences. Does the distribution of the paired differences follow a normal distribution?
d. Redo the analysis, this time using the Wilcoxon Sign Rank test. Do your conclusions change when you use this test?
e. Save your changes to the workbook and write a report summarizing your results and give your conclusions.

Have you accepted the null hypothesis, or is there evidence that reaction times do change from one round to another?

21. Return to your analysis of the results of the 1996 100-meter dash. This time, analyze the race times from the three rounds of the race.

   a. Open the **Race** workbook from the Chapter06 folder and save it as **Race Results Analysis**.
   b. Perform a paired *t* test of the race times (use the Race1, Race2, and Race3 variables), comparing the differences between Round 1 and Round 2 and then Round 2 and Round 3. Is there significant evidence that the race times decrease as the runner advances in the competition? Calculate the 95% confidence interval for the change in race time.
   c. Save your changes to the workbook and write a report summarizing your results, including any descriptive statistics and *p* values. Does your evidence suggest any difference in the competition level as a runner goes from Round 1 to Round 2 as compared to going from Round 2 to Round 3?

Chapter 7

TABLES

Objectives

In this chapter you will learn to:

- Create PivotTables of a single categorical variable
- Create PivotCharts as column and pie charts
- Relate two categorical variables with a two-way table
- Apply the chi-square test to a two-way table
- Compute expected values of a two-way table
- Combine or eliminate small categories to get valid tests
- Test for association between ordinal variables
- Create a custom sort order for your workbook

In this chapter you'll learn how to work with categorical data in the form of tables and ordinal variables. You'll learn how to use Excel's PivotTable feature to create tables, and you'll explore how to analyze these data in the table using StatPlus.

PivotTables

In the previous chapter you used *t* tests and nonparametric tests to analyze continuous variables. You can also apply hypothesis tests to categorical and ordinal data. These type of data are most commonly seen in surveys, which record counts broken down by categories. For example, you create a table of instructors broken down by title (assistant professor, associate professor, or full professor) and gender (male or female). Are there significantly more male full professors than female? How many female professors would you expect given the data? An analysis of categorical data addresses questions of this type.

To illustrate how to work with categorical variables, let's look at data from a survey of professors who teach statistics courses. The Survey workbook includes 392 responses to questions about whether the course requires calculus, whether statistical software is used, how the software is obtained by students, what kind of computer is used, and so on. The workbook contains the following variables shown in Table 7-1:

Table 7-1 Survey of Statistics Professors Data

Range Name	Range	Description
Computer	A2:A393	Computer used in the course
Dept	B2:B393	Department
Available	C2:C393	Type of computer system available to the student
Interest	D2:D393	The amount of interest in a supplementary statistics text
Calculus	E2:E393	The extent to which calculus is required for the course
Uses_Software	F2:F393	Whether the course uses software
Enroll_A	G2:G393	Categorical variable indicating semester course enrollment in the instructor's course (For example, 001-050 means that from 1 to 50 students are enrolled each semester).
Enroll_B	H2:H393	Categorical variable indicating annual course enrollment
Max_Cost	I2:I393	Maximum cost for a supplementary computer text

To open the Survey workbook:

1 Start Excel if necessary and maximize the Excel window.

2 Open the **Survey** workbook from the Chapter07 folder.

3 Save the workbook as **Survey Table Statistics**. The workbook appears as shown in Figure 7-1.

Figure 7-1
Survey data

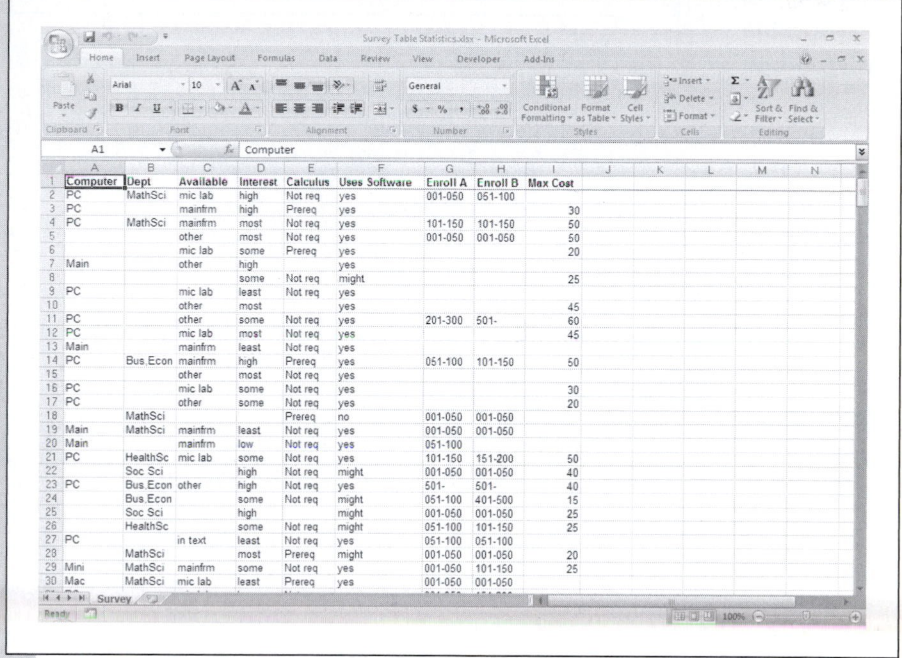

	Computer	Dept	Available	Interest	Calculus	Uses Software	Enroll A	Enroll B	Max Cost
2	PC	MathSci	mic lab	high	Not req	yes	001-050	051-100	
3	PC		mainfrm	high	Prereq	yes			30
4	PC	MathSci	mainfrm	most	Not req	yes	101-150	101-150	50
5			other	most	Not req	yes	001-050	001-050	50
6			mic lab	some	Prereq	yes			20
7	Main		other	high		yes			
8				some	Not req	might			25
9	PC		mic lab	least	Not req	yes			
10			other	most		yes			45
11	PC		other	some	Not req	yes	201-300	501-	60
12	PC		mic lab	most	Not req	yes			45
13	Main		mainfrm	least	Not req	yes			
14	PC	Bus,Econ	mainfrm	high	Prereq	yes	051-100	101-150	50
15			other	most	Not req	yes			
16	PC		mic lab	some	Not req	yes			30
17	PC		other	some	Not req	yes			20
18		MathSci			Prereq	no	001-050	001-050	
19	Main	MathSci	mainfrm	least	Not req	yes	001-050	001-050	
20	Main		mainfrm	low	Not req	yes	051-100		
21	PC	HealthSc	mic lab	some	Not req	yes	101-150	151-200	50
22		Soc Sci		high	Not req	might	001-050	001-050	40
23	PC	Bus,Econ	other	high	Not req	yes	501-	501-	40
24		Bus,Econ		some	Not req	might	051-100	401-500	15
25		Soc Sci		high		might	001-050	001-050	25
26		HealthSc		some	Not req	might	051-100	101-150	25
27	PC		in text	least	Not req	yes	051-100	051-100	
28		MathSci		most	Prereq	might	001-050	001-050	20
29	Mini	MathSci	mainfrm	some	Not req	yes	001-050	101-150	25
30	Mac	MathSci	mic lab	least	Prereq	yes	001-050	001-050	

You've been asked to determine the relationship between the department and whether calculus is required as a prerequisite for courses in that department. First you'll examine the distribution of professors in different department categories.

In Excel, you obtain category counts by generating a **PivotTable**, a worksheet table that summarizes data from the source data list (in this case the survey data). Excel PivotTables are interactive; that is, you can update them whenever you change the source data list, and you can switch row and column headings to view the data in different ways (hence the term *pivot*).

Try creating a PivotTable that summarizes the number of professors in each department. You can insert a PivotTable using the commands on the Insert tab.

To insert a PivotTable:

1 Click the **Insert** tab from the Excel ribbon and then click the **Pivot-Table** button from the Tables group.

Excel displays the Create PivotTable dialog box shown in Figure 7-2 with the data range A1:I393 already selected for you.

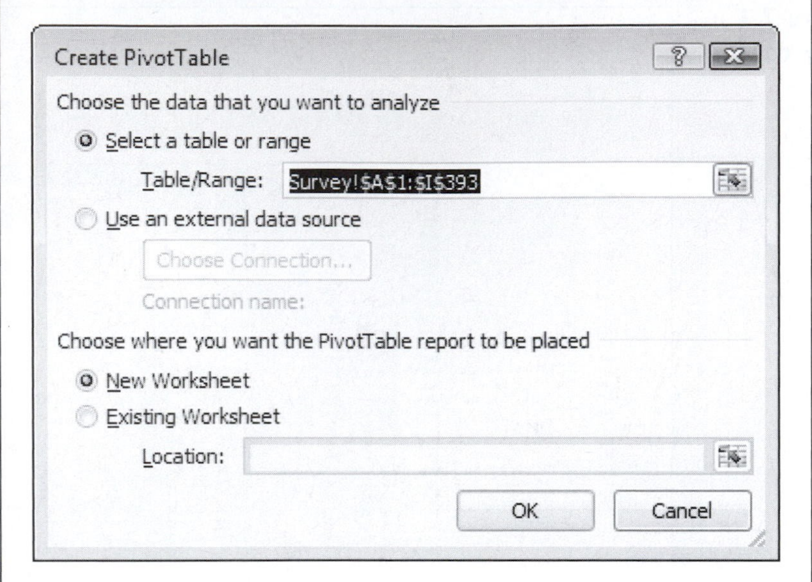

Figure 7-2 Create PivotTable dialog box

2 Verify that the New Worksheet option button is selected and then click the **OK** button.

Excel opens a new worksheet containing the PivotTable tools shown in Figure 7-3. Note that Excel has automatically added a new Pivot-Table Tools ribbon used for creating and editing PivotTables.

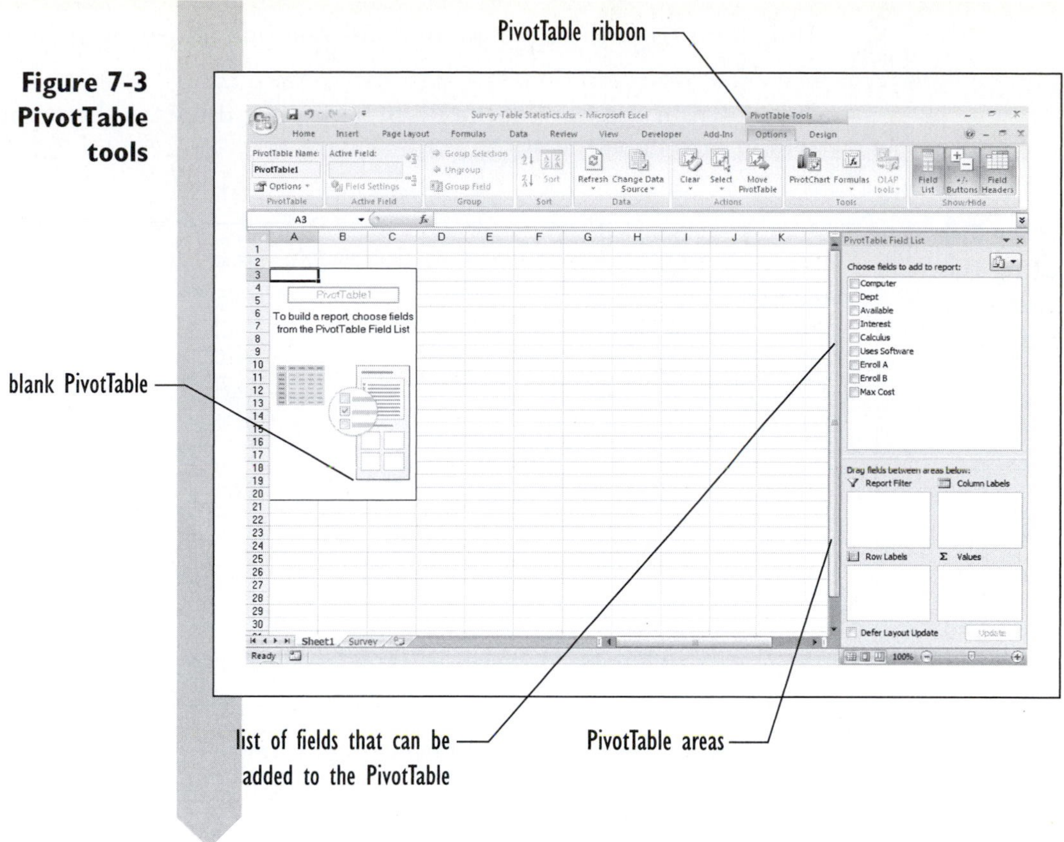

Figure 7-3 PivotTable tools

Now you control the layout of the PivotTable. A PivotTable has four areas. The **Row Labels** determine the categories that will appear in each row of the table. Similarly, the **Column Labels** control the categories for each table column. The **Values** area determines the values that will appear at each intersection of row and column categories. Finally, the **Report Filter** area is used to filter the PivotTable, showing only a subset of all of the data in the selected data set.

You can design your PivotTable by dragging fields from the PivotTable Field List box into these different areas. Try this now by creating a PivotTable showing the breakdown of department categories in your data set.

To create a PivotTable of department categories:

1. Click **Dept** from the PivotTable Field List box and drag it to the Row Labels area box.

   Excel adds the different department categories to the PivotTable. Now show the counts within each category.

2. Drag **Dept** from the PivotTable Field List box and drop it in the Values box.

   As shown in Figure 7-4, the PivotTable now displays the counts within each department.

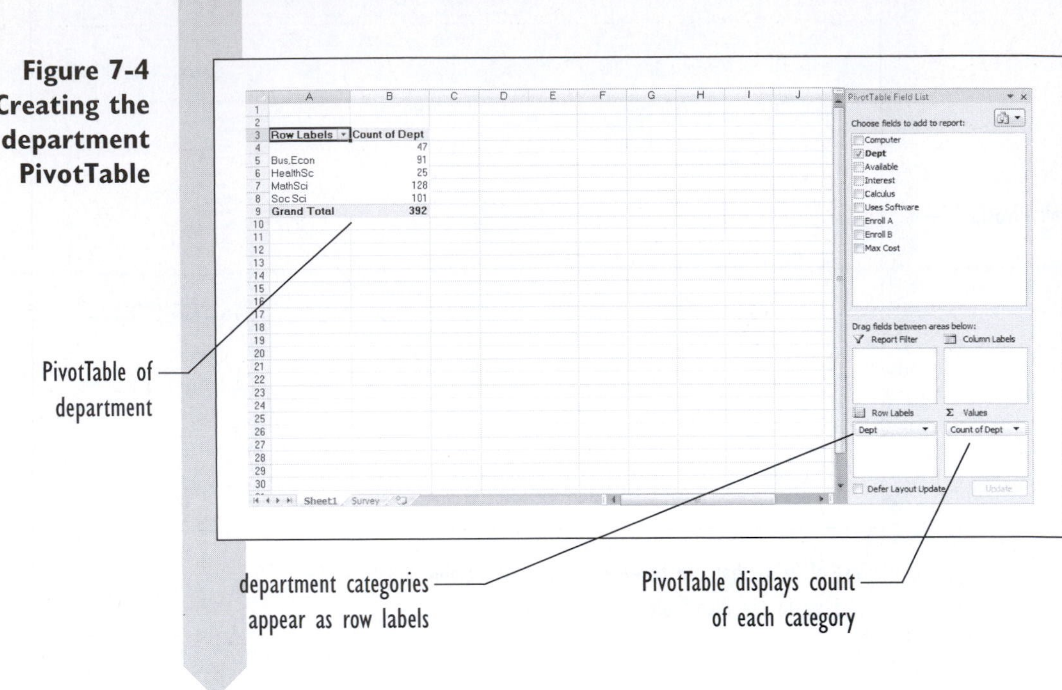

Figure 7-4 Creating the department PivotTable

The department PivotTable shows the count of different departments represented in the survey. From the table you can quickly see that there are 392 responses in the survey coming from 91 professors in business/economics, 25 professors in the health sciences, 128 professors in math and science, and 101 professors in the social sciences. There are also 47 respondents who did not specify a department.

You can edit the PivotTable to remove the respondents that did not specify a department.

Removing Categories from a PivotTable

PivotTables include drop-down list boxes that you can use to specify which categories are displayed in the table. Use a list box to remove respondents that did not specify a department.

To remove a category from the PivotTable:

1 Click the **Row Labels** drop-down box on the PivotTable.

2 Deselect the blank checkbox in the list of categories. See Figure 7-5.

Figure 7-5 Removing the blank category from the PivotTable

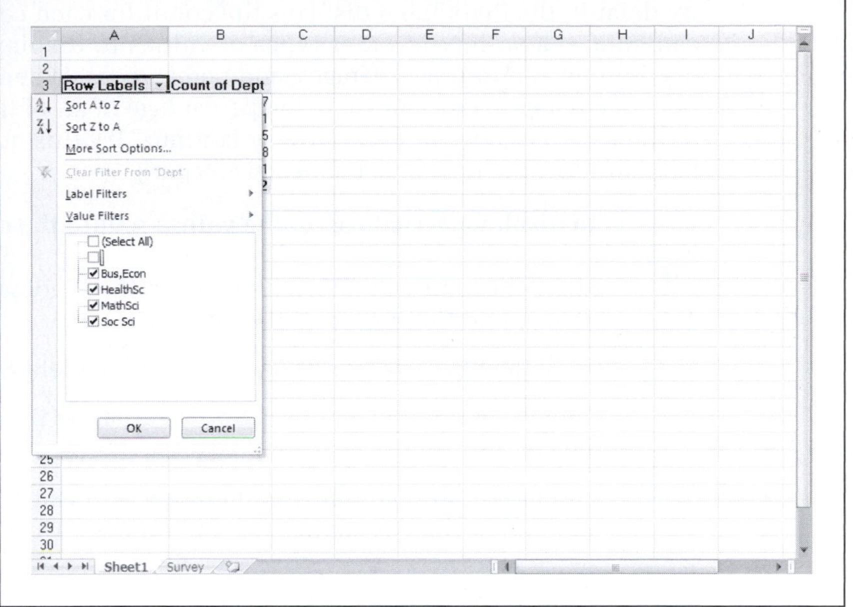

3 Click the **OK** button.

The PivotTable changes and no longer displays results from respondents that did not specify a department. See Figure 7-6.

Figure 7-6 The PivotTable with the blank category removed

	A	B	C
1			
2			
3	**Row Labels**	**Count of Dept**	
4	Bus,Econ	91	
5	HealthSc	25	
6	MathSci	128	
7	Soc Sci	101	
8	**Grand Total**	**345**	
9			

The PivotTable now shows a grand total of 345 respondents with the blank department category removed from the table.

Changing the Values Displayed by the PivotTable

By default, the PivotTable displays the count for each cell in the table. You can choose a variety of other types of values to display, including sums, maximums, minimums, averages, and percentages. When you choose to display percentages, you can show the percentage of all of the cells in the table or the percentage within each row or column. Try this now by changing the table to show the percentage for each category.

To display percentages of the values within the columns:

1. Right-click any of the count values in the PivotTable, and then click **Value Field Settings** from the pop-up menu.
2. Click the **Show values as** tab in the Value Field Settings dialog box.
3. Click % **of column** from the Show values as list box and then click the **OK** button.

The PivotTable is modified to show values as percentages of the total rather than as counts. See Figure 7-7.

Figure 7-7 PivotTable showing percentages

	A	B	C
1			
2			
3	**Row Labels**	**Count of Dept**	
4	Bus,Econ	26.38%	
5	HealthSc	7.25%	
6	MathSci	37.10%	
7	Soc Sci	29.28%	
8	**Grand Total**	**100.00%**	
9			

You can see at a glance that 26.38% of professors who listed their department came from business/economics, 7.25% came from the health sciences, 37.10% came from math and science, and 29.28% came from the social sciences.

To view the counts of the responses again:

1. Right-click any of the percentages in the PivotTable and then click **Value Field Settings** from the pop-up menu.
2. Click the **Show values as** tab, select **Normal** from the Show values as list box, and click the **OK** button.

Displaying Categorical Data in a Bar Chart

You can quickly display your PivotTable data in a PivotChart. The default PivotChart is a bar chart, in which the length of the bar is proportional to the number of counts in each cell.

To create a bar chart for department category:

1. With the PivotTable still selected, click the **PivotTable** button located in the Tools group of the Options tab on the PivotTable Tools ribbon.
2. Select the first column chart from the Insert Chart dialog box and click the **OK** button.

   As shown in Figure 7-8, Excel adds a column chart to the worksheet displaying the department counts from the PivotTable.

Figure 7-8 PivotChart of department affiliation

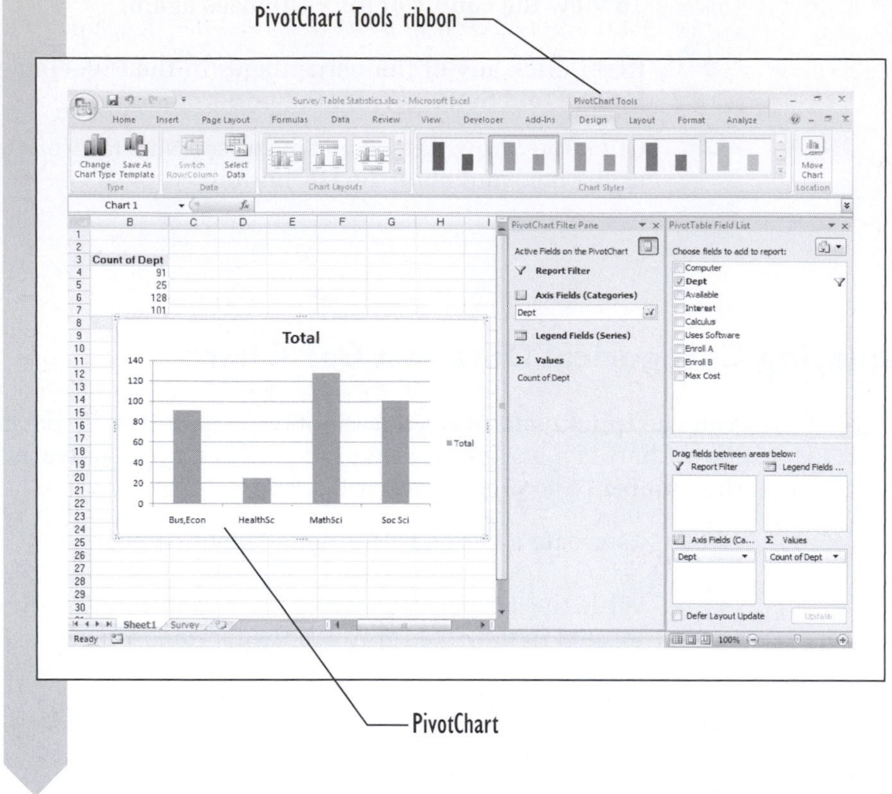

The PivotChart works the same as the PivotTable. For example, if you click the Axis Fields button in the PivotChart Filter Pane, you can add or remove categories from the chart. The pivot chart is also linked to the PivotTable, so any changes you make to layout or formatting of the chart are automatically reflected in the PivotTable.

Column charts are often used by statisticians when the need is to show the relative sizes of the groups. For example, it's quickly apparent from the chart that the mathematics and science departments have the highest representation of any category in the data set. What is not clear from the chart is the size of each group compared to the whole. For example, does the MathSci group make up more than half of the total respondents? It's not so easy to determine that kind of information from the column chart. To deal with that problem, we can have Excel add the counts to the chart.

To add counts to the column chart:

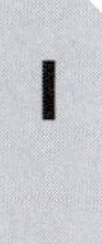

1 With the PivotChart still selected, click the **Data Labels** button from the Labels group on the Layout tab of the PivotChart Tools ribbon.

2 Click **Center** from the list of Data Label options.

Excel updates the chart, which now shows the counts for each column in the PivotChart. See Figure 7-9.

Figure 7-9 Column chart with counts

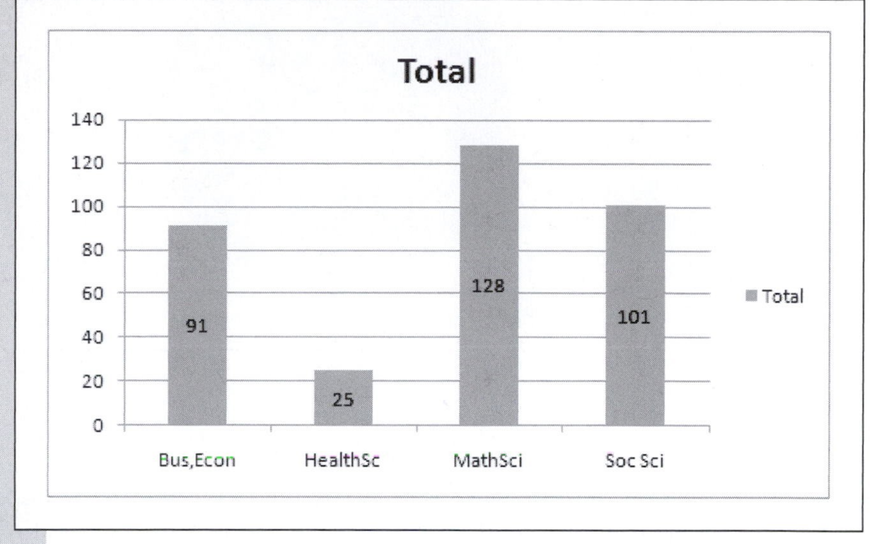

From this display we can see that 128 respondents come from the MathSci group, which corresponds to the information we saw earlier in the PivotTable.

Displaying Categorical Data in a Pie Chart

Another way of comparing the size of individual groups to the whole is the pie chart. A **pie chart** displays a circle (like a pie), in which each pie slice represents a different category. Let's see how the pie chart displays the department data. Rather than re-creating the chart from scratch, we'll simply change the chart type of the current graph.

To display a pie chart:

1 With the PivotChart still selected, click the **Change Chart Type** button located on the Type group of the Design tab on the PivotChart Tools ribbon.

2 Select the first Pie Chart subtype listed in the Change Chart Type dialog box and click the **OK** button.

Excel displays the pie chart in Figure 7-10.

Figure 7-10 Pie chart with counts

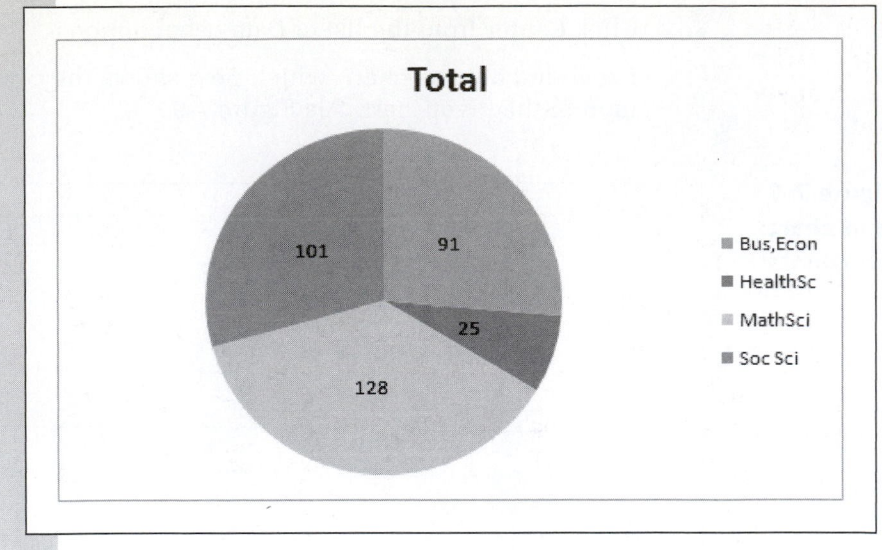

From the pie chart we can easily see a graphical comparison of the size of each department group relative to the entire collection of departments represented in the survey. The graph retained these count labels when we converted to the pie chart, but we can change that option as well. We can display the percentage of each slice, and we can add a label to the slice identifying which category it represents.

To change the pie chart's display options:

1. With the PivotChart still selected, click the **Data Labels** button again and then click **More Data Label Options** from the menu.

   Excel opens the Format Data Labels dialog box.

2. Click the **Percentage** check box and deselect the Value check box.
3. Click the **Category Name** check box.
4. Click the **Close** button.

   The chart changes as shown in Figure 7-11, displaying the percentages and labels for each pie slice.

Figure 7-11
Pie chart with percentages and labels

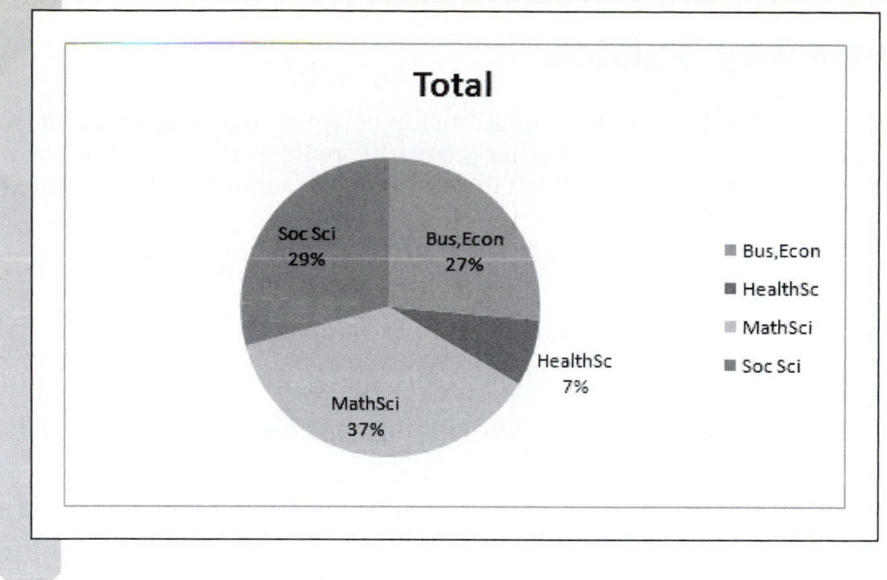

The pie chart is an effective tool for displaying categorical data, though it tends to be used more often in business reports than in statistical analyses.

EXCEL TIPS

- To view the source data for any particular cell in a PivotTable, double-click the cell. Excel will open a new worksheet containing the observations from the original data source that make up that cell.
- You can specify data sources other than the current workbook for your PivotTable data, such as databases and external files.
- If the values in the data source change, the PivotTable updates automatically to reflect those changes.
- You can create your own customized functions for the PivotTable. To do so, select any cell in the table and click the Formulas button located in the Tools group of the Options tab on the PivotTable Tools ribbon, and then click Calculated Field.

Two-Way Tables

What about the relationship between two categorical variables? You might want to see whether a calculus requirement for statistics varies by department. Excel's PivotTable feature can compile a table of calculus requirement by department.

To create a PivotTable for calculus requirement by department:

1. Return to the Survey worksheet.
2. Click the **PivotTable** button from the Tables group on the Insert tab.
3. Click the **OK** button to add a new worksheet containing the PivotTable tools.
4. Drag **Calculus** from the PivotTable Field List box and drop it into the Column Labels box.
5. Drag the **Dept** field to the Row Labels area.
6. Drag the **Dept** field onto the Values area.

Excel creates a PivotTable with counts of the Dept field broken down by the Calculus field. See Figure 7-12.

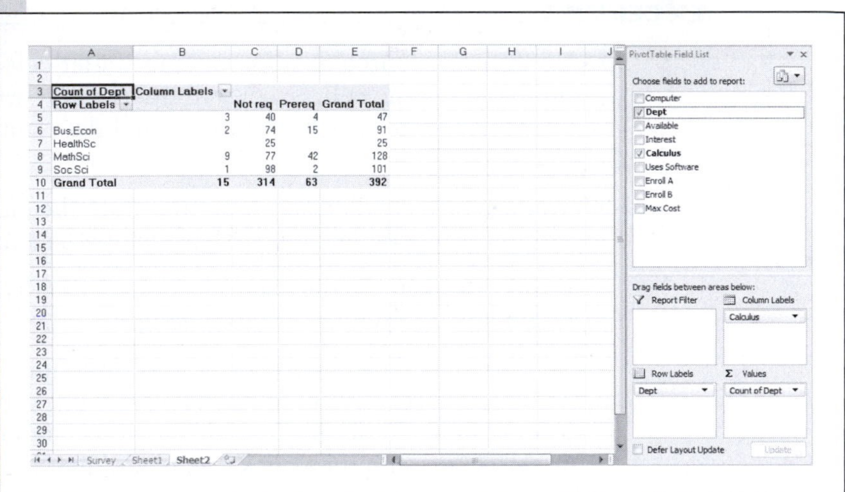

Figure 7-12 PivotTable of Dept versus Calculus

When you created the first PivotTable, you hid the blank category levels for the table. Do this as well for the two-way table.

To hide the missing values:

1. Click the **Row Labels** drop-down list arrow in the PivotTable and deselect the blank checkbox. Click **OK**.
2. Click the **Column Labels** drop-down list arrow and deselect the blank checkbox. Click **OK**.

Figure 7-13 shows the completed PivotTable.

Figure 7-13 Two-way table of department versus calculus requirement

	A	B	C	D	E
1					
2					
3	**Count of Dept**	**Column Labels**			
4	**Row Labels**	**Not req**	**Prereq**	**Grand Total**	
5	Bus,Econ	74	15	89	
6	HealthSc	25		25	
7	MathSci	77	42	119	
8	Soc Sci	98	2	100	
9	**Grand Total**	**274**	**59**	**333**	
10					

The table in Figure 7-13 shows frequencies of different combinations of department and calculus requirement. For example, cell B5, the intersection of Bus,Econ and Not req, shows that 74 professors in the business or economics departments do not require calculus as a prerequisite for their statistics course. There are a total of 333 responses. Note that missing combinations of Dept and Calculus are displayed as blanks in the PivotTable. For example, none of the statistics courses offered in the HealthSc category has calculus as a prerequisite. How do these values compare when viewed as percentages within each department? Let's modify the PivotTable to find out.

To show column percentages:

1. Right-click any of the count values in the table; then click **Value Field Settings** in the pop-up menu.
2. Click the **Show values as** tab and then select **% of row** from the Show values as list box.
3. Click the **OK** button. The revised PivotTable is shown in Figure 7-14.

Figure 7-14 Two-way table of percentages

	A	B	C	D	E
1					
2					
3	**Count of Dept**	**Column Labels**			
4	**Row Labels**	**Not req**	**Prereq**	**Grand Total**	
5	Bus,Econ	83.15%	16.85%	100.00%	
6	HealthSc	100.00%	0.00%	100.00%	
7	MathSci	64.71%	35.29%	100.00%	
8	Soc Sci	98.00%	2.00%	100.00%	
9	**Grand Total**	**82.28%**	**17.72%**	**100.00%**	
10					

On the basis of the percentages you can quickly observe that for most departments calculus is not a prerequisite for statistics, except for courses in the math or science departments in which more than one-third of the courses have such a requirement.

How do these percentages compare to the overall total number of respondents in the survey? Let's find out.

To calculate percentages of the total responses in the survey:

1. Right-click any cell in the PivotTable and click **Value Field Settings** from the pop-up menu.
2. Click the **Show values as** tab and select % **of total** from the Show values as list box. Click the **OK** button. Figure 7-15 shows the revised PivotTable.

Figure 7-15 Percentages of the total

	A	B	C	D	E
1					
2					
3	**Count of Dept**	**Column Labels**			
4	**Row Labels**	**Not req**	**Prereq**	**Grand Total**	
5	Bus,Econ	22.22%	4.50%	26.73%	
6	HealthSc	7.51%	0.00%	7.51%	
7	MathSci	23.12%	12.61%	35.74%	
8	Soc Sci	29.43%	0.60%	30.03%	
9	**Grand Total**	**82.28%**	**17.72%**	**100.00%**	
10					

On the basis of this table, almost 18% of the professors in the survey teach a class that has calculus as a prerequisite whereas more than 82% do not.

To reformat the table to show counts again:

1 Right-click any of the percentages in the PivotTable; then click **Value Field Settings** in the pop-up menu.

2 Click the **Show values as** tab and select **Normal** from the Show values as list box. Click the **OK** button.

Computing Expected Counts

If a calculus prerequisite were the same in each department, we would expect to find the column percentages (shown in Figure 7-14) to be about the same for each department. We would then say that department and calculus prerequisite are **independent** of each other, so that the pattern of usage does not depend on the department. It's the same for all of them. On the other hand, if there *is* a difference between departments, we would say that department and calculus prerequisite use are **related**. We cannot say anything about whether knowledge of calculus is usually required without knowing which department is being examined.

You've seen that there might be a relationship between the calculus variables and department. Is this difference significant? We could formulate the following hypotheses:

H_0: The calculus requirement is the same in all departments

H_a: The calculus requirement is related to the department

How can you test the null hypothesis? Essentially, you want a test statistic that will examine the calculus requirement across departments and then compare it to what we would expect to see if the calculus requirement and department type were independent variables.

How do you compute the expected counts? Under the null hypothesis, the percentage of courses requiring calculus should be the same across departments. Our best estimate of these percentages comes from the percentage of the grand total, shown in Figure 7-14. Thus we expect about 82.28% of the courses to require calculus and about 17.72% not to. To express this value in terms of counts, we multiply the expected percentage by the total number of courses in each department. For example, there are 119 courses in the MathSci departments, and if 17.72% of these had a calculus prerequisite, this would be 119×0.1772, or about 21.08, courses. Note that the actual observed value is 42 (cell C7 in Figure 7-13), so the number of courses that require calculus is higher than expected under the null hypothesis. The expected count can also be calculated using the formula

$$\text{Expected count} = \frac{(\text{Row total}) \times (\text{Column total})}{\text{Total observations}}$$

Thus the expected count of courses that have a calculus prerequisite in the MathSci departments could also be calculated as follows:

$$\text{Expected count} = \frac{59 \times 119}{333} = 21.08$$

To create a table of expected counts, you can either use Excel to perform the manual calculations or use the StatPlus add-in to create the table for you.

To create a table of expected counts:

1. Click **Descriptive Statistics** and then **Table Statistics** from the StatPlus menu.
2. Enter the range **A4:C8**.
3. Click the **Output** button; then click the **New Worksheet** option button and type the worksheet name, **Calculus Department Table**. Click **OK**.
4. Click the **OK** button to start generating the table of expected counts. See Figure 7-16.

Figure 7-16 Table of observed and expected counts

	A	B	C	D
1	Table Statistics			
2	*Observed Counts*	Not req	Prereq	
3	Bus,Econ	74	15	
4	HealthSc	25		
5	MathSci	77	42	
6	Soc Sci	98	2	
7				
8	*Expected Counts*	Not req	Prereq	
9	Bus,Econ	73.23	15.77	
10	HealthSc	20.57	4.43	
11	MathSci	97.92	21.08	
12	Soc Sci	82.28	17.72	

The command generates some output in addition to the table of expected counts shown in Figure 7-16, which we'll discuss later. The values in the Expected Counts table are the counts we would expect to see if a calculus requirement were independent of department.

The Pearson Chi-Square Statistic

With our tables of observed counts and expected counts, we need to calculate a single test statistic that will summarize the amount of difference between the two tables. In 1900, the statistician Karl Pearson devised such a test statistic, called the **Pearson chi-square**. The formula for the Pearson chi-square is

$$\text{Pearson chi-square} = \sum_{\text{all cells}} \frac{(\text{Observed count} - \text{Expected count})^2}{\text{Expected count}}$$

If the frequencies all agreed with their expected values, this total would be 0. If there is a substantial difference between the observed and expected counts, this value will be large. For the data in Figure 7-16, this value is

$$\text{Pearson chi-square} = \frac{(74 - 73.23)^2}{73.23} + \frac{(15 - 15.77)^2}{15.77} + \frac{(25 - 20.57)^2}{20.57} + \cdots + \frac{(2 - 17.72)^2}{17.72} = 47.592$$

Is this value large or small? Pearson discovered that when the null hypothesis is true, values of this test statistic approximately follow a distribution called the χ^2 **distribution** (pronounced "chi-squared"). Therefore, one needs to compare the observed value of the Pearson chi-square with the χ^2 distribution to decide whether the value is large enough to warrant rejection of the null hypothesis.

CONCEPT TUTORIALS
The χ^2 Distribution

To understand the χ^2 distribution better, use the explore workbook for Distributions.

To use the Distribution workbook:

1. Open the **Distributions** workbook, located in the Explore folder. Enable the macros in the workbook.
2. Click **Chi-squared** from the Table of Contents column. Review the material and scroll to the bottom of the worksheet. See Figure 7-17.

Figure 7-17
χ^2 distribution

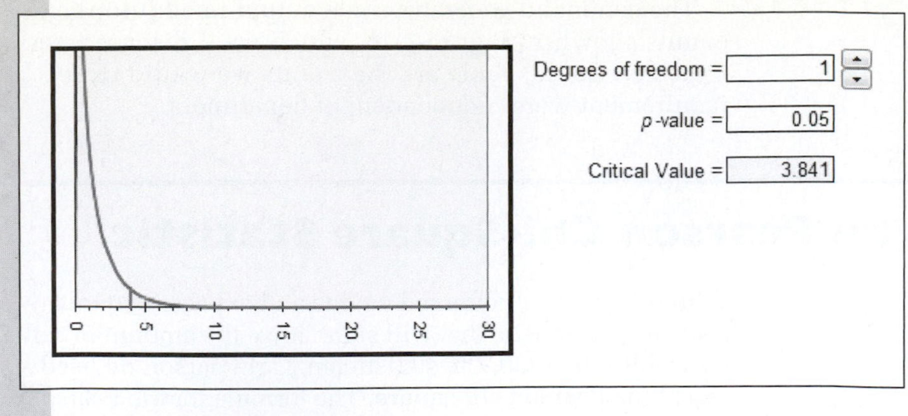

Unlike the normal distribution and *t* distribution, the X^2 distribution is limited to values ≥ 0. However, like the *t* distribution, the X^2 distribution involves a single parameter—the degrees of freedom. When the degrees of freedom are low, the distribution is highly skewed. As the degrees of freedom increase, the weight of the distribution shifts farther to the right and becomes less skewed. To see how the shape of the distribution changes, try changing the degrees of freedom in the worksheet.

To increase the degrees of freedom for the X^2 distribution:

1 Click the **Degrees of freedom** spin arrow and increase the degrees of freedom to **3**.

The distribution changes shape as shown in Figure 7-18.

Figure 7-18
χ^2 distribution with 3 degrees of freedom

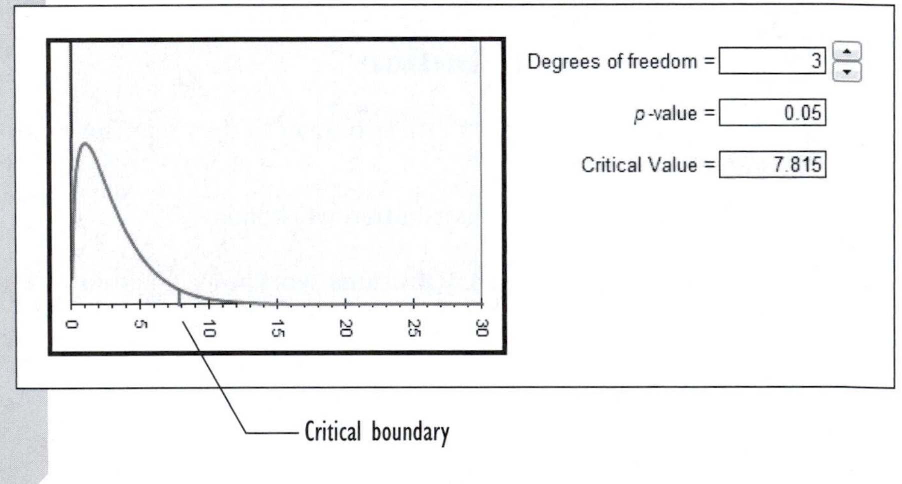

Like the normal and t distributions, the X^2 distribution has a critical boundary for rejecting the null hypothesis, but unlike those distributions, it's a one-sided boundary. There are a few situations where one might use upper and lower critical boundaries.

The critical boundary is shown in your chart with a vertical red line. Currently, the critical boundary is set for $\alpha = 0.05$. In Figure 7-18, this is equal to 7.815. You can change the value of α in this worksheet to see the critical boundary for other p values.

To change the critical boundary:

1 Click the ***p* value** box, type **0.10**, and press **Enter**.

The critical boundary changes, moving back to 6.251.

Experiment with other values for the degrees of freedom and the critical boundary.

When you're finished with the worksheet:

1 Close the Distributions workbook. Do not save any changes.

2 Return to the Survey Table Statistics workbook, displaying the Calculus Department Table worksheet.

The degrees of freedom for the Pearson chi-square are determined by the numbers of rows and columns in the table. If there are r rows and c columns, the number of degrees of freedom are $(r-1) \times (c-1)$. For our table of calculus requirement by department, there are 4 rows and 2 columns, and the number of degrees of freedom for the Pearson chi-square statistic is $(4-1) \times (2-1)$, or 3.

Where does the formula for degrees of freedom come from? The Pearson chi-square is based on the differences between the observed and expected counts. Note that the sum of these differences is 0 for each row and column in the table. For example, in the first column of the table, the expected and observed counts are as shown in Table 7-2:

Table 7-2 Counts for Calculus Requirement by Department

Observed	Expected	Difference
74	73.23	0.77
25	20.57	4.43
77	97.92	−20.92
98	82.28	15.72
	Sum	0.00

Because this sum is 0, the last difference can be calculated on the basis of the previous three, and there are only three cells that are free to vary in value. Applied to the whole table, this means that if we know 3 of the 8 differences, then we can calculate the values of the remaining 5 differences. Hence the number of degrees of freedom is 3.

Working with the χ^2 Distribution in Excel

Now that we know the value of the test statistic and the degrees of freedom, we are ready to test the null hypothesis. Excel includes several functions to help you work with the χ^2 distribution. Table 7-3 shows some of these.

Table 7-3 Excel Functions for χ^2 Distribution

Function	Description
CHIDIST(*x*, *df*)	Returns the *p* value for the χ^2 distribution for a given value of *x* and degrees of freedom *df*.
CHIINV(*p*, *df*)	Returns the χ^2 value from the χ^2 distribution with degrees of freedom *df* and *p* value *p*.
CHITEST(*observed*, *expected*)	Calculates the Pearson chi square, where *observed* is a range containing the observed counts and *expected* is a range containing the expected counts.
PEARSONCHISQ(*observed*)	Calculates the Pearson chi-square, where *observed* is the table containing the observed counts. *StatPlus required.*
PEARSONP(*observed*)	Calculates the *p* value of the Pearson chi-square, where *observed* is the table containing the observed counts. *StatPlus required.*

The output you generated earlier displays (among other things) the value for the Pearson chi-square statistic. The χ^2 value is 47.592 with a *p* value of less than 0.001. Because this probability is less than 0.05, you reject the null hypothesis that the calculus requirement does not differ on the basis of the department (not surprisingly).

Breaking Down the Chi-Square Statistic

The value of the Pearson chi-square statistic is built up from every cell in the table. You can get an idea of which cells contributed the most to the total value by observing the table of standardized residuals. The value of the **standardized residual** is

$$\text{Standardized residual} = \frac{\text{Observed count} - \text{Expected count}}{\sqrt{\text{Expected count}}}$$

Figure 7-19 displays the standardized residuals for the Calculus Requirement by Department table. Note that the highest standardized residual, 4.56, is found for the MathSci department under the Prereq column, leading us to believe that this count had the highest impact on rejecting the null hypothesis.

Figure 7-19 Table of standardized residuals

	A	B	C	D
14	*Std. Residuals*	Not req	Prereq	
15	Bus,Econ	0.09	-0.19	
16	HealthSc	0.98	-2.10	
17	MathSci	-2.11	4.56	
18	Soc Sci	1.73	-3.73	

Other Table Statistics

A common mistake is to use the value of X^2 to measure the degree of association between the two categorical variables. However, the X^2, along with the p value, measures only the *significance* of the association. This is because the value of X^2 is partly dependent on sample size and the size of the table. For example, in a 3×3 table, a X^2 value of 10 is significant with a p value of 0.04, but the same value in a 4×4 table is *not* significant with a p value of 0.35.

A **measure of association**, on the other hand, gives a value to the association between the row and column variables that is not dependent on the sample size or the size of the table. Generally, the higher the measure of association, the stronger the association between the two categorical variables.

Figure 7-20 shows other test statistics and measures of association created by the StatPlus Table Statistics command.

Figure 7-20
Other table statistics

	A	B	C	D	E
20	*Test Statistics*	Value	df	p-value	
21	Pearson Chi-Square	47.592	3	0.000	
22	Continuity Adjusted Chi-Square	44.153	3	0.000	
23	Likelihood Ratio Chi-Square	56.216	3	0.000	
24					
25	*Measures of Association*	Value	Std. Error	p-value	
26	Phi	0.378			
27	Contigency	0.354			
28	Cramer's V	0.378			
29	Goodman-Kruskal Gamma	-0.224	0.086	0.009	
30	Kendalls tau-b	-0.104	0.041	0.011	
31	Stuart's tau-c	-0.094	0.037	0.012	
32	Somer's D (C|R)	-0.067	0.027	0.012	
33	Somer's D (R|C)	-0.161	0.062	0.010	
34					

Table 7-4 summarizes these statistics and their uses.

Table 7-4 StatPlus Table Statistics

Statistic	Description
Pearson chi-square	Calculates the difference between the observed and expected counts. Approximately follows a X^2 distribution with $(r-1) \times (c-1)$ degrees of freedom, where r is the number of rows in the table and c is the number of table columns.
Continuity-adjusted chi-square	Similar to the Pearson chi square, except that it adjusts the X^2 value for the continuity of the X^2 distribution.
Likelihood ratio chi-square	It approximately follows a X^2 distribution with $(r-1) \times (c-1)$ degrees of freedom.
Phi	Measures the association between the row and column variables, varying from -1 to 1. A value near 0 indicates no association. Phi varies from 0 to 1 unless the table's dimension is 2×2.
Contingency	A measure of association ranging from 0 (no association) to a maximum of 1 (high association). The upper bound may be less than 1, depending on the values of the row and column totals.
Cramer's V	A variation of the contingency measure, modifying the statistic so that the upper bound is always 1.

(continued)

Goodman-Kruskal gamma	A measure of association used when the row and column values are ordinal variables. Gamma ranges from −1 to 1. A negative value indicates negative association, a positive value indicates positive association, and 0 indicates no association between the variables.
Kendall's tau-b	Similar to gamma, except that tau-b includes a correction for ties. Used only for ordinal variables.
Stuart's tau-c	Similar to tau-b, except that it includes a correction for table size. Used only for ordinal variables.
Somers' D	A modification of the tau-b statistic. Somers' D is used for ordinal variables in which one variable is used to predict the value of the other variable. Somers' D (R I C) is used when the column variable is used to predict the value of the row variable. Somers' D (C I R) is used when the row variable is used to predict the value of the column variable.

Because the χ^2 distribution is a continuous distribution and counts represent discrete values, some statisticians are concerned that the Pearson chi-square statistic is not appropriate. They recommend using the continuity-adjusted chi-square statistic instead. We feel that the Pearson chi-square statistic is more accurate and can be used without adjustment.

Among the other statistics in Table 7-4, the likelihood ratio chi-square statistic is usually close to the Pearson chi-square statistic. Many statisticians prefer using the likelihood ratio chi-square because it is used in log-linear modeling—a topic beyond the scope of this book.

All of the three test statistics shown in Figure 7-20 are significant at the 5% level. The association between the Calculus Requirement and Department variables ranges from 0.354 to 0.378 for the three measures of association (Phi, Contingency, and Cramer's V). The final four measures of association (gamma, tau-b, tau-c, and Somers' D) are used for ordinal data and are not appropriate for nominal data.

Validity of the Chi-Square Test with Small Frequencies

One problem you may encounter is that it might not be valid to use the Pearson chi-square test on a table with a large number of sparse cells. A **sparse cell** is defined as a cell in which the expected count is less than 5. The Pearson chi-square test requires large samples, and this means that cells with small counts can be a problem. You might get by with as many as one-fifth of the expected counts under 5, but if it's more than that, the *p* value returned by the Pearson chi square might lead you erroneously to reject or accept the null hypothesis. The StatPlus add-in will automatically display a warning if this occurs in your sample data. This warning was not displayed with the data from the survey; however, there was one cell that had a low expected count of 4.43.

If you wanted to remove a sparse cell from your analysis how would you go about it? You can either pool columns or rows together to increase the cell counts or remove rows or columns from the table altogether.

For these data you'll combine the counts from the Bus,Econ, HealthSc, and SocSci departments into a single group and compare that group to the counts in the MathSci department. Rather than editing the data in the worksheet, you can combine the groups in the PivotTable.

To compare the MathSci department to all other departments:

1. Return to the Survey worksheet and create another PivotTable on a new worksheet with Department in the row area and Calculus in the column area. Display the count of Department in the Values area.
2. Remove the blank Department entry from the row area of the PivotTable and remove the blank Calculus entry from the Column area of the table.
3. With the **Ctrl** key held down, click cells **A5, A6**, and **A8** in the PivotTable. This selects the three rows corresponding to the Bus,Econ, HealthSc, and SocSci departments respectively.
4. Click the **Group Selection** button from the Group on the Options tab of the PivotTable Tools ribbon. Excel adds grouping variables to the PivotTable rows as shown in Figure 7-21.

Figure 7-21 Grouping rows in the PivotTable

	A	B	C	D	E
1					
2					
3	Count of Dept	Column Labels			
4	Row Labels	Not req	Prereq	Grand Total	
5	Group1				
6	Bus,Econ	74	15	89	
7	HealthSc	25		25	
8	Soc Sci	98	2	100	
9	MathSci				
10	MathSci	77	42	119	
11	Grand Total	274	59	333	
12					

Next you want to give the two groups descriptive names and collapse the PivotTable around the two groups you created.

To rename and collapse the PivotTable groups:

1. Click cell **A5** in the PivotTable, type **Other Depts**, and press **Enter**.
2. Click the **minus** boxes in front of the group titles in cells A5 and A9 to collapse the groups. Figure 7-22 shows the collapsed PivotTable.

Figure 7-22 PivotTable with grouped rows

	A	B	C	D	E
1					
2					
3	Count of Dept	Column Labels			
4	Row Labels	Not req	Prereq	Grand Total	
5	⊞Other Depts	197	17	214	
6	⊞MathSci	77	42	119	
7	Grand Total	274	59	333	
8					

Notice that the expected cell counts are all much greater than 5. Grouping the rows in the PivotTable is a good way to remove problems with sparseness in the cell counts. Now that you've restructured the PivotTable, rerun your analysis, comparing the MathSci department to all of the other combined departments.

To check the table statistics on the revised table:

1. Click **Descriptive Statistics** from the StatPlus menu and then click **Table Statistics**.
2. Enter the range **A4:C6**.
3. Click the **Output** button and send the output to the new worksheet, Calculus **Department Table 2**.
4. Click the **OK** button to start generating the new batch of table statistics. See Figure 7-23.

Figure 7-23
Table statistics for the grouped PivotTable

	A	B	C	D	E
1	**Table Statistics**				
2	*Observed Counts*	Not req	Prereq		
3	Other Depts	197	17		
4	MathSci	77	42		
5					
6	*Expected Counts*	Not req	Prereq		
7	Other Depts	176.08	37.92		
8	MathSci	97.92	21.08		
9					
10	*Std. Residuals*	Not req	Prereq		
11	Other Depts	1.58	-3.40		
12	MathSci	-2.11	4.56		
13					
14	*Test Statistics*	Value	df	p-value	
15	Pearson Chi-Square	39.239	1	0.000	
16	Continuity Adjusted Chi-Square	37.386	1	0.000	
17	Likelihood Ratio Chi-Square	37.832	1	0.000	
18					
19	*Measures of Association*	Value	Std. Error	p-value	
20	Phi	0.343			
21	Contigency	0.325			
22	Cramer's V	0.343			
23	Goodman-Kruskal Gamma	0.727	0.075	0.000	
24	Kendalls tau-b	0.343	0.054	0.000	
25	Stuart's tau-c	0.251	0.044	0.000	
26	Somer's D (C|R)	0.274	0.048	0.000	
27	Somer's D (R|C)	0.431	0.065	0.000	
28					

There is little difference in our conclusion by grouping the table rows. Once again the Pearson chi-square test statistic is highly significant with a p value of less than 0.001. Thus we would reject the null hypothesis and accept the alternative hypothesis that the MathSci departments are more likely to require a calculus prerequisite.

Tables with Ordinal Variables

The two-way table you just produced was for two nominal variables. The Pearson chi-square test makes no assumption about the ordering of the categories in the table. Now let's look at categorical variables that have inherent order. For ordinal variables, there are more powerful tests than the Pearson chi square, which often fails to give significance for ordered variables.

As an example, consider the Calculus and Enroll B variables. Since the Calculus variable tells the extent to which calculus is required for a given statistics class (Not req or Prereq), it can be treated as either an ordinal or a nominal variable. When a variable takes on only two values, there really is no distinction between nominal and ordinal, because any two values can be regarded as ordered. The other variable, Enroll B, is a categorical variable that contains measures of the size of the annual course enrollment. In the survey, instructors were asked to check one of eight categories (0–50, 51–100, 101–150, 151–200, 201–300, 301–400, 401–500, and 501–) for the number of students in the course. You might expect that classes requiring calculus would have smaller enrollments.

Testing for a Relationship between Two Ordinal Variables

We want to test whether there is a relationship between a class requiring calculus as a prerequisite and the size of the class. Our hypotheses are

H_0: The pattern of enrollment is the same regardless of a calculus prerequisite.

H_a: The pattern of enrollment is related to a calculus prerequisite.

To test the null hypothesis, first form a two-way table for categorical variables, Calculus and Enroll B.

To form the table:

1. Return to the Survey worksheet and create another PivotTable on a new worksheet.
2. Place the **Enroll B** field in the Row Labels area of the PivotTable and **Calculus** in the Column Labels area. Also place **Calculus** in the Values area of the table.
3. Remove blank entries from both the row and column labels in the PivotTable. Figure 7-24 shows the final PivotTable.

Figure 7-24 Table of Enrollment versus Calculus Prerequisite

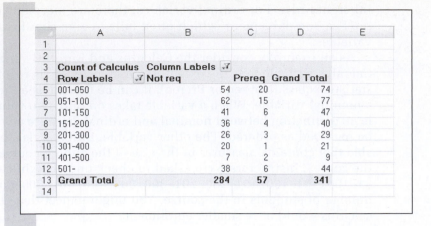

	A	B	C	D	E
1					
2					
3	Count of Calculus	Column Labels			
4	Row Labels	Not req	Prereq	Grand Total	
5	001-050	54	20	74	
6	051-100	62	15	77	
7	101-150	41	6	47	
8	151-200	36	4	40	
9	201-300	26	3	29	
10	301-400	20	1	21	
11	401-500	7	2	9	
12	501-	38	6	44	
13	Grand Total	284	57	341	
14					

In the table you just created, Excel automatically arranges the Enroll B levels in a combination of numeric and alphabetic order (alphanumeric order), so be careful with category names. For example, if 051–100 were written as 51–100 instead, Excel would place it near the bottom of the table because 5 comes after 1, 2, 3, and 4.

Remember that most of the expected values should exceed 5, so there is cause for concern about the sparsity in the second column between 201 and 500, but because only 4 of the 16 cells have expected values less than 5, the situation is not terrible. Nevertheless, let's combine enrollment levels from 200 to 500 of the PivotTable.

To combine levels, use the same procedure you did with the calculus requirement table:

1. Highlight **A9:A11**, the enrollment row labels for the categories from 200 through 500.
2. Click the **Group Selection** button from the Group on the Options tab of the PivotTable Tools ribbon.
3. Click cell **A4**, type **Enrollment**, and press **Enter**.
4. Click cell **A13**, type **201–500**, and press **Enter**.
5. Click the **minus** boxes in front of all of the row categories to collapse them. See Figure 7-25.

Figure 7-25 Enrollment table with pooled table rows

	A	B	C	D	E
1					
2					
3	Count of Calculus	Column Labels			
4	Enrollment	Not req	Prereq	Grand Total	
5	001-050	54	20	74	
6	051-100	62	15	77	
7	101-150	41	6	47	
8	151-200	36	4	40	
9	201-500	53	6	59	
10	501-	38	6	44	
11	Grand Total	284	57	341	
12					

Now generate the table statistics for the new table.

To generate table statistics:

1. Click **Descriptive Statistics** from the StatPlus menu and then click **Table Statistics**.
2. Select the range **A4:C10**.
3. Click the **Output** button and send the table to a new worksheet named **Enrollment Statistics**.
4. Click the **OK** button.

The statistics for this table are as shown in Figure 7-26.

Figure 7-26 Statistics for the Enrollment table

	A	B	C	D
26	*Test Statistics*	Value	df	p-value
27	Pearson Chi-Square	10.013	5	0.075
28	Continuity Adjusted Chi-Square	7.818	5	0.167
29	Likelihood Ratio Chi-Square	9.762	5	0.082
30				
31	*Measures of Association*	Value	Std. Error	p-value
32	Phi	0.171		
33	Contigency	0.169		
34	Cramer's V	0.171		
35	Goodman-Kruskal Gamma	-0.280	0.101	0.006
36	Kendalls tau-b	-0.133	0.049	0.006
37	Stuart's tau-c	-0.127	0.048	0.007
38	Somer's D (C|R)	-0.077	0.029	0.007
39	Somer's D (R|C)	-0.229	0.083	0.006

statistics that don't assume ordinal data (rows 27–29)

statistics that assume ordinal data (rows 35–39)

It's interesting to note that those statistics which do not assume that the data are ordinal (the Pearson chi-square, the continuity-adjusted X^2, and the likelihood ratio X^2) all fail to reject the null hypothesis at the 0.05 level. On the other hand, the statistics that take advantage of the fact that we're using ordinal data (the Goodman-Kruskal gamma, Kendall's tau-b, Stuart tau-c, and Somers' D) all reject the null hypothesis. This illustrates an important point: Always use the statistics test that best matches the characteristics of your data. Relying on the ordinal tests, we reject the null hypothesis, accepting the alternative hypothesis that the pattern of enrollment differs on the basis of whether calculus is a prerequisite.

To explore how that difference manifests itself, let's examine the table of expected values and standardized residuals in Figure 7-27.

Figure 7-27 Expected counts and standardized residuals for the enrollment table

	A	B	C
1	**Table Statistics**		
2	*Observed Counts*	Not req	Prereq
3	001-050	54	20
4	051-100	62	15
5	101-150	41	6
6	151-200	36	4
7	201-500	53	6
8	501-	38	6
9			
10	*Expected Counts*	Not req	Prereq
11	001-050	61.63	12.37
12	051-100	64.13	12.87
13	101-150	39.14	7.86
14	151-200	33.31	6.69
15	201-500	49.14	9.86
16	501-	36.65	7.35
17			
18	*Std. Residuals*	Not req	Prereq
19	001-050	-0.97	2.17
20	051-100	-0.27	0.59
21	101-150	0.30	-0.66
22	151-200	0.47	-1.04
23	201-500	0.55	-1.23
24	501-	0.22	-0.50

From the table, we see that the null hypothesis underpredicts the number of courses with class sizes in the 1–50 range that require knowledge of calculus. The null hypothesis predicts that 12.37 classes fit this classification, and there were 20 of them. As the size of the classes increases, the null hypothesis increasingly overpredicts the number of courses that require calculus. For example, if the null hypothesis were true, we would expect to see almost 10 courses with 201–500 students each require knowledge of calculus. The observed number from the survey was 6. From this we conclude that class size and a calculus prerequisite are not independent and that the courses that require knowledge of calculus are more likely to be smaller.

Custom Sort Order

With ordinal data you want the values to appear in the proper order when created by the PivotTable. If order is alphabetic or the variable itself is numeric, this is not a problem. However, what if the variable being considered has a definite order, but this order is neither alphabetic nor numeric? Consider the Interest variable from the Survey workbook, which measures the degree of interest in a supplementary statistics text. The values of this variable have a definite order (least, low, some, high, most), but this order is not alphabetic or numeric. You could create a numeric variable based on the values of Interest, such as 1 = least, 2 = low, 3 = some, 4 = high, and 5 = most. Another approach is to create a custom sort order, which lets you define a sort order for a variable.

You can define any number of custom sort orders. Excel already has some built in for your use, such as months of the year (Jan, Feb, Mar, ... Dec), so if your data set has a variable with month values, you can sort the data list by months (you can also do this with PivotTables). Try creating a custom sort order for the values of the Interest variable.

To create a custom sort order for the values of the Interest variable:

1. Click the **Office Button** and then click **Excel Options**.
2. Click **Popular** from the list of Excel options.
3. Click the **Edit Custom Lists** button.
4. Click the **List entries** list box.
5. Type **least** and press **Enter**.
6. Type **low** and press **Enter**.
7. Type **some** and press **Enter**.
8. Type **high** and press **Enter**.
9. Type **most** and click **Add**.

Your Custom Lists dialog box should look like Figure 7-28.

Figure 7-28
Custom sort order

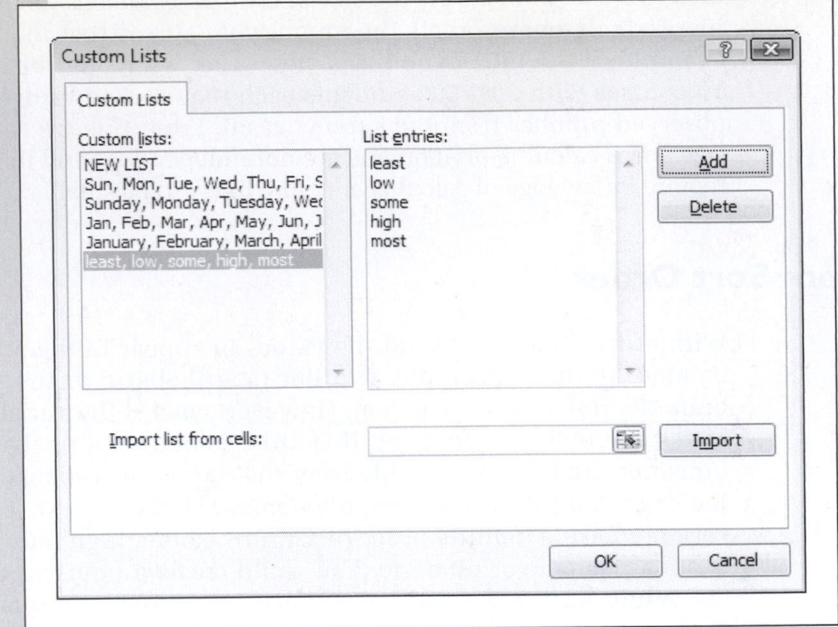

10 Click the **OK** button twice to return to the workbook.

Now create a PivotTable of Interest values to see whether Excel automatically applies the sort order you just created.

To create the PivotTable:

1 Return to the Survey worksheet and insert a new PivotTable on a new worksheet.

2 Drag the **Interest** field to the Row Labels and Values area of the PivotTable. Figure 7-29 shows the resulting PivotTable.

Figure 7-29
Custom sort order

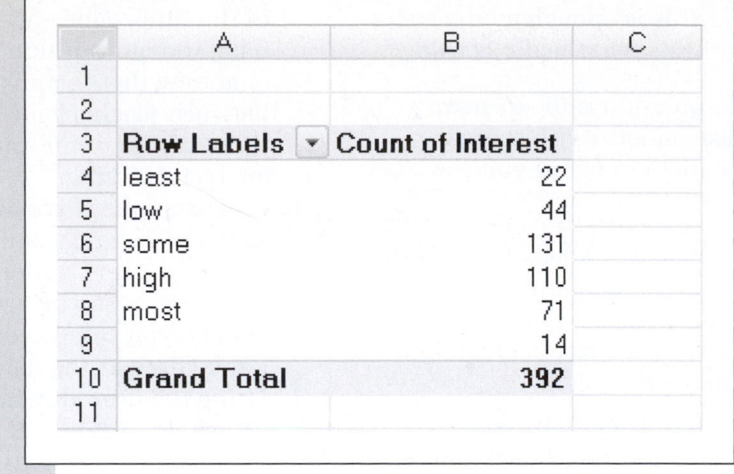

	A	B	C
1			
2			
3	**Row Labels**	**Count of Interest**	
4	least	22	
5	low	44	
6	some	131	
7	high	110	
8	most	71	
9		14	
10	**Grand Total**	**392**	
11			

Excel automatically sorts the Interest categories in the PivotTable in the proper order—least, low, some, high, and most—rather than alphabetically.

You've completed your work with categorical data with Excel. You can save your changes to the workbook now and close Excel if you want to take a break before starting on the exercises.

Exercises

1. Use Excel to calculate the following p values for the χ^2 distribution:
   a. $\chi^2 = 4$ with 4 degrees of freedom
   b. $\chi^2 = 4$ with 1 degree of freedom
   c. $\chi^2 = 10$ with 6 degrees of freedom
   d. $\chi^2 = 10$ with 3 degrees of freedom

2. Use Excel to calculate the following critical values for the χ^2 distribution:
   a. $\alpha = 0.10$, degrees of freedom $= 4$
   b. $\alpha = 0.05$, degrees of freedom $= 4$
   c. $\alpha = 0.05$, degrees of freedom $= 9$
   d. $\alpha = 0.01$, degrees of freedom $= 9$

3. *True or false, and why?* The Pearson chi-square test measures the degree of association between one categorical variable and another.

4. You are suspicious that a die has been tampered with. You decide to test it. After several tosses, you record the following results shown in Table 7-5:

Table 7-5 Die-Tossing Experiment

Number	Occurrences
1	32
2	20
3	28
4	14
5	23
6	15

Use Excel's CHITEST function to determine whether this is enough evidence to reject a hypothesis that the die is true.

5. Why should you not use the Pearson chi-square test statistic with ordinal data? What statistics should you use instead?

6. The Junior College workbook (discussed earlier in Chapter 3) contains information on hiring practices at a junior college. Analyze the data from PivotTables created for the workbook.
   a. Open the **Junior College** workbook from the Chapter07 data folder and save it as **Junior College Table Statistics**.
   b. Create a customized list of teaching ranks sorted in the following order: instructor, assistant professor, associate professor, full professor.
   c. Create a PivotTable with Rank Hired as the row label, Gender as the column label, and count of rank hired in the values area.
   d. Explore the question of whether there is a relationship between teaching rank and gender. What are your hypotheses? Generate the table statistics for this PivotTable. Which statistics are appropriate to use with the table? Is there any difficulty with the data in the table? How would you correct these problems?
   e. Group the associate professor and full professor groups together and redo your analysis.
   f. Group the three professor ranks into one and redo your analysis, relating gender to the instructor/professor split.
   g. Write a report summarizing your results, displaying the relevant tables and statistics. How do your three tables differ with respect to your conclusions? Discuss some of the problems one could encounter when trying to eliminate sparse data. Which of the three tables best describes the data, in your opinion, and would you conclude that there is a relationship between teacher rank and gender? What pieces of information is this analysis missing?
   h. Create another PivotTable with Degree as the page field, Rank Hired as the row field, and Gender as the column field. (Because you are obtaining counts, you can use either Gender or Rank Hired as the data field.)
   i. Using the drop-down arrows on the Page field button, display the table of persons hired with a Master's degree.
   j. Generate table statistics for this group. Is the rank when hired independent of gender for candidates with Master's degrees? Redo the analysis if necessary to remove sparse cells.
   k. Save your changes to the workbook and write a report summarizing your conclusions, including any relevant tables and statistics.

7. The Cold workbook contains data from a 1961 French study of 279 skiers during two 5–7 day periods. One group of skiers received a placebo (an ineffective saline solution), and another group received 1 gram of ascorbic acid per day. The study was designed to measure the incidence of the common cold in the two groups.
   a. State the null and alternative hypotheses of this study.
   b. Open the **Cold** workbook from the Chapter07 data folder and save it as **Cold Statistics**.
   c. Analyze the results of the study using the appropriate table statistics.
   d. Save your changes to the workbook and write a report summarizing your results.

8. The data in the Marriage workbook contain information on the heights of newly married couples. The study is designed to test whether people tend to choose marriage partners similar in height to themselves.
   a. State the null and alternative hypotheses of this study.
   b. Open the **Marriage** workbook from the Chapter07 folder and save it as **Marriage Analysis**.
   c. Analyze the data in the marriage table. What test statistics are appropriate to use with these data? Do you accept or reject the null hypothesis?
   d. Save your changes to the workbook, write a report summarizing your results, and print the relevant statistics supporting your conclusion.

9. The Gender Poll workbook contains polling data on how males and females respond to various social and political issues. Each worksheet in the workbook contains a table showing the responses to a particular question.
   a. Open the **Gender Poll** workbook from the Chapter07 folder and save it as **Gender Poll Statistics**.
   b. On each worksheet, calculate the table statistics for the opinion poll table.
   c. For which questions is the gender of the respondent associated with the outcome?
   d. Save your changes to the workbook and write a report summarizing your results, speculating on the types of issues that men and women might agree or disagree on.

10. The Race Poll workbook contains additional polling data on how different races respond to social and political questions.
    a. Open the **Race Poll** workbook from the Chapter07 folder and save it as **Race Poll Statistics**.
    b. On each worksheet, calculate the table statistics for the opinion poll table. Resolve any problem with sparse data by combining the Black and Other categories.
    c. For which questions is the race of the respondent associated with the outcome?
    d. Save your changes to the workbook and write a report summarizing your results, and speculating on the types of questions that blacks and members of other races might agree or disagree on.

11. The Home Sales workbook contains historic data on home prices in Albuquerque (see Chapter 4 for a discussion of this workbook.)
    a. Open the **Home Sales** workbook from the Chapter07 folder and save it as **Home Sales Statistics**.
    b. Analyze the data to determine whether there is evidence that houses in the NE sector are more likely to have offers pending than houses outside the NE sector.
    c. Save your changes to the workbook and write a report summarizing your conclusions.

12. The Montana workbook contains polling data from a survey in 1992 about the financial conditions in Montana. You've been asked to analyze what factors influence political affiliation in the state.
    a. Open the **Montana** workbook from the Chapter07 folder and save it as **Montana Political Statistics**.
    b. Create a PivotTable of Political Party versus Region (of the state). Remove any blank categories from the table. Is there evidence to suggest that different regions have different political leanings?

c. Create a PivotTable of Political Party versus Gender, removing any blank categories from the table. Is there evidence to suggest that political affiliation depends on gender?
d. Create a PivotTable of Political Party versus Age, removing any blank categories from the table. Age is an ordinal variable. Using the appropriate test statistic, analyze whether there is any significant relation between age and political affiliation.
e. Save your changes to the workbook and write a report summarizing your conclusions.

13. The Montana workbook discussed in Exercise 12 also contains information on the public's assessment of the financial status of the state. Analyze the results from this survey.

a. Open the **Montana** workbook from the Chapter07 folder and save it as **Montana Financial Statistics**.
b. Create a PivotTable comparing the Financial Status variable to the Gender variable. Is there statistical evidence that gender plays a roll in how individuals view the economy?
c. Create a PivotTable of Financial Status and Income. Note that both are ordinal variables. Is there statistical evidence of a relation between the two?
d. Create a PivotTable of Financial Status and Age. Once again note that both variables are ordinal. Using the appropriate statistical test, determine whether there is evidence of a relation between age and the assessment of the state's financial status.
e. Create a PivotTable comparing Financial Status and Political Party. Is there statistical evidence of a relation between the two?
f. Save your changes to the workbook and write a report summarizing your observations.

Chapter 8

Regression and Correlation

Objectives

In this chapter you will learn to:

- Fit a regression line and interpret the coefficients
- Understand regression statistics
- Use residuals to check the validity of the assumptions needed for statistical inference
- Calculate and interpret correlations and their statistical significance
- Understand the relationship between correlation and simple regression
- Create a correlation matrix and apply hypothesis tests to the correlation values
- Create and interpret a scatter plot matrix

This chapter examines the relationship between two variables using linear regression and correlation. Linear regression estimates a linear equation that describes the relationship, whereas correlation measures the strength of that linear relationship.

Simple Linear Regression

When you plot two variables against each other in a scatter plot, the values usually don't fall exactly in a perfectly straight line. When you perform a linear regression analysis, you attempt to find the line that best estimates the relationship between two variables (the *y*, or dependent, variable, and the *x*, or independent, variable). The line you find is called the **fitted regression line**, and the equation that specifies the line is called the **regression equation**.

The Regression Equation

If the data in a scatter plot fall approximately in a straight line, you can use linear regression to find an equation for the regression line drawn over the data. Usually, you will not be able to fit the data perfectly, so some points will lie above and some below the fitted regression line.

The regression line that Excel fits will have an equation of the form $y = a + bx$. Here *y* is the **dependent variable**, the one you are trying to predict, and *x* is the **independent**, or **predictor**, **variable**, the one that is doing the predicting. Finally, *a* and *b* are called **coefficients**. Figure 8-1 shows a line with $a = 10$ and $b = 2$. The short vertical line segments represent the errors, also called **residuals**, which are the gaps between the line and the points. The residuals are the differences between the observed dependent values and the predicted values. Because *a* is where the line intercepts the vertical axis, *a* is sometimes called the **intercept** or **constant term** in the model. Because *b* tells how steep the line is, *b* is called the **slope**. It gives the ratio between the vertical change and the horizontal change along the line. Here *y* increases from 10 to 30 when *x* increases from 0 to 10, so the slope is

$$b = \frac{\text{Vertical change}}{\text{Horizontal change}} = \frac{30 - 10}{10 - 0} = 2$$

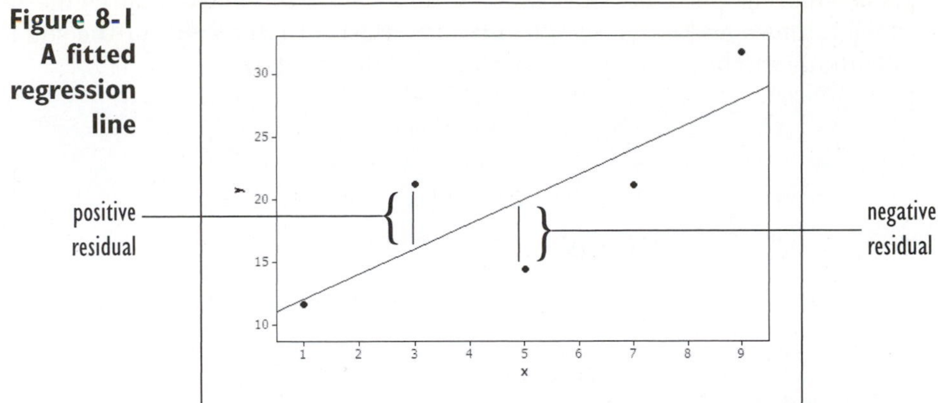

Figure 8-1 **A fitted regression line**

Suppose that x is years on the job and y is salary. Then the y intercept ($x = 0$) is the salary for a person with zero years' experience, the starting salary. The slope is the change in salary per year of service. A person with a salary above the line would have a positive residual, and a person with a salary below the line would have a negative residual.

If the line trends downward so that y decreases when x increases, then the slope is negative. For example, if x is age and y is price for used cars, then the slope gives the drop in price per year of age. In this example, the intercept is the price when new, and the residuals represent the difference between the actual price and the predicted price. All other things being equal, if the straight line is the correct model, a positive residual means a car costs more than it should, and a negative residual means a car costs less than it should (that is, it's a bargain).

Fitting the Regression Line

When fitting a line to data, you assume that the data follow the **linear model**:

$$y = \alpha + \beta x + \varepsilon$$

where α is the "true" intercept, β is the "true" slope, and ε is an error term. When you fit the line, you'll try to estimate α and β, but you can never know them exactly. The estimates of α and β, we'll label a and b. The predicted values of y using these estimates, we'll label $\hat{y}$, so that

$$\hat{y} = a + bx$$

To get estimates for α and β, we use values of a and b that result in a minimum value for the sum of squared residuals. In other words, if y_i is an observed value of y, we want values of a and b such that

$$\text{Sum of squared residuals} = \sum_{i=1}^{n} (y_i - \hat{y}_i)^2$$

is as small as possible. This procedure is called the **least-squares** method. The values a and b that result in the smallest possible sum for the squared residuals can be calculated from the following formulas:

$$b = \frac{\sum_{i=1}^{n}(x_i - \bar{x})(y_i - \bar{y})}{\sum_{i=1}^{n}(x_i - \bar{x})^2}$$

$$a = \bar{y} - b\bar{x}$$

These are called the **least-squares estimates**. For example, say our data set contains the values listed in Table 8-1:

Table 8-1 Data for Least-Squares Estimates

x	y
1	3
2	4
1	3
3	4
2	5

The sample averages for x and y are 1.8 and 3.4, and the estimates for a and b are

$$\begin{aligned} b &= \frac{\sum_{i=1}^{n}(x_i - \bar{x})(y_i - \bar{y})}{\sum_{i=1}^{n}(x_i - \bar{x})^2} \\ &= \frac{(1 - 1.8)(3 - 3.4) + (2 - 1.8)(4 - 3.4) + \cdots + (2 - 1.8)(5 - 3.4)}{(1 - 1.8)^2 + (2 - 1.8)^2 + \cdots + (2 - 1.8)^2} \\ &= 0.5 \\ a &= \bar{y} - b\bar{x} \\ &= 3.4 - 0.5 \times 1.8 \\ &= 2.5 \end{aligned}$$

Thus the least-squares estimate of the regression equation is $y = 2.5 + 0.5x$.

Regression Functions in Excel

Excel contains several functions to help you calculate the least-squares estimates. Two of these are shown in Table 8-2.

Table 8-2 Calculating Least-Squares Estimates

Function	Description
INTERCEPT(*y*, *x*)	Calculates the least-squares estimate, *a*, for known values *y* and *x*.
SLOPE(*y*, *x*)	Calculates the least-squares estimate, *b*, for known values *y* and *x*.

For example, if the *y* values are in the cell range A2:A11, and the *x* values are in the range B2:B11, then the function INTERCEPT(A2:A11, B2:B11) will display the value of *a*, and the function SLOPE(A2:A11, B2:B11) will display the value of *b*.

EXCEL TIPS

- You can also calculate linear regression values using the LINEST and LOGEST functions, but this is a more advanced topic. Both of these functions use arrays. You can learn more about these functions and about array functions in general by using Excel's online Help.

Exploring Regression

If you wish to explore the concepts behind linear regression, an Explore workbook has been provided for you.

To explore the regression concept:

1. Open the **Regression** workbook in the Explore folder.
2. Review the contents of the workbook. The Explore Regression worksheet shown in Figure 8-2 allows you to test the impact of different intercept and slope estimates on the sum of the squared differences.

Figure 8-2 Explore Regression worksheet

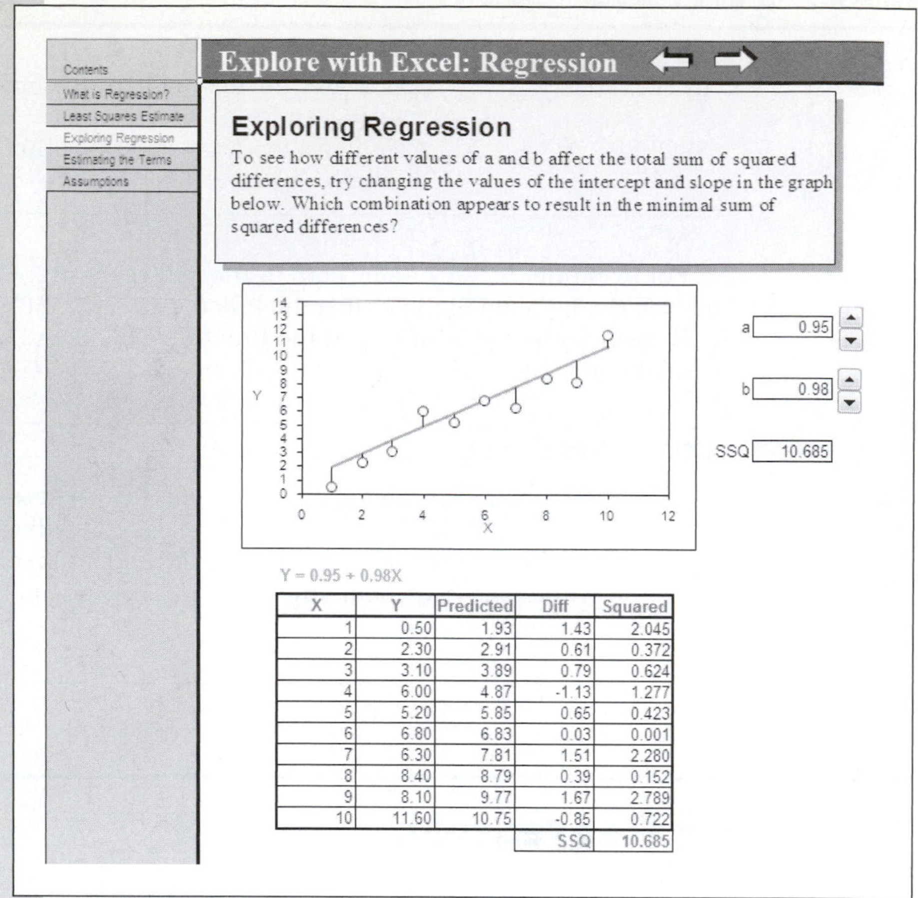

X	Y	Predicted	Diff	Squared
1	0.50	1.93	1.43	2.045
2	2.30	2.91	0.61	0.372
3	3.10	3.89	0.79	0.624
4	6.00	4.87	-1.13	1.277
5	5.20	5.85	0.65	0.423
6	6.80	6.83	0.03	0.001
7	6.30	7.81	1.51	2.280
8	8.40	8.79	0.39	0.152
9	8.10	9.77	1.67	2.789
10	11.60	10.75	-0.85	0.722
			SSQ	10.685

3 After you are finished working with the file, close it.

Performing a Regression Analysis

The Breast Cancer workbook contains data from a 1965 study analyzing the relationship between mean annual temperature and the mortality rate for women with a certain type of breast cancer. The subjects came from 16 different regions in Great Britain, Norway, and Sweden. Table 8-3 presents the data.

Table 8-3 Data for the Breast Cancer Workbook

Range Name	Range	Description
Region	A2:A17	A number indicating the region where the data have been collected
Temperature	B2:B17	The mean annual temperature of the region
Mortality	C2:C17	Mortality index for neoplasms of the female breast for the region

You've been asked to determine whether there is evidence of a linear relationship between the mean annual temperature in the region and the mortality index. Is the mortality index different for women who live in regions with different temperatures?

To open the Breast Cancer workbook:

1. Open the **Breast Cancer** workbook from the Chapter08 folder.
2. Save the workbook as **Breast Cancer Regression**. The workbook appears as shown in Figure 8-3.

Figure 8-3
The Breast Cancer workbook

	A	B	C	D	E
1	Region	Temperature	Mortality		
2	1	31.8	67.3		
3	2	34.0	52.5		
4	3	40.2	68.1		
5	4	42.1	84.6		
6	5	42.3	65.1		
7	6	43.5	72.2		
8	7	44.2	81.7		
9	8	45.1	89.2		
10	9	46.3	78.9		
11	10	47.3	88.6		
12	11	47.8	95.0		
13	12	48.5	87.0		
14	13	49.2	95.9		
15	14	49.9	104.5		
16	15	50.0	100.4		
17	16	51.3	102.5		

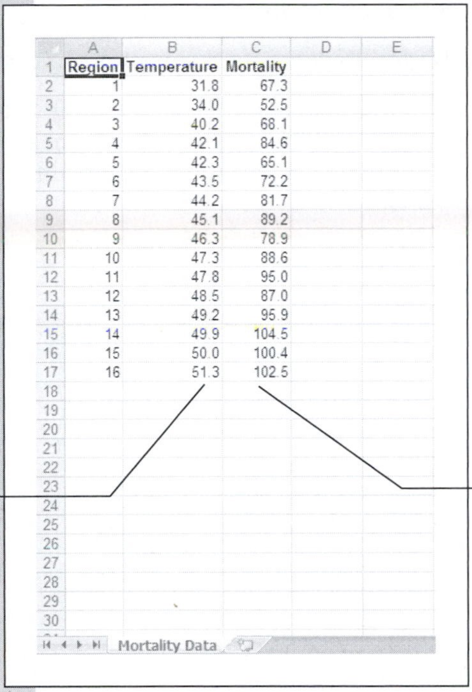

Mortality Data

annual mean temperature of the region in degrees Fahrenheit

breast cancer mortality index

Plotting Regression Data

Before you calculate any regression statistics, you should always plot your data. A scatter plot can quickly point out obvious problems in assuming that a linear model fits your data (perhaps the scatter plot will show that the data values do *not* fall along a straight line). Scatter plots in Excel also allow you to superimpose the regression line on the plot along with the regression equation. From this information, you can get a pretty good idea whether a straight line fits your data or not.

To create a scatter plot of the mortality data:

1. Click **Single Variable Charts** from the StatPlus menu and then click **Fast Scatter plot**.
2. Click the **x axis** button, select the **Temperature** range name and click **OK**. Click the **y axis** button, select the **Mortality** range name and click **OK**.
3. Click the **Chart Options** button, enter **Mortality vs. Temp** for the Chart title, **Temperature** for the x axis title, and **Mortality** for the y axis title. Click **OK**.
4. Click the **Output** button and save the chart to a new chart sheet named **Mortality Scatter plot**. Click **OK**.
5. Click the **OK** button to generate the scatter plot.
6. In the scatter plot that's created, resize the horizontal scale so that the lower boundary is **30**, and then resize the vertical scale so that the lower boundary is **50**. Figure 8-4 shows the final version of the scatter plot.

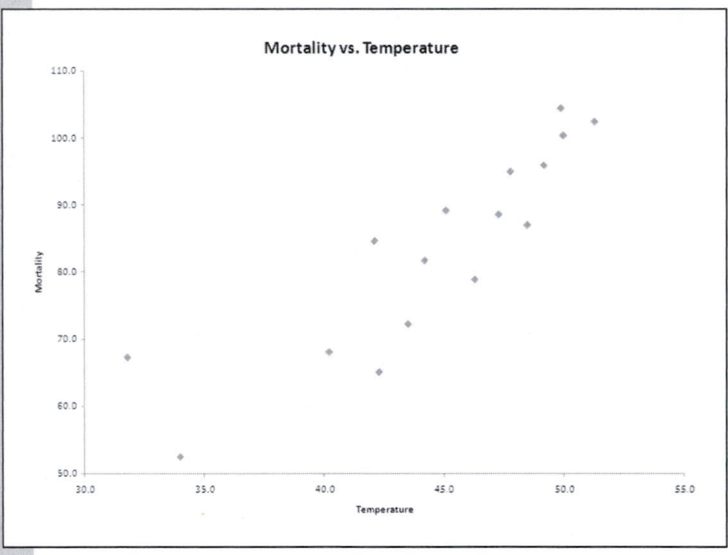

Figure 8-4 Scatter plot of the mortality index versus mean annual temperature

Now you'll add a regression line to the data.

To add a regression line:

1 Right-click any of the data points in the graph and click **Add Trendline** from the menu. See Figure 8-5.

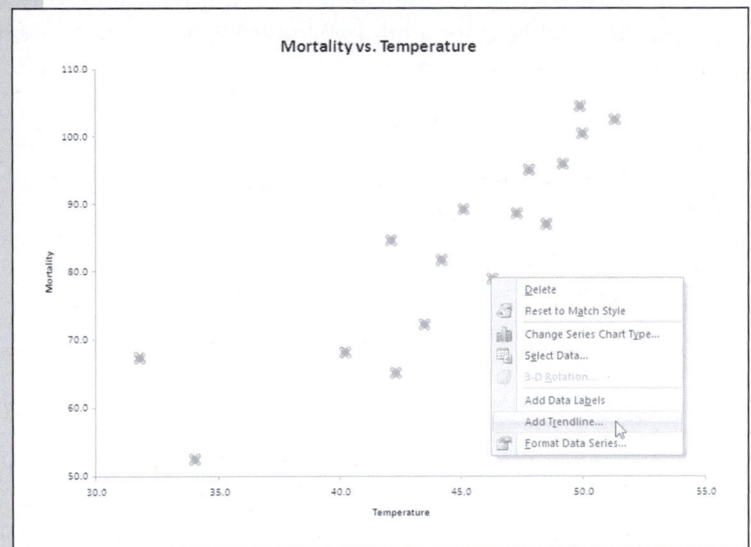

Figure 8-5 The Add Trendline command

2 Excel displays a list of possible regression and trend lines. Verify that the Linear regression option is selected as shown in Figure 8-6.

Figure 8-6 Choosing a trend line

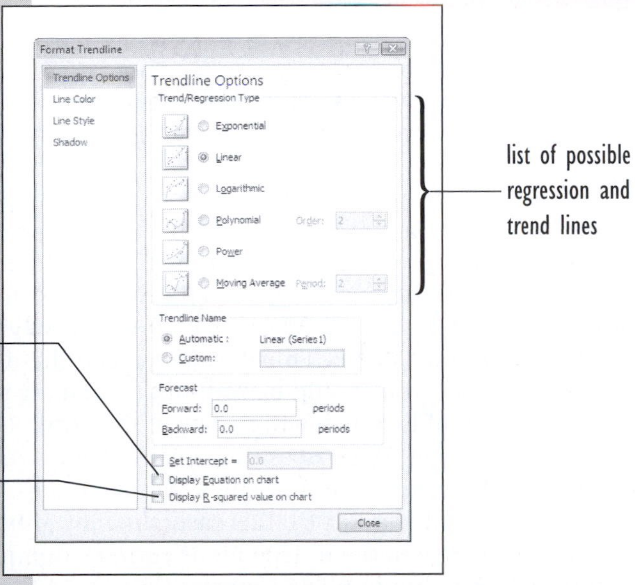

list of possible regression and trend lines

display the regression equation on the chart

display the regression's r^2 value

3 Click the **Display Equation on chart** and **Display R-squared value on chart** checkboxes and then click the **Close** button.

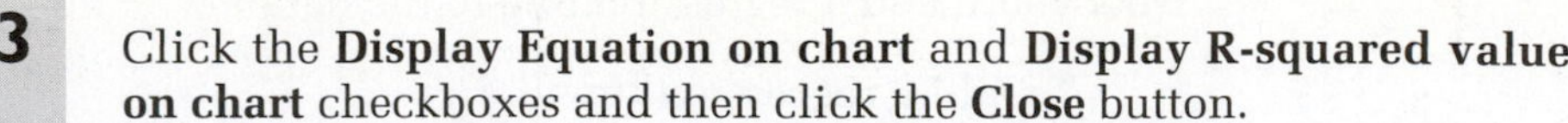

Excel adds a regression line to the plot along with the regression equation and R^2 value.

4 Drag the text containing the regression equation and R^2 value to a point above the plot. See Figure 8-7.

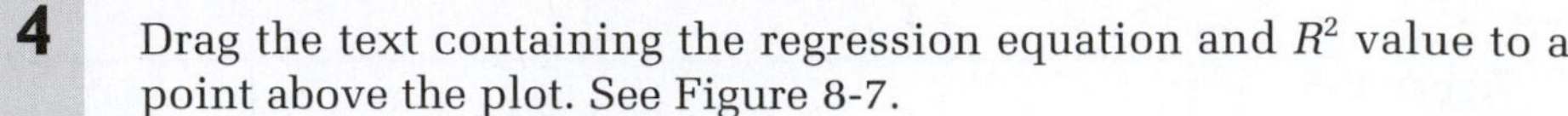

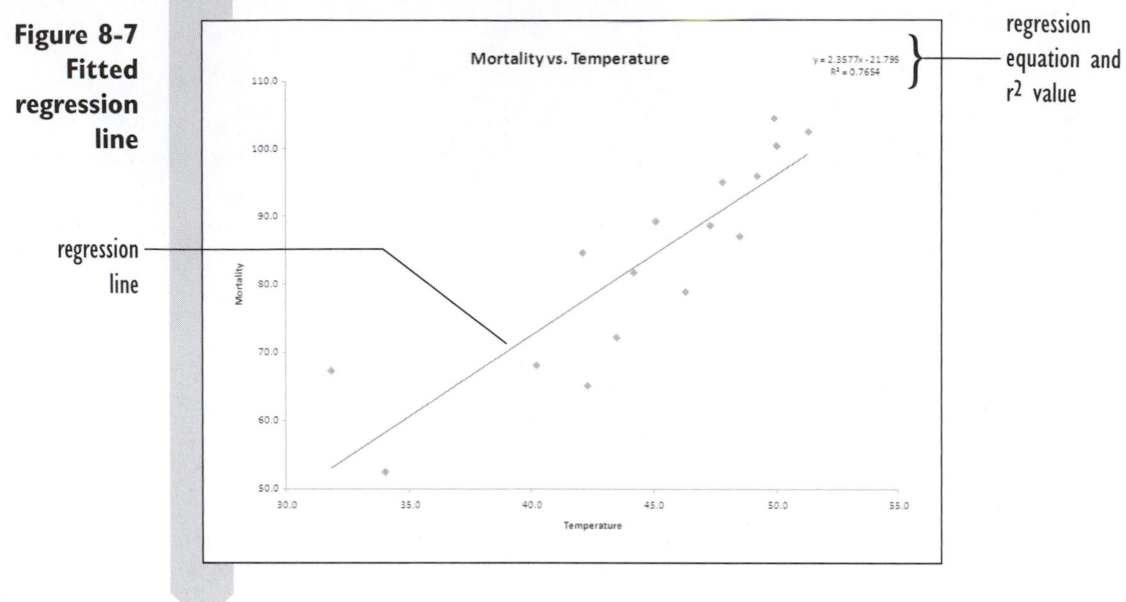

Figure 8-7 Fitted regression line

The regression equation for the mortality data is $y = -21.795 + 2.3577x$. This means that for every degree that the annual mean temperature increased in these regions, the breast cancer mortality index increased by about 2.3577 points.

How would you interpret the constant term in this equation (−21.795)? At first glance, this is the y intercept, and it means that if the mean annual temperature is 0, the value of the mortality index would be −21.795. Clearly this is absurd; the mortality index can't drop below zero. In fact, any mean annual temperature of less than 9.24 degrees Fahrenheit will result in a negative estimate of the mortality index. This does not mean that the linear equation is useless, but it means you should be cautious in making any predictions for temperature values that lie outside the range of the observed data.

The R^2 value is 0.7654. What does this mean? The $\boldsymbol{R^2}$ value, also known as the **coefficient of determination**, measures the percentage of variation in the values of the dependent variable (in this case, the mortality index) that can be explained by the change in the independent variable (temperature). R^2 values vary from 0 to 1. A value of 0.7654 means that 76.54% of the variation in the mortality index can be explained by the change in annual mean temperature. The remaining 23.46% of the variation is presumed to be due to random variability.

EXCEL TIPS

- You can use the Add Trendline command to add other types of least-squares curves, including logarithmic, polynomial, exponential, and power curves. For example, instead of fitting a straight line, you can fit a second-degree curve to your data.

Calculating Regression Statistics

The regression equation in the scatter plot is useful information, but it does not tell you whether the regression is statistically significant. At this point, you have two hypotheses to choose from.

H_0: There is no linear relationship between the mortality index and the mean annual temperature.

H_a: There is a linear relationship between the mortality index and the mean annual temperature.

The linear relationship we're testing is expressed in terms of the regression equation.

In order to analyze our regression, we need to use the Analysis ToolPak, an add-in that comes with Excel and provides tools for analyzing regression. If you do not have the Analysis ToolPak loaded on your system, you should install it now. Refer to Chapter 1 for information on installing the Data Analysis ToolPak.

To create a table of regression statistics:

1. Return to the **Mortality Data** worksheet.
2. Click **Data Analysis** from the Analysis group on the Data tab to open the Data Analysis dialog box.
3. Scroll down the list of data analysis tools and click **Regression**, and then click the **OK** button.
4. Enter the cell range **C1:C17** in the Input Y Range box.
5. Enter the cell range **B1:B17** in the Input X Range box.
6. Because the first cell in these ranges contains a text label, click the **Labels** checkbox.

7 Click the **New Worksheet Ply** option button and type **Regression Statistics** in the accompanying text box.

8 Click all four of the Residuals checkboxes.

Your Regression dialog box should appear as shown in Figure 8-8. Note that we did not select the Normal Probability Plots checkbox. This option creates a normal plot of the dependent variable. In most situations, this plot is not needed for regression analysis.

Figure 8-8 The Regression dialog box

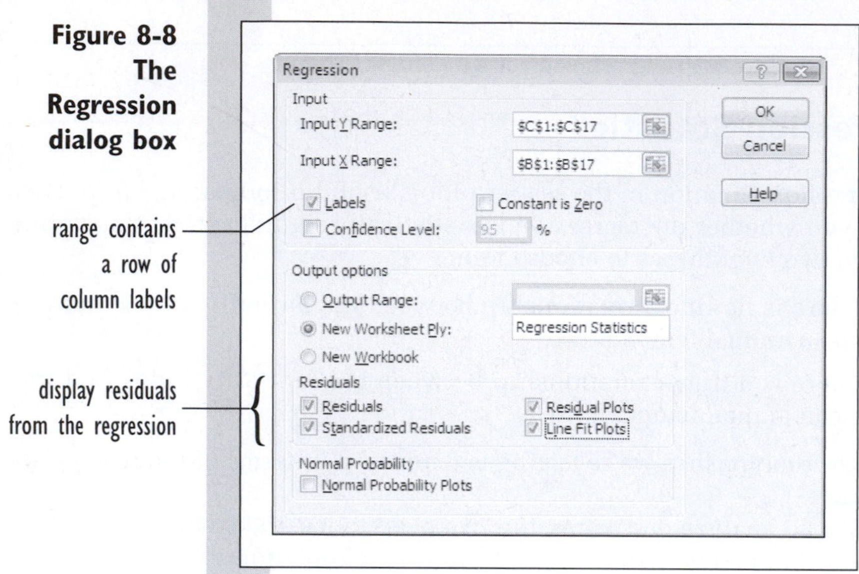

9 Click **OK.**

Excel generates the output shown in Figure 8-9.

Figure 8-9 Output from the Regression command

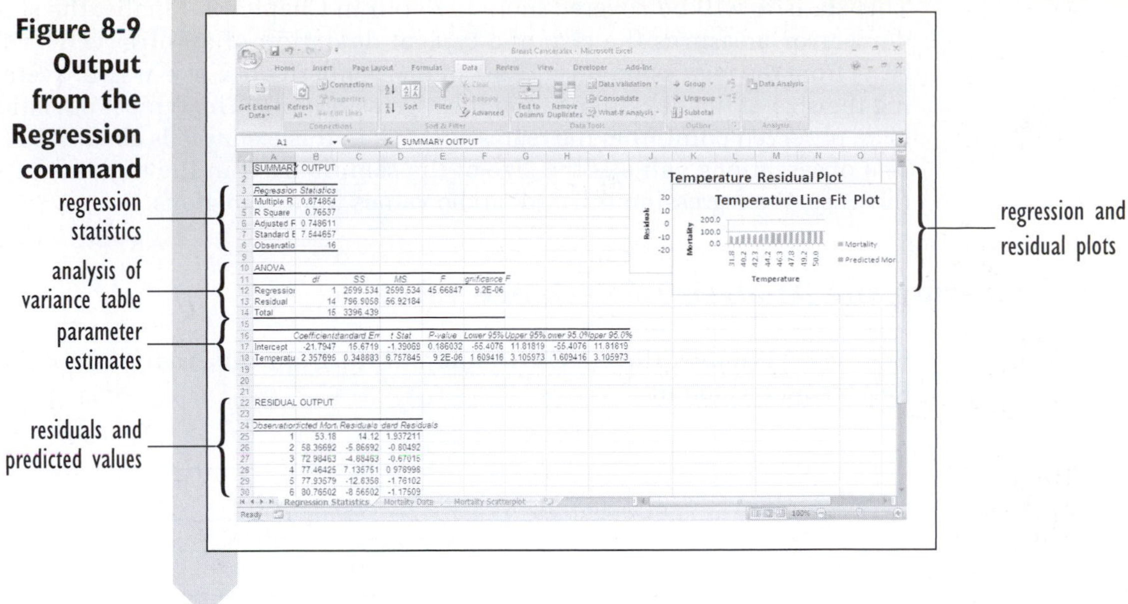

The output is divided into six areas: regression statistics, analysis of variance (ANOVA), parameter estimates, residual output, probability output (not shown in Figure 8-9), and plots. Let's examine these areas more closely. The Regression command doesn't format the output for us, so we may want to do that ourselves on the worksheet.

Interpreting Regression Statistics

Figure 8-10 Regression statistics

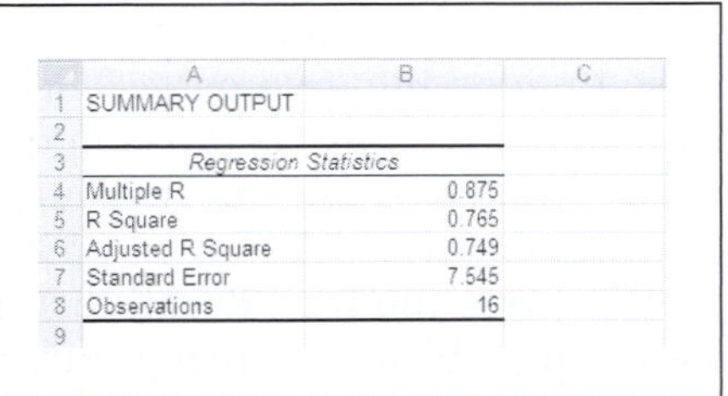

	A	B
1	SUMMARY OUTPUT	
2		
3	*Regression Statistics*	
4	Multiple R	0.875
5	R Square	0.765
6	Adjusted R Square	0.749
7	Standard Error	7.545
8	Observations	16
9		

You've seen some of the regression statistics shown in Figure 8-10 before. The R^2 value of 0.765 you've seen in the scatter plot. The multiple R value is equal to the square root of the R^2 value. The multiple R is equal to the absolute value of the correlation between the dependent variable and the predictor variable. You'll learn about correlation later in this chapter. The adjusted R^2 is used when performing a regression with several predictor variables.

This statistic will be covered more in depth in Chapter 9. Finally, the standard error measures the size of a typical deviation of an observed value (x, y) from the regression line. Think of the standard error as a way of averaging the size of the deviations from the regression line. The typical deviation of an observed point from the regression line in this example is about 7.5447. The observations value is the size of the sample used in the regression. In this case, the regression is based on the values from 16 regions.

Interpreting the Analysis of Variance Table

Figure 8-11 shows the ANOVA table output from the Analysis ToolPak Regression command.

Figure 8-11 Analysis of Variance (ANOVA) table

	ANOVA					
11		*df*	*SS*	*MS*	*F*	*Significance F*
12	Regression	1	2599.53	2599.53	45.67	0.000009
13	Residual	14	796.91	56.92		
14	Total	15	3396.44			

The ANOVA table analyzes the variability of the mortality index. The variability is divided into two parts: the first is the variability due to the regression line, and the second is random variability.

The values in the *df* column of the table indicate the number of degrees of freedom for each part. The total degrees of freedom are equal to the number of observations minus 1. In this case the total degrees of freedom are 15. Of those 15 degrees of freedom, 1 degree of freedom is attributed to the regression, and the remaining 14 degrees of freedom are attributed to random variability.

The *SS* column gives you the sums of squares. The total sum of squares is the sum of the squared deviations of the mortality index from the overall mean. This total is also divided into two parts. The first part, labeled in the table as the regression sum of squares, is the sum of squared deviations between the regression line and the overall mean. The second part, labeled the residual sum of squares, is equal to the sum of the squared deviations of the mortality index from the regression line. Recall that this is the value that we want to make as small as possible in the regression equation. In this example, the total sum of squares is 3,396.44, of which 2,599.53 is attributed to the regression and 796.91 is attributed to error.

What percentage of the total sum of squares can be attributed to the regression? In this case, it is 2,599.53/3,396.44 = 0.7654, or 76.54%. This is equal to the R^2 value, which, as you learned earlier, measures the percentage of variability explained by the regression. Note also that the total sum of squares (3,396.44) divided by the total degrees of freedom (15) equals 226.43, which is the variance of the mortality index. The square root of this value is the standard deviation of the mortality index.

The *MS* (mean square) column displays the sum of squares divided by the degrees of freedom. Note that the mean square for the residual is equal to the square of the standard error in cell B7 ($7.5447^2 = 56.9218$). Thus you can use the mean square for the residual to derive the standard error.

The next column displays the ratio of the mean square for the regression to the mean square error of the residuals. This value is called the ***F* ratio**. A large *F* ratio indicates that the regression may be statistically significant. In this example, the ratio is 45.7. The *p* value is displayed in the next column and equals 0.0000092. Because the *p* value is less than 0.05, the regression is statistically significant. You'll learn more about analysis of variance and interpreting ANOVA tables in an upcoming chapter.

Parameter Estimates and Statistics

The file output table created by the Analysis ToolPak Regression command displays the estimates of the regression parameters along with statistics measuring their significance. See Figure 8-12.

Figure 8-12 Parameter estimates and statistics

16		Coefficients	Standard Error	t Stat	P-value	Lower 95%	Upper 95%
17	Intercept	-21.795	15.672	-1.391	0.186	-55.408	11.818
18	Temperature	2.358	0.349	6.758	0.000	1.609	3.106
19							

As you've already seen, the constant coefficient, or intercept, equals about –21.79, and the slope based on the temperature variable is about 2.36. The standard errors for these values are shown in the Standard Error column and are 15.672 and 0.349, respectively. The ratio of the parameter estimates to their standard errors follows a *t* distribution with $n - 2$, or 14, degrees of freedom. The ratios for each parameter are shown in the *t* Stat column, and the corresponding two-sided *p* values are shown in the *P* value column. In this example, the *p* value for the intercept term is 0.186, and the *p* value for the slope term (labeled Temperature) is 9.2×10^{-6}, or 0.0000092 (note that this is the same *p* value that appeared in the ANOVA table).

The final part of this table displays the 95% confidence interval for each of the terms. In this case, the 95% confidence interval for the intercept term is about (−55.41, 11.82), and the 95% confidence interval for the slope is (1.61, 3.11).

Note: In your output the confidence intervals might appear twice. The first pair, a 95% interval, always appears. The second pair always appears, but with the confidence level you specify in the Regression dialog box. In this case, you used the default 95% value, so that interval appears in both pairs.

What have you learned from the regression statistics? First of all, you would decide to reject the null hypothesis and accept the alternative hypothesis that a linear relationship exists between the mortality index and temperature. On the basis of the confidence interval for the slope parameter, you can report with 95% confidence that for each degree increase in the mean annual temperature, the mortality index for the region increases between 1.61 to 3.11 points.

Residuals and Predicted Values

The last part of the output from the Analysis ToolPak's Regression command consists of the residuals and the predicted values. See Figure 8-13 (the values have been reformatted to make them easier to view).

Figure 8-13 Residuals and predicted values

RESIDUAL OUTPUT

Observation	Predicted Mortality	Residuals	Standard Residuals
1	53.180	14.120	1.937
2	58.367	-5.867	-0.805
3	72.985	-4.885	-0.670
4	77.464	7.136	0.979
5	77.936	-12.836	-1.761
6	80.765	-8.565	-1.175
7	82.415	-0.715	-0.098
8	84.537	4.663	0.640
9	87.367	-8.467	-1.162
10	89.724	-1.124	-0.154
11	90.903	4.097	0.562
12	92.553	-5.553	-0.762
13	94.204	1.696	0.233
14	95.854	8.646	1.186
15	96.090	4.310	0.591
16	99.155	3.345	0.459

As you've learned, the residuals are the differences between the observed values and the regression line (the predicted values). Also included in the output are the standardized residuals. From the values shown in Figure 8-13, you see that there is one residual that seems larger than the others, it is found in the first observation and has a standardized residual value of 1.937. Standardized residuals are residuals standardized to a common scale, regardless of the original unit of measurement. A standardized residual whose value is above 2 or below −2 is a potential outlier. There are many ways to calculate standardized residuals. Excel calculates using the following formula:

$$\text{Standardized residual} = \frac{\text{Residual}}{\sqrt{\text{Sum of squared residuals}/(n-1)}}$$

where n is the number of observations in the data set. In this data set, the value of the first standardized residual is

$$\frac{14.12}{\sqrt{796.9058/15}} = 1.937$$

You'll want to keep an eye on this observation as you continue to explore this regression model. As you'll see shortly, the residuals play an important role in determining the appropriateness of the regression model.

Checking the Regression Model

As in any statistical procedure, for statistical inference on a regression, you are making some important assumptions. There are four:

1. The straight-line model is correct.
2. The error term ε is normally distributed with mean 0.
3. The errors have constant variance.
4. The errors are independent of each other.

Whenever you use regression to fit a line to data, you should consider these assumptions. Fortunately, regression is somewhat robust, so the assumptions do not need to be perfectly satisfied.

One point that cannot be emphasized too strongly is that *a significant regression is not proof that these assumptions haven't been violated.* To verify that your data do not violate these assumptions is to go through a series of tests, called **diagnostics**.

Testing the Straight-Line Assumption

To test whether the straight-line model is correct, you should first create a scatter plot of the data to inspect visually whether the data depart from this assumption in any way. Figure 8-14 shows a classic problem that you may see in your data.

Figure 8-14
A curved relationship

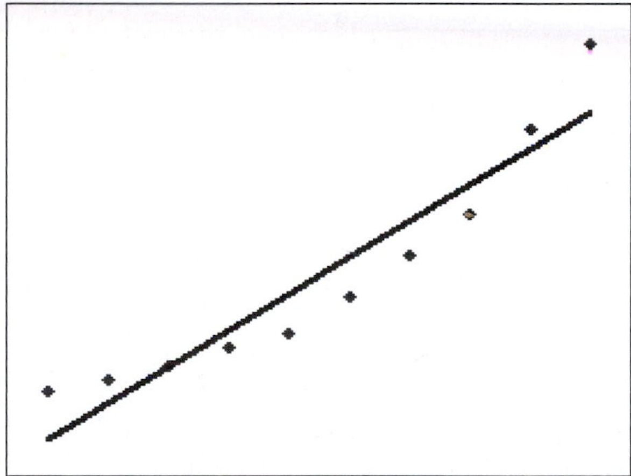

Another sharper way of seeing whether the data follow a straight line is to fit the regression line and then plot the residuals of the regression against the values of the predictor variable. A U-shaped (or upside-down U-shaped) pattern to the plot, as shown in Figure 8-15, is a good indication that the data follow a curved relationship and that the straight-line assumption is wrong.

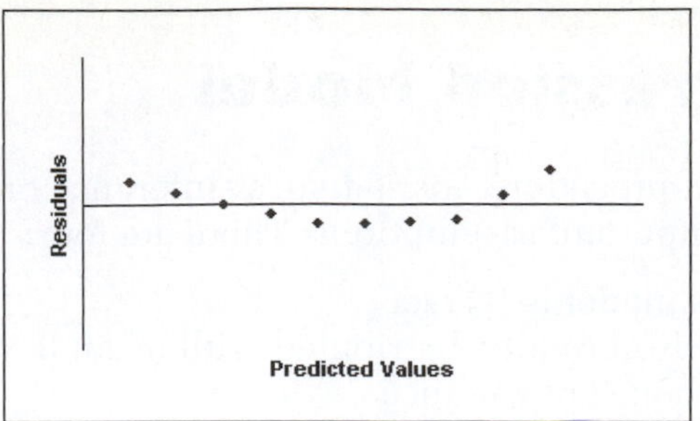

Figure 8-15 Residuals showing a curved relationship

Let's apply this diagnostic to the mortality index data. The Regression command creates this plot for you, but it can be difficult to read because of its size. We'll move the plot to a chart sheet and reformat the axes for easier viewing.

To create a plot of the residuals versus the predictor variable:

1. Scroll to cell J1 in the Regression Statistics worksheet and click the **Temperature Residual Plot**.
2. Click the **Move Chart** button located in the Location group of the Design tab on the Chart Tools ribbon.
3. Click the **As new sheet** option button and then type **Residuals vs. Temperature** and click **OK**.
4. Rescale the horizontal axes of the temperature variable, so that the lower boundary is **30**. The revised plot appears in Figure 8-16.

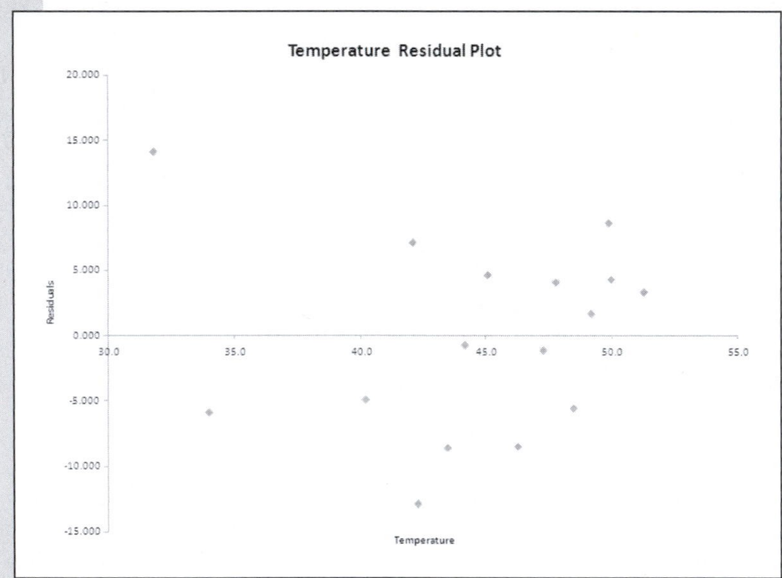

Figure 8-16 Residuals versus temperature values

The plot shows that most of the positive residuals tend be located at the lower and higher temperatures and most of the negative residuals are concentrated in the middle temperatures. This may indicate a curve in the data. The large first observation is influential here. Without it, there would be less indication of a curve.

Testing for Normal Distribution of the Residuals

The next diagnostic is a normal plot of the residuals. The Analysis ToolPak does not provide this chart, so we'll create one with StatPlus.

To create a Normal Probability Plot of the residuals:

1. Return to the Regression Statistics worksheet.
2. Click **Single Variable Charts** from the StatPlus menu and then click **Normal P-Plots**.
3. Click the **Data Values** button.
4. In the Input Options dialog box, click the **Use Range References** option button and then select the range **C24:C40**. Verify that the Range Include a Row of Column Labels checkbox is selected and click **OK**.
5. Click the **Output** button, verify that the As a New Chart sheet option button is selected, and type **Residual Normal Plot** in the accompanying text box. Click the **OK** button.
6. Click the **OK** button to start generating the Normal Probability plot. See Figure 8-17.

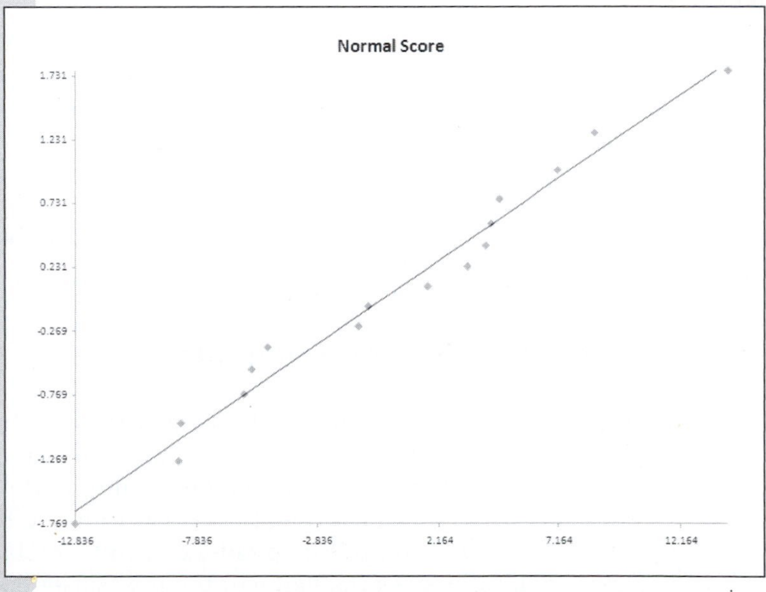

Figure 8-17 Normal probability plot of residuals

Recall that if the residuals follow a normal distribution, they should fall evenly along the superimposed line on the normal probability plot. Although the points in Figure 8-17 do not fall perfectly on the line, the departure is not strong enough to invalidate our assumption of normality.

Testing for Constant Variance in the Residuals

The next assumption you should always investigate is the assumption of constant variance in the residuals. A commonly used plot to help verify this assumption is the plot of the residuals versus the predicted values. This plot will also highlight any problems with the straight-line assumption.

Figure 8-18 Residuals showing nonconstant variance

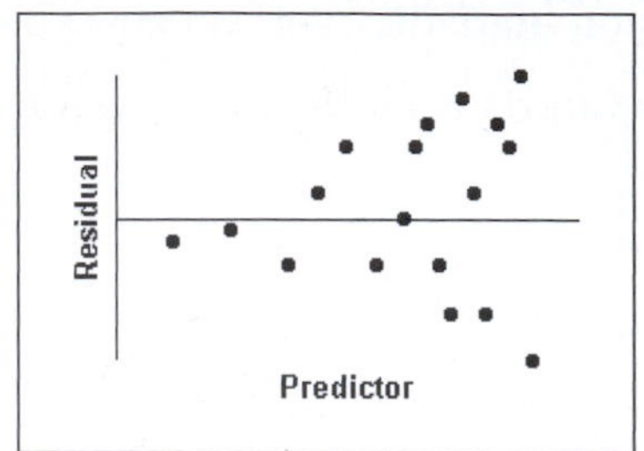

If the constant variance assumption is violated, you may see a plot like the one shown in Figure 8-18. In this example, the variance of the residuals is larger for larger predicted values. It's not uncommon for variability to increase as the value of the response variable increases. If that happens, you might remove this problem by using the log of the response variable and performing the regression on the transformed values.

With one predictor variable in the regression equation, the scatter plot of the residuals versus the predicted values is identical to the scatter plot of the residuals versus the predictor variable (shown earlier in Figure 8-16). The scatter plot indicates that there *may* be a decrease in the variability of the residuals as the predicted values increase. Once again, though, this interpretation is influenced by the presence of the possible outlier in the first observation. Without this observation, there might be no reason to doubt the assumption of constant variance.

Testing for the Independence of Residuals

The final regression assumption is that the residuals are independent of each other. This assumption is of concern only in situations where there is a defined order for the observations. For example, if we do a regression of a predictor variable versus time, the observations will follow a sequential order. The assumption of independence can be violated if the value of one observation influences the value of the next observation. For example, a large value might be followed by a small value, or large and small values could be clustered together (see Figure 8-19). In these cases, the residuals do not show independence, because you can predict what the sign of the next value will be on the basis of the current value.

Figure 8-19 Residuals versus predicted values

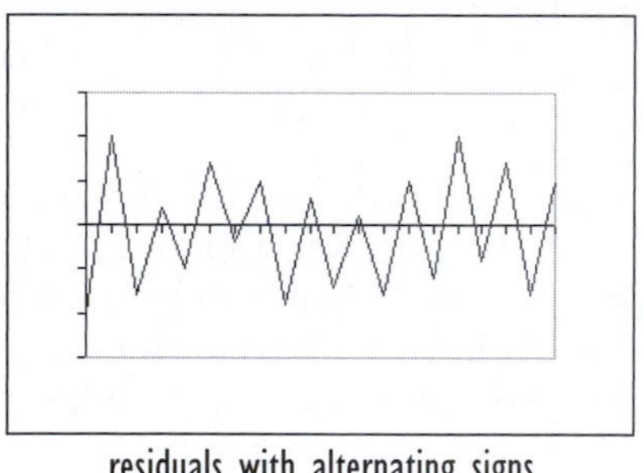

residuals with alternating signs

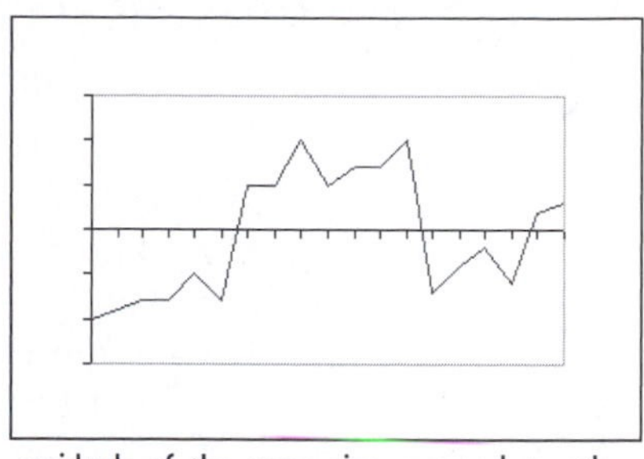

residuals of the same sign grouped together

In examining residuals, we can examine the sign of the values (either positive or negative) and determine how many values with the same sign are clustered together. These groups of similarly signed values are called runs. For example, consider a data set of 10 residuals containing 5 positive values and 5 negative values. The values could follow an order with only two runs, such as

$$+ + + + + - - - - -$$

In this case, we would suspect that the residuals were not independent, because the positives and negatives are clustered together in the sequence. On the other hand, we might have the opposite problem, where there could be as many as ten runs, such as

$$+ - + - + - + - + -$$

Here, we suspect the residuals are not independent, because the residuals are constantly switching sign. Finally, we might have something in-between, such as

$$+ + - - - + + - - +$$

which has five runs. If the number of runs is very large or very small, we would suspect that the residuals are not independent. How large (or how small) does this value have to be? Using probability theory, statisticians have calculated the p values for a **runs test**, associated with the number of runs observed for different sample sizes. If we let n be the sample size, n_+ be the number of positive values, and n_- be the number of negative values, the expected number of runs μ is

$$\mu = \frac{2n_+n_-}{n} + 1$$

and the standard deviation σ is

$$\sigma = \sqrt{\frac{2n_+n_-}{n(n-1)}\left(\frac{2n_+n_-}{n} - 1\right)}$$

If r is the observed number of runs, then the value

$$z = \frac{(r - \mu + \frac{1}{2})}{\sigma}$$

approximately follows a standard normal distribution for large sample sizes (where n_+ and n_- are both > 10). For example, if $n = 10$, $n_+ = 5$, and $n_- = 5$, then $\mu = 6$ and $\sigma = \sqrt{20/9} = 1.49$. If 5 runs have been observed, $z = -0.335$ and the p value is 0.368. This is very close to the exact p value of 0.357, so we would not find this an extremely unusual number of runs. On the other hand, if we observe only 3 runs, then $z = -2.012$ and the p value is .022 (the exact value is 0.04).

Another statistic used to test the assumption of independence is the **Durbin-Watson** test statistic. In this test, we calculate the value

$$DW = \frac{\sum_{i=2}^{n}(e_i - e_{i-1})^2}{\sum_{i=1}^{n} e_i^2}$$

where e_i is the ith residual in the data set. The value of DW is then compared to a table of Durbin-Watson values to see whether there is evidence of a lack of independence in the residuals. Generally, a value of DW approximately equal to 0 or 4 suggests that the residuals are not independent. A value of DW near 2 suggests independence. Values in between may be inconclusive.

Because the mortality index data are *not* sequential, you shouldn't apply the runs test or the Durbin-Watson test. Remember, these statistics are most useful when the residuals have a definite sequential order.

After performing the diagnostics on the residuals, you conclude that there is no hard evidence to suggest that the regression assumptions have been violated. On the other hand, there is a problematic large residual in the first observation to consider. You should probably redo the analysis without the first observation to see what effect (if any) this has on your model. You'll have a chance to do that in the exercises at the end of the chapter.

STATPLUS TIPS

- Excel does not include a function to perform the runs test, but you can use the Runs Test command from the Time Series submenu on the StatPlus menu on your time-ordered residuals to perform this analysis.

- Use the functions RUNS(*range*, *center*) and RUNSP(*range*, *center*) to calculate the number of runs in a data set and the corresponding *p* value for a set of data in the cell range, *range*, around the central line *center*. *StatPlus required.*
- Use the function DW(*range*) to calculate the Durbin-Watson test statistic for the values in the cell range *range*. *StatPlus required.*

Correlation

The value of the slope in our regression equation is a product of the scale in which we measure our data. If, for example, we had chosen to express the temperature values in degrees Centigrade, we would naturally have a different value for the slope (though, of course, the statistical significance of the regression would not change). Sometimes, it's an advantage to express the strength of the relationship between one variable and another in a dimensionless number, one that does not depend on scale. One such value is the correlation. The **correlation** expresses the strength of the relationship on a scale ranging from −1 to 1.

A positive correlation indicates a strong positive relationship, in which an increase in the value of one variable implies an increase in the value of the second variable. This might occur in the relationship between height and weight. A negative correlation indicates that an increase in the first variable signals a decrease in the second variable. An increase in price for an object could be negatively correlated with sales. See Figure 8-20. A correlation of zero does *not* imply there is no relationship between the two variables. One can construct a nonlinear relationship that produces a correlation of zero.

Figure 8-20
Correlations

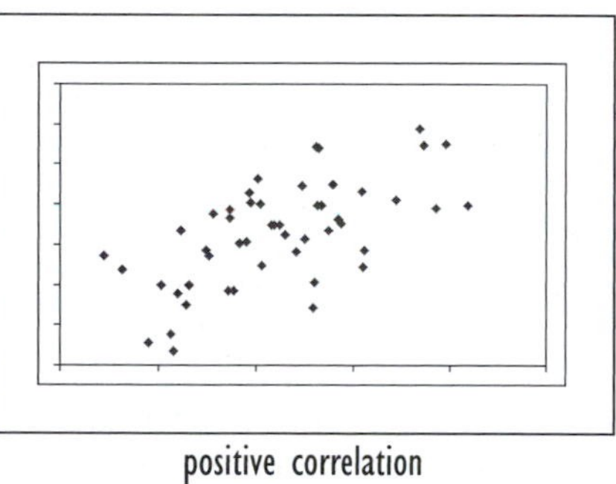

positive correlation

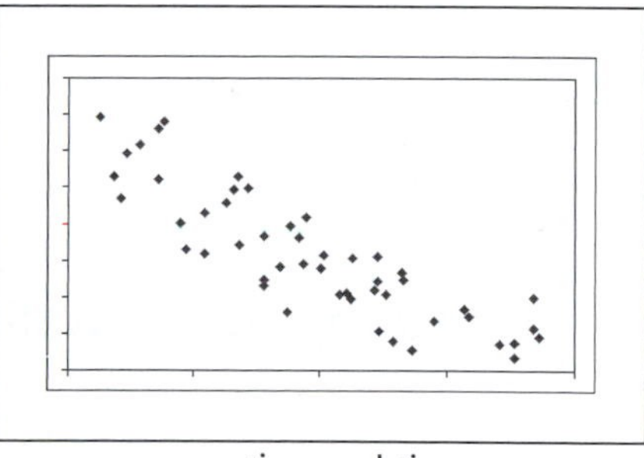

negative correlation

The most often used measure of correlation is the **Pearson Correlation coefficient**, which is usually identified with the letter r. The formula for r is

$$r = \frac{\sum_{i=1}^{n}(x_i - \overline{x})(y_i - \overline{y})}{\sqrt{\sum_{i=1}^{n}(x_i - \overline{x})^2}\sqrt{\sum_{i=1}^{n}(y_i - \overline{y})^2}}$$

For example, the correlation of the data in Table 8-1 is

$$\begin{aligned} r &= \frac{(1 - 1.8)(3 - 3.4) + (2 - 1.8)(4 - 3.4) + \cdots + (2 - 1.8)(5 - 3.4)}{\sqrt{(1 - 1.8)^2 + \cdots + (2 - 1.8)^2} \times \sqrt{(3 - 3.4)^2 + \cdots + (5 - 3.4)^2}} \\ &= \frac{1.4}{\sqrt{2.8} \times \sqrt{1.2}} \\ &= 0.763 \end{aligned}$$

which indicates a high positive correlation.

Correlation and Slope

Notice that the numerator in the equation for r is exactly the same as the numerator for the slope b in the regression equation shown earlier. This is important because it means that the slope $= 0$ when $r = 0$ and that the sign of the slope is the same as the sign of the correlation. The slope can be any real number, but the correlation must always be between -1 and $+1$. A correlation of $+1$ means that all of the data points fall perfectly on a line of positive slope. In such a case, all of the residuals would be 0 and the line would pass right through the points; it would have a perfect fit.

In terms of hypothesis testing, the following statements are equivalent:

H_0: There is no linear relationship between the predictor variable and the dependent variable.

H_0: There is no population correlation between the two variables.

In other words, the correlation is zero if the slope is zero, and vice versa. When you do a statistical test for correlation, the assumptions are the same as the assumptions for linear regression.

Correlation and Causality

Correlation indicates the relationship between two variables without assuming that a change in one causes a change in the other. For example, if you learn of a correlation between the number of extracurricular activities and grade-point average (GPA) for high school students, does this imply that if you raise a student's GPA, he or she will participate in more after-school activities? Or that if you ask the student to get more involved in extracurricular activities, his or her grades will improve as a result? Or is it more likely that if this correlation is true, the type of people who are good students also tend to be the type of people who join after-school groups? You should therefore be careful never to confuse correlation with cause and effect, or causality.

Spearman's Rank Correlation Coefficient *s*

Pearson's correlation coefficient is not without problems. It can be susceptible to the influence of outliers in the data set, and it assumes that a straight-line relationship exists between the two variables. In the presence of outliers or a curved relationship, Pearson's *r* may not detect a significant correlation. In those cases, you may be better off using a nonparametric measure of correlation, **Spearman's rank correlation**, which is usually denoted by the symbol *s*. As with the nonparametric tests in Chapter 7, you replace observed values with their ranks and calculate the value of *s* on the ranks. Spearman's rank correlation, like many other nonparametric statistics, is less susceptible to the influence of outliers and is better than Pearson's correlation for nonlinear relationships. The downside to the Spearman correlation is that it is not as powerful as the Pearson correlation in detecting significant correlations in situations where the parametric assumptions are satisfied.

Correlation Functions in Excel

To calculate correlation values in Excel, you can use some of the functions shown in Table 8-4. Note that Excel does not include functions to calculate Spearman's rank correlation or the *p* values for the two types of correlation measures.

Table 8-4 Calculating Correlation Values

Function	Description
CORREL(*x*, *y*)	Calculates Pearson's correlation *r* for the values in *x* and *y*.
CORRELP(*x*, *y*)	Calculates the two-sided *p* value of Pearson's correlation for the values in *x* and *y*. *StatPlus required.*
SPEARMAN(*x*, *y*)	Calculates Spearman's rank correlation *s* for the values in *x* and *y*. *StatPlus required.*
SPEARMANP(*x*, *y*)	Calculates the two-sided *p* value of Spearman's rank correlation for the values in *x* and *y*. *StatPlus required.*

Let's use these functions to calculate the correlation between the mortality index and the mean annual temperature for the breast cancer data.

To calculate the correlations and p values:

1 Return to the Mortality Data worksheet.

2 Enter the labels **Pearson's r**, ***p* value**, **Spearman's s**, and ***p* value** in the cell range **A19:A22**. Enlarge the width of column A to fit the size of the new labels.

3 Click cell **B19**, type **=CORREL(temperature, mortality)**, and press **Enter**.

4 In cell B20, type **=CORRELP(temperature, mortality)** and press **Enter**.

5 In cell B21, type **=SPEARMAN(temperature, mortality)** and press **Enter**.

6 In cell B22, type **=SPEARMANP(temperature, mortality)** and press **Enter**.

The correlation values are shown in Figure 8-21.

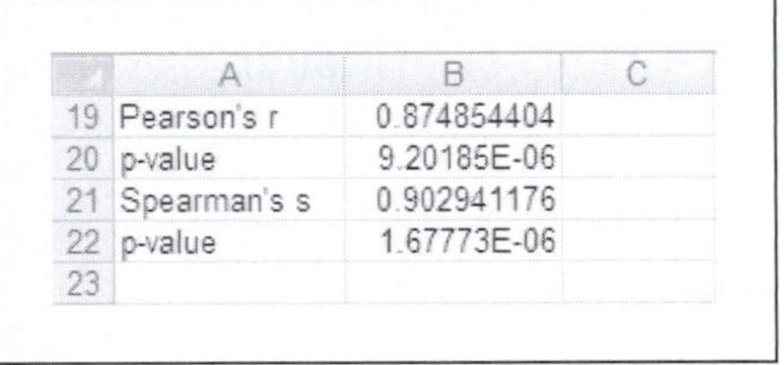

	A	B	C
19	Pearson's r	0.874854404	
20	p-value	9.20185E-06	
21	Spearman's s	0.902941176	
22	p-value	1.67773E-06	
23			

Figure 8-21 Correlations and *p* value

The values in Figure 8-21 indicate a strong positive correlation between the mortality index and the mean annual temperature. The p values for both measures are also very significant, indicating that this correlation is statistically different from zero. Note that the p value for Pearson's r is equal to the p value for the linear regression shown earlier in Figure 8-11. One more important point: the value of r, 0.875, is equal to the square root of the R^2 statistic, computed earlier in Figure 8-10. It will always be the case that R^2 is equal to the square of Pearson's correlation coefficient between two variables.

You can close the Breast Cancer Regression Analysis workbook now. You've completed your analysis of the data, but you'll return to it in the chapter exercises.

Creating a Correlation Matrix

When you have several variables to study, it's useful to calculate the correlations between the variables. In this way, you can get a quick picture of the relationships between the variables, determining which variables are highly correlated and which are not. One way of doing this is to create a **correlation matrix**, in which the correlations (and associated *p* values) are laid out in a square grid.

To illustrate the use of a correlation matrix, consider the Calculus workbook. This file contains data collected to see how performance in a freshman calculus class is related to various predictors (Edge and Friedberg, 1984). Table 8-5 describes the variables in the Calculus workbook.

Table 8-5 Calculus Workbook Variables

Range Name	Range	Description
Calc_HS	A2:A81	Indicates whether calculus was taken in high school (0 = no; 1 = yes)
ACT_Math	B2:B81	The student's score on the ACT mathematics exam
Alg_Place	C2:C81	The student's score on the algebra placement exam given in the first week of classes
Alg2_Grade	D2:D81	The student's grade point in second-year high school algebra
HS_Rank	E2:E81	The student's percentile rank in high school
Gender	F2:F81	The student's gender
Gender_Code	G2:G81	The student's gender code (0 = female; 1 = male)
Calc	H2:H81	The student's grade in calculus

To open the Calculus workbook:

1 Open the **Calculus** workbook from the Chapter08 data folder.

2 Save the workbook as **Calculus Correlation Analysis** to the same folder. The workbook appears as shown in Figure 8-22.

Figure 8-22
Calculus
workbook

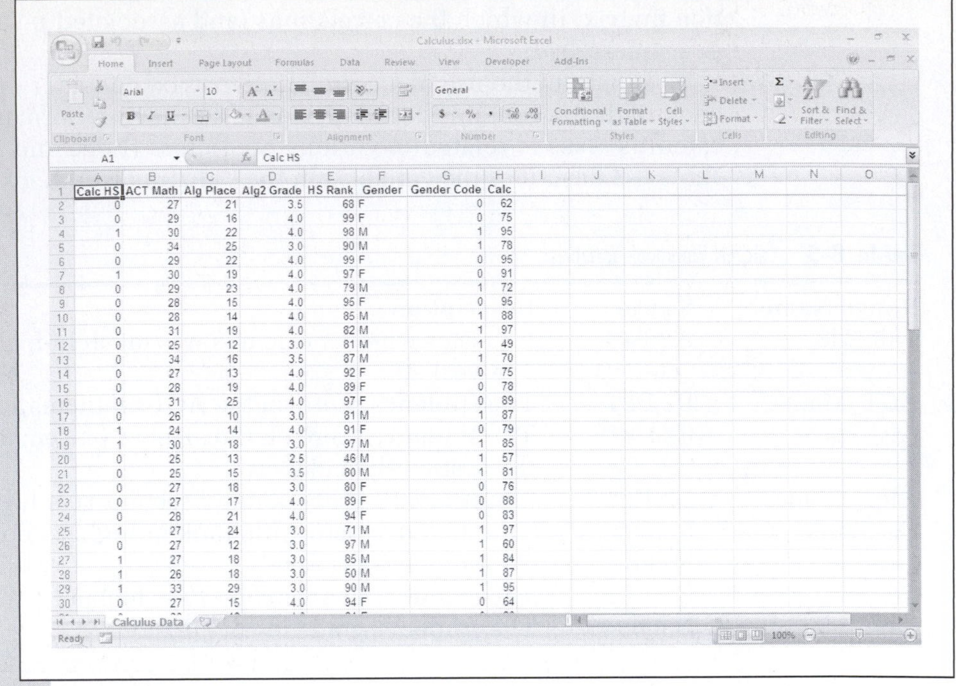

Calc HS	ACT Math	Alg Place	Alg2 Grade	HS Rank	Gender	Gender Code	Calc
0	27	21	3.5	68	F	0	62
0	29	16	4.0	99	F	0	75
1	30	22	4.0	98	M	1	95
0	34	25	3.0	90	M	1	78
0	29	22	4.0	99	F	0	95
1	30	19	4.0	97	F	0	91
0	29	23	4.0	79	M	1	72
0	28	15	4.0	95	F	0	95
0	28	14	4.0	85	M	1	88
0	31	19	4.0	82	M	1	97
0	25	12	3.0	81	M	1	49
0	34	16	3.5	87	M	1	70
0	27	13	4.0	92	F	0	75
0	28	19	4.0	89	F	0	78
0	31	25	4.0	97	F	0	89
0	26	10	3.0	81	M	1	87
1	24	14	4.0	91	F	0	79
1	30	18	3.0	97	M	1	85
0	25	13	2.5	46	M	1	57
0	25	15	3.5	80	M	1	81
0	27	18	3.0	80	F	0	76
0	27	17	4.0	89	F	0	88
0	28	21	4.0	94	F	0	83
1	27	24	3.0	71	M	1	97
0	27	12	3.0	97	M	1	60
1	27	18	3.0	85	M	1	84
1	26	18	3.0	50	M	1	87
1	33	29	3.0	90	M	1	95
0	27	15	4.0	94	F	0	64

Now let's create a matrix of the correlations for all of the variables in the workbook.

To create a correlation matrix of the numeric variables:

1. Click **Multivariate Analysis** from the StatPlus menu and then click **Correlation Matrix.**
2. Click the **Data Values** button and select all of the variables in the workbook *except* the Gender variable.
3. Click the **Output** button and send the output to a new worksheet named **Corr Matrix**. Click the **OK** button. Figure 8-23 shows the completed dialog box.

Figure 8-23 Create Correlation dialog box

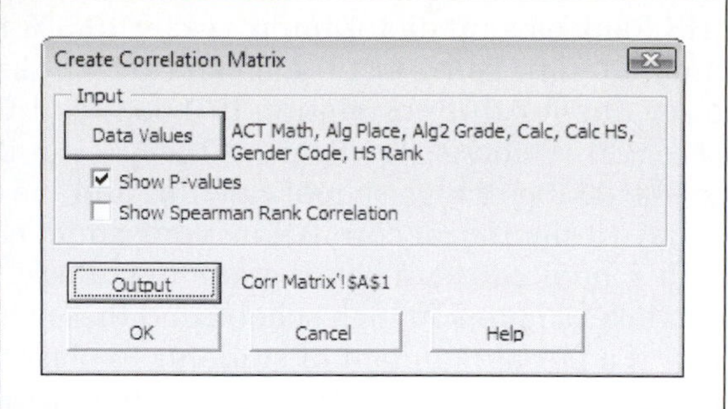

4 Click the **OK** button.

Excel generates the matrix of correlations as displayed in Figure 8-24.

Figure 8-24 Correlation matrix

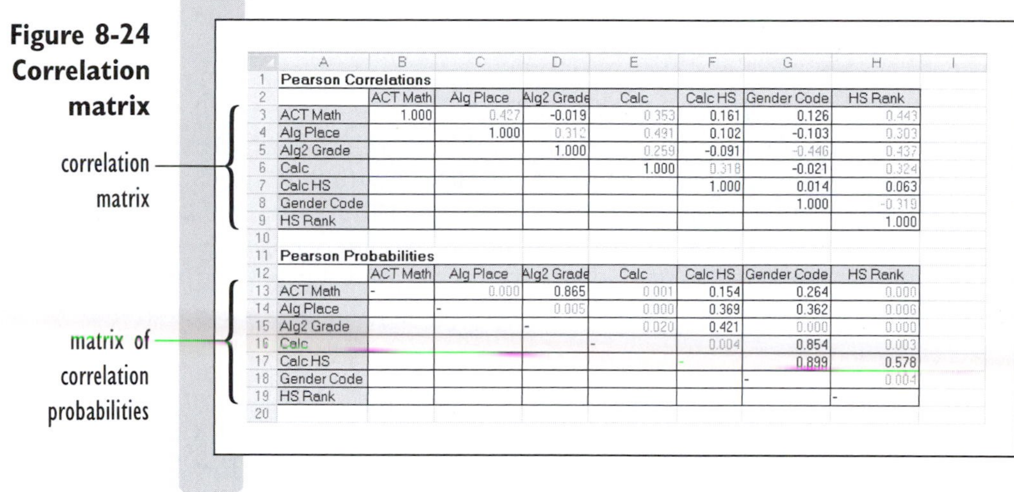

Pearson Correlations

	ACT Math	Alg Place	Alg2 Grade	Calc	Calc HS	Gender Code	HS Rank
ACT Math	1.000	0.427	-0.019	0.353	0.161	0.126	0.443
Alg Place		1.000	0.312	0.491	0.102	-0.103	0.303
Alg2 Grade			1.000	0.259	-0.091	-0.446	0.437
Calc				1.000	0.318	-0.021	0.324
Calc HS					1.000	0.014	0.063
Gender Code						1.000	-0.319
HS Rank							1.000

Pearson Probabilities

	ACT Math	Alg Place	Alg2 Grade	Calc	Calc HS	Gender Code	HS Rank
ACT Math	-	0.000	0.865	0.001	0.154	0.264	0.000
Alg Place		-	0.005	0.000	0.369	0.362	0.006
Alg2 Grade			-	0.020	0.421	0.000	0.000
Calc				-	0.004	0.854	0.003
Calc HS					-	0.899	0.578
Gender Code						-	0.004
HS Rank							-

Figure 8-24 shows two matrices. The first, in cells A1:H9, is the correlation matrix, which shows the Pearson correlations. The second, the matrix of probabilities in cells A11:H19, gives the corresponding two-sided *p* values. *P* values less than 0.05 are highlighted in red.

The most interesting numbers here are the correlations with the calculus score, because the object of the study was to predict this score. The highest correlation appears in cell E4 (0.491), with Alg Place, the algebra placement test score. The other correlations, ACT Math, HS Rank, and Calc HS, are not impressive predictors when you consider that the squared correlation gives R^2, the percentage of variance explained by the variable as a regression predictor.

For example, the correlation between Calc and HS Rank is 0.324 (cell H6); the square of this is 0.105, so regression on HS Rank would account for only 10.5% of the variation in the calculus score. Another way of saying this is that using HS Rank as a predictor improves by 10.5% the sum of squared errors, as compared with using just the mean calculus score as a predictor. Note that the p value for this correlation is 0.003 (cell H16), which is less than 0.05, so the correlation is significant at the 5% significance level.

Just because the taking of high school calculus and the subsequent college calculus score have a significant correlation, you cannot conclude that taking calculus in high school *causes* a better grade in college. The stronger math students tend to take calculus in high school, and these students also do well in college. Only if a fair assignment of students to classes could be guaranteed (so that the students in high school calculus would be no better or worse than others) could the correlation be interpreted in terms of causation.

Correlation with a Two-Valued Variable

You might reasonably wonder about using Calc HS here. After all, it assumes only the two values 0 and 1. Does the correlation between Calc and Calc HS make sense? The positive correlation of 0.324 indicates that if the student has taken calculus in high school, the student is more likely to have a high calculus grade.

Another categorical variable in this correlation matrix is Gender Code, which has a significant negative correlation with the Alg2 Grade ($r = -0.446$, p value $= 0.000$) and HS Rank ($r = -0.319$, p value $= 0.004$). Recall that in the gender code, 0 = female and 1 = male. A negative correlation here means that females tended to have higher grades in second-year algebra and were ranked higher in high school.

Adjusting Multiple *p* Values with Bonferroni

The second matrix in Figure 8-24 gives the p values for the correlations. Except for Gender, all of the correlations with Calc are significant at the 5% level because all the p values are less than 0.05.

Some statisticians believe that the p values should be adjusted for the number of tests, because conducting several hypothesis tests raises above 5% the probability of rejecting at least one true null hypothesis. The **Bonferroni** approach to this problem is to multiply the p value in each test by the total number of tests conducted. With this approach, the probability of rejecting one or more of the true hypotheses is less than 5%.

Let's apply this approach to correlations of Calc with the other variables. Because there are six correlations, the Bonferroni approach would have us multiply each p value by 6 (equivalent to decreasing the p value required for statistical significance to $0.05/6 = 0.0083$). Alg2 Grade has a p value of 0.020, and because $6 \times (0.020) = 0.120$, the correlation is no longer significant from this point of view. Instead of focusing on the individual correlation tests, the Bonferroni approach rolls all of the tests into one big package, with 0.05 referring to the whole package.

Bonferroni makes it much harder to achieve significance, and many researchers are reluctant to use it because it is so conservative. In any case, this is a controversial area, and professional statisticians argue about it.

EXCEL TIPS

- To create a correlation matrix with Excel, you can click the Data Analysis button from the Analysis group on the Data tab and then click Correlation from the Data Analysis dialog box. Complete the Correlation dialog box to create the matrix.

Creating a Scatter Plot Matrix

The Pearson correlation measures the extent of the linear relationship between two variables. To see whether the relationship between the variables is really linear, you should create a scatter plot of the two variables. In this case, that would mean creating 15 different scatter plots, a time-consuming task! To speed up the process, you can create a scatter plot matrix. In a **scatter plot matrix**, or **SPLOM**, you can create a matrix containing the scatter plots between the variables. By viewing the matrix, you can tell at a glance the nature of the relationships between the variables.

To create a scatter plot matrix:

1. Click **Multi-variable Charts** from the StatPlus menu and then click **Scatter plot Matrix.**
2. Click the **Data Values** button and select the range names **ACT Math**, **Alg Place**, **Alg2_Grade**, **Calc**, and **HS Rank**.
3. Click the **Output** button, and send the output to the worksheet **SPLOM**. Click the **OK** button twice.

Excel generates the scatter plot matrix shown in Figure 8-25.

Figure 8-25
Scatter plot matrix

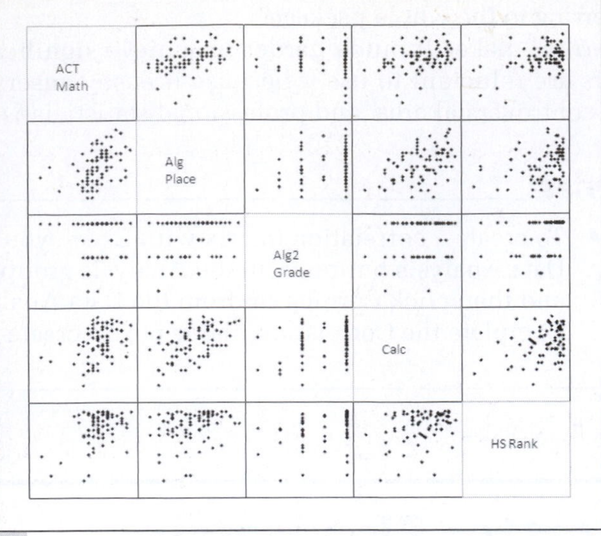

Depending on the number of variables you are plotting, SPLOMs can be difficult to view on the screen. If you can't see the entire SPLOM on your screen, consider reducing the value in the Zoom Control box. You can also reduce the SPLOM by selecting it and dragging one of the resizing handles to make it smaller.

How should you interpret the SPLOM? Each of the five variables is plotted against the other four variables, with the four plots displayed in a row. For example, ACT Math is plotted as the *y* variable against the other four variables in the first row of the SPLOM. The first plot in the first row is ACT Math versus Alg Place, and so on. On the other hand, the first plot in the first column displays Alg Place as the *y* variable and is plotted against the *x* variable, ACT Math. The scales of the plot are not shown in order to save space. If you find a plot of interest, you can recreate it using Excel's Chart Wizard to show more details and information.

Carefully consider the plots in the second to last row, which show Calc against the other variables. Each plot shows a roughly linear upward trend. It would be reasonable to conclude here that correlation and linear regression are appropriate when predicting Calc from ACT Math, Alg2 Grade, Alg Place, and HS Rank.

Recall from Figure 8-24 that Alg Place had the highest correlation with Calc. How is that evident here? A good predictor has good accuracy, which means that the range of *y* is small for each *x*. Of the four plots in the fourth row, the plot of Calc against Alg Place has the narrowest range of *y* values for each *x*. However, Alg Place is the best of a weak lot. None of the plots shows that really accurate prediction is possible. None of these plots shows a relationship anywhere near as strong as the relationship between mortality index and temperature that you worked with earlier in the chapter.

Save your work and close the Calculus Correlation Analysis workbook.

Exercises

1. *True or false, and why:* If the slope of a regression line is large, the correlation between the variables will also be large.
2. *True or false, and why:* If the correlation between two variables is near 1, the slope will be a large positive number.
3. *True or false, and why:* If the *p* value of the Pearson's correlation coefficient is low, the *p* value of the slope parameter of the regression equation will also be low.
4. *True or false, and why:* A correlation of zero means that the two variables are unrelated.
5. *True or false, and why:* The runs test is one of the diagnostic tests you should always apply to the residuals in your regression analysis.
6. In a time-ordered study, you have 25 residuals from the regression model. There are 10 negative residuals and 15 positive ones. There are a total of 10 runs. Is this an unusual number of runs? What is the level of statistical significance?
7. Using the following ANOVA table for the regression of variable *y* on variable *x*, answer the questions below.

Table 8-6 Regression of Variable *y* on *x*

ANOVA	*df*	*SS*	*MS*	*F*	Significance *F*
Regression	1	129.6	129.6	4.91	0.057
Residual	8	210.9	26.4		
Total	9	340.5			

   a. How many observations are in the data set?
   b. What is the variance of *y*?
   c. What is the value of R^2?
   d. What percentage of the variability in *y* is explained by the regression?
   e. What is the absolute value of the correlation of *x* and *y*?
   f. What is the *p* value of the correlation of *x* and *y*?
   g. What is the standard error (the typical deviation of an observed point from the regression line)?

8. Return to the Breast Cancer Mortality study discussed in this chapter. There may be an outlier in the data set. Perform the following analysis to determine the effect of this outlier on the regression analysis:
   a. Open the **Breast Cancer** workbook from the Chapter08 folder and save it as **Breast Cancer Outlier Regression** to the same folder.
   b. Remove the observation for the first region from the data set.

c. Create a scatter plot of mortality versus temperature and add a linear trend line to the plot, showing both the R^2 value and the regression equation.
d. Calculate the regression statistics for the new data set.
e. Create scatter plots of the residuals of the regression equation versus temperature and the predicted values. Also create a normal probability plot of the residuals.
f. Calculate the Pearson and Spearman correlation coefficients, including the p values.
g. Save your changes to the workbook and write a report summarizing your results, including a description of the diagnostic tests you performed. How does this regression compare with the regression you performed earlier that included the possible outlier? How do the diagnostic plots compare?

9. You've been given an Excel workbook containing nutritional information on 10 wheat products. Perform the following analysis:
   a. Open the **Wheat** workbook from the Chapter08 folder and save it as **Wheat Regression Analysis**.
   b. Plot Calories versus ServingGrams, adding a regression line, equation, and R^2 value to the plot. How does the serving size (in grams) predict the calories of the different wheat products?
   c. Compute Pearson's correlation and the corresponding p value between Calories and ServingGrams.
   d. Use the Data Analysis ToolPak to calculate the statistics for the regression equation.
   e. Create diagnostic plots of residuals versus ServingGrams, and the normal probability plot of the residuals. Do the regression assumptions seem to be satisfied?
   f. In the plot of residuals versus predicted values, label each point with the food type (pretzel, bagel, bread, etc). Where do the residuals for the breads appear?
   g. Breads are often low in calories because of high moisture content. One way of removing the moisture content from the equation is to create a new variable that sums up the total of the nutrient weights. With this in mind, create a new variable, total, which is the sum of the weights of carbohydrates, proteins, and fats. From this total subtract the value of the Fiber variable since fiber does not contribute to the calorie total. Plot Calories versus Total on a new chart sheet.
   h. Redo your regression equation, regressing the Calories variable on the new variable Total. How does the diagnostic plot of residuals versus predicted values compare to the earlier plot? How do the R^2 values compare? Where are the residuals for the bread values located?
   i. Save your changes to the workbook and write a report summarizing your observations.

10. Continue to investigate the nutritional data in the Wheat workbook by performing the following analysis:
    a. Open the **Wheat** workbook from the Chapter08 folder and save it as **Wheat Correlation Matrix**.
    b. Create scatter plot and correlation matrices (Pearson correlation only) for the variables serving grams, calories, protein, carbohydrate, and fat.
    c. Why is fat so weakly related to the other variables? Given that fat is supposed to be very important in calories, why is the correlation so weak here?
    d. Would the relationship between the fat and calories variables be stronger

if we used foods that cover a wider range of fat-content values?

e. Save your changes to the workbook and write a report summarizing your observations.

11. You've been given a workbook containing the ages and prices of used Mustangs from Cars.com in 2002. Perform the following analysis:

a. Open the **Mustang** workbook from the Chapter08 folder and save it as **Mustang Regression Analysis**.

b. Compute the Pearson and Spearman correlations (and p values) between age and price.

c. Plot price against age. Does this scatter plot cause you any concern about the validity of the correlations?

d. How do the correlations change if you concentrate only on cars that are less than 20 years old?

e. Excluding the old classic cars (older than 19 years), perform a regression of price against age and find the drop in price per year of age.

f. Do you see any problems in the diagnostic plots of the residuals?

g. Save your changes to the workbook and write a report summarizing your observations.

12. Return to the Calculus data set you examined in this chapter and perform the following analysis:

a. Open the **Calculus** workbook from the Chapter08 folder and save it as **Calculus Regression Analysis**.

b. Regress Calc on Alg Place and obtain a 95% confidence interval for the slope.

c. Interpret the slope in terms of the increase in final grade when the placement score increases by 1 point.

d. Do the residuals give you any cause for concern about the validity of the model?

e. Save your changes to the workbook and write a report summarizing your observations.

13. The Booth workbook gives total assets and net income for 45 of the largest American U.S. banks in 1973. Open the workbook and perform the following analysis of this historical economic data set:

a. Open the **Booth** workbook from the Chapter08 folder and save it as **Booth Regression Analysis**.

b. Plot net income against total assets and notice that the points tend to bunch up toward the lower left, with just a few big banks dominating the upper part of the graph. Add a linear trend line to the plot.

c. Regress net income against total assets and plot the standard residuals against the predictor values. (The standardized residuals appear with the regression output when you select the Standardized Residuals check box in the Regression dialog box.)

d. Given that the residuals tend to be bigger for the big banks, you should be concerned about the assumption of constant variance. Try taking logs of both variables. Now repeat the plot of one against the other, repeat the regression, and again look at the plot of the residuals against the predicted values. Does the transformation help the relationship? Is there now less reason to be concerned about the assumptions? Notice that some banks have strongly positive residuals, indicating good performance, and some banks have strongly negative residuals, indicating below-par performance. Indeed, bank 20, Franklin National Bank, has the second most negative residual and failed the following year. Booth (1985) suggests that regression is a good way to locate problem banks before it is too late.

e. Save your changes to the workbook and write a report summarizing your observations.

14. You've been given a workbook which contains mass and volume measurements on eight chunks of aluminum from a high school chemistry class.

a. Open the **Aluminum** workbook from the Chapter08 folder and save it as **Aluminum Regression Analysis**.

b. Plot mass against volume, and notice the outlier.

c. After excluding the outlier, regress mass on volume, without the constant term (select the Constant is Zero checkbox in the Regression dialog box), because the mass should be 0 when the volume is 0. The slope of the regression line is an estimate of the density (not a statistical word here but a measure of how dense the metal is) of aluminum.

d. Give a 95% confidence interval for the true density. Does your interval include the accepted true value, which is 2.699?

e. Save your changes to the workbook and write a report summarizing your observations.

15. You've been given data containing health statistics from 2007 for the 50 states of the United States. The data set contains two variables: Diabetes and FluPneum. The Diabetes variable contains the death rates (per 100,000) for diabetes while the FluPneum variable contains the death rates for causes related to the flu or pneumonia. You've been asked to determine if there is any correlation between these two measures.

a. Open the **Health** workbook from the Chapter08 folder and save it as **Health Correlation Analysis**.

b. Compute the Pearson correlation and the Spearman rank correlation between them. How does the Spearman rank correlation differ from the Pearson correlation? How do the *p* values compare? Are both tests significant at the 5% level?

c. Create the corresponding scatter plot. Label each point on the scatter plot with the name of the state. Which state is a possible outlier on the lower left of the plot?

d. Copy the data to a new worksheet, removing the most extreme outlier. Redo the correlations and your scatter plot.

e. How are the size and significance of the correlations influenced by removing that one state? Make a case for the deletions on the basis of the plot and some geography. Does the original correlation give an exaggerated notion of the relationship between the two variables? Does the nonparametric correlation coefficient solve the problem? Explain. Would you say that a correlation without a plot can be deceiving?

f. Save your workbook and write a report summarizing your observations.

16. The Fidelity workbook contains financial data from 1989, 1990, and 1991 for 33 Fidelity sector funds. The source is the Morningside Mutual Fund Sourcebook 1992, Equity Mutual Funds. You've been asked to explore the relationships between some of the financial variables in this data set. The name of the fund is given in the Sector column. The TOTL90 column is the percentage total return during the year 1990, and TOTL91 is the percentage total return for the year 1991. NAV90 is the percentage increase in net asset value during 1990, and similarly, NAV91 is the percentage change in net asset value during 1991. INC90 is the percentage net income for 1990, and similarly, INC91 is the percentage net income for 1991. CAPRET90 is the percentage capital gain for 1990, and CAPRET91 is the percentage capital gain for 1991.

a. Open the **Fidelity** workbook from the Chapter08 folder and save it as **Fidelity Financial Analysis**.
b. What is the correlation between the percentage capital gains for 1990 and 1991? Do your analysis using both the Pearson and Spearman correlations, calculating the p value for both. Is there evidence to support the supposition that the percentage capital gains from 1990 are highly correlated with the percent capital gains for 1991?
c. What is the correlation between the percentage net income for 1990 and 1991? Use both the Pearson and Spearman correlation coefficents and include the p values. Is net income from 1990 highly correlated with net income from 1991?
d. Create a scatter plot for the two correlations in parts a and b. Label each point on the scatter plot with labels from the Sector column.
e. You should get a stronger correlation for income than for capital gains. How do you explain this?
f. Calculate the correlation between the percentage increase in net asset value in 1990 to 1991 using the NAV90 and NAV91 variables and then generate the scatter plot, labeling the points with the sector names. Note that the Biotechnology Fund stands out in the plot. It was the only fund that performed well in both years.
g. Compare the Pearson and Spearman correlation values for NAV90 and NAV91. Are they the same sign? What could account for the different correlation values? Which do you think is more representative of the scatter plot you created?
h. If the correlation is this weak, what does it suggest about using fund performance in one year as a guide to fund performance in the following year?
i. Save your changes to the workbook and write a report summarizing your observations.

17. The Draft workbook contains information on the 1970 military draft lottery. Draft numbers were determined by placing all 366 possible birth dates in a rotating drum and selecting them one by one. The first birth date drawn received a draft number of 1 and men born on that date were drafted first, the second birth date entered received a draft number of 2, and so forth. Is there any relationship between the draft number and the birth date?

a. Open the **Draft** workbook from the Chapter08 folder and save it as **Draft Correlation Analysis**.
b. Using the values in the Draft Numbers worksheet, calculate the Pearson and Spearman correlation coefficients and p value between the Day_of_the_Year and the Draft number. Is there a significant correlation between the two? Using the value of the correlation, would you expect higher draft numbers to be assigned to people born earlier in the year or later?
c. Create a scatter plot of Number versus Day_of_the_Year. Is there an obvious relationship between the two in the scatter plot?
d. Add a trend line to your scatter plot and include both the regression equation and the R^2 value. How much of the variation in draft number is explained by the Day_of_the_Year variable?
e. Calculate the average draft number for each month and then calculate

the correlation between the month number and the average draft number. How do the values of the correlation in this analysis compare with those of the correlation you performed earlier?

f. Create a scatter plot of average draft number versus month number. Add a trend line and include the regression equation and R^2 value. How much of the variability in the average draft number per month is explained by the month?

g. Save your changes to the workbook and write a report summarizing your conclusions. Which analysis (looking at daily values or looking at monthly averages) better describes any problem with the draft lottery?

18. The Emerald health care providers claim that components of their health plan cause it to rise significantly more slowly than overall health costs. You decide to investigate to see whether there is evidence for Emerald's claim. You have recorded Emerald costs over the past seven years, along with the consumer price index (CPI) for all urban consumers and the medical component of the CPI.

a. Open the **Emerald** workbook from the Chapter08 folder and save it as **Emerald Regression Analysis**.

b. Using the Analysis ToolPak's Regression command, calculate the regression equation for each of the three price indexes against the year variable. What are the values for the three slopes? Express the slope in terms of the increase in the index per year. How does Emerald's change in cost compare to the other two indexes?

c. Look at the 95% confidence intervals for the three slopes. Do the confidence intervals overlap? Does there appear to be a significant difference in the rate of increase under the Emerald plan as compared to the increases under the other two indexes?

d. Summarize your conclusions. Do you see evidence to substantiate Emerald's claim?

e. Save your changes and write a report summarizing your observations.

19. The Teacher workbook contains data on the relationship between teachers' salaries and the spending on public schools per pupil in 1985. Perform the following analysis on this data set:

a. Open the **Teacher** workbook from the Chapter08 folder and save it as **Teacher Salary Analysis**.

b. Create a scatter plot of spending per pupil versus teacher salary. Add a trend line containing the R^2 value and regression to the plot.

c. Compute the regression statistics for the data, and then create the diagnostic plots discussed in this chapter. Is there any evidence of a problem in the diagnostic plots?

d. Copy the spending per pupil versus teacher salary scatter plot to a new chart sheet and then break down the points in the plot on the basis of the values of the area variable. For each of the three series in the chart, add a linear trend line and compute the R^2 value and regression equation. How do the least-squares lines compare among the three regions? What do you think accounts for any difference in the trend lines?

e. Redo the regression statistics, performing three regressions, one for each of the three areas in the data set. Compare the regression equations. What are the 95% confidence intervals for the slope parameters in the three areas?

f. Save your changes to the workbook and write a report summarizing your observations.

20. The Highway workbook contains data on highway fatalities per million vehicle miles from 1945 to 1984 for the United States and the state of New Mexico. You've been asked to use regression analysis to analyze and compare the trend in the fatality rates.

  a. Open the **Highway** workbook from the Chapter08 folder and save it as **Highway Regression Analysis**.
  b. Create a scatter plot that shows the New Mexico and U.S. fatality rates versus the Year variable. For each data series, display the linear regression line, along with the regression equation and R^2 value. How much of the variation in highway fatalities is explained by the linear regression line for the two data sets? Do the trend lines appear to be the same? What problems would you see for this trend line if it is extended out for many years into the future?
  c. Calculate the regression statistics for both data sets and create residual plots for both regressions. Do the residual plots indicate any possible violations of the regression assumptions?
  d. Since these are time-ordered data, perform a runs test on the standardized residuals for both the New Mexico and U.S. data. Calculate the Durbin-Watson test statistic for both sets of residuals. Does your analysis lead you to believe that one of the regression assumptions has been violated?
  e. Save your changes to the workbook and write a report summarizing your conclusions.

21. The HomeTax workbook contains data on home prices and property taxes for houses in Albuquerque, New Mexico, sold back in 1993. Many factors were involved in assessing the property tax for a home during that time. You've been asked to do a general analysis comparing the price of the home to its assessed property tax.

  a. Open the **HomeTax** workbook from the Chapter08 folder and save it as **HomeTax Regression Analysis**.
  b. Create a scatter plot of the tax on each home versus that home's price. Add a trend line to the scatter plot and include the regression equation and R^2 value. How much of the variation in property taxes is explained by the price of the house?
  c. Calculate the regression statistics, comparing property tax to home price, and create a plot of the residuals.
  d. Create a Normal plot of the residuals. Is there anything in the two residual plots that may violate the regression assumptions?
  e. Create two new variables in the workbook named log(price) and log(tax) that contain the Base10 logarithms of the price and tax data. Redo steps b through d on these transformed data. Has the transformation solved any problems with the regression assumptions on the untransformed values? What problems, if any, still remain?
  f. Save your changes to the workbook and write a report summarizing your conclusions.

Chapter 9

MULTIPLE REGRESSION

Objectives

In this chapter you will learn to:

- Use the F distribution
- Fit a multiple regression equation and interpret the results
- Use plots to help understand a regression relationship
- Validate a regression using residual diagnostics

Regression Models with Multiple Parameters

In Chapter 8, you used simple linear regression to predict a dependent variable (y) from a single independent variable (x, a predictor variable). In multiple regression, you predict a dependent variable from several independent variables. For three predictors, x_1, x_2, and x_3, the multiple regression model takes the form:

$$y = \beta_0 + \beta_1 x_1 + \beta_2 x_2 + \beta_3 x_3 + \varepsilon$$

where the coefficients are unknown parameter values that you can estimate and ε is random error, which follows a normal distribution with mean 0 and variance σ^2. Note that the predictors can also be functions of variables. The following are also examples of models whose parameters you can estimate with multiple regression:

Polynomial: $y = \beta_0 + \beta_1 x + \beta_2 x^2 + \beta_3 x^3 + \varepsilon$

Trigonometric: $y = \beta_0 + \beta_1 \sin x + \beta_2 \cos x + \varepsilon$

Logarithmic: $y = \beta_0 + \beta_1 \log x_1 + \beta_2 \log x_2 + \varepsilon$

Note that all of these equations are examples of linear models, even though they use various trigonometric and logarithmic functions. The *linear* in *linear model* refers to the error term ε and the parameters β_i. The equations are linear in those terms. For example, one could create new variables $l = \sin x$ and $k = \cos x$, and then the second model is the linear equation $y = \beta_0 + \beta_1 l + \beta_2 k + \varepsilon$.

After computing estimated values for the β coefficients, you can plug them into the equation to get predicted values for y. The estimated regression model is expressed as

$$y = b_0 + b_1 x_1 + b_2 x_2 + b_3 x_3$$

where the b_i's are the estimated parameter values, and the residuals correspond to the error term ε.

CONCEPT TUTORIALS
The *F* distribution

The *F distribution* is basic to regression and analysis of variance as studied in this chapter and the next. An example of the *F* distribution is shown in the Distributions workbook.

To view the *F* distribution:

1 Open the **Distributions** workbook located in the Explore folder of your Student files. Enable the macros in the workbook.

2 Click **F** from the Table of Contents column. Review the material and scroll to the bottom of the worksheet. See Figure 9-1.

Figure 9-1 The *F* distribution

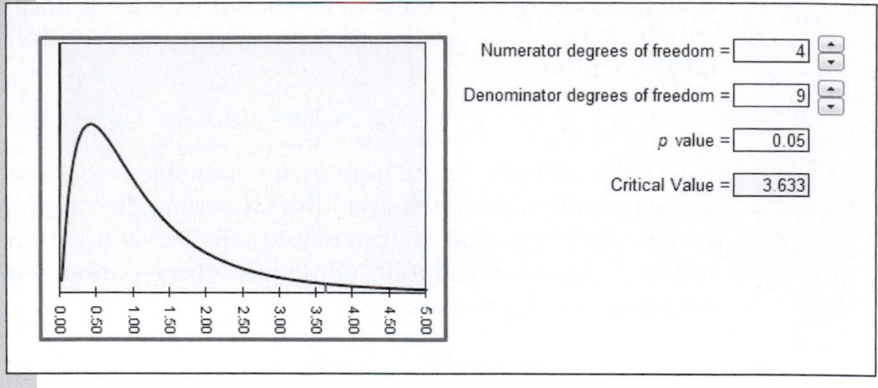

The *F* distribution has two degree-of-freedom parameters: the numerator and denominator degrees of freedom. The distribution is usually referred to as $F(m, n)$—that is, the *F* distribution with *m* numerator degrees of freedom and *n* denominator degrees of freedom. The Distribution workbook opens with an $F(4,9)$ distribution.

Like the χ^2 distribution, the *F* distribution is skewed. To help you better understand the shape of the *F* distribution, the worksheet lets you vary the degrees of freedom of the numerator and the denominator by clicking the degrees-of-freedom scroll arrows. Experiment with the worksheet to view how the distribution of *F* changes as you increase the degrees of freedom.

To increase the degrees of freedom in the numerator and denominator:

1 Click the up spin arrow to increase the numerator degrees of freedom to 10.

2 Click the up spin arrow to increase the denominator degrees of freedom to 15. Then watch how the distribution changes.

In this book, hypothesis tests based on the *F* distribution always use the area under the upper tail of the distribution to determine the *p* value.

To change the *p* value:

1. Click the **Critical Value** box, type 0.10, and then press **Enter**. This gives you the critical value for the *F* test at the 10% significance level.

Notice that the critical value shifts to the left, telling you that 10% of the values of the *F* distribution lie to the right of this point.

Continue working with the *F* distribution worksheet, trying different parameter values to get a feel for the *F* distribution. Close the workbook when you're finished. You do not need to save any changes you may have inadvertently made to the document.

Using Regression for Prediction

One of the goals of regression is prediction. For example, you could use regression to predict what grade a student would get in a college calculus course. (This is the dependent variable, the one being predicted.) The predictors (the independent variables) might be ACT or SAT math score, high school rank, and a placement test score from the first week of class. Students with low predictions might be asked to take a lower-level class.

However, suppose the dependent variable is the price of a four-unit apartment building and the independent variables are the square footage, the age of the building, the total current rent, and a measure of the condition of the building. Here you might use the predictions to find a building that is undervalued, with a price that is much less than its prediction. This analysis was actually carried out by some students, who found that there was a bargain building available. The owner needed to sell quickly as a result of cash flow problems.

You can use multiple regression to see how several variables combine to predict the dependent variable. How much of the variability in the dependent variable is accounted for by the predictors? Do the combined independent variables do better or worse than you might expect, on the basis of their individual correlations with the dependent variable? You might be interested in the individual coefficients and in whether they seem to matter in the prediction equation. Could you eliminate some of the predictors without losing much prediction ability?

When you use regression in this way, the individual coefficients are important. Rosner and Woods (1988) compiled statistics from baseball box scores, and they regressed runs on singles, doubles, triples, home runs, and walks (walks are combined with hit by pitched ball). Their estimated prediction equation is

$$\text{Runs} = -2.49 + 0.47 \text{ singles} + 0.76 \text{ doubles} + 1.14 \text{ triples} + 1.54 \text{ home runs} + 0.39 \text{ walks}$$

Notice that walks have a coefficient of 0.39, and singles have a coefficient of 0.47, so a walk has more than 80% of the weight of a single. This is in contrast to the popular slugging percentage used to measure the offensive production of players, which gives weight 0 to walks, 1 to singles, 2 to doubles, 3 to triples, and 4 to home runs. The Rosner-Woods equation gives relatively more weight to singles, and the weight for doubles is less than twice as much as the weight for singles. Similar comparisons are true for triples and home runs. Do baseball general managers use equations like the Rosner-Woods equation to evaluate ball players? If not, why not?

You can also use regression to see whether a particular group is being discriminated against. A company might ask whether women are paid less than men with comparable jobs. You can include a term in a regression to account for the effect of gender. Alternatively, you can fit a regression model for just men, apply the model to women, and see whether women have salaries that are less than would be predicted for men with comparable positions. It is now common for such arguments to be offered as evidence in court, and many statisticians have experience in legal proceedings.

Regression Example: Predicting Grades

For a detailed example of a multiple regression, consider the Calculus workbook first discussed in Chapter 8, which examined how scores in first-semester calculus were related to various measures of student achievement in high school (Edge and Friedberg, 1984).

To open the Calculus workbook:

1. Start Excel and open the **Calculus** workbook from the Chapter09 data folder.
2. 

Save the file as **Calculus** Multiple **Regression**.

In Chapter 8, it appeared from the correlation matrix and scatter plot matrix that the algebra placement test is the best individual predictor of the first semester calculus score (although it is not very successful). Multiple regression gives a measure of how good the predictors are when used together. The model is

$$\text{Calculus score} = \beta_0 + \beta_1(\text{Calc HS}) + \beta_2(\text{ACT Math}) + \beta_3(\text{Alg Place}) + \beta_4(\text{Alg2 Grade}) + \beta_5(\text{HS Rank}) + \beta_6(\text{Gender Code}) + \varepsilon$$

You can use the Analysis ToolPak Regression command to perform a multiple regression on the data, but the predictor variables must occupy a contiguous range. You will be using columns A, B, C, D, E, and G as your predictor variables, so you need to move column G, Gender Code, next to columns A:E.

To move column G next to columns A:E:

1. Click the **G** column header to select the entire column.
2. Right-click the selection to open the shortcut menu; then click **Cut**.
3. Click the F column header.
4. Right-click the selection to open the shortcut menu; then click **Insert Cut Cells**. You can now identify the contiguous range of columns A:F as your predictor variables.

To perform a multiple regression on the calculus score based on the predictor variables Calc HS, ACT Math, Alg Place, Alg2 Grade, HS Rank, and Gender Code, use the Regression command found in the Analysis ToolPak provided with Excel.

To perform the multiple regression:

1. Click **Data Analysis** from the Analysis group on the Data tab, select **Regression** from the Analysis Tools list box, and click **OK**.
2. Type **H1:H81** in the Input *Y* Range text box, press **Tab**, and then type **A1:F81** in the Input *X* Range text box.
3. Click the **Labels** checkbox and the **Confidence Level** checkbox to select them, and then verify that the **Confidence Level** box contains 95.
4. Click the **New Worksheet Ply** option button, click the corresponding text box, and then type Multiple **Reg**.
5. Click the **Residuals, Standardized Residuals, Residual Plots,** and **Line Fit Plots** checkboxes to select them. Your Regression dialog box should look like Figure 9-2.

Figure 9-2 The completed Regression dialog box

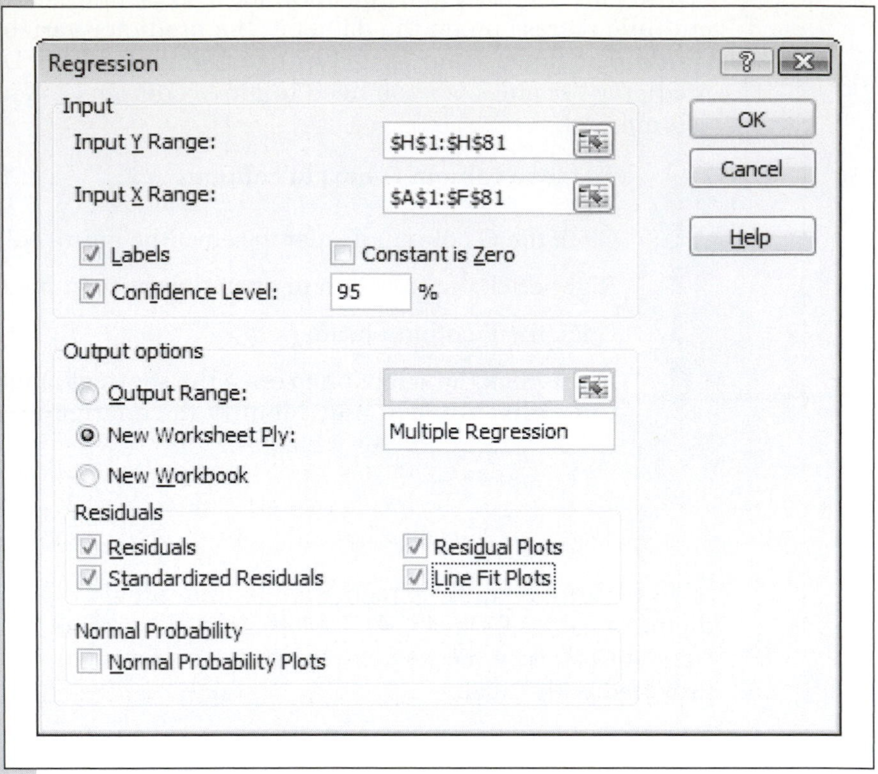

6 Click **OK**.

Excel creates a new sheet, Multiple Reg, which contains the summary output and the residual plots.

Interpreting the Regression Output

To interpret the output, look first at the analysis of variance (ANOVA) table found in cells A10:F14. Figure 9-3 shows this range with the columns widened to display the labels and the values reformatted. The analysis of variance table shows you whether the fitted regression model is significant.

The analysis of variance table helps you choose between two hypotheses.

H_0: The population coefficients of all six predictor variables = 0

H_a: At least one of the six population coefficients $\neq 0$

Figure 9-3 Multiple regression ANOVA table

	A	B	C	D	E	F
10	ANOVA					
11		*df*	*SS*	*MS*	*F*	*Significance F*
12	Regression	6	3840.164	640.027	7.197	4.696E-06
13	Residual	73	6492.036	88.932		
14	Total	79	10332.200			
15						

There are many different parts to an ANOVA table. At this point, you should just concentrate on the *F* ratio and its *p* value, which tell you whether the regression is significant. This ratio is large when the predictor variables explain much of the variability of the response variable, and hence it has a small *p* value as measured by the *F* distribution. A small value for this ratio indicates that much of the variability in *y* is due to random error (as estimated by the residuals of the model) and is not due to the regression. The next chapter, on analysis of variance, contains a more detailed description of the ANOVA table.

The *F* ratio, 7.197, is located in cell E12. Under the null hypothesis, you assume that there is no relationship between the six predictors and the calculus score. If the null hypothesis is true, the *F* ratio in the ANOVA table follows the *F* distribution, with 6 numerator degrees of freedom and 73 denominator degrees of freedom. You can test the null hypothesis by seeing whether this observed *F* ratio is much larger than you would expect in the *F* distribution. If you want to get a visual picture of this hypothesis test, use the *F* distribution worksheet from the Distributions workbook and display the $F(6, 73)$ distribution.

The Significance *F* column gives a *p* value of 4.69×10^{-6} (cell F12), representing the probability that an *F* ratio with 6 degrees of freedom in the numerator and 73 in the denominator has a value 7.197 or more. This is much less than .05, so the regression is significant at the 5% level. You could also say that you reject the null hypothesis at the 5% level and accept the alternative that at least one of the coefficients in the regression is not zero. If the *F* ratio were not significant, there would not be much interest in looking at the rest of the output.

Multiple Correlation

The regression statistics appear in the range A3:B8, shown in Figure 9-4 (formatted to show column labels and the values to three decimal places).

Figure 9-4 Multiple regression statistics

	A	B
1	SUMMARY OUTPUT	
2		
3	*Regression Statistics*	
4	Multiple R	0.610
5	R Square	0.372
6	Adjusted R Square	0.320
7	Standard Error	9.430
8	Observations	80
9		

The R Square value in cell B5 (.372) is the coefficient of determination R^2 discussed in the previous chapter. This value indicates that 37% of the variance in calculus scores can be attributed to the regression. In other words, 37% of the variability in the final calculus score is due to differences among students (as quantified by the values of the predictor variables) and the rest is due to random fluctuation. Although this value might seem low, it is an unfortunate fact that decisions are often made on the basis of weak predictor variables, including decisions about college admissions and scholarships, freshman eligibility in sports, and placement in college classes.

The Multiple R (0.610) in cell B4 is just the square root of the R^2; this is also known as the **multiple correlation**. It is the correlation among the response variable, the calculus score, and the linear combination of the predictor variables as expressed by the regression. If there were only one predictor, this would be the absolute value of the correlation between the predictor and the dependent variable. The Adjusted R Square value in cell B6 (0.320) attempts to adjust the R^2 for the number of predictors. You look at the adjusted R^2 because the unadjusted R^2 value either increases or stays the same when you add predictors to the model. If you add enough predictors to the model, you can reach some very high R^2 values, but not much is to be gained by analyzing a data set with 200 observations if the regression model has 200 predictors, even if the R^2 value is 100%. Adjusting the R^2 compensates for this effect and helps you determine whether adding additional predictors is worthwhile.

The standard error value, 9.430 (cell B7), is the estimated value of σ, the standard deviation of the error term ε, in other words, the standard deviation of the calculus score once you compensate for differences in the predictor variables. You can also think of the standard error as the typical error for prediction of the 80 calculus scores. Because a span of 10 points corresponds to a difference of one letter grade (A vs. B, B vs. C, and so on), the typical error of prediction is about one letter grade.

Coefficients and the Prediction Equation

At this point you know the model is statistically significant and accounts for about 37% of the variability in calculus scores. What is the regression equation itself and which predictor variables are most important?

You can read the estimated regression model from cells A16:I23, shown in Figure 9-5, where the first column contains labels for the predictor variables.

Figure 9-5 Parameter estimates and *p* values

	A	B	C	D	E	F	G
16		*Coefficients*	*Standard Error*	*t Stat*	*P-value*	*Lower 95%*	*Upper 95%*
17	Intercept	27.943	12.438	2.247	0.028	3.155	52.732
18	Calc HS	7.192	2.488	2.891	0.005	2.233	12.151
19	ACT Math	0.352	0.430	0.817	0.417	-0.506	1.209
20	Alg Place	0.827	0.268	3.092	0.003	0.294	1.360
21	Alg2 Grade	3.683	2.441	1.509	0.136	-1.182	8.548
22	HS Rank	0.111	0.116	0.953	0.344	-0.121	0.342
23	Gender Code	2.627	2.469	1.064	0.291	-2.294	7.548
24							

The Coefficients column (B16:B23) gives the estimated coefficients for the model. The corresponding prediction equation is

$$\text{Calc} = 27.943 + 7.192(\text{Calc HS}) + 0.352(\text{ACT Math}) + 0.827(\text{Alg Place})$$
$$+ 3.683(\text{Alg2 Grade}) + 0.111(\text{HS Rank}) + 2.627(\text{Gender Code})$$

The coefficient for each variable estimates how much the calculus score will change if the variable is increased by 1 and the other variables are held constant. For example, the coefficient 0.352 of ACT Math indicates that the calculus score should increase by 0.352 point if the ACT math score increases by 1 point and all other variables are held constant.

Some variables, such as Calc HS, have a value of either 0 or 1, in this case to indicate the absence or presence of calculus in high school. The coefficient 7.192 is the estimated effect on the calculus score of taking high school calculus, other things being equal. Because 10 points correspond to one letter grade, the coefficient 7.192 for Calc HS is almost one letter grade.

Using the coefficients of this regression equation, you can forecast what a particular student's calculus score may be, given background information on the student. For example, consider a male student who did not take calculus in high school, scored 30 on his ACT Math exam, scored 23 on his algebra placement test, had a 4.0 grade in second-year high school algebra, and was ranked in the 90th percentile in his high school graduation class. You would predict that his calculus score would be

$$\text{Calc} = 27.943 + 7.192(0) + 0.352(30) + 0.827(23) + 3.683(4.0)$$
$$+ 0.111(90) + 2.627(1) = 74.87, \text{ or about 75 points}$$

Notice the Gender Code coefficient, 2.627, which shows the effect of gender if the other variables are held constant. Because the males are coded 1 and the females are coded 0, if the regression model is true, a male student will score 2.627 points higher than a female student, even when the backgrounds of both students are equivalent (equivalent in terms of the predictor variables in the model).

Whether you can trust that conclusion depends partly on whether the coefficient for Gender Code is significant. For that you have to determine the precision with which the value of the coefficient has been determined. You can do this by examining the estimated standard deviations of the coefficients, displayed in the Standard Error column.

t Tests for the Coefficients

The *t* Stat column shows the ratio between the coefficient and the standard error. If the population coefficient is 0, then this has the *t* distribution with degrees of freedom $n - p - 1 = 80 - 6 - 1 = 73$. Here n is the number of cases (80) and p is the number of predictors (6). The next column, *P* value, is the corresponding p value—the probability of a *t* value this large or larger in absolute value. For example, the *t* value for Alg Place is 3.092, so the probability of a *t* this large or larger in absolute value is about .003. The coefficient is significant at the 5% level because this is less than .05. In terms of hypothesis testing, you would reject the null hypothesis that the coefficient is 0 at the 5% level and accept the alternative hypothesis. This is a two-tailed test—it rejects the null hypothesis for either large positive or large negative values of *t*—so your alternative hypothesis is that the coefficient is not zero. Notice that only the coefficients for Alg Place and Calc HS are significant. This suggests that you not devote a lot of effort to interpreting the others. In particular, it would not be appropriate to assume from the regression that male students perform better than equally qualified female students.

The range F17:G23 indicates the 95% confidence intervals for each of the coefficients. You are 95% confident that having calculus in high school is associated with an increase in the calculus score of at least 2.233 points and not more than 12.151 points in this particular regression equation.

Is it strange that the ACT math score is nowhere near significant here, even though this test is supposed to be a strong indication of mathematics achievement? Looking back at the correlation matrix in Chapter 8, you can see that it has correlation 0.353 with Calc, which is highly significant ($p = .001$). Why is it not significant here? The answer involves other variables that contain some of the same information. In using the *t* distribution to test the significance of the ACT Math term, you are testing whether you can get away with deleting this term. If the other predictors can take up the slack and provide most of its information, then the test says that this term is not significant and therefore is not needed in the model. If each of the predictors can be predicted from the others, any single predictor can be eliminated without losing much.

You might think that you could just drop from the model all the terms that are not significant. However, it is important to bear in mind that the individual tests are correlated, so each of them changes when you drop one of the terms. If you drop the least-significant term, others might then become significant. A frequently used strategy for reducing the number of predictors involves the following steps:

1. Eliminate the least-significant predictor if it is not significant.
2. Refit the model.
3. Repeat Steps 1 and 2 until all predictors are significant.

In the exercises, you'll get a chance to rerun this model and eliminate all non significant variables. For now, examine the model and see whether any assumptions have been violated.

Testing Regression Assumptions

There are a number of useful ways to look at the results produced by multiple linear regression. This section reviews the four common plots that can help you assess the success of the regression.

1. Plotting dependent variables against the predicted values shows how well the regression fits the data.
2. Plotting residuals against the predicted values magnifies the vertical spread of the data so you can assess whether the regression assumptions are justified. A curved pattern to the residuals indicates that the model does not fit the data. If the vertical spread is wider on one side of the plot, it suggests that the variance is not constant.
3. Plotting residuals against individual predictor variables can sometimes reveal problems that are not clear from a plot of the residuals versus the predicted values.
4. Creating a normal plot of the residuals helps you assess whether the regression assumption of normality is justified.

Observed versus Predicted Values

How successful is the regression? To see how well the regression fits the data, plot the actual Calculus values against the predicted values stored in B29:B109. (You can scroll down to view the residual output.) To plot the observed calculus scores versus the predicted scores, you must first place the data on the same worksheet.

To copy the observed scores:

1. Select the range **B29:B109** and click the **Copy** button from the Clipboard group on the Home tab.
2. Click the **Calculus Data** sheet tab.
3. Select the range **H1:H81**; right-click the selection and click **Insert Copied Cells** from the popup menu to paste the predicted values into column H.
4. Click the **Shift Cells Right** option button to move the observed calculus scores into column I; then click **OK**.

The predicted calculus scores appear in column H, as shown in Figure 9-6 (formatted to show the column labels).

Figure 9-6 Predicted and observed calculus scores

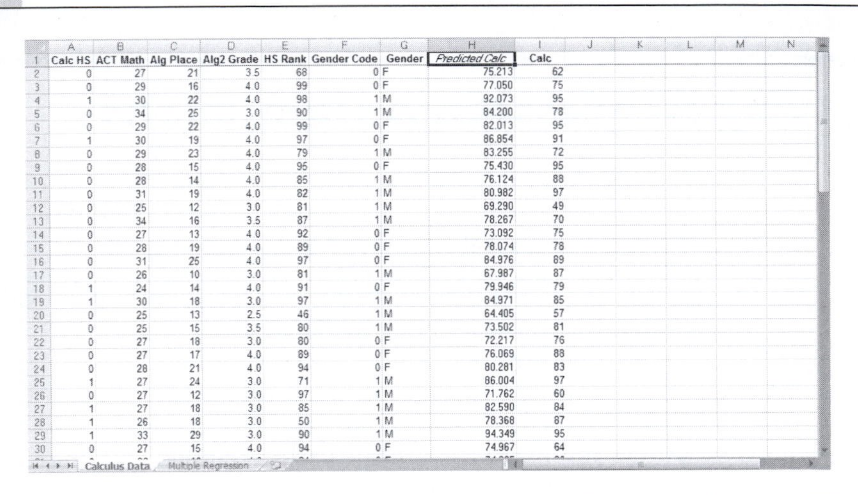

	A	B	C	D	E	F	G	H	I
1	Calc HS	ACT Math	Alg Place	Alg2 Grade	HS Rank	Gender Code	Gender	*Predicted Calc*	Calc
2	0	27	21	3.5	68	0	F	75.213	62
3	0	29	16	4.0	99	0	F	77.050	75
4	1	30	22	4.0	98	1	M	92.073	95
5	0	34	25	3.0	90	1	M	84.200	78
6	0	29	22	4.0	99	0	F	82.013	95
7	1	30	19	4.0	97	0	F	86.854	91
8	0	29	23	4.0	79	1	M	83.255	72
9	0	28	15	4.0	95	0	F	75.430	95
10	0	28	14	4.0	85	1	M	76.124	88
11	0	31	19	4.0	82	1	M	80.982	97
12	0	25	12	3.0	81	1	M	69.290	49
13	0	34	16	3.5	87	1	M	78.267	70
14	0	27	13	4.0	92	0	F	73.092	75
15	0	28	19	4.0	89	0	F	78.074	78
16	0	31	25	4.0	97	0	F	84.976	89
17	0	26	10	3.0	81	1	M	67.987	87
18	1	24	14	4.0	91	0	F	79.946	79
19	1	30	18	3.0	97	1	M	84.971	85
20	0	25	13	2.5	46	1	M	64.405	57
21	0	25	15	3.5	80	1	M	73.502	81
22	0	27	18	3.0	80	0	F	72.217	76
23	0	27	17	4.0	89	0	F	76.069	88
24	0	28	21	4.0	94	0	F	80.281	83
25	1	27	24	3.0	71	1	M	86.004	97
26	0	27	12	3.0	97	1	M	71.762	60
27	1	27	18	3.0	85	1	M	82.590	84
28	1	26	18	3.0	50	1	M	78.368	87
29	1	33	29	3.0	90	1	M	94.349	95
30	0	27	15	4.0	94	0	F	74.967	64

Calculus Data | Multiple Regression

Now create a scatter plot of the data in the range H1:I81.

To create the scatter plot of the observed scores versus the predicted scores:

1. Click **Single Variable Charts** from the StatPlus menu and then click **Fast Scatterplot**.
2. Click the **x-axis** button and then click the **Use Range References** option button and select the range **H1:H81** from the worksheet. Click the **OK** button.

3 Click the **y-axis** button and select **Calc** from the list of range names and click the **OK** button.

4 Click the **Chart Options** button and enter **Calculus Scores** for the chart title, **Predicted** for the *x* axis title, and **Observed** for the *y* axis title. Click the **OK** button.

5 Click the **Output** button and save the chart to the **Observed vs. Predicted** chart sheet. Click **OK**.

6 Click the **OK** button to generate the scatter plot.

7 Rescale the *x* axis and *y* axis in the plot so that the ranges go from 40 to 100 rather than from 0 to 100 or 0 to 120.

The final form of the scatter plot should look like Figure 9-7.

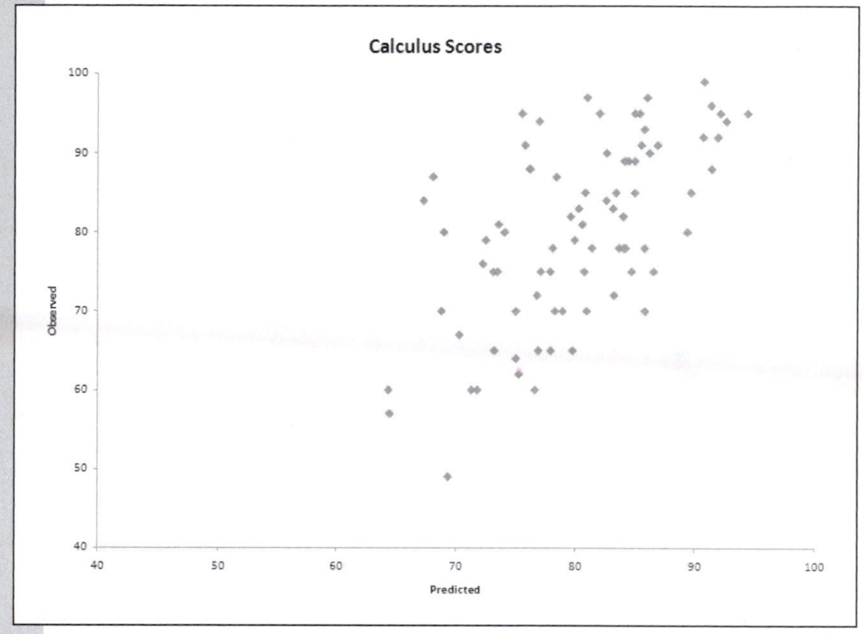

Figure 9-7 Scatter plot of observed and predicted scores

How good is the prediction shown here? Is there a narrow range of observed values for a given predicted value? This plot is a slight improvement on the plot of Calc versus Alg Place from the scatter plot matrix in Chapter 8. Figure 9-7 should be better because Alg Place and five other predictors are being used here.

Does it appear that the range of values is narrower for large values of predicted calculus score? If the error variance were lower for students with high predicted values, it would be a violation of the third regression assumption, which requires a constant error variance. Consider the students predicted to have a grade of 80 in calculus. These students have actual grades of around 65 to around 95, a wide range. Notice that the variation is lower for students predicted to have a grade of 90. Their actual scores are all in the 80s and 90s. There is a barrier at the top—no score can be above 100—and this limits the possible range. In general, when a barrier limits the range of the dependent variable, it can cause nonconstant error variance. This issue is considered further in the next section.

Plotting Residuals versus Predicted Values

The plot of the residuals versus the predicted values shows another view of the variation in Figure 9-7 because the residuals are the differences between the actual calculus scores and the predicted values.

To make the plot:

1. Click the **Multiple Regression** sheet tab to return to the regression output.
2. Create a scatter plot of the Residuals in the cell range C29:C109 versus Predicted Values in the cell range B29:B109 using either the StatPlus Fast Scatterplot command or using Excel's built-in commands to create a scatter plot.
3. Specify a chart title of **Residual Plot**, and label the *x* axis **Predicted Calculus Scores** and the *y* axis **Residuals**. Save the scatter plot to a chart sheet named **Residuals vs. Predicted**.
4. Change the scale of the *x* axis from 0–100 to 60–100. Your chart sheet should look like Figure 9-8.

Figure 9-8
Scatter plot of residuals and predicted scores

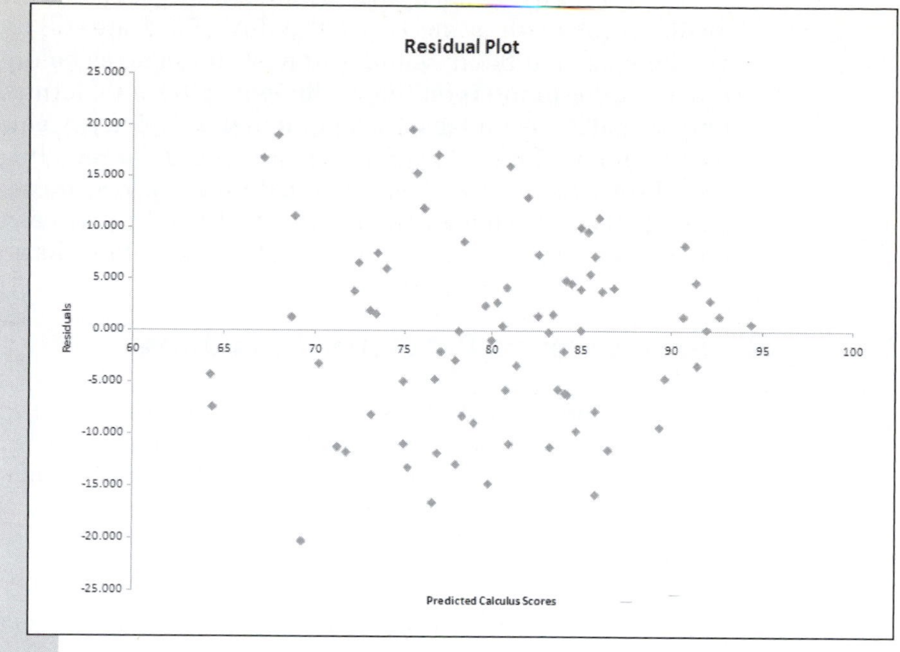

This plot is useful for verifying the regression assumptions. For example, the first assumption requires that the form of the model be correct. A violation of this assumption might be seen in a curved pattern. No curve is apparent here.

If the assumption of constant variance is not satisfied, then it should be apparent in Figure 9-8. Look for a trend in the vertical spread of the data. For example, the data may widen out as the predicted value increases. There appears to be a definite trend toward a narrower spread on the right, and it is cause for concern about the validity of the regression—although regression does have some robustness with respect to the assumption of constant variance.

For data that range from 0 to 100 (such as percentages), the arcsine–squareroot transformation sometimes helps fix problems with nonconstant variance. The transformation involves creating a new column of transformed calculus scores where

$$\text{Transformed calc score} = \sin^{-1}\sqrt{\text{calculus score}/100}$$

Using Excel, you would enter the formula

$$= \text{ASIN}(\text{SQRT}(x/100))$$

where x is the value or cell reference of a value you want to transform.

If you were to apply this transformation here and use the transformed calculus score in the regression in place of the untransformed score, you would find that it helps to make the variance more constant, but the regression results are about the same. Calc HS and Alg Place are still the only significant coefficients, and the R^2 value is almost the same as before. Of course, it is much harder to interpret the coefficients after transformation. Who would understand if you said that each point in the algebra placement score is worth 0.012 point in the arcsine of the square root of the calculus score divided by 100? From this perspective, the transformed regression is useful mainly to validate the original regression. If it is valid and it gives essentially the same results as the original regression, then the original results are valid.

Plotting Residuals versus Predictor Variables

It is also useful to look at the plot of the residuals against each of the predictor variables because a curve might show up on only one of those plots or there might be an indication of nonconstant variance. Such plots are created automatically with the Analysis ToolPak Add-Ins.

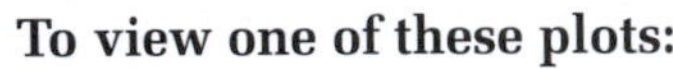

To view one of these plots:

1 Click the **Multiple Regression** sheet tab to return to the regression output.

The plots generated by the add-in start in cell J1 and extend to cell Z32. Two types of plots are generated: scatter plots of the regression residuals versus each of the regression variables, and the observed and predicted values of the response variable (calculus score) against each of the regression variables. See Figure 9-9. (You might have to scroll up and right to see the charts.)

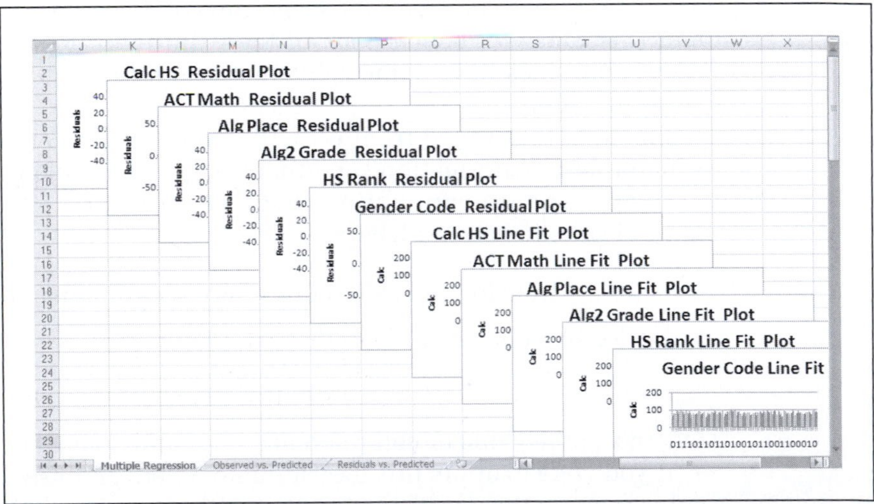

Figure 9-9 Plots created with the Regression command

The plots are shown in a cascading format, in which the plot title is often the only visible element of a chart. When you click a chart title, the chart goes to the front of the stack. The charts are small and hard to read, however. You can better view each chart by placing it on a chart sheet of its own. Try doing this with the plot of the residuals versus Alg Place.

To view the chart:

1. Click the chart **Alg Place Residual Plot** (located in the range L5:Q14).
2. Click the **Move Chart** button from the Location group on the Design tab of the Chart Tools ribbon.
3. Click the **As new sheet** option button and type **Alg Place Residual Plot** in the accompanying text box. Click **OK**.

The scatter plot is moved to a chart sheet shown in Figure 9-10.

Figure 9-10 Alg Place residual plot

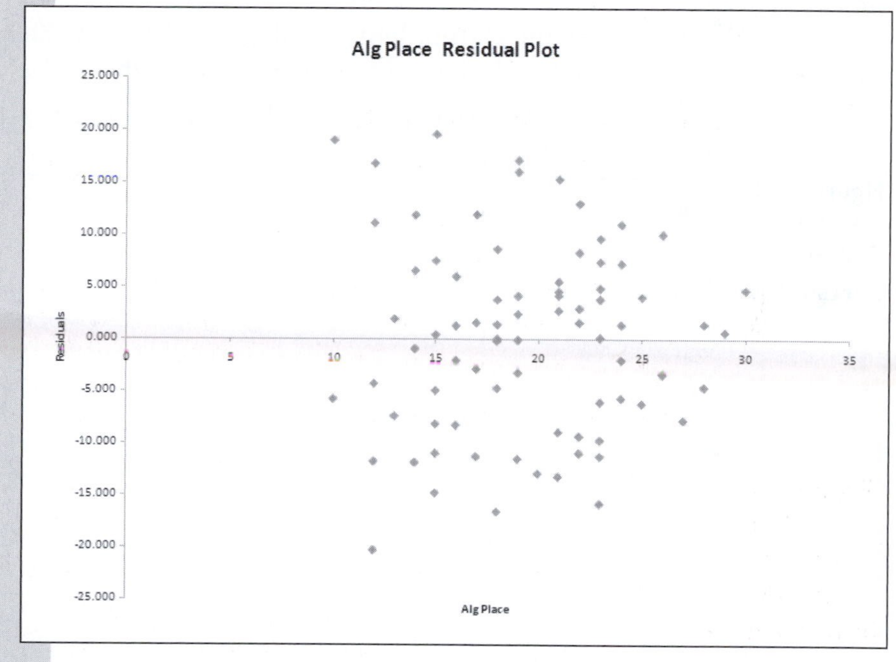

Does the spread of the residual values appear constant for differing values of the algebra placement score? It appears that the spread of the residuals is wider for lower values of Alg Place. This might indicate that you have to transform the data, perhaps using the arcsine transformation just discussed.

Normal Errors and the Normal Plot

What about the assumption of normal errors? Usually, if there is a problem with non normal errors, extreme values show up in the plot of residuals versus predicted values. In this example there are no residual values beyond 25 in absolute value, as shown in Figure 9-8.

How large should the residuals be if the errors are normal? You can decide whether these values are reasonable with a normal probability plot.

To make a normal plot of the residuals:

1. Return to the Multiple Regression worksheet.
2. Click **Single Variable Charts** from the StatPlus menu and then click **Normal P-plots**.
3. Click the **Data Values** button, click the **Use Range References** option button, and select the range **C29:C109**. Click **OK**.
4. Click the **Output** button and specify the new chart sheet **Residual P-plot** as the output destination. Click **OK**.
5. Click **OK** to start generating the plot. See Figure 9-11.

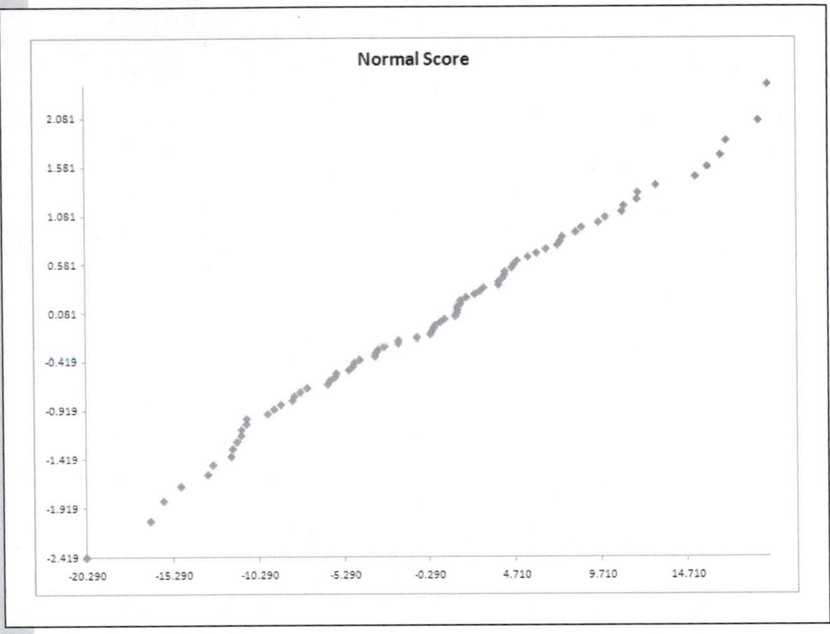

Figure 9-11 Normal P-plot of the residuals

The plot is quite well behaved. It is fairly straight, and there are no extreme values (either in the upper right or lower left corners) at either end. It appears there is no problem with the normality assumption.

Summary of Calc Analysis

What main conclusions can you make about the calculus data, now that you have done a regression, examined the regression residual file, and plotted some of the data? With an R^2 of 0.37 and an adjusted R^2 of 0.320, the regression accounts for only about one-third of the variance of the calculus score. This is disappointing, considering all the weight that college scholarships, admissions, placement, and athletics place on the predictors. Only the algebra placement score and whether calculus was taken in high school have significant coefficients in the regression. There is a slight problem with the assumption of a constant variance, but that does not affect these conclusions. You can close your workbook now, saving your changes.

Regression Example: Sex Discrimination

In this next example, you use regression analysis to determine whether a particular group is being discriminated against. For example, some of the female faculty at a junior college felt underpaid, and they sought statistical help in proving their case. The college collected data for the variables that influence salary for 37 females and 44 males. The data are stored in the Discrimination workbook.

To open the file:

1 Open **Discrimination** from the Chapter09 data folder.

2 Save the workbook as **Discrimination Multiple Regression**.

Table 9-1 shows the variables in the workbook.

Table 9-1 The Discrim Workbook

Range Name	Range	Description
Gender	A2:A82	Gender of faculty member (F = female, M = male)
MS_Hired	B2:B82	1 for Master's degree when hired, 0 for no Master's degree when hired
Degree	C2:C82	Current degree: 1 for Bachelor's, 2 for Master's, 3 for Master's plus 30 hours, and 4 for PhD
Age_Hired	D2:D82	Age when hired
Years	E2:E82	Number of years the faculty member has been employed at the college
Salary	F2:F82	Current salary of faculty member

In this example, you use salary as the dependent variable, using four other quantitative variables as predictors. One way to see whether female faculty have been treated unfairly is to do the regression using just the male data and then apply the regression to the female data. For each female faculty member, this predicts what a male faculty member would make with the same years, age when hired, degree, and Master's degree status. The residuals are interesting because they are the difference between what each woman makes and her predicted salary if she were a man. This assumes that all of the relevant predictors are being used, but it would be the college's responsibility to point out all the variables that influence salary in an important way. When there is a union contract, which is the case here, it should be clear which factors influence salary.

Regression on Male Faculty

To do the regression on just the male faculty and then look at the residuals for the females, use Excel's AutoFilter capability and copy the male rows to a new worksheet.

To create a worksheet of salary information for male faculty only:

1. Right-click the **Salary Data** sheet tab to open the pop-up menu and then click **Insert** from the menu.
2. Click **Worksheet** from the General sheet of the Insert dialog box and click **OK**.
3. Double-click the new sheet tab and type **Male Faculty**. Return to the Salary Data worksheet.
4. Click the **Filter** button from the Sort & Filter group on the Data tab. Excel adds drop-down arrows to all of the column headers in the list.
5. Click the **Gender** drop-down arrow; then deselect all of the checkboxes except for **M** and click the **OK** button. Excel displays only the data for the male faculty.
6. Select the range **A1:F82**; then click the **Copy** button from the Clipboard group on the Home tab.
7. Go to cell A1 on the Male Faculty worksheet and click the **Paste** button from the Clipboard group on the Home tab. The salary data for male faculty now occupy the range A1:F45 on the Male Faculty worksheet.

Using a SPLOM to See Relationships

To get a sense of the relationships among the variables for the male faculty, it is a good idea to compute a correlation matrix and plot the corresponding scatter plot matrix.

To create the SPLOM:

1. Click **Multi-variable Charts** from the StatPlus menu and then click **Scatterplot Matrix**.
2. Click the **Data Values** button, click the **Use Range References** option button, and select the range **B1:F45**. Click **OK**.
3. Click the **Output** button, click the **New Worksheet** option button, and type **Male SPLOM** in the accompanying text box. Click **OK**.
4. Click **OK** to start generating the scatter plot matrix. See Figure 9-12 for the completed SPLOM.

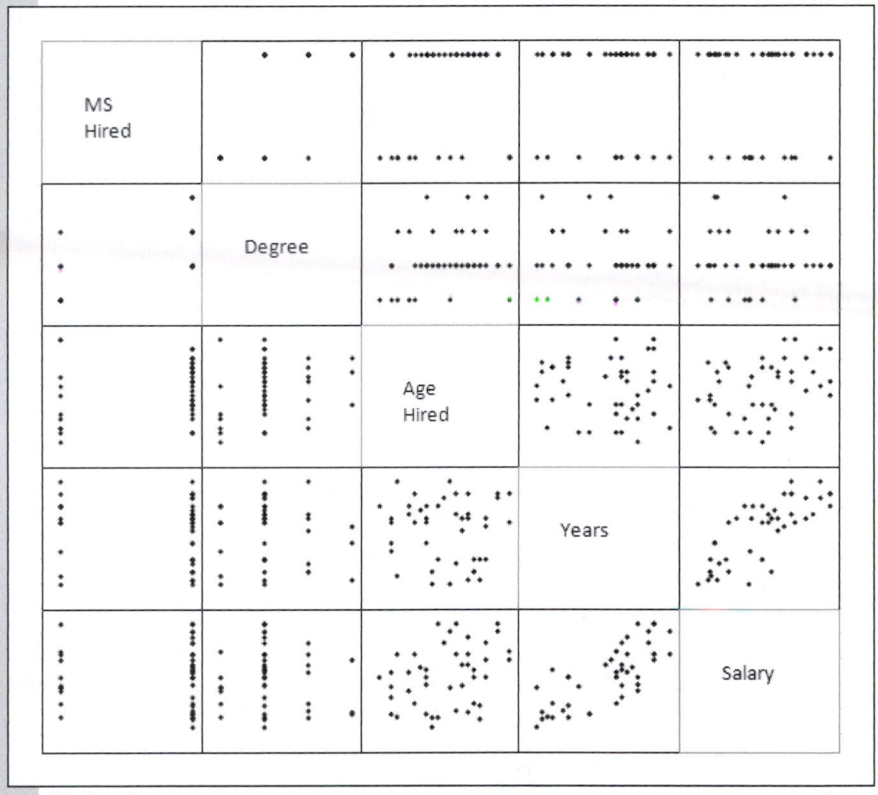

Figure 9-12 SPLOM of variables for male faculty salary data

Focus on the last row because it shows the relationships of the other variables to salary. Years employed is a good predictor because the range of salary is fairly narrow for each value of years employed (although the relationship is not perfectly linear). Age at which the employee was hired is not a very good predictor because there is a wide range of salary values for each value of age hired. There is not a significant relationship between the two predictors years employed and age hired. What about the other two predictors? Looking at the plots of salary against degree and MS hired makes it clear that neither of them is closely related to salary. The people with higher degrees do not seem to be making higher salaries. Those with a Master's degree when hired do not seem to be making much more either. Therefore, the correlations of degree and MS hired with salary should be low.

You might have some misgivings about using Degree as a predictor. After all, it is only an ordinal variable. There is a natural order to the four levels, but it is arbitrary to assign the values 1, 2, 3, and 4. This says that the spacing from Bachelor's to Master's (1 to 2) is the same as the spacing from Master's plus 30 hours to PhD (3 to 4). You could instead assign the values 1, 2, 3, and 5, which would mean greater space from Master's plus 30 hours to the PhD. In spite of this arbitrary assignment, ordinal variables are frequently used as regression predictors. Usually, it does not make a significant difference whether the numbers are 1, 2, 3, and 4 or 1, 2, 3, and 5. In the present situation, you can see from Figure 9-12 that salaries are about the same in all four degree categories, which implies that the correlation of salary and degree is close to 0. This is true no matter what spacing is used.

Correlation Matrix of Variables

The SPLOM shows the relationships between salary and the other variables. To quantify this relationship, create a correlation matrix of the variables.

To form the correlation matrix:

1. Click the **Male Faculty** sheet tab.
2. Click **Multivariate Analysis** from the StatPlus menu and then click **Correlation Matrix**.
3. Click the **Data Values** button, click the **Use Range References** option button, select the range **B1:F45**, and click **OK**.

4 Click the **Output** button, click the New Sheet option button, and type **Male Corr Matrix** in the New Sheet text box; then click **OK** twice. The resulting correlation matrix appears on its own sheet, as shown in Figure 9-13.

Figure 9-13 Correlation matrix for male faculty salary data

	A	B	C	D	E	F	G
1	**Pearson Correlations**						
2		MS Hired	Degree	Age Hired	Years	Salary	
3	MS Hired	1.000	0.520	0.219	-0.099	0.009	
4	Degree		1.000	0.215	-0.103	-0.072	
5	Age Hired			1.000	-0.064	0.325	
6	Years				1.000	0.765	
7	Salary					1.000	
8							
9	**Pearson Probabilities**						
10		MS Hired	Degree	Age Hired	Years	Salary	
11	MS Hired	-	0.000	0.153	0.525	0.952	
12	Degree		-	0.161	0.505	0.643	
13	Age Hired			-	0.681	0.032	
14	Years				-	0.000	
15	Salary					-	
16							

You might wonder why the variable Age Hired is used instead of employee age. The problem with using employee age is one of collinearity. **Collinearity** means that one or more of the predictor variables are highly correlated with each other. In this case, the age of the employee is highly correlated with the number of years employed because there is some overlap between the two. (People who have been employed more years are likely to be older.) This means that the information those two variables provide is somewhat redundant. However, you can tell from Figure 9-13 that the relationship between years employed and age when hired is negligible because the p value is .681 (cell E13). Using the variable age hired instead of age gives the advantage of having two nearly uncorrelated predictors in the model. When predictors are only weakly correlated, it is much easier to interpret the results of a multiple regression.

The correlations for Salary show a strong relationship to the number of years employed and some relationship to age when hired, but there is little relationship to a person's degree. This is in agreement with the SPLOM in Figure 9-12.

Multiple Regression

What happens when you throw all four predictors into the regression pot?

To specify the model for the regression:

1 Click the **Male Faculty** sheet tab.

2 Click the **Data Analysis** button from the Analysis group on the Data tab, select **Regression** from the list of Analysis Tools, and click **OK**. The Regression dialog box might contain the options you selected for the previous regression.

3 Type **F1:F45** in the Input *Y* Range text box, press **Tab**, and then type **B1:E45** in the Input *X* Range text box.

4 Verify that the **Labels** checkbox is selected and that the **Confidence Level** checkbox is selected and contains a value of 95.

5 Click the **New Worksheet Ply** option button and type **Male Faculty Regression** in the corresponding text box (replace the current contents if necessary).

6 Verify that the **Residuals, Standardized Residuals, Residual Plots,** and **Line Fit Plots** checkboxes are selected.

7 Click **OK**.

The first portion of the summary output is shown in Figure 9-14, with columns resized and values reformatted.

Figure 9-14 Regression output for male faculty data

	A	B	C	D	E	F	G
1	SUMMARY OUTPUT						
2							
3	*Regression Statistics*						
4	Multiple R	0.856					
5	R Square	0.732					
6	Adjusted R Square	0.705					
7	Standard Error	3168.434					
8	Observations	44					
9							
10	ANOVA						
11		*df*	*SS*	*MS*	*F*	*Significance F*	
12	Regression	4	1.07E+09	2.68E+08	2.67E+01	1.05518E-10	
13	Residual	39	3.92E+08	1.00E+07			
14	Total	43	1.46E+09				
15							
16		*Coefficients*	*Standard Error*	*t Stat*	*P-value*	*Lower 95%*	*Upper 95%*
17	Intercept	12900.669	3178.168	4.059	0.000	6472.218	19329.119
18	MS Hired	744.482	1304.009	0.571	0.571	-1893.125	3382.089
19	Degree	-783.529	746.088	-1.050	0.300	-2292.633	725.576
20	Age Hired	373.735	83.198	4.492	0.000	205.451	542.020
21	Years	606.176	64.531	9.393	0.000	475.649	736.703

Interpreting the Regression Output

The R^2 of 0.732 shows that the regression explains 73.2% of the variance in salary. However, when this is adjusted for the number of predictors (four), the adjusted R^2 is about 0.705=70.5%. The standard error is 3,168.434, so salaries vary roughly plus or minus $3,000 from their predictions. The overall F ratio is about 26.67, with a p value in cell F12 of 1.063×10^{-10}, which rules out the hypothesis that all four population coefficients are 0. Looking at the coefficient values and their standard errors, you see that the coefficients for the variables Degree and MS Hired have values that are not much more than 1 times their standard errors. Their t statistics are much less than 2, and their p values are much more than .05; therefore, they are not significant at the 5% level. On the other hand, Years employed and Age Hired do have coefficients that are much larger than their standard errors, with t values of 9.39 and 4.49, respectively. The corresponding p values are significant at the 0.1% level.

The coefficient estimate of 606 for years employed indicates that each year on the job is worth $606 in annual salary if the other predictors are held fixed. Correspondingly, because the coefficient for Age Hired is about $374, all other factors being equal, an employee who was hired at an age 1 year older than another employee will be paid an additional $374.

Residual Analysis of Discrimination Data

Now check the assumptions under which you performed the regression.

To create a plot of residuals versus predicted salary values:

1. Using either the StatPlus Fast Scatterplot command or the Excel's Scatter button on the Insert menu, create a scatter plot of the Residual values in the cell range C27:C71 versus Predicted Values in the range B27:B71.
2. Enter a chart title of **Residual Plot**, and label the x axis **Predicted Salaries** and the y axis **Residuals**. Save the scatter plot to a chart sheet named **Residuals vs. Predicted**.
3. Change the scale of the x axis from 0–45,000 to 20,000–45,000. Your chart sheet should look like Figure 9-15.

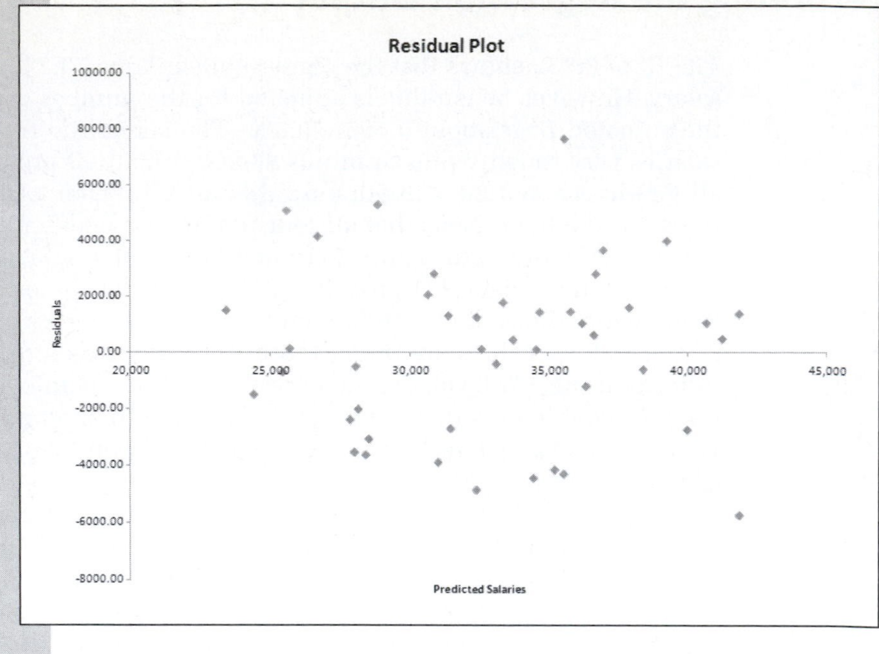

Figure 9-15 Residuals versus predicted values for the male salary data

There does not appear to be a problem with nonconstant variance. At least, there is not a big change in the vertical spread of the residuals as you move from left to right. However, there are two points that look questionable. The one at the top has a residual value near 8,000 (indicating that this individual is paid $8,000 more than predicted from the regression equation), and at the bottom of the plot an individual is paid about $6,000 less than predicted from the regression.

Except for these two, the points have a somewhat curved pattern—high on the ends and low in the middle—of the kind that is sometimes helped by a log transformation. As it turns out, the log transformation would straighten out the plot, but the regression results would not change much. For example, if log(salary) is used in place of salary, the R^2 value changes only from 0.732 to 0.733. When the results are unaffected by a transformation, it is best not to bother because it is much easier to interpret the untransformed regression.

Normal Plot of Residuals

What about the normality assumption? Are the residuals reasonably in accord with what is expected for normal data?

To create a normal probability plot of the residuals:

1. Click the **Male Faculty Regression** sheet tab.
2. Click **Single Variable Charts** from the StatPlus menu and then click **Normal P-plots.**
3. Click the **Data Values** button, click the **Use Range References** option button, and select the range **C27:C71**. Click **OK.**
4. Click the **Output** button and specify the new chart sheet **Male Residual P-plot** as the output destination. Click **OK.**
5. Click **OK** to start generating the plot. See Figure 9-16.

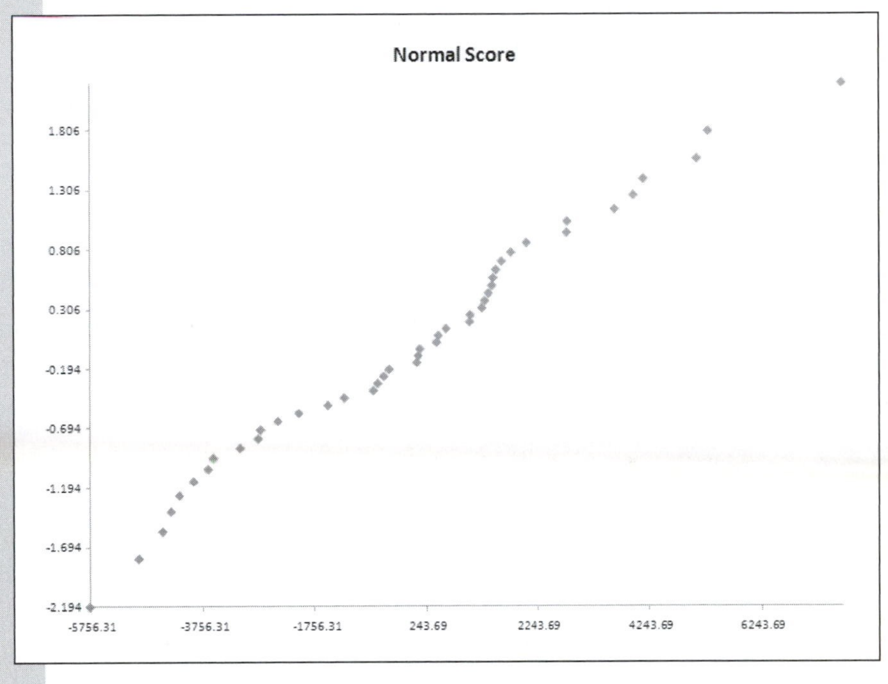

Figure 9-16 Normal P-plot of residuals for the male faculty data

The plot is reasonably straight, although there is a point at the upper right that is a little farther to the right than expected. This point belongs to the employee whose salary is $8,000 more than predicted, but it does not appear to be too extreme. You can conclude that the residuals seem consistent with the normality assumption.

Are Female Faculty Underpaid?

Being satisfied with the validity of the regression on males, let's go ahead and apply it to the females to see whether they are underpaid. The idea is to look at the differences between what female faculty members were paid and what we would predict they would be paid on the basis of the regression model for male faculty. Your ultimate goal is to choose between two hypotheses.

H_0: The mean population salaries of females are equal to the salaries predicted from the population model for males.

H_a: The mean population salaries of females are lower than the salaries predicted from the population model for males.

To obtain statistics on the salaries for females relative to males, you must create new columns of predicted values and residuals.

To create new columns of predicted values and residuals:

1. Create a new blank worksheet named **Female Faculty** and then go to the Salary Data worksheet.
2. Click the Gender drop-down list arrow and select only the **F** checkbox.

   Verify that the range A1:F38 displaying data on only the female faculty is displayed in the worksheet.
3. Copy the selection and paste it to cell A1 on the Female Faculty worksheet.
4. In the Female Data worksheet, click cell **G1**, type **Predicted Salary** press **Tab**, type **Residuals** and then press **Enter**.
5. Select the range **G2:H38**.
6. In cell G2, type **=12900.67+744.4821*B2−783.529*C2+373.7354*D2+606.1759*E2** (the regression equation for males), and press **Tab**.
7. Type **=F2–G2** in cell H2, and press **Enter**.
8. Select the cell range **F2:H38** and then click the **Fill** button from the Editing group on the Home tab and click **Down**. Excel inserts the formula into the remaining cells in the two columns. The data should appear as in Figure 9-17.

Figure 9-17 Predicted salaries and data for the female faculty

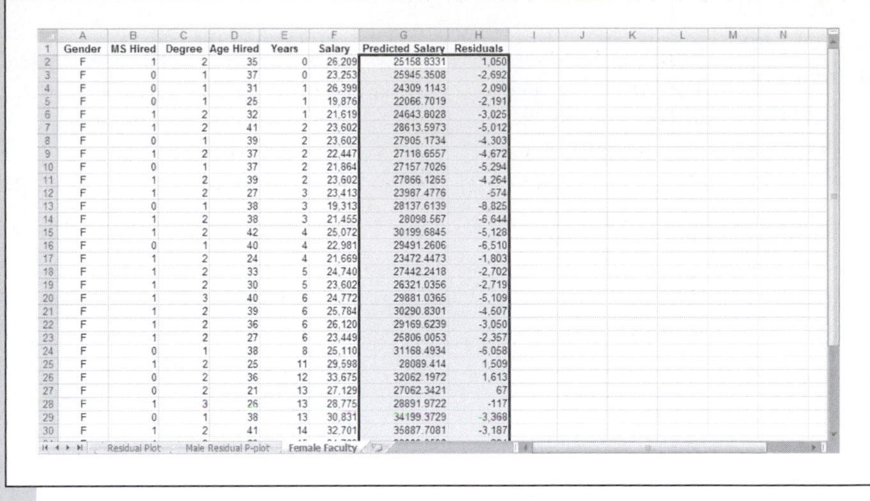

	A	B	C	D	E	F	G	H
1	Gender	MS Hired	Degree	Age Hired	Years	Salary	Predicted Salary	Residuals
2	F	1	2	35	0	26,209	25158.8331	1,050
3	F	0	1	37	0	23,253	25945.3508	-2,692
4	F	0	1	31	1	26,399	24309.1143	2,090
5	F	0	1	25	1	19,876	22066.7019	-2,191
6	F	1	2	32	1	21,619	24643.8028	-3,025
7	F	1	2	41	2	23,602	28613.5973	-5,012
8	F	0	1	39	2	23,602	27905.1734	-4,303
9	F	1	2	37	2	22,447	27118.6557	-4,672
10	F	0	1	37	2	21,864	27157.7026	-5,294
11	F	1	2	39	2	23,602	27866.1265	-4,264
12	F	1	2	27	3	23,413	23987.4776	-574
13	F	0	1	38	3	19,313	28137.6139	-8,825
14	F	1	2	38	3	21,455	28098.567	-6,644
15	F	1	2	42	4	25,072	30199.6845	-5,128
16	F	0	1	40	4	22,981	29491.2606	-6,510
17	F	1	2	24	4	21,669	23472.4473	-1,803
18	F	1	2	33	5	24,740	27442.2418	-2,702
19	F	1	2	30	5	23,602	26321.0356	-2,719
20	F	1	3	40	6	24,772	29881.0365	-5,109
21	F	1	2	39	6	25,784	30290.8301	-4,507
22	F	1	2	36	6	26,120	29169.6239	-3,050
23	F	1	2	27	6	23,449	25806.0053	-2,357
24	F	0	1	38	8	25,110	31168.4934	-6,058
25	F	1	2	25	11	29,598	28089.414	1,509
26	F	0	2	36	12	33,675	32062.1972	1,613
27	F	0	2	21	13	27,129	27062.3421	67
28	F	1	3	26	13	28,775	28891.9722	-117
29	F	0	1	38	13	30,831	34199.3729	-3,368
30	F	1	2	41	14	32,701	35887.7081	-3,187

Residual Plot | Male Residual P-plot | Female Faculty

To see whether females are paid about the same salary that would be predicted if they were males, create a scatter plot of residuals versus predicted salary.

To create the scatter plot:

1. Using either the StatPlus Fast Scatterplot command or the Scatterplot button on the Insert tab, create a scatterplot of the Residuals (H1:H38) versus Predicted Salary (G1:G38).

2. Give a chart title of **Female Residual Plot**, and label the *x* axis **Predicted Salaries** and the *y* axis **Residuals**. Save the chart to a chart sheet named **Female Residual Plot**.

3. Change the scale of the *x* axis from 0–45,000 to 20,000–40,000. Your chart sheet should resemble that shown in Figure 9-18.

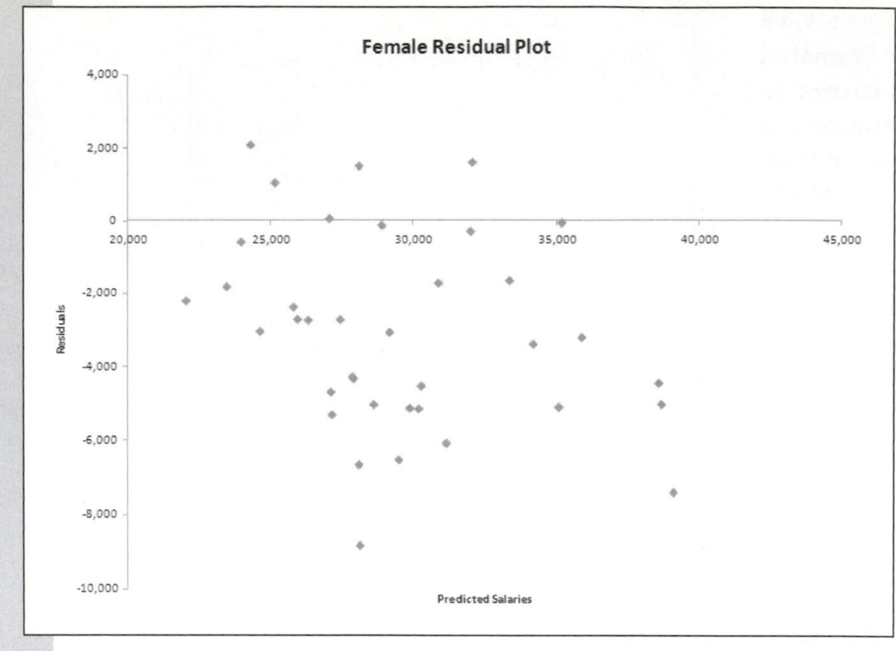

Figure 9-18 Scatter plot of residuals versus predicted values for female faculty

Out of 37 female faculty, only 5 have salaries greater than what would be predicted if they were males, whereas 32 have salaries less than predicted. Calculate the descriptive statistics for the residuals to determine the average discrepancy in salary.

To calculate descriptive statistics for female faculty's salaries:

1. Click the **Female Faculty** worksheet tab.
2. Click **Descriptive Statistics** from the StatPlus menu and then click **Univariate Statistics**.
3. Click the **All summary statistics** and **All variability statistics** checkboxes.
4. Click the **Input** button, click the **Use Range References** option button, and select the range **H1:H38**. Click **OK**.
5. Click the **Output** button, and select the new worksheet **Female Residual Stats** as the output destination. Click **OK**.
6. Click **OK** to generate the table of descriptive statistics shown in Figure 9-19.

Figure 9-19 Descriptive statistics of the residuals for female faculty

	A	B
1		**Univariate Statistics**
2		Residuals
3	Count	37
4	Sum	-113,355
5	Average	-3,063.64
6	Median	-3,049.6
7	Mode	#N/A
8	Trimmed Mean (0.2)	-3,087.60
9	Minimum	-8,825
10	Maximum	2,090
11	Range	10,914
12	Standard Deviation	2,662.033
13	Variance	7,086,418.085
14	Standard Error	437.635
15	Skewness	0.130
16	Kurtosis	-0.503

On the basis of the descriptive statistics, you can conclude that the female faculty are paid, on average, $3,063.64 less than what would be expected for equally qualified male faculty members (as quantified by the predictor variables). The largest discrepancy is for a female faculty member who is paid $8,825 less than expected (cell B9). Of those with salaries greater than predicted, there is a female faculty member who is paid $2,090 more than expected (cell B10).

To understand the salary deficit better, you can plot residuals against the relevant predictor variables. Start by plotting the female salary residuals versus age when hired. (You could plot residuals versus years employed, but you would see no particular trend in the pattern of the residuals.)

To plot the residuals against Age Hired:

1. Click the **Female Faculty** sheet tab to return to the data worksheet.
2. Using either the StatPlus Fast Scatterplot command or the Scatter-chart button on the Insert tab of the Excel ribbon, create a scatter plot of Residuals (H1:H38) versus Age Hired (D1:D38).

3 Give a chart title of **Residuals vs. Age Hired**, and label the *x* axis **Age Hired** and the *y* axis **Residuals**. Save the chart in a new chart sheet named **Female Resid vs. Age Hired**.

4 Change the scale of the *x* axis from 0–60 to 20–50. Your chart should look like Figure 9-20.

Figure 9-20 Scatter plot of residuals versus Age Hired for female faculty

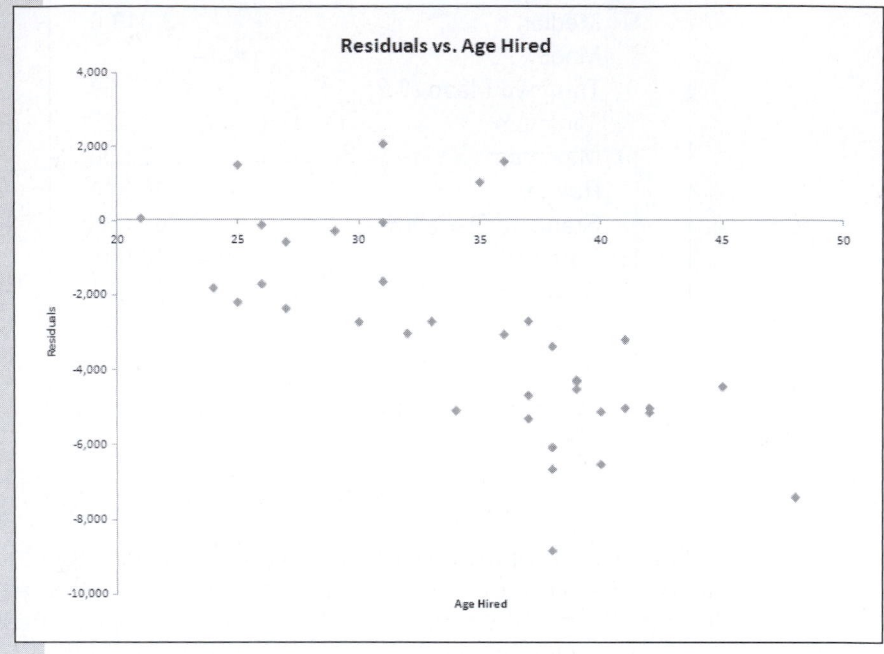

There seems to be a downward trend to the scatterplot, indicating that the greater discrepancies in salaries occur for older female faculty. Add a linear regression line to the plot, regressing residuals versus age when hired.

To add a linear regression line to the plot:

1 Right-click the data series (any one of the data points in Figure 9-20), and click **Add Trendline** in the shortcut menu.

2 Verify that the Linear Trend/Regression Type option is selected, and then click **Close**. Your plot should now look like Figure 9-21.

Figure 9-21 Scatter plot with trend line added

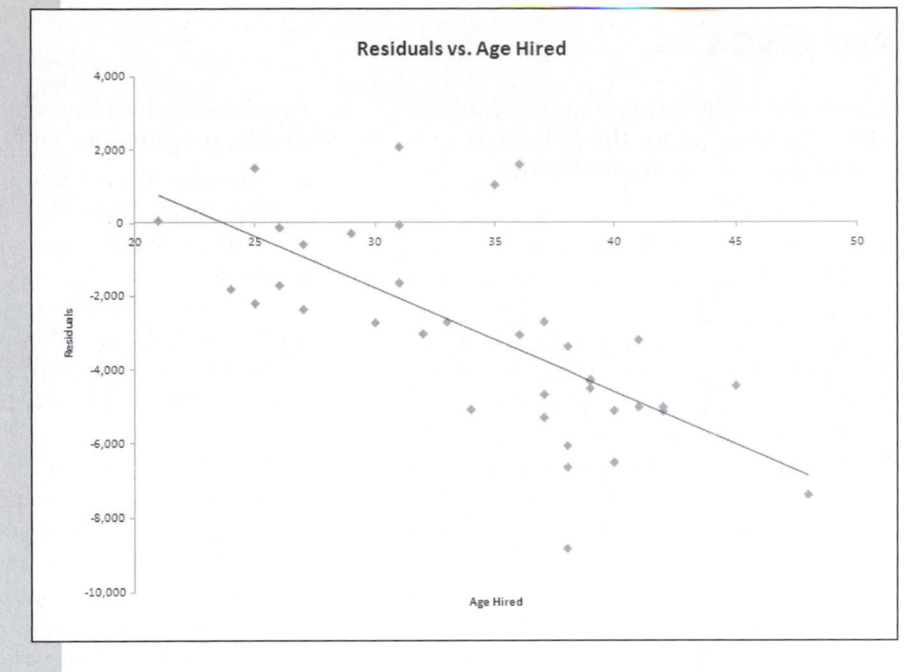

This plot shows a salary deficiency that depends very much on the age at which a female was hired. Those who were hired under the age of 25 have residuals that average around 0 or a little below. Those who were hired over the age of 40 are underpaid by more than $5,000 on average. The most underpaid female has a deficit of nearly $9,000.

Drawing Conclusions

Why should age make a difference in the discrepancies? One possibility is that women are more likely than men to take time off from their careers to raise their children. If this is the case, an older male faculty member would have more job experience and thus be paid more. However, this might not be true of all women, yet all of the females who were hired over the age of 36 were underpaid.

To summarize, the female faculty are underpaid an average of about $3,000. However, there is a big difference depending on how old they were when hired. Those who were hired after the age of 40 have an average deficit of more than $5,000. It should be noted that when the case was eventually settled out of court, each woman received the same compensation, regardless of age. You can now close the workbook, saving your changes.

Exercises

1. Use Excel's FINV function to calculate the critical value for the following F distributions (assume that the p value = .05):
   a. Numerator degrees of freedom = 1; denominator degrees of freedom = 9.
   b. Numerator degrees of freedom = 2; denominator degrees of freedom = 9.
   c. Numerator degrees of freedom = 3; denominator degrees of freedom = 9.
   d. Numerator degrees of freedom = 4; denominator degrees of freedom = 9.
   e. Numerator degrees of freedom = 5; denominator degrees of freedom = 9.

2. Use Excel's FDIST function to calculate the p value for the following F distributions (assume that the critical value = 3.5):
   a. Numerator degrees of freedom = 1; denominator degrees of freedom = 9.
   b. Numerator degrees of freedom = 2; denominator degrees of freedom = 9.
   c. Numerator degrees of freedom = 3; denominator degrees of freedom = 9.
   d. Numerator degrees of freedom = 4; denominator degrees of freedom = 9.
   e. Numerator degrees of freedom = 5; denominator degrees of freedom = 9.

3. Which of the following models can be solved using linear regression? Justify your answers.
   a. $y = \beta_0 + \beta_1 x_1 + \beta_2 x_2 + \varepsilon$
   b. $y = \dfrac{\beta_0 x}{\beta_1 + x} + \varepsilon$
   c. $y = \beta_0 + \beta_1 \sin x + \beta_2 \cos x + \varepsilon$

4. What is collinearity?

5. The Wheat workbook contains nutritional data on ten different wheat products. You've been asked to determine the relationship between calories, carbohydrates, protein, and fat.
   a. Open the **Wheat** workbook from the Chapter09 folder and save it as **Wheat Multiple Regression**.
   b. Generate the correlation matrix for the variables Calories, Carbo-Fiber (the carbohydrate value minus the fiber value), Protein, and Fat. Also create the corresponding scatterplot matrix.
   c. Regress Calories on the other three variables and obtain the residual output. How successful is the regression? It is known that carbohydrates (once adjusted for fiber content) have 4 calories per gram, protein has 4 calories per gram, and fats have 9 calories per gram. How do the coefficients compare with the known values?
   d. Explain why the coefficient for fat is inaccurate, in terms of its standard error and in comparison with the known value of 9. (*Hint:* Examine the data and notice that the fat content is specified with the least precision.)
   e. Plot the residuals against the predicted values. Is there an outlier? Label the points of the scatter plot by food brand to see which case is most extreme. Do the calories add up correctly for this case? That is, when you multiply the carbohydrate content (adjusted for fiber content) by 4, the protein content by 4, and the fat content by 9, does it add up to more calories than are stated on the package? Notice also that another case has close to the same values of Carbo-Fiber, Protein, and Fat, but the Calories value is 10 higher. How do you explain this? Would

a company understate the calorie content?

f. Save your changes to the workbook and write a report summarizing your observations.

6. The Fritos workbook is a slight modification of the Wheat workbook with data added about Fritos corn chips. It is included because it has a substantial fat content, in contrast to the other foods in the data set. Because none of the foods there have much fat, it is impossible to see from the Wheat workbook how much fat contributes to the calories in the foods.

   a. Open the **Fritos** workbook from the Chapter09 folder and save it as **Fritos Multiple Regression**.

   b. Repeat the regression of the previous exercise and see whether the coefficient for fat is now estimated more accurately. Use both the known value of 9 for comparison and the standard error of the regression that is printed in the output.

   c. Save your changes to the workbook and write a report summarizing your observations.

7. The Baseball workbook contains team statistics for each of the major league teams from the 2001–2007 baseball seasons. You've been asked to derive an equation that predicts the number of runs per game on the basis of the number of singles, doubles, triples, home runs, bases on balls, and strike outs.

   a. Open the **Baseball** workbook from the Chapter09 folder and save it as **Baseball Multiple Regression**.

   b. Create six new columns in the Baseball Stats worksheet and calculate the average number of singles, doubles, triples, home runs, bases on balls, and strikeouts per game for each of the teams in the data set.

   c. Regress Runs per Game on the six variables you created to derive an equation for the average number of runs per game on the basis of the average number of singles, doubles, triples, home runs, bases on balls, and strike outs. Are all of the variables in your equation significant? Remove any insignificant variables from your model and rerun the regression. Compare your results with the results obtained by Rosner and Woods (1988), as quoted in the beginning of this chapter. Can the differences be explained in terms of the standard errors of the coefficients?

   d. Do the Rosner-Woods coefficients make more sense in terms of which should be largest and which should be smallest?

   e. Save your changes to the workbook and write a report summarizing your results.

8. The Toyota workbook contains price, age, and mileage data for used car sales of Toyota Corollas from 2009. You've been asked to analyze the data to model the effect of age and mileage on the used car price.

   a. Open the **Toyota** workbook from the Chapter09 data folder and save it as **Toyota Multiple Regression**.

   b. Regress price on age and miles. What impact do age and miles have on the sale price of the car? Are both variables significant in your regression equation?

   c. Create plots of Residuals versus Miles, Residuals versus Age, and Residuals versus Predicted Price. Do you notice any pattern in the graphs that would indicate a problem with the constant variance assumption or the linearity assumption?

d. Create a normal plot of the residuals. Does your plot support a conclusion that the residuals are normally distributed?
e. Save your changes to the workbook and write a report summarizing your observations.

9. The regression performed in the previous exercise assumed that prices would change linearly with miles and age. It could also be the case that the prices will change instead as a percentage so that instead of dropping $1000 per year, the price would drop 10% per year. You can check this assumption by performing a logarithm of the used car sales price.

a. Open the **Toyota** workbook from the Chapter09 folder and save it as **Toyota Log Regression**.
b. Create a new variable named **LogPrice** equal to the log(price) value.
c. Repeat the regression from the last exercise using the log(price) rather than price.
d. Does this improve the multiple correlation? Have the *p* values associated with the miles and age coefficients become more significant?
e. When log(price) is used as the dependent variable, the regression can be interpreted in terms of percentage drop in Price per year of age, instead of a fixed drop per year of Age when Price is used as the dependent variable. Does it make more sense to have the price drop by 16.5% each year or to have the price drop by $721 per year? In particular, would an old car lose as much value per year as it did when it was young?
f. Save your changes to the workbook and write a report summarizing your conclusions.

10. The Cars workbook contains data based on reviews published in *Consumer Reports®*, 2003–2008. See Exercise 10 of Chapter 2. The workbook includes observations from 275 car models on the variables Price, MPG (miles per gallon), Cyl (number of cylinders), Eng size (engine displacement in liters), Eng type (normal, hybrid, turbo, turbodiesel), HP (horsepower), Weight (vehicle weight in pounds), Time0–60 (time to accelerate from 0 to 60 miles per hour in seconds), Date (month of publication), and Region (United States, Europe, or Asia). There is an additional variable Eng type01 that is 1 for hybrids and diesels and 0 otherwise.

a. Open the **Cars** workbook from the Chapter09 folder and save it as **Cars Multiple Regression**.
b. Create a correlation matrix (excluding Spearman's rank correlation) and a scatter plot matrix of the seven quantitative variables Price, MPG, Cyl, Eng size, HP, Weight, and Eng type01.
c. Regress MPG on Cyl, Eng size, HP, Weight, Price, and Eng type01.
d. Note that the regression coefficients for Cyl and Eng size are not significant at the .05 level. Compare this to the *p* values for these variables in the correlation matrix. What accounts for the lack of significance? (*Hint:* Look at the correlations among Cyl, Eng size, HP, Price, and Weight.)
e. Create a scatter plot of the regression residuals versus the predicted values. Judging by the scatter plot, do the assumptions of the regression appear to be violated? Why or why not?
f. Create a new variable, GPM100, that displays 100 divided by the miles per gallon. This measures the fuel necessary to go 100 miles. Some statisticians and car magazines use this

because it gives a direct measure of the energy consumption. Redo your regression model with this new dependent variable in place of MPG. How does the residual versus predicted value plot compare to the earlier one? Compare the regression results with the previous results for MPG (including *R* squared).

g. Save your changes to the workbook and write a report summarizing your conclusions.

11. Return to the Cars workbook and perform the following analysis:

a. Open the **Cars** workbook from the Chapter09 folder and save it as **Cars Reduced Model**.

b. Recreate the GPM100 variable described in the previous exercise and then regress GPM100 on the same numeric variables. Try to reduce the number of variables in the model using the following algorithm:

i. Perform the regression.

ii. If any coefficients in the regression are nonsignificant, redo the regression with the least significant variable removed.

iii. Continue until all coefficients remaining are significant.

To do this, you may have to move the columns around because the Regression command requires that all predictor variables lie in adjacent columns.

c. How does the R^2 value for this reduced model compare to the full model with six predictors?

d. Report your final model and save your changes to the workbook.

12. Perform the following final analysis on the Cars data:

a. Reopen the **Cars** workbook from the Chapter09 folder and save it as **Cars Final Analysis**.

b. Regress the variable GPM100 described in Exercise 11 on Cyl, Eng size, HP, Weight, Price, and Eng type01 for only the U.S. cars. (You will have to copy the data to a new worksheet using the AutoFilter function.)

c. Analyze the residuals of the model. Do they follow the assumptions reasonably well?

d. In the Car Data worksheet, add a new column containing the predicted GPM100 values for all car models using the regression equation you created for only the U.S. cars. Create another new column containing the residuals.

e. Plot the residuals against the predicted values for all of the cars, and then break down the scatter plot into categories on the basis of origin. Rescale the *x* axis so that it ranges from 3 to 8.

f. Calculate descriptive statistics (include the summary, variability, and 95% *t*-confidence intervals) for the residuals column, broken down by region.

g. Save your changes to the workbook and write a report summarizing your conclusions, including a discussion of whether Asian and European cars appear to have better MPG after correction for the other factors. Because the model was developed for U.S. cars, the average residual for U.S. cars will be 0. If a car has a negative residual for GPM100, then the car uses less energy than was predicted for it, and therefore it gets better gas mileage than predicted (the value predicted for a U.S. car).

13. The Temperatures workbook contains average January temperatures for 56 cities in the United States, along with the cities' latitude and longitude. Perform the following analysis of the data:

a. Open the **Temperatures** workbook from the Chapter09 folder and save it as **Temperatures Regression**.
b. Create a chart sheet containing a scatterplot of latitude vs. longitude. Modify the scales for the horizontal and vertical axes to go from 60 to 120 degrees in longitude and from 20 to 50 degrees in latitude. Reverse the orientation of the *x*-axis so that it starts from 120 degrees on the left and goes down to 60 degrees on the right. Add labels to the points, showing the temperature for each city.
c. Construct a regression model that relates average temperature to latitude and longitude.
d. Examine the results of the regression. Are both predictor variables statistically significant at the 5% level? What is the R^2 value? How much of the variability in temperature is explained by longitude and latitude?
e. Format the regression values generated by the Analysis ToolPak to display residual values as integers. Copy the map chart from part b to a new chart sheet, and delete the temperature labels. Now label the points using the residual values.
f. Interpret your findings. Where are the negative values clustered? Where do you usually find positive residuals?
g. Write a report summarizing your findings, discussing where the linear model fails and why.

14. The Housing Price workbook contains information on home prices in Albuquerque, New Mexico.
    a. Open the **Housing Price** workbook from the Chapter09 folder and save it as **Housing Price Regression**.
    b. Regress the price of the houses in the sample on three predictor variables: Square Feet, Age, and number of features.
    c. Examine the plot of residuals versus predicted values. Is there any violation of the regression assumptions evident in this plot?
    d. Redo the regression analysis, this time regressing the Log Price on the three predictor variables. How does the plot of residuals versus predicted values appear in this model? Did the logarithm correct the problem you noted earlier?
    e. There is an outlier in the plot. Identify the point and describe what this tells us about the price of the house if the model is correct.
    f. Save your changes to the workbook and write a report summarizing your observations.

15. The Unemployment workbook contains the U.S. unemployment rate, Federal Reserve Board index of industrial production, and year of the decade 1950–1959. Unemployment is the dependent variable; Industrial Production and Year of the Decade are the predictor variables.
    a. Open the **Unemployment** workbook from the Chapter09 folder and save it as **Unemployment Regression**.
    b. Create a chart sheet showing the scatter plot of Unemployment versus FRB_Index. Add a linear trend line to the chart. Does unemployment appear to rise along with production?
    c. Using the Analysis ToolPak, run a simple linear regression of Unemployment versus FRB_Index. What is the regression equation? What is the R^2 value? Does the regression explain much of the variability in unemployment values during the 1950s?
    d. Rerun the regression, adding Years to the regression equation. How does the R^2 value change with the addition of the Years factor? What is the regression equation?

e. Compare the parameter value for FRB_Index in the first equation with that in the second. How are they different? Does your interpretation of the effect of production on unemployment change from one regression to the other?
f. Calculate the correlation between FRB_Index and Year. How significant is the correlation?
g. Save your changes to the workbook and write a report summarizing your observations.

17. The Beer Rating workbook contains ratings from ratebeer.com along with values of IBU (international bittering units, a measure of bitterness) and ABV (alcohol by volume) for 25 beers.
    a. Open the **Beer Rating** workbook from the Chapter09 folder and save it as **Beer Rating Regression**.
    b. Create a correlation matrix and scatterplot matrix for Rating, IBU, and ABV.
    c. Which beer has the highest rating? The lowest?
    d. Describe the relationships among the three variables. Given how the variables are related, do the correlations fully describe the strengths of the relationships? Explain.
    e. Regress Rating on IBU and ABV. Notice that, although both predictors have strongly significant correlations with Rating, they do not both have significant regression coefficients. How do you explain this?
    f. Plot the residuals from (e) to check the assumptions. Which of the assumptions is clearly not satisfied? Why should this be expected based on (d)?
    g. Repeat the multiple regression in (e) with the square of IBU as a third predictor. Check the assumptions again.
    h. How effective is the regression in (g)? Interpret the coefficients with regard to statistical significance and sign. In particular, discuss the coefficient for the square of IBU.
    i. Save your changes to the workbook and then write a report summarizing your conclusions.

Chapter 10

ANALYSIS OF VARIANCE

Objectives

In this chapter you will learn to:

- Compare several groups graphically
- Compare the means of several groups using analysis of variance
- Correct for multiple comparisons using the Bonferroni test
- Find which pairs of means differ significantly
- Compare analysis of variance to regression analysis
- Perform a two-way analysis of variance
- Create and interpret an interaction plot
- Check the validity of assumptions

One-Way Analysis of Variance

Earlier we used the *t* test to compare two treatment groups, such as two groups taught by two different methods. What if there are four treatment groups? We might have 40 subjects split into four groups, with each group receiving a different treatment. The treatments might be four different drugs, four different diets, or four different advertising videos. **Analysis of variance**, or **ANOVA**, provides a test to determine whether to accept or reject the hypothesis that all of the group means are equal.

The model we'll use for analysis of variance, called a means model, is

$$y = \mu_i + \varepsilon$$

Here, μ_i is the mean of the *i*th group, and ε is a random error following a normal distribution with mean 0 and variance σ^2. If there are *P* groups, the null and alternative hypotheses for the means model are

$$H_0\text{: } \mu_1 = \mu_2 = \cdots = \mu_P.$$

$$H_a\text{: Not all of the } \mu_i \text{ are equal.}$$

Note that the assumptions for the means model are similar to those used for regression analysis.

- The errors are normally distributed.
- The errors are independent.
- The errors have constant variance σ^2.

The similarity to regression is no accident. As you will see later in this chapter, analysis of variance can be thought of as a special case of regression.

To verify analysis of variance assumptions, it is helpful to make a plot that shows the distribution of observations in each of the treatment groups. If the plot shows large differences in the spread among the treatment groups, there might be a problem of nonconstant variance. If the plot shows outliers, there might be a problem with the normality assumption. Independence could also be a problem if time is important in the data collection, in which case consecutive observations might be correlated. However, there are usually no problems with the independence assumption in the analysis of variance.

Analysis of Variance Example: Comparing Hotel Prices

Some professional associations are reluctant to hold meetings in New York City because of high hotel prices and taxes. Are hotels in New York City more expensive than hotels in other major cities?

To answer this question, let's look at hotel prices in four major cities: New York City, Chicago, Denver, and San Francisco. For each city, a random sample of eight hotels was taken from the *TripAdvisor.com* website (February 2008) and stored in the Hotel workbook. The workbook contains the following variables as shown in Table 10-1:

Table 10-1 The Hotel Workbook

Range Name	Range	Description
City	A2:A33	City of each hotel
Hotel	B2:B33	Name of hotel
Stars	C2:C33	*TripAdvisor.com* rating (February 2008), on a scale from 1 to 5
Price	D2:D33	The room price

To open the Hotel workbook:

1 Open the **Hotel** workbook from the Chapter10 folder.

2 Save the workbook as **Hotel ANOVA**. The workbook appears as shown in Figure 10-1.

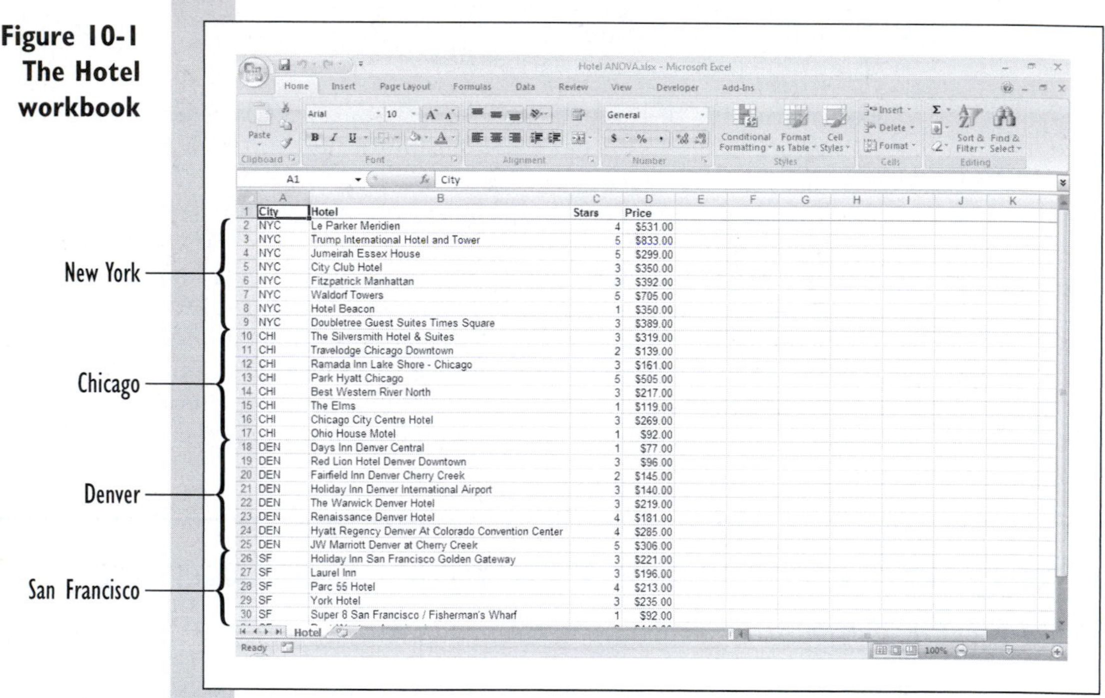

City	Hotel	Stars	Price
NYC	Le Parker Meridien	4	$531.00
NYC	Trump International Hotel and Tower	5	$833.00
NYC	Jumeirah Essex House	5	$299.00
NYC	City Club Hotel	3	$350.00
NYC	Fitzpatrick Manhattan	3	$392.00
NYC	Waldorf Towers	5	$705.00
NYC	Hotel Beacon	1	$350.00
NYC	Doubletree Guest Suites Times Square	3	$389.00
CHI	The Silversmith Hotel & Suites	3	$319.00
CHI	Travelodge Chicago Downtown	2	$139.00
CHI	Ramada Inn Lake Shore - Chicago	3	$161.00
CHI	Park Hyatt Chicago	5	$505.00
CHI	Best Western River North	3	$217.00
CHI	The Elms	1	$119.00
CHI	Chicago City Centre Hotel	3	$269.00
CHI	Ohio House Motel	1	$92.00
DEN	Days Inn Denver Central	1	$77.00
DEN	Red Lion Hotel Denver Downtown	3	$96.00
DEN	Fairfield Inn Denver Cherry Creek	2	$145.00
DEN	Holiday Inn Denver International Airport	3	$140.00
DEN	The Warwick Denver Hotel	3	$219.00
DEN	Renaissance Denver Hotel	4	$181.00
DEN	Hyatt Regency Denver At Colorado Convention Center	4	$285.00
DEN	JW Marriott Denver at Cherry Creek	5	$306.00
SF	Holiday Inn San Francisco Golden Gateway	3	$221.00
SF	Laurel Inn	3	$196.00
SF	Parc 55 Hotel	4	$213.00
SF	York Hotel	3	$235.00
SF	Super 8 San Francisco / Fisherman's Wharf	1	$92.00

Figure 10-1 The Hotel workbook

We have to decide between two hypotheses.

H_0: The mean hotel population price is the same for each city.

H_a: The mean hotel population prices are not the same.

Graphing the Data to Verify ANOVA Assumptions

It is best to begin with a graph that shows the distribution of hotel prices in each of the four cities. To do this, you can use the multiple histograms command available in StatPlus.

To create the graphs:

1. Click **Multi-variable Charts** from the StatPlus menu and then click **Multiple Histograms.**
2. Because the workbook is laid out with the variable values in one column and the categories in another, verify that the **Use a column of category levels** option button is selected.
3. Click the **Data Values** button, and select **Price** from the list of range names. Click **OK.**
4. Click the **Categories** button, and select **City** from the range names list. Click **OK.**
5. Click the **Display normal curve** checkbox, and verify that the **Frequency** option button is selected.
6. Click the **Output** button, and send the output to a new worksheet named **Histograms.** Click **OK.**

Your completed dialog box should appear as shown in Figure 10-2.

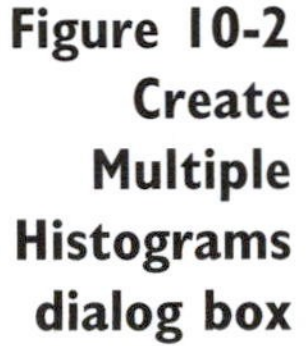

Figure 10-2 Create Multiple Histograms dialog box

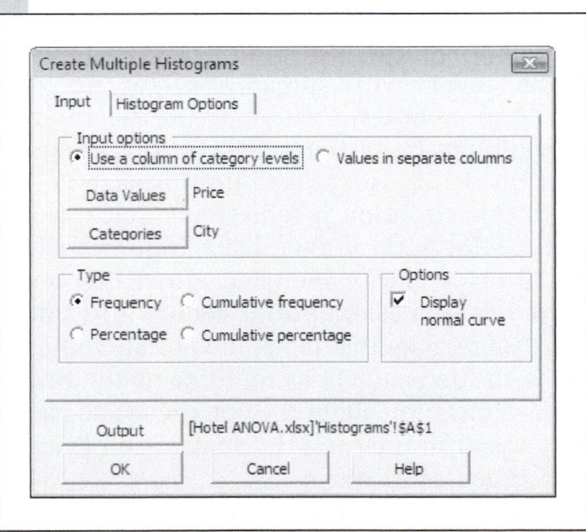

7 Click **OK**. StatPlus generates the histograms shown in Figure 10-3.

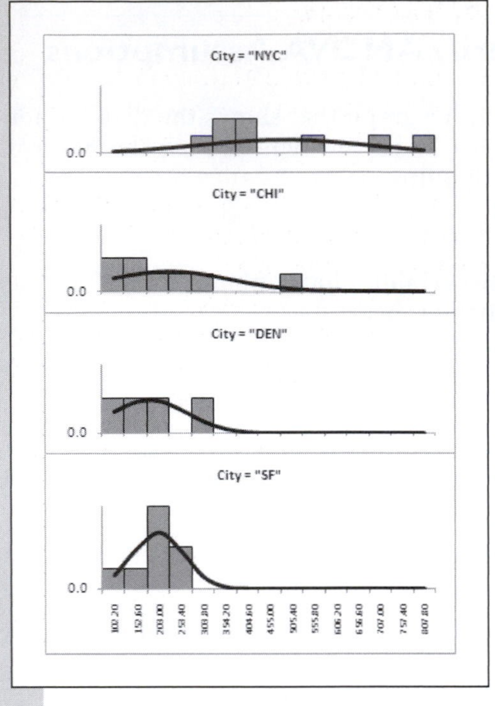

Figure 10-3 Multiple histograms of prices in each city

What do these histograms tell you about the analysis of variance assumptions? One of the assumptions states that the population variance is the same in each group. If one city has prices that are all bunched together and another has a very wide spread of prices, unequal variances can be a problem. The plot shows a tendency for the spread to be larger when the prices are higher. In particular, New York seems to have the highest mean price and the biggest spread, and Chicago is second in both mean and spread. In this situation it sometimes helps to replace the response variable with its logarithm. Exercise 7 requests that you replace Price with the log of price and to see what effect this has on the analysis. If you find there that the assumptions seem valid but that the results are essentially the same, then that tends to validate the original analysis. Here we continue with the analysis using Price as the response variable, even though there is a question about the equal variance assumption. Generally speaking, it is easier to interpret the analysis on Price rather than its logarithm or some other transform.

What about the assumption of normal data? The analysis of variance is robust to the normality assumption, so only major departures from normality would cause trouble. In any case, eight observations in each group may be too few to determine whether the normality assumption is violated.

Computing the Analysis of Variance

From the histograms, it appears that New York has the highest mean hotel price. Still, there is some overlap between the New York prices and the others. Do you think that New York City is significantly more expensive than the other cities? We'll soon find out by performing an analysis of variance. To do so, we'll have to use the Analysis of Variance command available from the Analysis ToolPak, the statistical add-in supplied with Excel. The Analysis ToolPak requires that the group values be placed in separate columns. In this workbook, groups are identified by a category variable, so you'll have to unstack the price values on the basis of the City variable, creating four separate price columns.

To unstack the Price column:

1 Click **Manipulate Columns** from the StatPlus menu and then click **Unstack Column**.

2 Click the **Data Values** button, and select **Price** from the range names list. Click **OK**.

3 Click the **Categories** button, select **City** from the range name list, and click **OK**.

4 Deselect the **Sort the Columns** checkbox.

5 Click the **Output** button and send the unstacked values to a new worksheet named **Price Data**. Click **OK**.

Figure 10-4 shows the completed Unstack Column dialog box. Click **OK**.

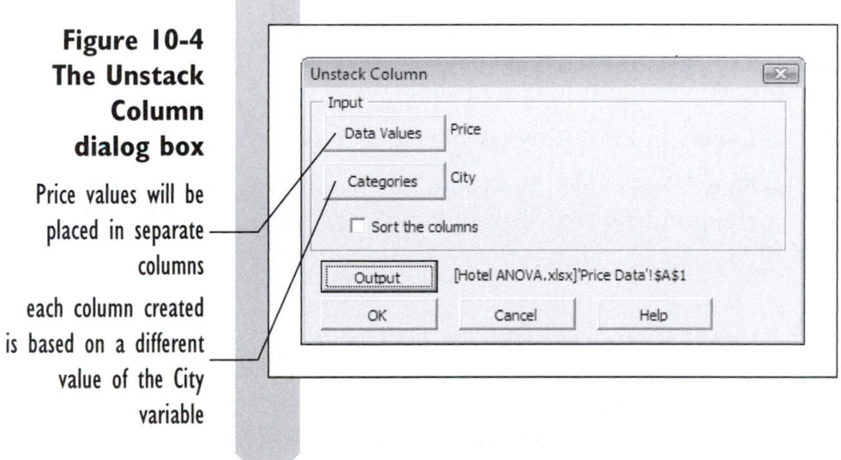

Figure 10-4 The Unstack Column dialog box

The unstacked data are shown in Figure 10-5.

Figure 10-5 The unstacked data

	A	B	C	D	E
1	NYC	CHI	DEN	SF	
2	531	319	77	221	
3	833	139	96	196	
4	299	161	145	213	
5	350	505	140	235	
6	392	217	219	92	
7	705	119	181	149	
8	350	269	285	223	
9	389	92	306	274	
10					

STATPLUS TIPS

- You can use the StatPlus' Stack Columns to stack a series of columns. The resulting data set will contain two columns: a column of data values and a column containing the category labels.

Now you can perform the analysis of variance on the price data.

To perform the analysis of variance:

1 Click **Data Analysis** from the Analysis group on the Data tab, then click **Anova: Single Factor** in the Analysis Tools list box, and then click **OK**.

2 Type **A1:D9** in the Input Range text box, and verify that the **Grouped By Columns** option button is selected.

3 Click the **Labels in First Row** checkbox to select it.

4 Click the **New Worksheet Ply** option button and type **Price ANOVA** in the corresponding text box. Your dialog box should look like Figure 10-6.

Figure 10-6 The Anova: Single Factor dialog box

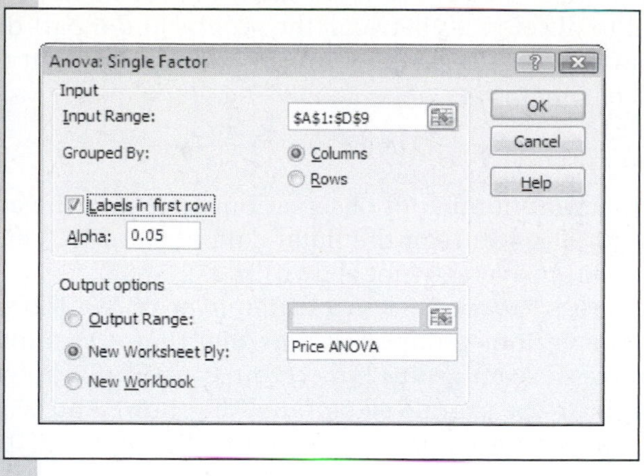

Anova: Single Factor

Input
Input Range: $A$1:$D$9
Grouped By: Columns / Rows
Labels in first row
Alpha: 0.05

Output options
Output Range:
New Worksheet Ply: Price ANOVA
New Workbook

OK
Cancel
Help

5 Click **OK**.

Figure 10-7 shows the resulting analysis of variance output, with some minor formatting.

Figure 10-7 Analysis of variance output

	A	B	C	D	E	F	G
1	Anova: Single Factor						
2							
3	SUMMARY						
4	*Groups*	*Count*	*Sum*	*Average*	*Variance*		
5	NYC	8	3849	481.125	37201.554		
6	CHI	8	1821	227.625	18513.982		
7	DEN	8	1449	181.125	6986.125		
8	SF	8	1603	200.375	3159.982		
9							
10							
11	ANOVA						
12	*Source of Variation*	*SS*	*df*	*MS*	*F*	*P-value*	*F crit*
13	Between Groups	472716.4	3	157572.13	9.570	0.000163	2.947
14	Within Groups	461031.5	28	16465.41			
15							
16	Total	933747.9	31				
17							

Interpreting the Analysis of Variance Table

In performing an analysis of variance, you determine what part of the variance you should attribute to randomness and what part you can attribute to other factors. Analysis of variance does this by splitting the total sum of squares (the sum of squared deviations from the mean) into two parts: a part attributed to differences between the groups and a part due to random error or random chance. To see how this is done, recall that the formula for the total sum of squares is

$$\text{Total } SS = \sum_{i=1}^{n} (y_i - \bar{y})^2$$

Here, the total number of observations is n, and the average of *all* observations is $\bar{y}$. The value for the hotel data is 933,747.9 and is shown in cell B16. The sample average (not shown) is 272.5625.

Let's express the total SS in a different way. We'll break the calculation down by the various groups. Assume that there are a total of P groups and that the size of each group is n_i (groups need not be equal in size, so n_i would indicate the sample size of the ith group), and we calculate the total sum of squares for each group separately. We can write this as

$$\text{Total } SS = \sum_{i=1}^{P} \sum_{j=1}^{n_i} (y_{ij} - \bar{y})^2$$

Here, y_{ij} identifies the jth observation from the ith group (for example, y_{23} would mean the third observation from the second group). Notice that we haven't changed the value; all we've done is specify the order in which we'll calculate the total sum of squares. We'll calculate the sum of squares in the first group, and then in the second and so forth, adding up all of the sums of squares in each group to arrive at the total sum of squares.

Next we'll calculate the sample average for *each* group, labeling it $\bar{y}_i$, which is the sample average of the ith group. For example, in the hotel data, the values (shown in cells D5:D8) are

NYC	481.125
CHI	227.625
DEN	181.125
SF	200.375

Using the group averages, we can calculate the total sum of squares within each group. This is equal to the sum of the squared deviations, where the deviation is from each observation to its group average. We'll call this value the **error sum of squares**, or **SSE**, and express it as

$$\text{SSE} = \sum_{i=1}^{P} \sum_{j=1}^{n_i} (y_{ij} - \bar{y}_i)^2$$

Another term for this value is the **within-groups sum of squares** because it is the sum of squares within each group. The value for SSE in the hotel data is 461,031.5 (shown in cell B14).

The final piece of the analysis of variance is to calculate the sum of squares between each of the group averages and the overall average. This value, called the **between-groups sum of squares** and otherwise known as the **treatment sum of squares**, or **SST**, is

$$\text{SST} = \sum_{i=1}^{P} n_i (\bar{y}_i - \bar{y})^2$$

Note that we take each squared difference and multiply it by the number of observations in the group. In this hotel data set, each group has eight observations, so the value of n_i is always eight. The between-groups sum of squares for the hotel data is equal to 472,716.4 (cell B13).

But note that the total sum of squares is equal to the within-groups sum of squares plus the between-groups sum of squares, because 933,747.9 = 461,031.5 + 472,716.4. In general terms,

$$\text{Total SS} = \text{SSE} + \text{SST}$$

Let's try to relate this to the price of staying at hotels in various cities. If the average prices in the various cities are very different, the between-groups sum of squares will be a large value. However, if the city averages are close in value, the between-groups sum of squares will be near zero. The argument goes the other way, too; a large value for the between-groups sum of squares could indicate that the city averages are very different, whereas a small value might show that they are not so different.

A large value for the between-groups sum of squares could also be due to a large number of groups, so you have to adjust for the number of groups in the data set. The **degrees of freedom** (*df*) column in the ANOVA table (cells C13: C16) tells you that. The *df* for the city factor (in this case the between-groups term) is the number of groups minus 1, or 4 − 1 = 3 (cell C13). The degrees of freedom for the total sum of squares is the total number of observations minus 1, or 32 − 1 = 31 (cell C16). The remaining degrees of freedom are assigned to the error term (the within-groups term) and are equal to 31 − 3 = 28 (cell C14).

The **Mean Square** (MS) column (cells D13:D14) shows the sum of squares divided by the degrees of freedom; you can think of the entries in this column as variances. The first value, 157,572.3 (cell D13), measures the variance in hotel cost between the various cities; the second value, 16,465.1 (cell D14), measures the variance of the cost within cities. The within-groups mean square also estimates the value of σ^2—the variance of the error term ε shown in the means model earlier in the chapter. If the variability in hotel prices between cities is large relative to the variability of hotel prices within the cities, then we might conclude that mean hotel price is not the same for each city.

To test this, we calculate the ratio of the two variances. Under the null hypothesis, this value should follow an F distribution with n, m degrees of freedom, where n is the degrees of freedom for the between-groups variance and m is the degrees of freedom for the within-groups variance.

In the hotel data, the F value is 9.570 (cell E13) and follows an $F(3,28)$ distribution. Excel calculates the p value to be .000163 (cell F13), which is less than .05. We reject the null hypothesis, accepting the alternative that there is a difference in the mean hotel price.

Although the output does not show it, you can use the values in the ANOVA table to derive some of the same statistics you used in regression analysis. For example, the ratio of the between-groups sum of squares to the total sum of squares equals R^2, the coefficient of determination discussed in some depth in Chapters 8 and 9. In this case R^2 = 472,716.4/933,747.9 = 0.50626. Thus about 50% of the variability in hotel price is explained by the city of origin.

Comparing Means

The ANOVA table has led you to reject the hypothesis that the mean single-room price is the same in all four cities and to accept the alternative that the four means are not all the same. Looking at the mean values, you might be tempted to conclude that the high price for New York City hotel rooms is the cause and leave it at that. This assumption would be unwarranted because you haven't tested for this specific hypothesis. Are there significant differences between the other cities as well? To find out, you need to calculate the differences in mean value between all pairs of cities and then test the differences to discover their statistical significance.

Excel does not provide a function to test pairwise mean differences, but one has been provided for you with StatPlus.

To create a matrix of paired differences:

1. Click **Multivariate Analysis** from the StatPlus menu and then click **Means Matrix**.
2. Click the **Data Values** button, and select **Price** from the list of range names. Click **OK**.
3. Click the **Categories** button, and select **City** from the range names list. Click **OK**.
4. Click the **Use Bonferroni Correction** checkbox.
5. Click the **Output** button, and direct the output to a new worksheet named **Means Matrix**. Click **OK**.

Figure 10-8 shows the completed dialog box.

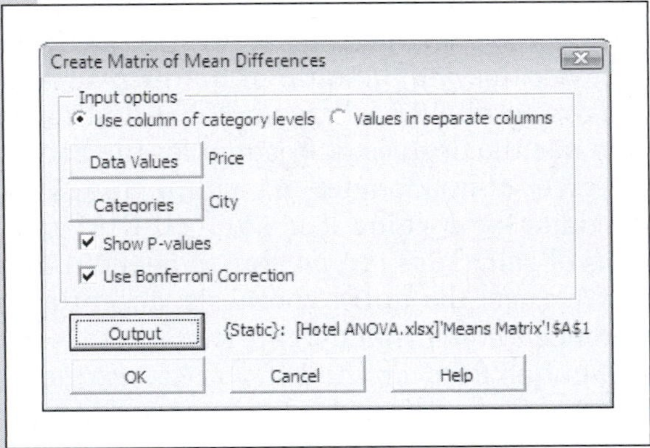

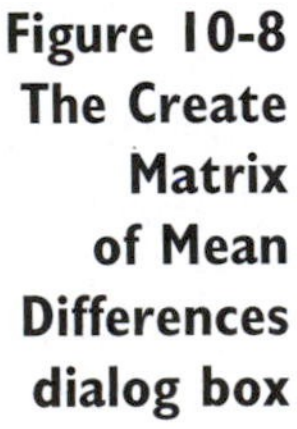
Figure 10-8 The Create Matrix of Mean Differences dialog box

6 Click **OK**. Excel generates the output shown in Figure 10-9.

Figure 10-9 Pairwise mean difference

	A	B	C	D	E	F
1			Descriptive Statistics			
2		City = "NYC"	City = "CHI"	City = "DEN"	City = "SF"	
3		Price	Price	Price	Price	
4	Count	8	8	8	8	
5	Average	481.1250	227.6250	181.1250	200.3750	
6	Standard Deviation	192.87704	136.06609	83.58304	56.21372	
7	Sum of Squares	2,112,261	544,103	311,353	343,321	
8						
9	Pairwise Mean Difference (row - column)					
10		"NYC"	"CHI"	"DEN"	"SF"	
11	"NYC"	0.000	253.500	300.000	280.750	
12	"CHI"		0.000	46.500	27.250	
13	"DEN"			0.000	-19.250	
14	"SF"				0.000	
15	MSE = 16465.4107142857					
16						
17	Pairwise Probabilities (Bonferroni Correction)					
18		"NYC"	"CHI"	"DEN"	"SF"	
19	"NYC"	-	0.003	0.000	0.001	
20	"CHI"		-	1.000	1.000	
21	"DEN"			-	1.000	
22	"SF"				-	
23						

You can tell from the Pairwise Mean Difference table that the mean cost for a single hotel room in Los Angeles is $27.25 less than the mean cost in Chicago. The largest difference is between Denver and New York City, with a single room in Denver hotels costing $300 less than a single room in New York City hotels. Note that the output includes the mean squared error value from the ANOVA table, 16,465.41, which is the estimate of the variance of hotel prices.

Using the Bonferroni Correction Factor

You also requested in the dialog box a table of *p* values for these mean differences using the Bonferroni correction factor. Recall from Chapter 8 that the Bonferroni procedure is a conservative method for calculating the probabilities by multiplying the *p* value by the total number of comparisons. Because the *p* values are much higher than you would see if you compared the cities with *t* tests, it is harder to get significant comparisons with the Bonferroni procedure. However, the Bonferroni procedure has the advantage of giving fewer false positives than *t* tests would give.

With the Bonferroni procedure, the chances of finding at least one significant difference among the means is less than 5% if all of the four population means are the same. On the other hand, if you do six *t* tests to compare the four cities at the 5% level, there is much more than a 5% chance of getting significance in at least one of the six tests if all four population means are the same. Other methods are available to help you adjust the p value for multiple comparisons, including Tukey's and Scheffé's, but the Bonferroni method is the easiest to implement in Excel, which does not provide a correction procedure.

Note: Essentially, the difference between the Bonferroni procedure and a t test is that for the Bonferroni procedure, the 5% applies to all six comparisons together but for t tests, the 5% applies to each of the six comparisons separately. In statistical language, the Bonferroni procedure is testing at the 5% level experimentwise, whereas the t test is testing at the 5% level pairwise.

The pairwise comparison probabilities show that the three biggest differences are significant (highlighted in red). The New York city room price is higher than the room price in the other three cities, but none of those three cities are significantly different in price from each other.

When to Use Bonferroni

As the size of the means matrix increases, the number of comparisons increases as well. Consequently, the p values for the pairwise differences are greatly inflated. As you can imagine, there might be a point where there are so many comparisons in the matrix that it is nearly impossible for any one of the comparisons to be statistically significant using the Bonferroni correction factor. Many statisticians are concerned about this problem and feel that although the Bonferroni correction factor does guard well against incorrectly finding significant differences, it is also too conservative and misses true differences in pairs of mean values.

In such situations, statisticians make a distinction between paired comparisons that are planned before the data are analyzed and those that occur only after we look at the data. For example, the planned comparisons here are the differences in hotel room price between New York City and the others. You should be careful with new comparisons that you come up with after you have collected the data. You should hold these comparisons to a much higher standard than the comparisons you've planned to make all along. This distinction is important in order to ward off the effects of data "snooping" (unplanned comparisons). Some statisticians recommend that you do the following when analyzing the paired means differences in your analysis of variance:

1. Conduct an F test for equal means.
2. If the F statistic is significant at the 5% level, make any planned comparisons you want without correcting the p value. For data snooping, use a correction factor such as Bonferroni's on the p value.
3. If the F statistic for equal means is not significant, you can still consider any planned comparisons, but only with a correction factor to the p value. Do not analyze any unplanned comparisons (Milliken and Johnson 1984).

It should be emphasized that although some statisticians embrace this approach, others question its validity.

Comparing Means with a Boxplot

Earlier you used multiple histograms to compare the distribution of hotel prices among the different cities. The boxplot is also very useful for this task because it shows the broad outline of the distributions and displays the medians for the four cities. Recall that if the data are very badly skewed, the mean might be strongly affected by outlying values. The median would not have this problem.

To create a boxplot of price versus city:

1. Click **Single Variable Charts** from the StatPlus menu and then click **Boxplots**.
2. Click the **Data Values** button, and select **Price** from the list of range names. Click **OK**.
3. Click the **Categories** button, and select **City** from the range names list. Click **OK**.
4. Click the **Output** button, and direct the output to a new chart sheet named **Boxplots**. Click **OK**.
5. Click **OK**. The resulting boxplots are shown in Figure 10-10.

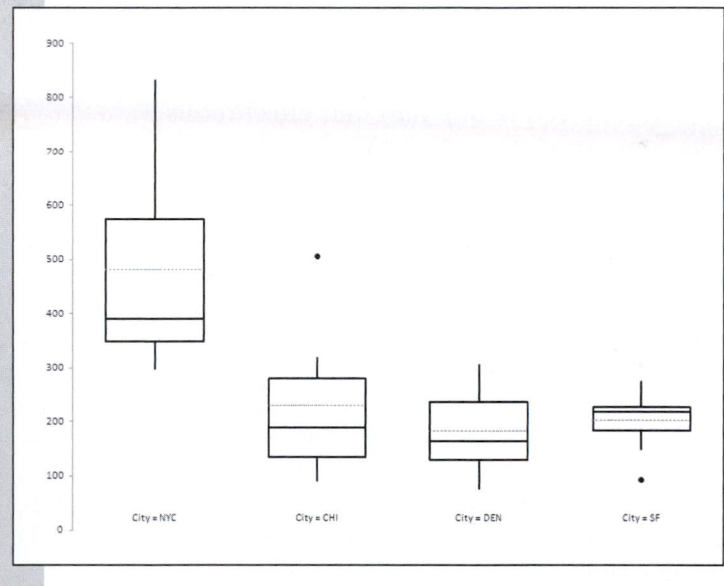

Figure 10-10 Boxplot of room prices for each city

Compare the medians, indicated by the middle horizontal line (not dotted) in each box. The median for San Francisco is above the median for Chicago even though you discovered from the pairwise mean difference matrix that the mean price in Chicago is $27.25 above the mean in San Francisco. The reason for the difference is an outlier in the sample of Chicago room prices. This outlier has a big effect on the Chicago mean price, but not on the median. The median is much more robust to the effect of outliers.

One-Way Analysis of Variance and Regression

You can think of analysis of variance as a special form of regression. In the case of analysis of variance, the predictor variables are discrete rather than continuous. Still, you can express an analysis of variance in terms of regression and, in doing so, can get additional insights into the data. To do this, you have to reformulate the model.

Earlier in this chapter you were introduced to the means model

$$y = \mu_i + \varepsilon$$

for the *i*th treatment group. An equivalent way to express this relationship is with the **effects model**

$$y = \mu + \alpha_i + \varepsilon$$

Here μ is a mean term, α_i is the effect from the *i*th treatment group, and ε is a normally distributed error term with mean 0 and variance σ^2.

Let's apply this equation to the hotel data. In this data set there are four groups representing the four cities, so you would expect the effects model to have a mean term μ and four effect terms α_1, α_2, α_3 and α_4 representing the four cities. There is a problem, however: You have five parameters in your model, but you are estimating only four mean values. This is an example of an **overparametrized model**, where you have more parameters than response values. As a result, an infinite number of possible values for the parameters will solve the equation. To correct this problem, you have to reduce the number of parameters. Statistical packages generally do this in one of two ways: Either they constrain the values of the effect terms so that the sum of the terms is zero, or they define one of the effect terms to be zero (Milliken and Johnson, 1984). Let's apply this second approach to the hotel data and perform the analysis of variance using regression modeling.

Indicator Variables

To perform the analysis of variance using regression modeling, you can create indicator variables for the data. **Indicator variables** take on values of either 1 or 0, depending on whether the data belong to a certain treatment group or not. For example, you can create an indicator variable where the variable value is 1 if the observation comes from a hotel in San Francisco or 0 if the observation comes from a hotel not in San Francisco.

You'll use the indicator variables to represent the terms in the effects model.

To create indicator variables for the hotel data:

1. Click the **Hotel** worksheet tab (you might have to scroll to see it) to return to the worksheet containing the hotel data.
2. Click **Manipulate Columns** from the StatPlus menu and then click **Create Indicator Columns.**
3. Click the **Categories** button, and select **City** from the list of range names. Click **OK.**
4. Click the **Output** button, click the **Cell** option button, and select cell **F1**. Click **OK.**
5. Click **OK.**

Excel generates the four new columns shown in Figure 10-11.

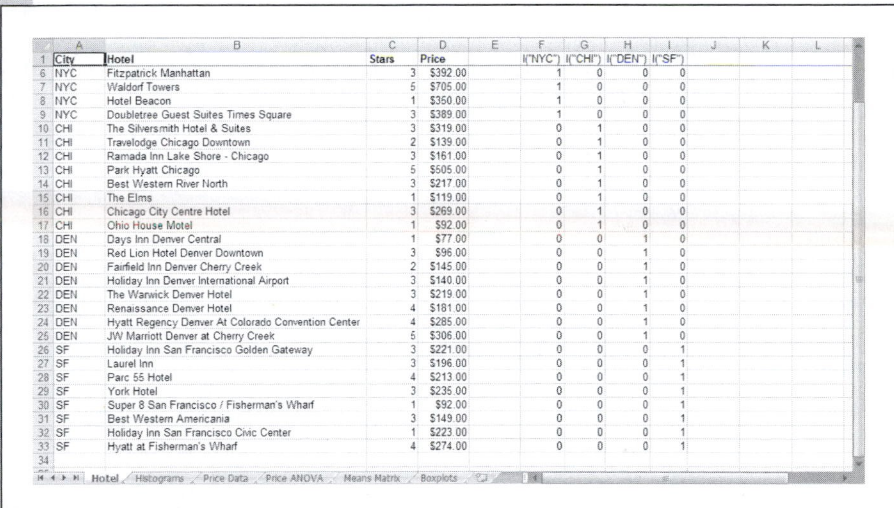

	A	B	C	D	E	F	G	H	I
1	City	Hotel	Stars	Price		I("NYC")	I("CHI")	I("DEN")	I("SF")
6	NYC	Fitzpatrick Manhattan	3	$392.00		1	0	0	0
7	NYC	Waldorf Towers	5	$705.00		1	0	0	0
8	NYC	Hotel Beacon	1	$350.00		1	0	0	0
9	NYC	Doubletree Guest Suites Times Square	3	$389.00		1	0	0	0
10	CHI	The Silversmith Hotel & Suites	3	$319.00		0	1	0	0
11	CHI	Travelodge Chicago Downtown	2	$139.00		0	1	0	0
12	CHI	Ramada Inn Lake Shore - Chicago	3	$161.00		0	1	0	0
13	CHI	Park Hyatt Chicago	5	$505.00		0	1	0	0
14	CHI	Best Western River North	3	$217.00		0	1	0	0
15	CHI	The Elms	1	$119.00		0	1	0	0
16	CHI	Chicago City Centre Hotel	3	$269.00		0	1	0	0
17	CHI	Ohio House Motel	1	$92.00		0	1	0	0
18	DEN	Days Inn Denver Central	1	$77.00		0	0	1	0
19	DEN	Red Lion Hotel Denver Downtown	3	$96.00		0	0	1	0
20	DEN	Fairfield Inn Denver Cherry Creek	2	$145.00		0	0	1	0
21	DEN	Holiday Inn Denver International Airport	3	$140.00		0	0	1	0
22	DEN	The Warwick Denver Hotel	3	$219.00		0	0	1	0
23	DEN	Renaissance Denver Hotel	4	$181.00		0	0	1	0
24	DEN	Hyatt Regency Denver At Colorado Convention Center	4	$285.00		0	0	1	0
25	DEN	JW Marriott Denver at Cherry Creek	5	$306.00		0	0	1	0
26	SF	Holiday Inn San Francisco Golden Gateway	3	$221.00		0	0	0	1
27	SF	Laurel Inn	3	$196.00		0	0	0	1
28	SF	Parc 55 Hotel	4	$213.00		0	0	0	1
29	SF	York Hotel	3	$235.00		0	0	0	1
30	SF	Super 8 San Francisco / Fisherman's Wharf	1	$92.00		0	0	0	1
31	SF	Best Western Americania	3	$149.00		0	0	0	1
32	SF	Holiday Inn San Francisco Civic Center	1	$223.00		0	0	0	1
33	SF	Hyatt at Fisherman's Wharf	4	$274.00		0	0	0	1
34									

Figure 10-11 Indicator variables in columns F:I

The values in column F, labeled I ("NYC"), are equal to 1 if the values in the row come from a hotel in New York City or 0 if they do not. Similarly, the values for the next three columns are 1 if the observations come from Chicago, Denver, and San Francisco, respectively, or 0 otherwise.

Fitting the Effects Model

With these columns of indicator variables, you can now fit the effects model to the hotel pricing data.

To fit the effects model using regression analysis:

1. Click the **Data Analysis** button from the Analysis group on the Data tab, then click **Regression** in the Analysis Tools list box, and click **OK**.
2. Type **D1:D33** in the Input Y Range text box, press **[Tab]**, and then type **F1:H33** in the Input X Range text box.

   Recall that you have to remove one of the effect terms to keep from overparametrizing the model. For this example, remove the New York effect term. (You could have removed any one of the four city effect terms.)
3. Click the **Labels** checkbox to select it because the range includes a header row.
4. Click the **New Worksheet Ply** option button; then type **Effects Model** in the corresponding text box.
5. Verify that all four Residuals checkboxes are deselected; then click **OK**.

   The regression output appears as in Figure 10-12. (The columns are resized to show the labels.)

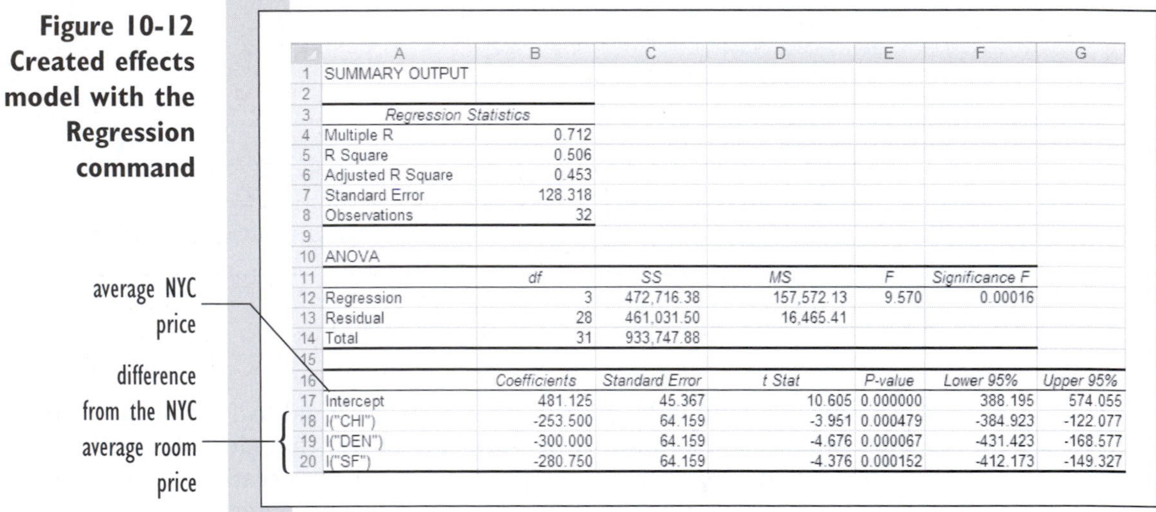

	A	B	C	D	E	F	G
1	SUMMARY OUTPUT						
2							
3	*Regression Statistics*						
4	Multiple R	0.712					
5	R Square	0.506					
6	Adjusted R Square	0.453					
7	Standard Error	128.318					
8	Observations	32					
9							
10	ANOVA						
11		*df*	*SS*	*MS*	*F*	*Significance F*	
12	Regression	3	472,716.38	157,572.13	9.570	0.00016	
13	Residual	28	461,031.50	16,465.41			
14	Total	31	933,747.88				
15							
16		*Coefficients*	*Standard Error*	*t Stat*	*P-value*	*Lower 95%*	*Upper 95%*
17	Intercept	481.125	45.367	10.605	0.000000	388.195	574.055
18	I("CHI")	-253.500	64.159	-3.951	0.000479	-384.923	-122.077
19	I("DEN")	-300.000	64.159	-4.676	0.000067	-431.423	-168.577
20	I("SF")	-280.750	64.159	-4.376	0.000152	-412.173	-149.327

Figure 10-12 Created effects model with the Regression command

The analysis of variance table produced by the regression (cells A10:F14) and shown in Figure 10-12 should appear familiar to you because it is equivalent to the ANOVA table created earlier and shown in Figure 10-7. There are two differences: the Between Groups row from the earlier ANOVA table is the Regression row in this table, and the Within Groups row is now termed the Residual row.

The parameter values of the regression are also familiar. The intercept coefficient 481.125 (cell B17) is the same as the mean price in New York. The values of the CHI, DEN, and SF effect terms now represent the difference between the mean hotel price in these cities and the price in New York. Note that this is exactly what you calculated in the matrix of paired mean differences shown in Figure 10-9. The *p* values for these coefficients are the uncorrected *p* values for comparing the paired mean differences between these cities and New York. If you multiplied these *p* values by 6 (the number of paired comparisons in the paired mean differences matrix), you would have the same *p* values shown in Figure 10-9.

Can you see how the use of indicator variables allowed you to create the effects model? Consider the values for I ("CHI"). For any non-Chicago hotel, the value of the indicator variable is 0, so the effect term is multiplied by 0, and therefore has no impact on the estimate of the hotel price. It is only for Chicago hotels that the effect term is present.

As you can see, using regression analysis to fit the effects model gives you much of the same information as the one-way analysis of variance.

The model you've considered suggests that the average price for a single room at a hotel in New York City is significantly higher than that for a single room in a hotel in Chicago, Denver, or San Francisco. You can expect to pay about an average of $481 for a single room in New York City, and $253.50 less than this in Chicago, $300 less in Denver, and $280.75 less in San Francisco. You've completed your study of the hotel data in this workbook. You can close the Hotel ANOVA workbook now, saving your changes.

EXCEL TIPS

- You can use the Regression command to calculate the means model instead of the effects model. To do this, run the Analysis ToolPak's Regression command, choose *all* of the indicator variables in the Input X Range text box, and select the Constant Is Zero checkbox. This will remove the constant term from the model. The parameter estimates will correspond to mean values of the different groups.

Two-Way Analysis of Variance

One-way analysis of variance compares several groups corresponding to a single categorical variable, or factor. A **two-way analysis of variance** uses two factors. In agriculture, for example, you might be interested in the effects of both potassium and nitrogen on the growth of potatoes. In medicine you might want to study the effects of medication and dose on the duration of headaches. In education you might want to study the effects of grade level and gender on the time required to learn a skill. A marketing experiment might consider the effects of advertising dollars and advertising medium (television, magazines, and so on) on sales.

Recall that earlier in the chapter you looked at the means model for a one-way analysis of variance. Two-way analysis of variance can also be expressed as a means model:

$$y_{ijk} = \mu_{ij} + \varepsilon_{ijk}$$

where y is the response variable and μ_{ij} is the mean for the ith level of one factor and the jth level of the second factor. Within each combination of the two factors, you might have multiple observations called **replicates**. Here ε_{ijk} is the error for the ith level of the first factor, the jth level of the second factor, and the kth replicate, following a normal distribution with mean 0 and variance σ^2.

The model is more commonly presented as an effects model where

$$y_{ijk} = \mu + \alpha_i + \beta_j + \alpha\beta_{ij} + \varepsilon_{ijk}$$

Here y is the response variable, μ is the overall mean, α_i is the effect of the ith treatment for the first factor, and β_j is the effect of the jth treatment for the second factor. The term $\alpha\beta_{ij}$ represents the interaction between the two factors, that is, the effect that the two factors have on each other. For example, in an experiment where the two factors are advertising dollars and advertising medium, the effect of an increase in sales might be the same regardless of what advertising medium (radio, newspaper, or television) is used, or it might vary depending on the medium. When the increase is the same regardless of the medium, the interaction is 0; otherwise, there is an interaction between advertising dollars and medium.

A Two-Factor Example

To see how different factors affect the value of a response variable, consider an example of the effects of four different assembly lines (A, B, C, or D) and two shifts (a.m. or p.m.) on the production of microwave ovens for an appliance manufacturer. Assembly line and shift are the two factors; the assembly line factor has four levels, and the shift factor has two levels. Each combination of the factors line and shift is called a **cell**, so there are $4 \times 2 = 8$ cells. The response variable is the total number of microwaves assembled in a week for one assembly line operating on one particular shift. For each of the eight combinations of assembly line and shift, six separate weeks' worth of data are collected.

You can describe the mean number of microwaves created per week with the effects model where

$$\text{Mean number of microwaves} = \text{overall mean} + \text{assembly line effect} + \text{shift effect} + \text{interaction} + \text{error}$$

Now let's examine a possible model of how the mean number of microwaves produced could vary between shifts and assembly lines. Let the overall mean number of microwaves produced for all shifts and assembly lines be 240 per week. Now let the four assembly line effects be A, +66 (that is, assembly line A produces on average 66 more microwaves than the overall mean); B, −2; C, −100; and D, +36. Let the two shift effects be p.m., −6, and a.m., +6. Notice that the four assembly line effects add up to zero, as do the two shift effects. This follows from the need to constrain the values of the effect terms to avoid overparametrization, as was discussed with the one-way effects model earlier in this chapter.

If you exclude the interaction term from the model, the population cell means (the mean number of microwaves produced) look like this.

	A	**B**	**C**	**D**
p.m.	300	232	134	270
a.m.	312	244	146	282

These values are obtained by adding the overall mean + the assembly line effect + the shift effect for each of the eight cells. For example, the mean for the p.m. shift on assembly line A is

$$\text{Overall mean} + \text{assembly line effect} + \text{shift effect} = 240 + 66 - 6 = 300$$

Without interaction, the difference between the a.m. and the p.m. shifts is the same (12) for each assembly line. You can say that the difference between a.m. and p.m. is 12 no matter which assembly line you are talking about. This works the other way, too. For example, the difference between line A and line C is the same (166) for both the p.m. shift (300 − 134) and the a.m. shift (312 − 146). You might understand these relationships better from a graph. Figure 10-13 shows a plot of the eight means with no interaction (you don't have to produce this plot).

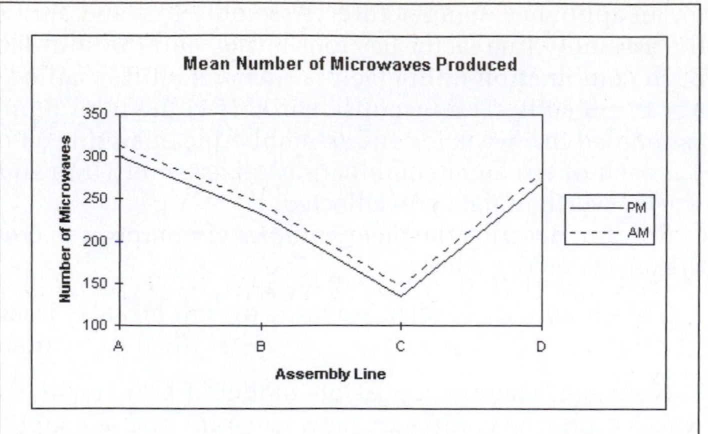

Figure 10-13 Means plot without an interaction effect

The cell means are plotted against the assembly line factor using separate lines for the shift factor. This is called an interaction plot; you'll create one later in this chapter.

Because there is a constant spacing of 12 between the two shifts, the lines are parallel. The pattern of ups and downs for the p.m. shift is the same as the pattern of ups and downs for the a.m. shift.

What if interaction is allowed? Suppose that the eight cell population means are as follows:

	A	B	C	D
p.m.	295	235	175	200
a.m.	317	241	142	220

In this situation, the difference between the shifts varies from assembly line to assembly line, as shown in Figure 10-14. This means that any inference on the shift effect must take into account the assembly line. You might claim that the a.m. shift generally produces more microwaves, but this is not true for assembly line C.

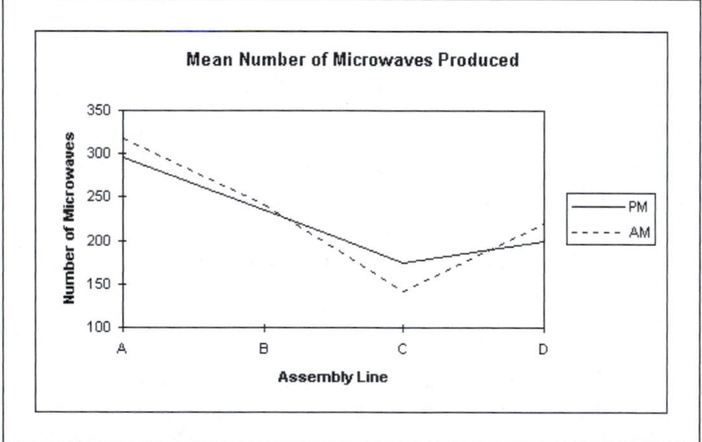

Figure 10-14 Means plot with an interaction effect

The assumptions for two-way ANOVA are essentially the same as those for one-way ANOVA. For one-way ANOVA, all observations on a treatment were assumed to have the same mean, but here all observations in a cell are assumed to have the same mean. The two-way ANOVA assumes independence, constant variance, and normality, just as the one-way ANOVA (and regression).

Two-Way Analysis Example: Comparing Soft Drinks

The Cola workbook contains data describing the effects of cola (Coke, Pepsi, Shasta, or generic) and type (diet or regular) on the foam volume of cola soft drinks. Cola and type are the factors; cola has four levels, and type has two levels. There are, therefore, eight combinations, or cells, of cola brand and soft drink type. For each of the eight combinations, the experimenter purchased and cooled a six-pack, so there are 48 different cans of soda. Then the experimenter chose a can at random, poured it in a standard way into a standard glass, and measured the volume of foam.

Why would it be wrong to test all of the regular Coke first, then the diet Coke, and so on? Although the experimenter might make every effort to keep everything standardized, trends that influence the outcome could appear. For example, the temperature in the room or the conditions in the refrigerator might change during the experiment. There could be subtle trends in the way the experimenter poured and measured the cola. If there were such trends, it would make a difference which brand was poured first, so it is best to pour the 48 cans in random order.

The Cola workbook contains the variables shown in Table 10-2.

Table 10-2 Data for Cola Workbook

Range Name	Range	Description
Can_No	A2:A49	The number of the can (1–6) in the six-pack
Cola	B2:B49	The cola brand
Type	C2:C49	Type of cola: regular or diet
Foam	D2:D49	The foam content of the cola
Cola_Type	E2:E49	The brand and type of the cola

To open the Cola workbook:

1 Open the **Cola** workbook from the Chapter10 data folder.

2 Save the workbook as **Cola ANOVA**. The workbook appears as shown in Figure 10-15.

Figure 10-15
Cola workbook

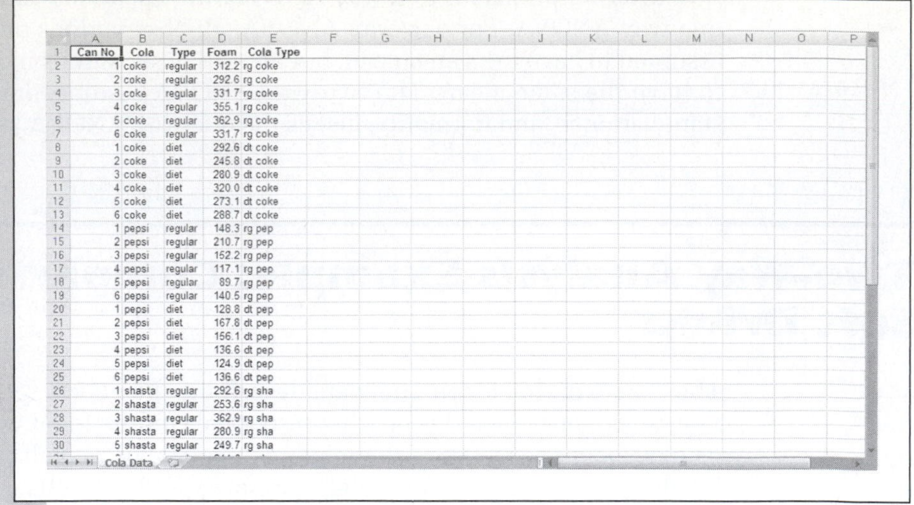

	A	B	C	D	E
1	Can No	Cola	Type	Foam	Cola Type
2	1	coke	regular	312.2	rg coke
3	2	coke	regular	292.6	rg coke
4	3	coke	regular	331.7	rg coke
5	4	coke	regular	355.1	rg coke
6	5	coke	regular	362.9	rg coke
7	6	coke	regular	331.7	rg coke
8	1	coke	diet	292.6	dt coke
9	2	coke	diet	245.8	dt coke
10	3	coke	diet	280.9	dt coke
11	4	coke	diet	320.0	dt coke
12	5	coke	diet	273.1	dt coke
13	6	coke	diet	288.7	dt coke
14	1	pepsi	regular	148.3	rg pep
15	2	pepsi	regular	210.7	rg pep
16	3	pepsi	regular	152.2	rg pep
17	4	pepsi	regular	117.1	rg pep
18	5	pepsi	regular	89.7	rg pep
19	6	pepsi	regular	140.5	rg pep
20	1	pepsi	diet	128.8	dt pep
21	2	pepsi	diet	167.8	dt pep
22	3	pepsi	diet	156.1	dt pep
23	4	pepsi	diet	136.6	dt pep
24	5	pepsi	diet	124.9	dt pep
25	6	pepsi	diet	136.6	dt pep
26	1	shasta	regular	292.6	rg sha
27	2	shasta	regular	253.6	rg sha
28	3	shasta	regular	362.9	rg sha
29	4	shasta	regular	280.9	rg sha
30	5	shasta	regular	249.7	rg sha

Cola Data

Graphing the Data to Verify Assumptions

Before performing a two-way analysis of variance on the data, you should plot the data values to see whether there are any major violations of the assumptions of equal variability in the different cells. Note that you can use the Cola_Type variable to identify the eight cells.

To create multiple histograms of the foam data:

1. Click **Multi-variable Charts** from the StatPlus menu and then click **Multiple Histograms**.
2. Click the **Data Values** button, select **Foam** from the range names list, and click **OK**.
3. Click the **Categories** button, select **Cola_Type** from the range names list, and click **OK**.
4. Click the **Display normal curve** checkbox.
5. Click the **Output** button, send the charts to a new worksheet named **Histograms**, and click **OK**.
6. Click **OK** to generate the histograms shown in Figure 10-16.

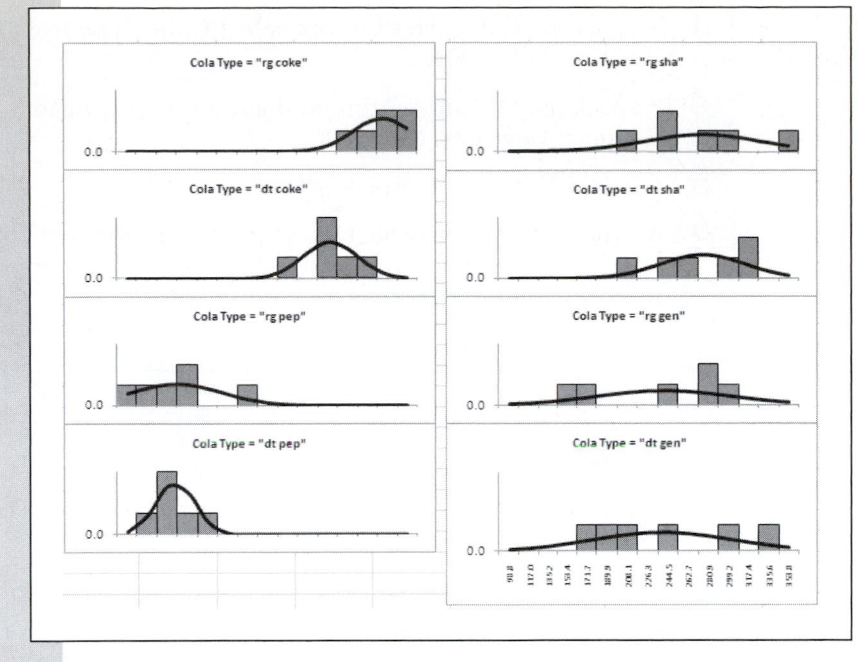

Figure 10-16 Multiple histograms of cola type

Because of the number of charts, you must either reduce the zoom factor on your worksheet or scroll vertically through the worksheet to see all the plots. Do you see major differences in spread among the eight groups? If so, it would suggest a violation of the equal-variances assumption, because all of the groups are supposed to have the same population variance. The histograms seem to indicate a greater variability in the generic colas and the Shasta brand, whereas less variability is indicated for the Coke and Pepsi brands. Once again, the two-way ANOVA is fairly robust with respect to the constant variance assumption, so this might not invalidate the analysis.

You should also look for outliers because extreme observations can make a big difference in the results. An outlier could be the result of a strange can of cola, a wrong observation, a recording error, or an error in entering the data. To gain further insight into the distribution of the data, create a boxplot of each of the eight combinations of brand and type.

To create boxplots of the foam data:

1. Click **Single Variable Charts** from the StatPlus menu and then click **Boxplots**.
2. Click the **Data Values** button, select **Foam** from the range names list, and click **OK**.

3 Click the **Categories** button, select **Cola_Type** from the range names list, and click **OK**.

4 Click the **Output** button, and send the output to a new chart sheet named **Boxplots**. Click **OK**.

5 Click **OK** to create the boxplots.

6 You improve the chart by editing the labels at the bottom of the boxplot, removing the text string Cola_Type= from each label, and increasing the font size. See Figure 10-17.

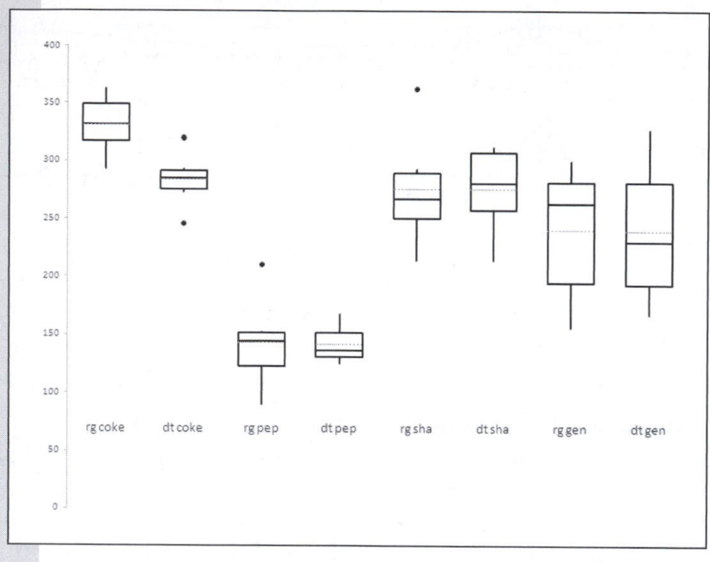

Figure 10-17 Boxplots of foam versus cola type

From the boxplots, you can see that there are no extreme outliers evident in the data, but there are several moderate outliers; perhaps most noteworthy are the outliers for regular Pepsi and regular Shasta. An advantage of the boxplot over the multiple histograms is that it is easier to view the relative change in foam volume from diet to regular for each brand of cola. The first two boxplots represent the range of foam values for regular and diet Coke, respectively, after which come the Pepsi values, Shasta values, and, finally, the generic values. Notice in the plot that the same pattern occurs for both the diet and the regular colas. Coke is the highest, Pepsi is the lowest, and Shasta and generic are in the middle. The difference in the foam between the diet and the regular sodas does not depend much on the cola brand. This suggests that there is no interaction between the cola effect and the type effect.

On the basis of this plot, can you draw preliminary conclusions regarding the effect of type (diet or regular) on foam volume? Does it appear that there is much difference due to cola type (diet or regular)? Because the foam levels do not appear to differ much between the two types, you can expect that the test for the type effect in a two-way analysis of variance will not be significant. However, look at the differences among the four brands of colas. The foamiest can of Pepsi is below the least foamy can of Coke, so you might expect that there will be a significant cola effect.

The Interaction Plot

The histograms and boxplots give us an idea of the influence of cola type and cola brand on foam volume. How do we graphically examine the interaction between the two factors? We can do so by creating an **interaction plot**, which displays the average foam volume for each combination of factors. To do this, we take advantage of Excel's pivot table feature.

To set up the pivot table:

1. Click the **Cola Data** sheet tab to return to the data.
2. Click the **PivotTable** button from the Tables group of the Insert tab.
3. Verify that the **New Worksheet** option button is selected and then click the **OK** button.

   Excel opens a new worksheet displaying the fields from the list in a PivotTable Field list pane.
4. Drag the **Type** field from the field list and drop it in the Column Labels box.
5. Drag the **Cola** field from the list of fields and drop it into the Row Labels box.
6. Drag the **Foam** field from the field list and drop it in the Values box.
7. Click **Sum of Foam** in the Values box and select **Value Field Settings** from the pop-up menu.
8. Click **Average** in the Summarize Value field by List box then click **OK.**
9. Click the **Grand Totals** button from the Layout group on the Design tab of the PivotTable Tools ribbon and select **Off for Rows and Columns** to run off the grand total for the rows and columns of the PivotTable.

Next you can create a line chart based on the values in the PivotTable.

To create a line chart of the cell average:

1. Click the **PivotChart** button from the Tools group on the Options tab of the PivotTable Tools ribbon.
2. Click **Line** from the list of chart types.
3. Click the first chart sub type in the list and click **OK**.

   Excel creates the PivotChart of the PivotTable data as a line chart.
4. With the PivotChart selected, click the **Move Chart** button from the Location group on the Design tab of the PivotChart Tools ribbon and move the chart to a new chart sheet named **Interaction Plot**.

   Figure 10-18 displays the chart.

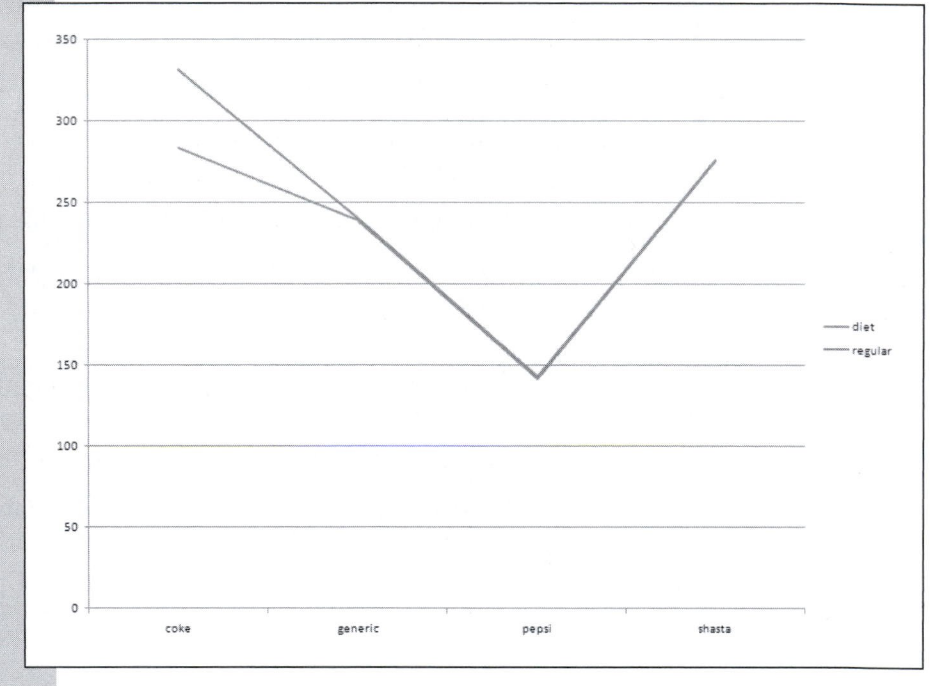

Figure 10-18 Interaction plot of cola type versus cola brand

The plot shows that the foam volumes of diet and regular colas are very close, except for Coke. If there is no interaction between cola brand and cola type, the difference in foam volume for diet and regular should be the same for each cola brand. This means that the lines would move in parallel, always with the same vertical distance. Of course, there is a certain amount of random variation, so the lines will usually not be perfectly parallel. The plot would seem to indicate that there is no interaction between cola brand and cola type. To confirm our visual impression, we'll have to perform a two-way analysis of variance.

Using Excel to Perform a Two-Way Analysis of Variance

The Analysis ToolPak provides two versions of the two-way analysis of variance. One is for situations in which there is no replication of each combination of factor levels. That would be the case in this example if the experimenter had tested only one can of soda for each cola brand and type. However, the experiment has been done with six cans, so you should perform a two-way analysis of variance with replication.

Note that the number of cans for each cell of brand and type must be the same. Specifically, you cannot use data that have five cans of diet Coke and six cans of regular Coke. Data with the same number of replications per cell are called **balanced data**. If the number of replicates is different for different combinations of brand and type, you cannot use the Analysis ToolPak's two-way analysis of variance command.

Finally, to use the Analysis ToolPak on this data set, it must be organized in a two-way table. Figure 10-19 shows this table for the cola data. The data are formatted so that the first factor (the four cola brands) is displayed in the columns, and the second factor (diet or regular) is shown in the rows of the table. Replications (the six cans in each pack) occupy six successive rows. Each cell in the two-way table is the value of the foam volume for a particular can. You can create this table using the Create Two-Way Table command included with StatPlus.

Figure 10-19 Two-way table of foam values

Foam	Cola			
Type	coke	generic	pepsi	shasta
diet	292.6	167.8	128.8	292.6
	245.8	249.7	167.8	253.6
	280.9	187.3	156.1	214.6
	320.0	210.7	136.6	269.2
	273.1	292.6	124.9	312.2
	288.7	327.8	136.6	312.2
regular	312.2	156.1	148.3	292.6
	292.6	253.6	210.7	253.6
	331.7	273.1	152.2	362.9
	355.1	175.6	117.1	280.9
	362.9	284.8	89.7	249.7
	331.7	300.4	140.5	214.6

To create a two-way table:

1. Return to the Cola Data worksheet.
2. Click **Manipulate Columns** from the StatPlus menu and then click **Create Two-Way Table**.
3. Click the **Data Values** button, select **Foam** from the range name list, and click **OK**.
4. Click the **Column Levels** button, select **Cola** from the list of range names, and click **OK**.
5. Click the **Row Levels** button, and select **Type** from the range names. Click **OK**.
6. Click the **Output** button, and direct the output to a new worksheet named **Two-Way Table**. Figure 10-20 shows the completed dialog box.

Figure 10-20 The Make a Two-way Table dialog box

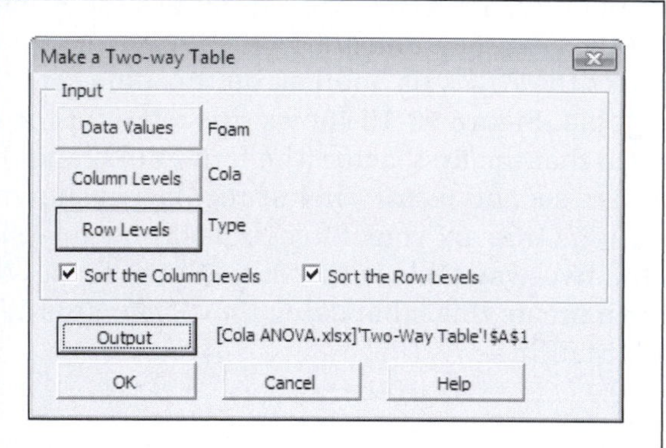

7. Click **OK**.

The structure of the data on the Two-Way Table worksheet now resembles Figure 10-19, and you can now use the Analysis ToolPak to compute the two-way ANOVA.

To calculate the two-way analysis of variance:

1. Click the **Data Analysis** button from the Analysis group on the Data tab, then click **Anova: Two-Factor With Replication** in the Analysis Tools list box, and then click **OK**.
2. Type **A2:E14** in the Input Range text box, press **[Tab]**.

You have to indicate the number of replicates in the two-way table for this command.

3. Type **6** in the Rows per sample text box.

4. Click the **New Worksheet Ply** option button, and type **Two-Way ANOVA** in the corresponding text box.

   Your dialog box should look like Figure 10-21.

Figure 10-21 Anova: Two-Factor With Replication dialog box

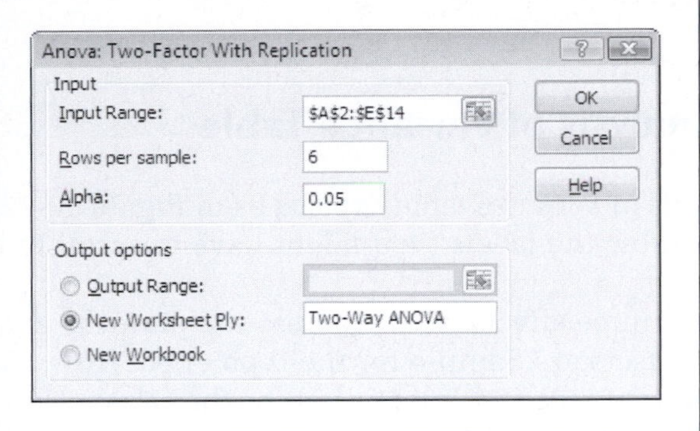

5. Click **OK**.

EXCEL TIPS

- If there is only one observation for each combination of the two factors, use Anova: Two-Factor Without Replication.
- If there is more than one observation for each combination of the two factors, use Anova: Two-Factor With Replication.
- If there are blanks for one or more of the factor combinations, you cannot use the Analysis ToolPak to perform two-way ANOVA.
- You can calculate the *p* value for the *F* distribution using Excel's FDIST(*F, df1, df2*), where *F* is the value of the *F* statistic, *df1* is the degrees of freedom for the factor, and *df2* is the degrees of freedom for the error term.

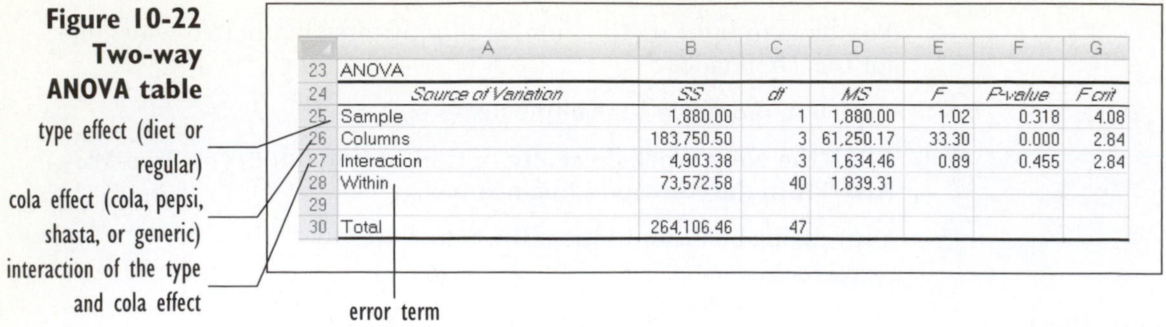

	A	B	C	D	E	F	G
23	ANOVA						
24	*Source of Variation*	*SS*	*df*	*MS*	*F*	*P-value*	*F crit*
25	Sample	1,880.00	1	1,880.00	1.02	0.318	4.08
26	Columns	183,750.50	3	61,250.17	33.30	0.000	2.84
27	Interaction	4,903.38	3	1,634.46	0.89	0.455	2.84
28	Within	73,572.58	40	1,839.31			
29							
30	Total	264,106.46	47				

Figure 10-22 Two-way ANOVA table

Interpreting the Analysis of Variance Table

The Analysis of Variance table appears as in Figure 10-22, with the columns resized to show the labels (you might have to scroll to see this part of the output).

There are three effects now, whereas the one-way analysis had just one. The three effects are Sample for the type effect (row 25), Columns for the cola effect (row 26), and Interaction for the interaction between type and cola (row 27). The Within row (row 28) displays the within sum of squares, also known as the error sum of squares.

As we saw earlier with the one-way ANOVA, the two-way ANOVA breaks the total sum of squares into different parts. If we designate SST as the sum of squares for the cola type, SSC as the sum of squares for cola brand, SSI for the interaction between brand and type, and SSE for random error, then

$$\text{Total} = \text{SST} + \text{SSC} + \text{SSI} + \text{SSE}$$

In this data set, the values for the various sums of squares are

SST	1,880.00
SSC	183,750.50
SSI	4,903.38
SSE	73,572.58

The degrees of freedom for each factor are equal to the number of levels in the factor minus 1. There are two cola types, diet and regular, so the degrees of freedom are 1. There are 3 degrees of freedom in the four cola brands (Coke, Pepsi, Shasta, and generic). The degrees of freedom for the interaction term are equal to the product of the degrees of freedom for the two factors. In this case, that would be $1 \times 3 = 3$. Finally, there are $n - 1$, or 47, degrees of freedom for the total sum of squares, leaving $47 - (1 + 3 + 3) = 40$ degrees of freedom for the error sum of squares. Note that the total degrees of freedom are equal to the sum of the degrees of freedom for each term in the model. In other words, if DFT are the degrees of freedom for the cola type, DFC are the degrees of freedom for cola brand, DFI are the interaction degrees of freedom, and DFE are the degrees of freedom for the error term, then

$$\text{Total degrees of freedom} = \text{DFT} + \text{DFC} + \text{DFI} + \text{DFE}$$

The next column of the two-way ANOVA table displays the mean square of each of the factors (equal to the sum of squares divided by the degrees of freedom). These are

Type	1,880.00
Cola	61,250.17
Interaction	1,634.46
Error	1,839.31

These values are the variances in foam volume within the various factors. The largest variance is displayed in the cola factor; this indicates that this is where the greatest difference in foam volume lies. The mean square value for the error term 1839.31 is an estimate of σ^2, the variance in foam volume after accounting for the factors of cola brand, type, and the interaction between the two. In other words, after accounting for these effects in your model, the typical deviation—or standard deviation—in foam volume is about $\sqrt{1840} = 42.9$.

As with one-way ANOVA, the next column of the table displays the ratio of each mean square to the mean square of the error term. These ratios follow a $F(m, n)$ distribution, where m is the degrees of freedom of the factor (type, cola or interaction) and n is the degrees of freedom of the error term. By comparing these values to the F distribution, Excel calculates the p values (cells F25:F27) for each of the three effects in the model. Examine first the interaction p value, which is .455 (cell F27)—much greater than .05 and not even close to indicating significance at the 5% level. This confirms what we suspected from viewing the interaction plot. Now let's look at the type and cola factors.

The column or cola effect is highly significant, with a p value of 5.84×10^{-11} (cell F26). This is less than .05, so there is a significant difference among colas at the 5% level (because the p value is less than .001, there is significance at the 0.1% level, too). However, the p value is .318 for the sample or type effect (cell F25), so there is no significant difference between diet and regular.

These quantitative conclusions from the analysis of variance are in agreement with the qualitative conclusions drawn from the boxplot: There is a significant difference in foam volume between cola brands, but not between cola types. Nor does there appear to be an interaction between cola brand and type in how they influence foam volume.

Finally, how much of the total variation in foam volume has been explained by the two-way ANOVA model? Recall that the coefficient of determination (R^2 value) is equal to the fraction of the total sum of squares that is explained by the sums of squares of the various factors. In this case that value is

$$\frac{(1880.00 + 183{,}750.50 + 4903.38)}{264{,}106.46} = 0.721$$

Thus about 72% of the total variation in foam volume can be attributed to differences in cola brand, cola type, and the interaction between cola brand and type. Only about 28% of the total variation can be attributed to random causes.

Summary

To summarize the results from the plots and the analysis of variance, we conclude the following:

1. There is no reason to reject the hypothesis that foam volume is the same regardless of cola type (diet or regular).
2. There is a significant difference among the four cola brands (Coke, Pepsi, Shasta, and generic) with respect to foam volume. Coke has the highest volume of foam, Pepsi has the lowest, and the other two brands fall in the middle.
3. There is no significant interaction between cola type and cola brand. In other words, we don't reject the null hypothesis that the difference in foam volume between diet and regular is the same for all four brands.

You can save and close the Cola ANOVA workbook now.

Exercises

1. Define the following terms:
   a. Error sum of squares
   b. Within-groups sum of squares
   c. Between-groups sum of squares
   d. Mean square error

2. Which value in the ANOVA table gives an estimate of σ^2, the variance of the error term in the means model?

3. If the between-groups mean square error = 7,000 and the within-groups mean square = 2,000, what is the value of the *F* ratio? If the degrees of freedom for the between-groups and within-groups are 4 and 14, respectively, what is the *p* value of the *F* ratio?

4. What is the Bonferroni correction factor and when should you use it?

5. Use Excel to calculate the *p* value for the following:
   a. $F = 2.5$; numerator $df = 10$; denominator $df = 20$
   b. $F = 3.0$; numerator $df = 10$; denominator $df = 20$
   c. $F = 3.5$; numerator $df = 10$; denominator $df = 20$
   d. $F = 4.0$; numerator $df = 10$; denominator $df = 20$
   e. $F = 4.5$; numerator $df = 10$; denominator $df = 20$

6. You're performing a two-way ANOVA on an education study to evaluate a new

teaching method. The two factors are region (East, Midwest, South, or West) and teaching method (standard or experimental). Schools are entered into the study, and their average test scores are recorded. There are five replicates for each combination of the region and method factors.

a. Using the information about the design of the study, complete the following ANOVA table:

Term	SS	df	MS	F
Region	9,305	?	?	?
Method	12,204	?	?	?
Interaction	6,023	?	?	?
Error	?	?	?	
Total SS	60,341	?		

b. What is the R^2 value of the ANOVA model?

c. Use Excel's FDIST function to calculate the p values for each of the factors and the interaction term in the model.

d. State your conclusions. What factors have a significant impact on the test scores? Is there an interaction between region and teaching method?

7. In analyzing the hotel data there appeared to be a problem of unequal population variances. Does it help to use the logarithm of price in place of price?

a. Open the **Hotel** workbook from the Chapter10 data folder and save it as **Hotel Log ANOVA**.

b. Compute a new variable LogPrice, the natural log of price.

c. Repeat the one-way ANOVA using LogPrice in place of Price (remember, you will have to unstack the data to use the Analysis ToolPak). Does there now appear to be a problem of unequal population variances?

d. Recalculate the matrix of paired differences (use the Bonferroni correction in calculating the p values).

e. Save your workbook and write a report summarizing your results. Do your conclusions differ in any important way from what was obtained for Price?

8. The Hotel Two-Way workbook is taken from the same source as the Hotel workbook, except that the data are balanced for a two-way ANOVA. This means that the random sample was forced to have the same number of hotels in each of 20 cells of city and stars (four levels of city and five levels of stars). For each of the 20 cells specified by a level of city and a level of stars, a random sample of two hotels was taken. Therefore, the sample has 40 hotels. Included in the file is a variable, city stars, which indicates the combination of city and stars. Perform the following analysis:

a. Open the **Hotel Two-Way** workbook from the Chapter10 folder and save it as **Hotel Two-Way? ANOVA**.

b. Using Excel's PivotTable feature, create an interaction plot of the average hotel price for the different combinations of city and stars. Is there evidence of an interaction apparent in the plot?

c. Do a two-way ANOVA for price versus stars and city. (You will have to create a two-way table that has stars as the row variable and city as the column variable.) Is there a significant interaction? Are the main effects significant?

d. On the basis of the means for the five levels of stars, give an approximate figure for the additional cost per star.

e. Compare the city effect in this model to the one-way analysis, which did not take into account the rating for each hotel.

f. As the number of stars increases, the mean price increases approximately linearly. Graph price versus stars. Break down the chart into categories on the basis of the city variable and then add trend lines to each of the four cities. Include the four regression equations on the chart. Do the slopes appear to be the same for the different cities?

   g. Save your changes to the workbook and write a report summarizing your observations.

9. Continue to explore the data from the Cola workbook discussed in this chapter by performing the following analysis:

   a. Open the **Cola** workbook from the Chapter10 folder and save it as **Cola Oneway ANOVA**.
   b. Create boxplots and multiple histograms of the foam variable for the different cola brands.
   c. Because the two-way ANOVA performed in the chapter showed that the interaction term and the type effect were not signficant, redo your analysis as a one-way ANOVA with cola as the single factor.
   d. Create a matrix of paired differences, using the Bonferroni correction. Which pairs of colas are different in terms of their foam volume?
   e. Save your changes to the workbook and write a report summarizing your results.

10. You've been given a workbook that contains information on 32 colleges from the 2008 edition of *U.S. News and World Report's* "America's Best Colleges," which lists 248 national liberal arts colleges in four tiers. The first two tiers are combined in a list of 125 colleges, and there are 61 in tier three and 62 in tier four. Splitting the 125 into 62 and 63, you have four tiers with 62, 63, 61, and 62 colleges, respectively. A random sample of eight was drawn from each of the four tiers, excluding nonprivate colleges. The data set includes Tier (from 1 to 4), College, Expenses (including tuition and fees but not room and board), and InState (the percentage of students who come from within the state). Perform the following analysis on the data set:

    a. Open the **Four Year** workbook from the Chapter10 folder and save it as **Four Year ANOVA**.
    b. Create a multiple histogram of the tuition for different tier levels. Are there apparent problems with the normality and constant variance assumptions?
    c. Perform a one-way ANOVA to compare expenses in the four tiers. Does the tier affect the cost of attending a private college?
    d. Create a matrix of paired mean differences. Does it cost significantly more to attend a college in a more prestigious tier?
    e. Notice that the means for expenses decrease roughly linearly as the tier number increases. Accordingly, regress Expenses on Tier. Interpret the tier regression coefficient in terms of the drop in expenses when you move to a higher tier number. Conversely, how much more does it cost to attend a college in a more prestigious tier (with a lower tier number)?
    f. Save your changes to the workbook and write a report summarizing your results, stating whether it is more expensive to attend a highly rated college and, if so, how the cost is related to the rating. Compare the regression and the ANOVA.

11. The Four Year workbook of Exercise 10 includes InState, the percentage of students coming from within the state. How does the InState variable depend on tier?

    a. Open the **Four Year** workbook from the Chapter10 data folder and save it as **Four Year Instate ANOVA**.
    b. Create a boxplot of InState broken down by tier. Notice the outlier in tier 4. Which college is it, and how

can you explain it? As a hint, consider that the college is in Vermont, which has a small population. Why is that relevant here?

c. Perform a one-way analysis of variance to compare tiers. Are there significant differences among tiers in the percentage of instate students?

d. Create a matrix of paired mean differences. Does the first tier have significantly fewer instate students in comparison to the other three tiers?

e. Redo the analysis in parts b and c but this time do not include the outlier. How does this affect the results?

f. Save your changes to the workbook and write a report summarizing your results.

12. The Infield workbook data set has statistics on 120 major league baseball infielders at the start of the 2007 season. The data include Salary, logSalary (the logarithm of salary), and Position.

a. Open the **Infield** workbook and save it as **Infield Salary ANOVA**.

b. Create multiple histograms and boxplots to see the distribution of Salary for each position. How would you describe the shape of the distribution?

c. Make the same plots for logSalary. How does the shape of the distribution change with the logarithm of the salary?

d. Perform a one-way ANOVA of logSalary on Position to see whether there is any significant difference of salary among positions.

e. Save your changes to the workbook and write a report summarizing your results.

13. The Infield workbook also contains the SLG variable, the slugging percentage of each infield player. Analyze the relationship between slugging percentage and position.

a. Open the **Infield** workbook from the Chapter10 data folder and save it as **Infield SLG ANOVA**.

b. Create multiple histograms and boxplots of the SLG variable against Position. Describe the shape of the distributions. Is there any reason to doubt the validity of the ANOVA assumptions?

c. Perform a one-way ANOVA of SLG against Position.

d. Create a matrix of paired mean differences to compare infield positions (use the Bonferroni correction factor). Which positions differ significantly? Can you explain why?

e. Save your changes to the workbook and write a report summarizing your results.

14. The Honda25 workbook contains the prices of used Hondas and indicates the age (in years) and whether the transmission is 5 speed or automatic.

a. Open the **Honda25** workbook from the Chapter10 data folder and save it as **Honda25 ANOVA**.

b. Perform a two-sample *t* test for the price data on the basis of the transmission type.

c. Perform a one-way ANOVA with price as the dependent variable and transmission as the grouping variable.

d. Compare the value of the *t* statistic in the *t* test to the value of the *F* ratio in the *F* test. Do you find that the *F* ratios for ANOVA are the same as the squares of the *t* values from the *t* test and that the *p* values are the same?

e. Use one-way ANOVA to compare the ages of the Hondas for the two types of transmissions. Does this explain why the difference in price is so large?

f. Perform two regressions of price vs. age—the first for automatic transmissions and the second for 5-speed transmissions. Compare the two linear regression lines. Do they appear to be the same? What problems do you see with this approach?

g. Save your changes to the workbook and write a report summarizing your results.

15. The Honda12 workbook contains a subset of the Honda workbook in which the age variable is made categorical and has the values 1–3, 4–5, and 6 or more. Some observations of the workbook have been removed to balance the data. The variable Trans indicates the transmission, and the variable Trans Age indicates the combination of transmission and age class.

    a. Open the **Honda12** workbook from the Chapter10 folder and save it as **Honda12 ANOVA**.
    b. Create a multiple histogram and boxplot of Price versus Trans Age. Does the constant variance assumption for a two-way analysis of variance appear justified?
    c. Create an interaction plot of Price versus Trans and Trans Age (you will need to create a pivot table of means for this). Does the plot give evidence for an interaction between the Trans and Trans Age factors?
    d. Perform a two-way analysis of variance of Price on Trans Age and Trans (you will have to create a two-way table using Trans Age as the row variable and Trans as the column variable).
    e. Save your changes to the workbook and write a report summarizing your observations.

16. At the Olympics, competitors in the 100-meter dash go through several rounds of races, called heats, before reaching the finals. The first round of heats involves over a hundred runners from countries all over the globe. The heats are evenly divided among the premier runners so that one particular heat does not have an overabundance of top racers. You decide to test this assumption by analyzing data from the 1996 Summer Olympics in Atlanta, Georgia.

    a. Open the **Race** workbook from the Chapter10 folder and save it as **Race Times ANOVA**.
    b. Create a boxplot of the race times broken down by heats. Note any large outliers in the plot and then rescale the plot to show times from 9 to 13 seconds. Is there any reason not to believe, based on the boxplot, that the variation of race times is consistent between heats?
    c. Perform a one-way ANOVA to test whether the mean race times among the 12 heats are significantly different.
    d. Create a pairwise means matrix of the race times by heat.
    e. Save your workbook and summarize your conclusions. Are the race times different between the heats? What is the significance level of the analysis of variance?

17. Repeat Exercise 16, this time looking at the reaction times among the 12 heats and deciding whether these reaction times vary. Write your conclusions and save your workbook as **Race Reaction ANOVA**.

18. Another question of interest to race observers is whether reaction times increase as the level of competition increases. Try to answer this question by analyzing the reaction times for the 14 athletes who competed in the first three rounds of heats of the men's 100-meter dash at the 1996 Summer Olympics.

    a. Open the **Race Rounds** workbook from the Chapter10 data folder and save it as **Race Rounds ANOVA**.
    b. Use the Analysis ToolPak's ANOVA: Two-Factor Without Replication command to perform a two-way analysis of variance on the data in the Reaction Times worksheet. What are the two factors in the ANOVA table?

c. Examine the ANOVA table. What factors are significant in the analysis of variance? What percentage of the total variance in reaction time can be explained by the two factors? What is the R^2 value?
d. Examine the means and standard deviations of the reaction times for each of the three heats. Using these values, form a hypothesis for how you think reaction times vary with rounds.
e. Test your hypothesis by performing a paired *t* test on the difference in reaction times between each pair of rounds (1 vs. 2, 2 vs. 3, and 1 vs. 3). Which pairs show significant differences at the 5% level? Does this confirm your hypothesis from the previous step?
f. Because there is no replication of a racer's reaction time within a round, you cannot add an interaction term to the analysis of variance. You can still create an interaction plot, however. Create an interaction plot with round on the *x*-axis and the reaction time for each racer as a separate line in the chart. On the appearance of the chart, do you believe that there is an interaction between round and the racer involved? What impact does this have on your overall conclusions as to whether reaction time varies with round?
g. Save your changes to the workbook and report your observations.

19. Researchers are examining the effect of exercise on heart rate. They've asked volunteers to exercise by going up and down a set of stairs. The experiment has two factors: step height and rate of stepping. The step heights are 5.75 inches (coded as 0) and 11.5 inches (coded as 1). The stepping rates are 14 steps/min (coded as 0), 21 steps/min (coded as 1), and 28 steps/min (coded as 2). The experimenters recorded both the resting heart rate (before the exercise) and the heart rate afterward. Analyze their findings.

a. Open the **Heart** workbook from the Chapter10 data folder and save it as **Heart ANOVA**.
b. Create a two-way table using StatPlus. Place frequency in the row area of the table, place height in the column area of the table, and use heart rate after the exercise as the response variable.
c. Analyze the values in the two-way table with a two-way ANOVA (with replication). Is there a significant interaction between the frequency at which subjects climb the stairs and the height of the stairs as it affects the subject's heart rate?
d. Create an interaction plot. Discuss why the interaction plot supports your findings from the previous step.
e. Create a new variable named Change, which is the change in heart rate due to the exercise. Repeat parts a–c for this new variable and answer the question of whether there is an interaction between frequency and height in affecting the change in heart rate.
f. Save your changes to the workbook and write a report summarizing your conclusions.

20. The Noise workbook contains data from a statement by Texaco, Inc. to the Air and Water Pollution Subcommittee of the Senate Public Works Committee on June 26, 1973. Mr. John McKinley, president of Texaco, cited an automobile filter developed by Associated Octel Company as effective in reducing pollution. However, questions had been raised about the effects of filters on vehicle performance, fuel consumption, exhaust gas backpressure, and silencing. On the last question, he referred to the data included here as evidence that the silencing properties of the Octel filter were at least equal to those of standard silencers.

a. Open the **Noise** workbook from the Chapter10 data folder and save it as **Noise ANOVA**.
b. Create boxplots and histograms of the Noise variable, broken down by the Size_Type variable (you should edit the labels in the boxplot to make the plot easier to read).
c. Create an interaction plot of the Noise variable for different levels of the Size and Type factors. Is there evidence of an interaction from the plot?
d. Create a two-way table of the Noise data for the Size and Type factors.
e. Using the two-way table, perform a two-way ANOVA on the data. What factors are significant?
f. Save your changes to the workbook and write a report summarizing your conclusions.

21. The Waste workbook contains data from a clothing manufacturer. The firm's quality control department collects weekly data on percent waste, relative to what can be achieved by computer layouts of patterns on cloth. A negative value indicates that the plant employees beat the computer in controlling waste. Your job is to determine whether there is a significant difference among the five plants in their percent waste values.

a. Open the **Waste** workbook from the Chapter10 folder and save it as **Waste ANOVA**.
b. Create boxplots of the waste value for the five plants. Are there any extreme outliers in the data that you should be concerned about?
c. Perform a one-way analysis of variance on the data.
d. Create a matrix of paired mean differences for the data. State your tentative conclusions.
e. Copy the waste data to another worksheet in the workbook, and delete any observations that were identified as extreme outliers on the boxplots.
f. Redo your one-way ANOVA and means matrix on the revised data. Have your conclusions changed?
g. Save your workbook and write a report summarizing your findings.

22. Cuckoos lay their eggs in the nests of other host birds. The host birds adopt and then later hatch the eggs. The Eggs workbook contains data on the lengths of eggs found in the nest of host birds. One theory holds that cuckoos lay their eggs in the nests of a particular host species and that they mate within a defined territory. If true, this would cause a geographical subspecies of cuckoos to develop and natural selection would ensure the survival of cuckoos most fitted to lay eggs that would be adopted by a particular host. If cuckoo eggs differed in length between hosts, this would lend some weight to that hypothesis. You've been asked to compare the length of the eggs placed in the different nests of the host birds.

a. Open the **Eggs** workbook from the Chapter10 folder and save it as **Eggs ANOVA**.
b. Perform a one-way ANOVA on the egg lengths for the six species. Is there evidence that the egg lengths differ between the species?
c. Create a boxplot of the egg lengths.
d. Analyze the pairwise differences between the species by creating a means matrix. Use the Bonferroni correction on the *p* values. What, if any, differences do you see between the species?
e. Save your changes to the workbook and write a report summarizing your observations.

Chapter 11

TIME SERIES

Objectives

In this chapter you will learn to:

- Plot a time series
- Compare a time series to lagged values of the series
- Use the autocorrelation function to determine the relationship between past and current values
- Use moving averages to smooth out variability
- Use simple exponential smoothing and two-parameter exponential smoothing
- Recognize seasonality and adjust data for seasonal effects
- Use three-parameter exponential smoothing to forecast future values of a time series
- Optimize the simple exponential smoothing constant

Time Series Concepts

A **time series** is a sequence of observations taken at evenly spaced time intervals. The sequence could be daily temperature measurements, weekly sales figures, monthly stock market prices, quarterly profits, or yearly power-consumption data. Time series analysis involves looking for patterns that help us understand what is happening with the data and help us predict future observations. For some time series data (for example, monthly sales figures), you can identify patterns that change with the seasons. This seasonal behavior is important in forecasting.

Usually the best way to start analyzing a time series is by plotting the data against time to show trends, seasonal patterns, and outliers. If the variability of the series changes with time, the series might benefit from a transformation that stabilizes the variance. Constant variance is assumed in much of time series analysis, just as in regression and analysis of variance, so it pays to see first whether a transformation is needed. The logarithmic transformation is one such example that is especially useful for economic data. For example, if there is growth in power consumption over the years, then the month-to-month variation might also increase proportionally. In this case, it might be useful to analyze either the log or the percentage change, which should have a variance that changes little over time.

Time Series Example: The Rise in Global Temperatures

To illustrate these ideas, you've been provided the Global Temperature workbook (Source: *http://data.giss.nasa.gov/gistemp/tabledata/GLB.Ts.txt*). The workbook contains average annual temperature readings compiled by NASA, covering the years 1880 through 1997. The NASA data are often used by climatologists investigating climate change and global warming. Table 11-1 describes the range names and data contained in the workbook.

Table 11-1 Global Temperature Workbook

Range Name	Range	Description
Year	A2:A129	The year
Decade	B2:B129	The decade
Celsius	C2:C129	The average annual global temperature in degrees Celsius
Fahrenheit	D2:D129	The average annual global temperature in degrees Fahrenheit

To open the Global Temperature workbook:

1. Open the **Global Temperature** workbook from the Chapter11 data folder.
2. Save the workbook as **Global Temperature Analysis**. The workbook appears as shown in Figure 11-1.

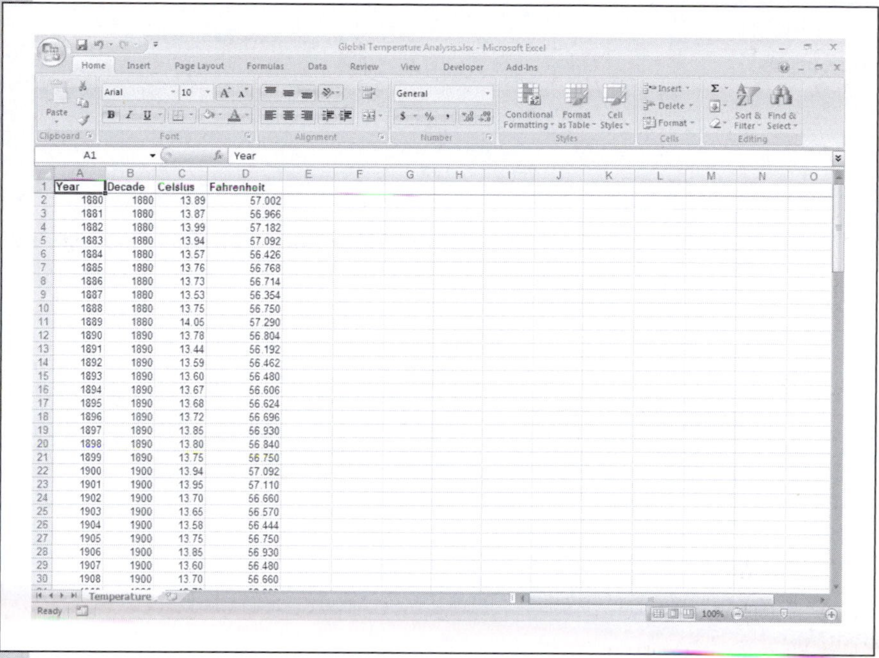

	A	B	C	D
1	Year	Decade	Celsius	Fahrenheit
2	1880	1880	13.89	57.002
3	1881	1880	13.87	56.966
4	1882	1880	13.99	57.182
5	1883	1880	13.94	57.092
6	1884	1880	13.57	56.426
7	1885	1880	13.76	56.768
8	1886	1880	13.73	56.714
9	1887	1880	13.53	56.354
10	1888	1880	13.75	56.750
11	1889	1880	14.05	57.290
12	1890	1890	13.78	56.804
13	1891	1890	13.44	56.192
14	1892	1890	13.59	56.462
15	1893	1890	13.60	56.480
16	1894	1890	13.67	56.606
17	1895	1890	13.68	56.624
18	1896	1890	13.72	56.696
19	1897	1890	13.85	56.930
20	1898	1890	13.80	56.840
21	1899	1890	13.75	56.750
22	1900	1900	13.94	57.092
23	1901	1900	13.95	57.110
24	1902	1900	13.70	56.660
25	1903	1900	13.65	56.570
26	1904	1900	13.58	56.444
27	1905	1900	13.75	56.750
28	1906	1900	13.85	56.930
29	1907	1900	13.60	56.480
30	1908	1900	13.70	56.660

Figure 11-1
The Global Temperature workbook

Plotting the Global Temperature Time Series

Before doing any computations, it is best to explore the time series graphically. You'll plot the average annual temperatures in degrees Fahrenheit.

To plot the annual average temperature readings:

1. Select the nonadjacent range **A1:A129;D1:D129**.
2. Click the **Scatter** button from the Charts group on the Insert tab.
3. Click the third scatter chart subtype (Scatter with Smooth lines).
4. Scroll up to the top of the window, and with the chart still selected, click the **Move Chart** button from the Location group on the Design tab of the Chart Tools ribbon. Move the chart to a new chart sheet named **Temperature Chart**.

5 Remove the legend and gridlines from the chart.

6 Change the chart title to **Average Global Temperature**. Change the *x* axis title to **Year** and the *y* axis title to **Temperature (F)**. Figure 11-2 shows the edited chart.

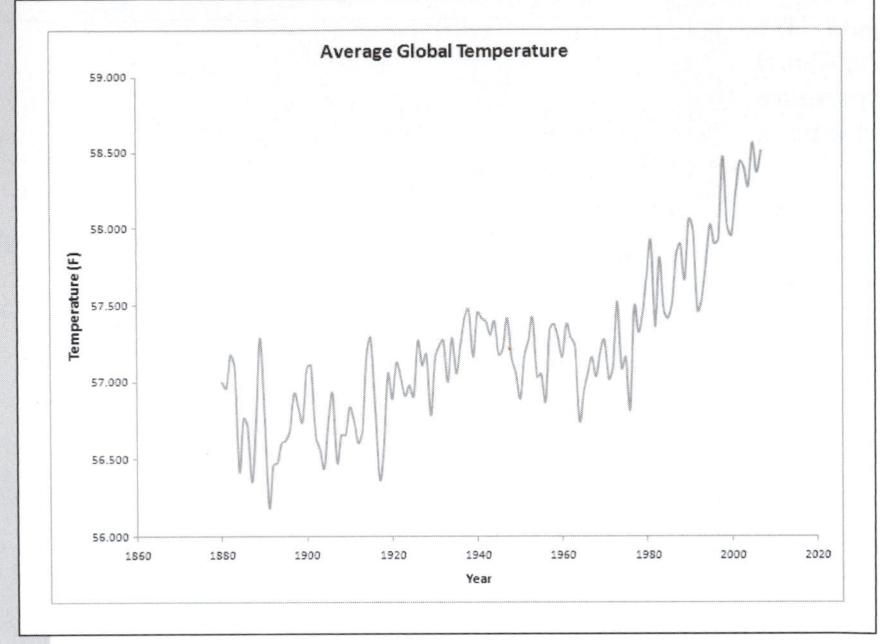

Figure 11-2 Time plot of annual global temperatures

The plot in Figure 11-2 shows a noticeable increase in mean annual temperature over the second half of the twentieth century. While the trend is striking, there is still a great deal of variability in the average annual temperature from year to year. We can smooth out this variability by plotting the averages per decade.

To calculate the average global temperature per decade:

1 Click the **Temperature** sheet tab.

2 Click **Descriptive Statistics** from the StatPlus menu and then click **Univariate Statistics.**

3 Click the **Summary** tab and select the **Count** and **Average** checkboxes.

4 Click the **Variability** tab and select the **Range** and **Std. Deviation** checkboxes.

5 Click the **General** tab and click the **Columns** option button to display the statistics by columns rather than by rows.

6 Click the **Input** button and select **Fahrenheit** from the list of range names. Click **OK**.

Now you'll break down the statistics by decade.

7 Click the **By** button and select **Decade** from the range names list. Click **OK**.

8 Click the **Output** button and direct the output to a new worksheet named **Temps by Decade**.

9 Click **OK**.

Excel generates the table shown in Figure 11-3.

Figure 11-3 Temperature statistics by decade

	A	B	C	D	E	F	G
1			Univariate Statistics				
2			Count	Average	Range	Standard Deviation	
3	Fahrenheit	Decade = 1880	10	56.85440	0.936	0.309079	
4		Decade = 1890	10	56.63840	0.738	0.217462	
5		Decade = 1900	10	56.73560	0.666	0.235732	
6		Decade = 1910	10	56.82020	0.918	0.282121	
7		Decade = 1920	10	57.02180	0.486	0.150945	
8		Decade = 1930	10	57.22880	0.468	0.145789	
9		Decade = 1940	10	57.30080	0.396	0.136610	
10		Decade = 1950	10	57.17300	0.540	0.197955	
11		Decade = 1960	10	57.12080	0.630	0.183800	
12		Decade = 1970	10	57.22700	0.702	0.226455	
13		Decade = 1980	10	57.66080	0.558	0.206316	
14		Decade = 1990	10	57.92180	1.008	0.284282	
15		Decade = 2000	8	58.34525	0.612	0.193373	
16		Overall	128	57.21716	2.376	0.500372	
17							

Now that you've calculated the per-decade averages create a scatter plot of the values.

To create a scatter plot of the decade averages:

1 Select the nonadjacent cell range **B3:B15;D3:D15** from the table of decade statistics.

2 Click the **Line** button from the Charts group on the Insert tab and click the first chart subtype (Line).

3 Move the line chart to a new chart sheet named **Decades Chart**.

4 Remove the legend from the chart.

5 Add the chart title **Average Global Temperature**. Set the *x* axis title to **Decade** and the *y* axis title to **Temperature (F)**.

6 Change the number format of the values on the *y* axis to one decimal place. Figure 11-4 shows the formatted scatter chart.

Figure 11-4 Average temperature by decade

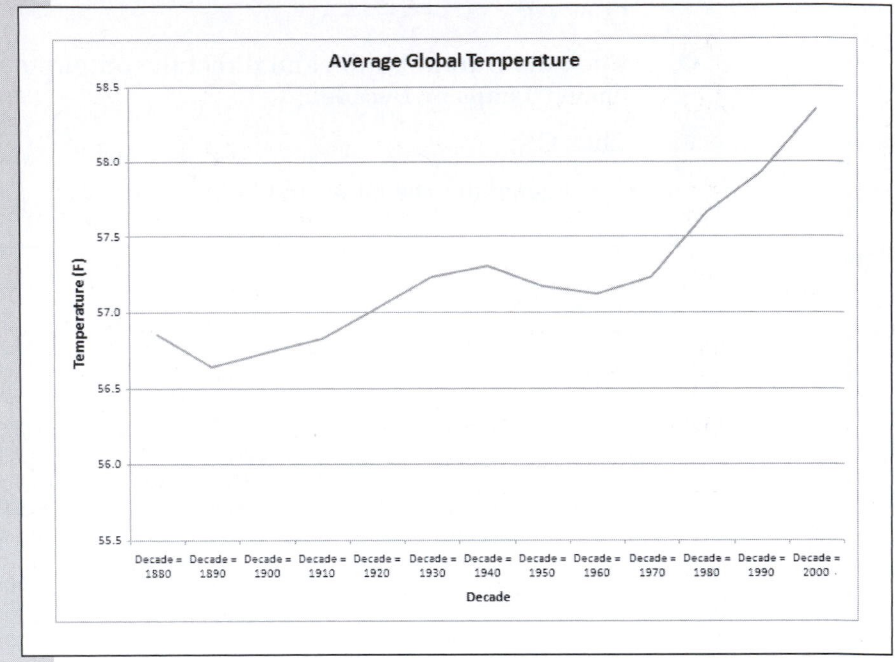

The chart clearly shows a minor dip in the decade average from 1940 to 1960 after which there is a steady increase in global temperature up through the decade of the 2000s.

Analyzing the Change in Global Temperature

The changes in the temperature average from year to year are important. Your next step in examining these values is to analyze annual average temperature change.

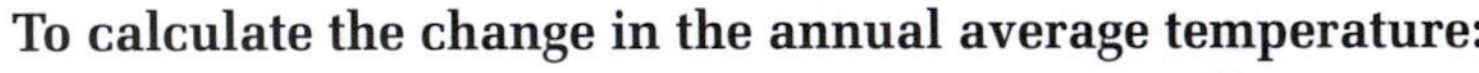

To calculate the change in the annual average temperature:

1 Click the **Temperature** sheet tab to return to the data.

2 Click cell **E1**, type **Change**, and then press **Enter**.

3 Select the range **E3:E129** (*not* E2:E129).

4 Type **=D3-D2** in cell **E3**; then press **Enter**.

5 Press the **Fill** button on the Editing group of the Home tab and then click **Down**. Excel fills the difference formula down the remaining cells in the column, displaying the change in mean annual temperature from one year to the next.

Now that you have calculated the differences in the mean annual temperature from one year to the next, you can plot those differences.

To plot the change in the temperature versus time:

1 Select the range **A1:A129**, press and hold the **CTRL** key, and then select the range **E1:E129**.

2 Click the **Scatter** button from the Charts group on the Insert tab and click the first chart subtype (Scatter with only markers).

3 Move the chart to a new chart sheet named **Yearly Change**.

4 Remove the legend and gridlines from the plot.

5 Enter the chart title **Yearly Temperature Change**. Set the title of the *x* axis to **Year** and the title of the *y* axis to **Change in Temperature (F)**. Figure 11-5 shows the formatted scatter chart.

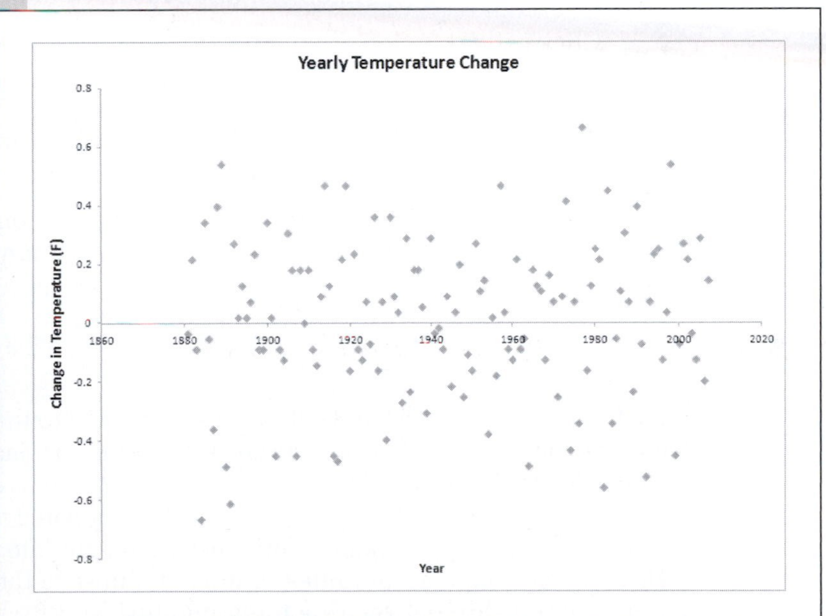

Figure 11-5 Yearly differences in mean global temperature

There does not appear to be any trend in the change in mean annual temperature over the 128 years represented in the data set. The changes in temperature values appear as scattered in recent years as they do in years from the beginning from the chart.

Looking at Lagged Values

Often in time series you will want to compare a value observed at one time point to the value observed one or more time points earlier. In the temperature data, for example, you might be interested in whether the mean temperature from one year can be used to predict the mean temperature of the following year. Such prior values are known as **lagged values**. Lagged values are an important concept in time series analysis. You can lag observations for one or more time points. In the example of the global temperature data, the lag 1 value is the temperature value one year prior, the lag 2 value is the temperature value two years prior, and so forth.

You can calculate lagged values by letting the values in rows of the lagged column be equal to values one or more rows above in the unlagged column. Let's add a new column to the Temperature worksheet, consisting of annual temperature averages lagged one year.

To create a column of lag 1 values for the global temperature data:

1. Click the **Temperature** sheet tab to return to the data.
2. Right-click the **D** column header so that the entire column is selected and the pop-up menu opens. Click **Insert** in the pop-up menu.
3. Click cell **D1**, type **Lag1 Temps (F)**, and press **Enter**.
4. Select the range **D3:D129** (not D2:D129).
5. Type **=E2** in cell D3 (this is the value from the previous year); then press **Enter**.
6. Click the **Fill** button from the Editing group on the Home tab and click **Down**. Excel fills in the rest of the column with the one-year lagged values.

Each row of the lagged temperature values is equal to the temperature value of the previous year. You could have created a column of lag 2 values by selecting the range D4:D129 and letting D4 be equal to E2, and so on. Note that for the lag 2 values you have to start two rows down, as compared to one row down for the lag 1 values. The lag 3 values would have been put into the range D5:D129.

How do the temperature values compare to those of the previous year? To see the relationship between each temperature and its one-year lag value, create a scatterplot.

To create a scatterplot of temperature versus one-year lagged temperatures:

1 Select the range **D1:E129**.

2 Click the **Scatter** button from Charts group on the Insert tab and then select the first chart subtype (Scatter).

3 Move the chart to the **Lag 1 Chart** chart sheet.

4 Remove the gridlines and legends from the plot. Name the chart title **Lagged Temperatures**, the *x* axis title **Prior Year Temperature (F)**, and the *y*-axis title **Temperature (F)**. See Figure 11-6.

Figure 11-6 Temperature and Lag1 temperature values

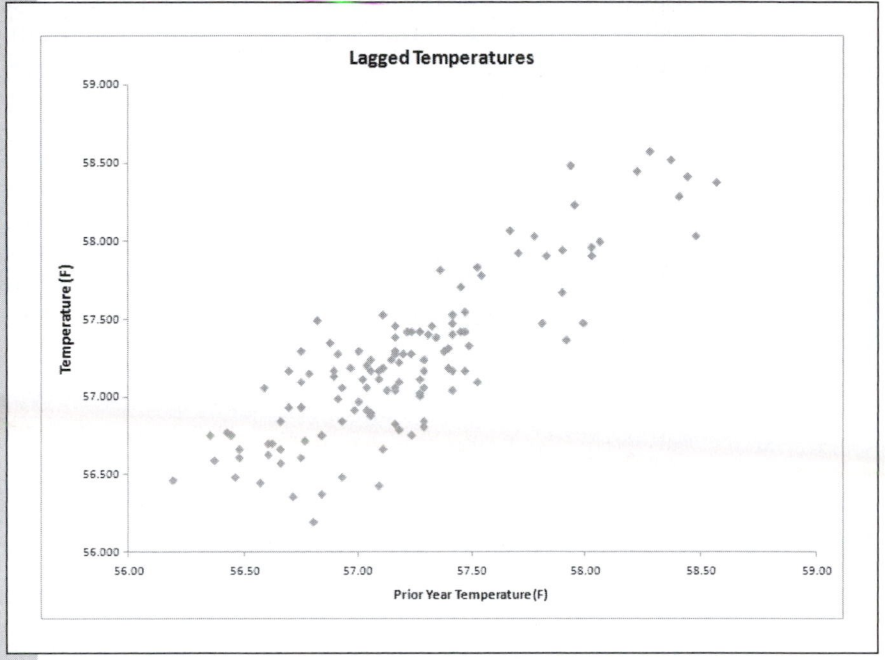

As shown in the chart, there is a strong positive relationship between temperature value in one year and the temperature value from the previous year. This means that a high temperature value in one year implies a high (or above average) value in the following year; a low value in one year indicates a low (or below average) value in the next year. In time series analysis, we study the correlations among observations, and these relationships are sometimes helpful in predicting future observations. In this example, the annual temperature value appears to be strongly correlated with the temperature value from the previous year. Temperatures might also be correlated with observations two, three, or more years earlier. To discover the relationship between a time series and other lagged values of the series, statisticians calculate the autocorrelation function.

The Autocorrelation Function

If there is some pattern in how the values of your time series change from observation to observation, you could use it to your advantage. Perhaps a below-average value in one year makes it more likely that the series will be high in the next year, or maybe the opposite is true—a year in which the series is low makes it more likely that the series will continue to stay low for a while.

The **autocorrelation function (ACF)** is useful in finding such patterns. It is similar to a correlation of a data series with its lagged values. The ACF value for lag 1 (denoted by r_1) calculates the relationship between the data series and its lagged values. The formula for r_1 is

$$r_1 = \frac{(y_2 - \bar{y})(y_1 - \bar{y}) + (y_3 - \bar{y})(y_2 - \bar{y}) + \cdots + (y_n - \bar{y})(y_{n-1} - \bar{y})}{(y_1 - \bar{y})^2 + (y_2 - \bar{y})^2 + \cdots + (y_n - \bar{y})^2}$$

Here, y_1 represents the first observation, y_2 the second observation, and so forth. Finally, y_n represents the last observation in the data set. Similarly, the formula for r_2, the ACF value for lag 2, is

$$r_2 = \frac{(y_3 - \bar{y})(y_1 - \bar{y}) + (y_4 - \bar{y})(y_2 - \bar{y}) + \cdots + (y_n - \bar{y})(y_{n-2} - \bar{y})}{(y_1 - \bar{y})^2 + (y_2 - \bar{y})^2 + \cdots + (y_n - \bar{y})^2}$$

The general formula for calculating the autocorrelation for lag k is

$$r_k = \frac{(y_{k+1} - \bar{y})(y_1 - \bar{y}) + (y_{k+2} - \bar{y})(y_2 - \bar{y}) + \cdots + (y_n - \bar{y})(y_{n-k} - \bar{y})}{(y_1 - \bar{y})^2 + (y_2 - \bar{y})^2 + \cdots + (y_n - \bar{y})^2}$$

Before considering the autocorrelation of the temperature data, let's apply these formulas to a smaller data set, as shown in Table 11-2.

Table 11-2 Sample Autocorrelation Data

Observation	Values	Lag 1 Values	Lag 2 Values
1	6		
2	4	6	
3	8	4	6
4	5	8	4
5	0	5	8
6	7	0	5

The average of the values is 5, y_1 is 6, y_2 is 4, y_3 is 8, and so forth, through y_n, which is equal to 7. To find the lag 1 autocorrelation, use the formula for r_1 so that

$$r_1 = \frac{(4-5)(6-5) + (8-5)(4-5) + \cdots + (7-5)(0-5)}{(6-5)^2 + (4-5)^2 + \cdots + (7-5)^2}$$

$$= \frac{-14}{40} = -0.35$$

In the same way, the value for r_2, the lag 2 ACF value, is

$$r_2 = \frac{(8-5)(6-5) + (5-5)(4-5) + \cdots + (7-5)(5-5)}{(6-5)^2 + (4-5)^2 + \cdots + (7-5)^2}$$

$$= \frac{-12}{40} = -0.30$$

The values for r_1 and r_2 imply a negative correlation between the current observation and its lag 1 and lag 2 values (that is, the previous two values). So a low value at one time point indicates high values for the next two time points. Now that you've seen how to compute r_1 and r_2, you should be able to compute r_3, the lag 3 autocorrelation. Your answer should be 0.275, a positive correlation, indicating that values of this series are positively correlated with observations three time points earlier.

Recall from earlier chapters that a constant variance is needed for statistical inference in simple regression and also for correlation. The same holds true for the autocorrelation function. The ACF can be misleading for a series with unstable variance, so it might first be necessary to transform for a constant variance before using the ACF.

Applying the ACF to Annual Mean Temperature

Now apply the ACF to the temperature data. You can use StatPlus to compute and plot the autocorrelation values for you.

To compute the autocorrelation function for the annual mean temperatures:

1. Click the **Temperature** sheet tab.
2. Click **Time Series** from the StatPlus menu and then click **ACF Plot**.
3. Click the **Data Values** button and select **Fahrenheit** from the range names list. Click **OK**.
4. Enter **20** in the Calculate ACF up through lag spin box to calculate the autocorrelations between the mean annual temperature values and mean temperatures up to 20 years earlier.

5 Click the **Output** button, and send the output to a new sheet named **Temperature ACF**. Click **OK** twice to close the dialog box and calculate the ACF function. Figure 11-7 shows the output from the ACF command.

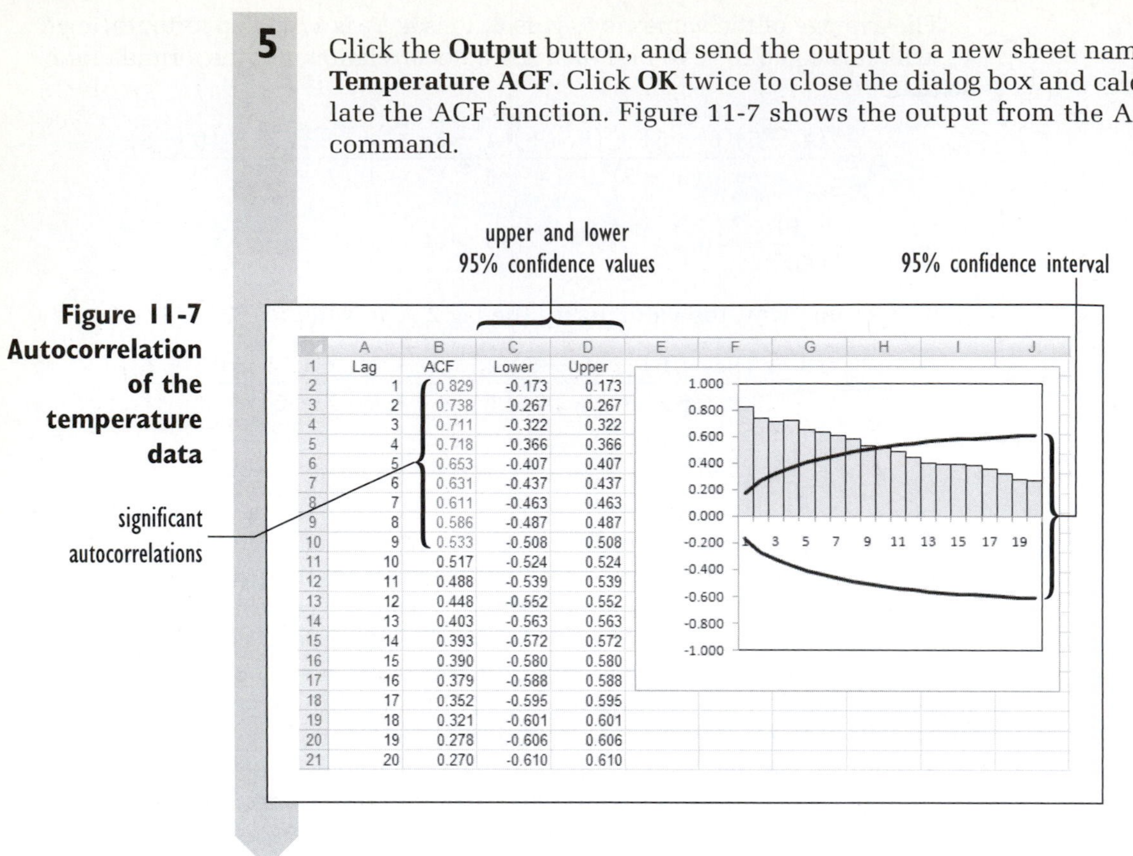

	A	B	C	D
1	Lag	ACF	Lower	Upper
2	1	0.829	-0.173	0.173
3	2	0.738	-0.267	0.267
4	3	0.711	-0.322	0.322
5	4	0.718	-0.366	0.366
6	5	0.653	-0.407	0.407
7	6	0.631	-0.437	0.437
8	7	0.611	-0.463	0.463
9	8	0.586	-0.487	0.487
10	9	0.533	-0.508	0.508
11	10	0.517	-0.524	0.524
12	11	0.488	-0.539	0.539
13	12	0.448	-0.552	0.552
14	13	0.403	-0.563	0.563
15	14	0.393	-0.572	0.572
16	15	0.390	-0.580	0.580
17	16	0.379	-0.588	0.588
18	17	0.352	-0.595	0.595
19	18	0.321	-0.601	0.601
20	19	0.278	-0.606	0.606
21	20	0.270	-0.610	0.610

Figure 11-7 Autocorrelation of the temperature data

The output shown in Figure 11-7 lists the lags from 1 to 20 and gives the corresponding autocorrelations in the next column.

The lower and upper ranges of the autocorrelations are shown in the next two columns and indicate how low or high the correlation needs to be for statistical significance at the 5% level. Autocorrelations that lie outside this range are shown in red in the worksheet. The plot of the ACF values and confidence widths gives a visual picture of the patterns in the data. The two curves indicate the width of the 95% confidence interval of the autocorrelations.

The autocorrelations are very high for the lower lag numbers, and they remain significant (that is, they lie outside the 95% confidence width boundaries) through lag 9. Specifically, the correlation between the mean annual temperature and the mean annual temperature of the previous year is 0.829 (cell B2). The correlation between the current temperature and the lag 2 value is 0.738 (cell B3), and so forth. This is typical for a series that has a strong trend upward or downward. Given the increase in the global temperatures during the latter half of the twentieth century, it shouldn't be surprising that high temperatures are correlated with the high temperatures of the previous year. In such a series, if an observation is above the mean, then its neighboring observations are also likely to be above the mean, and the autocorrelations with nearby observations are high. In fact, when there is a trend, the autocorrelations tend to remain high even for high lag numbers.

STATPLUS TIPS

- You can use StatPlus's ACF(*range*, *lag*) function to compute autocorrelations for specific lag values. Here, *range* is the range of cells containing the time series data, and *lag* is the number of observations to lag. Note that values must be placed within a single column.

Other ACF Patterns

Other time series show different types of autocorrelation patterns. Figure 11-8 shows four examples of time series (trend, cyclical, oscillating, and random), along with their associated autocorrelation functions.

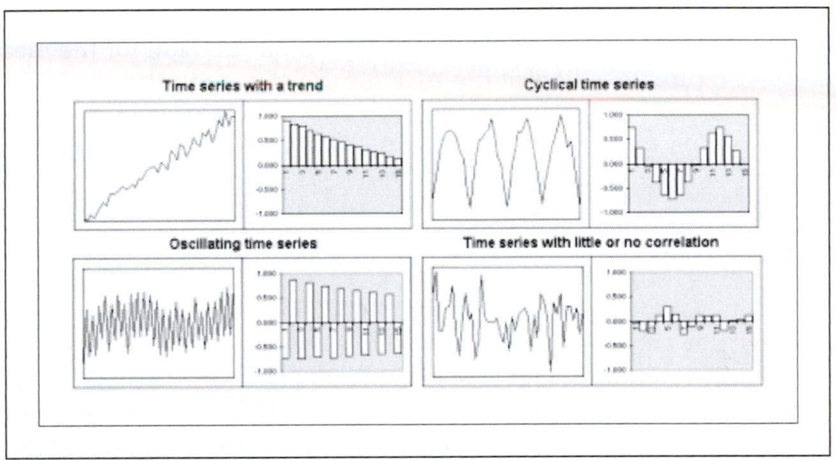

Figure 11-8 Four sample time series with corresponding ACF patterns

You have already seen the first example with the temperature data. The trend need not be increasing; a decreasing trend also produces the type of ACF pattern shown in the first example in Figure 11-8.

The seasonal or cyclical pattern shown in the second example is common in weather data that follows a seasonal pattern (such as monthly average temperature). The length of the cycle in this example is 12, indicated in the ACF by the large positive autocorrelation for lag 12. Because the data in the time series follow a cycle of length 12, you would expect that values 12 units apart would be highly correlated with each other. Seasonal time series models are covered more thoroughly later in this chapter.

The third example shows an oscillating time series. In this case, a large value is followed by a low value and then by another large value. An example of this might be winter and summer sales over the years for a large retail toy company. Winter sales might always be above average because of the holiday season, whereas summer sales might always be below average. This pattern of oscillating sales might continue and could follow the pattern shown in Figure 11-8. The ACF for this time series has an alternating pattern of positive and negative autocorrelations.

Finally, if the observations in the time series are independent or nearly independent, there is no discernible pattern in the ACF and all the autocorrelations should be small, as shown in the fourth example. This is characteristic of a **random walk model** in which current values are independent of previous values and thus you cannot use current values to predict future ones.

There are many other possible patterns of behavior for time series data besides the four examples shown here.

Applying the ACF to the Change in Average Global Temperature

Having looked at the autocorrelation function for the mean annual temperature, let's look at the ACF for the change in the average global temperature. Does an increase in temperature in one year imply that the next year will also show an increase? Or is the opposite more likely, where years that show a large increase in temperature are followed by years in which the temperature increase is smaller or is even a decrease? Let's find out.

To calculate the autocorrelation for the change in annual temperature:

1. Click the **Temperature** sheet tab.
2. Click **Time Series** from the StatPlus menu and then click **ACF Plot**.
3. Click the **Data Values** button, click the **Use Range References** option button, and select the range **F3:F129**.
4. Deselect the **Range includes a row of column labels** checkbox and click **OK**.

   You want to *deselect* this checkbox because this selection does not include a header row.

5 Enter **20** in the Calculate ACF up through the lag spin box.

6 Click the **Output** button, and send the output to a new sheet named **Change ACF**. Click **OK** twice.

Figure 11-9 shows the output from the command.

Figure 11-9 ACF for the change in the annual temperature

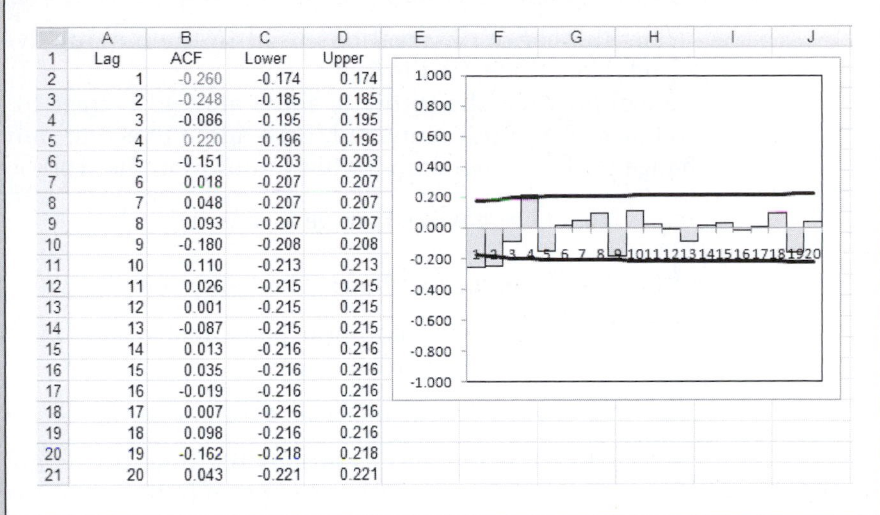

	A	B	C	D
1	Lag	ACF	Lower	Upper
2	1	-0.260	-0.174	0.174
3	2	-0.248	-0.185	0.185
4	3	-0.086	-0.195	0.195
5	4	0.220	-0.196	0.196
6	5	-0.151	-0.203	0.203
7	6	0.018	-0.207	0.207
8	7	0.048	-0.207	0.207
9	8	0.093	-0.207	0.207
10	9	-0.180	-0.208	0.208
11	10	0.110	-0.213	0.213
12	11	0.026	-0.215	0.215
13	12	0.001	-0.215	0.215
14	13	-0.087	-0.215	0.215
15	14	0.013	-0.216	0.216
16	15	0.035	-0.216	0.216
17	16	-0.019	-0.216	0.216
18	17	0.007	-0.216	0.216
19	18	0.098	-0.216	0.216
20	19	-0.162	-0.218	0.218
21	20	0.043	-0.221	0.221

The autocorrelations for change in average temperature are not as strong as you saw earlier using the yearly temperature values. However note that the lag 1 and lag 2 correlations are both statistically significant and negative. This indicates a negative correlation between the current change in temperature and changes from one or two years prior. Apparently an increase in temperature in one year is associated with a smaller increase or even a decrease in the next two years.

Moving Averages

As you saw earlier in Figure 11-5, the change in average temperature can vary unpredictably from one year to another. One way of smoothing out this fluctuation is to take the average change over an entire decade as you did for the yearly temperature values. Another way of smoothing your data is to calculate a moving average.

For example, if you calculate the average change for each of the last five years and you do this every year, you are forming a **moving average** of those values as you move forward in time. Specifically, to calculate the five-year moving average for values prior to the observation y_n, you define the moving average $y_{\text{ma}(5)}$ such that

$$y_{\text{ma}(5)} = \frac{y_{n-1} + y_{n-2} + y_{n-3} + y_{n-4} + y_{n-5}}{5}$$

The number of observations used in the moving average is called the **period**. Here the period is 5.

Excel provides the ability to add a moving average to a scatterplot using the Insert Trendline command. Let's add a five-year moving average to the change in the temperature for the values from the workbook.

To add a moving average to a chart:

1. Click the **Yearly Change** chart sheet tab.
2. Right-click the data series (any data value) on the chart to select it and open the shortcut menu.
3. Click **Add Trendline** in the shortcut menu.
4. Click the **Trendline Options** list item if it is not already selected, click the **Moving Average** option button, and then click the **Period** up spin arrow until **5** appears as the period. Your dialog box should look like Figure 11-10.

Figure 11-10
The Format Trendline dialog box

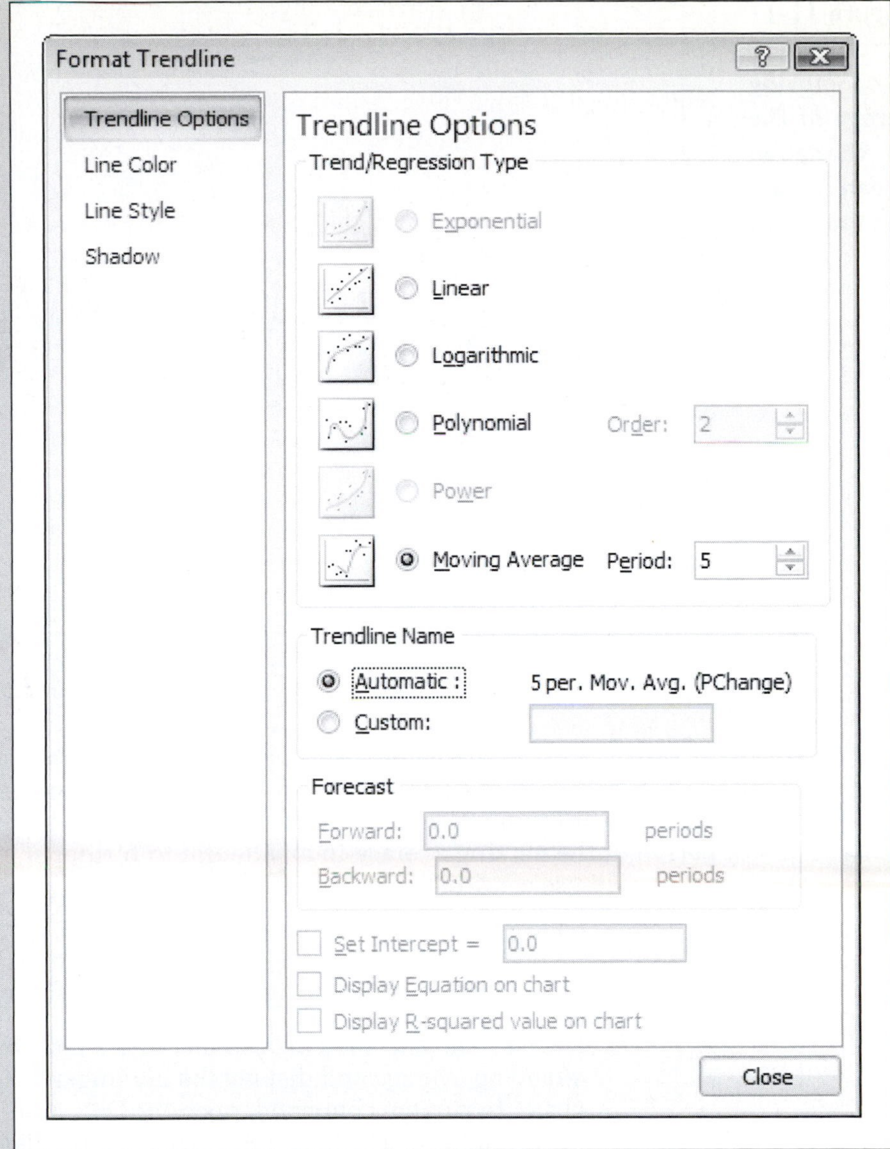

5 Click the **Close** button and then click outside the chart to deselect the data series. The moving-average curve appears as in Figure 11-11.

Figure 11-11 Five year moving average of the change in mean annual temperature

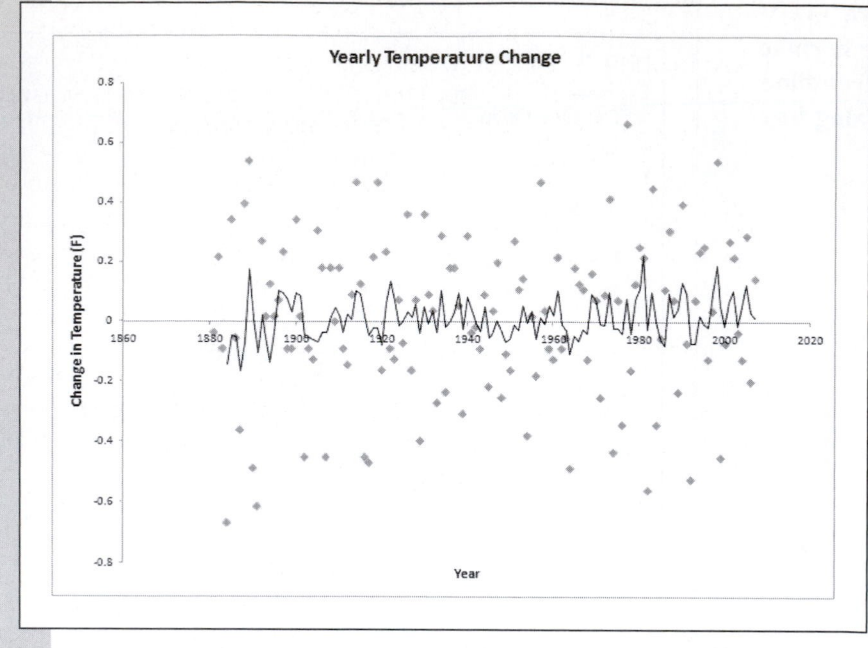

Taking the five-year moving average smoothes out the data a bit; however, even the smoothed values show so much fluctuation that it is difficult to spot a clear trend (if one exists). We can edit the trendline to increase the period of the moving average in an attempt to further smooth the data, but we should use caution because in smoothing the data some crucial information could be lost.

EXCEL TIPS

- Excel's Analysis ToolPak also includes a command to calculate a moving average and display the moving-average values in a chart. To run the command, open the Data Analysis ToolPak dialog box and select Moving Average from the list of analysis tools.

Simple Exponential Smoothing

The moving average gives equal weight to all previous values in the moving average period. Thus with a five-year period, a value recorded five years ago is given as much weight as the value from the previous year. Some feel that an approach that gives equal weight to all observations within the period is not always reasonable. For example, if the belief that increased industrialization is accelerating the effects of human-made global warming and we want to predict future temperature values, we may want to give greater weight to the most recent observations and lesser weight to observations further in the past.

Many analysts advocate a moving average that gives greater weight to more recent values and one in which the value of the weights drops off exponentially. This kind of moving average is not limited to a set period of values but gives some weight to all observations in the data set. The most recent observation gets weight w, where the value of w ranges from 0 to 1. The next most recent observation gets weight $w(1 - w)$, the one before that gets weight $w(1 - w)^2$, and so on. In general, the weight assigned to an observation k units prior to the current observation is equal to $w(1 - w)^{k-1}$. The exponentially weighted moving average is therefore

$$\text{Exponentially weighted average} = wy_{n-1} + w(1 - w)y_{n-2} + w(1 - w)^2 y_{n-3} + \cdots$$

Here w is called a **smoothing factor** or **smoothing constant**. This technique is called **exponential smoothing** or, specifically, **one-parameter exponential smoothing**. Table 11-3 gives the weights for prior observations under different values of w.

Table 11-3 Exponential Weights

y_{n-1}	y_{n-2}	y_{n-3}	y_{n-4}	y_{n-5}	y_{n-6}
w	$w(1-w)$	$w(1-w)^2$	$w(1-w)^3$	$w(1-w)^4$	$w(1-w)^5$
0.01	0.0099	0.0098	0.0097	0.0096	0.0095
0.15	0.1275	0.1084	0.0921	0.0783	0.0666
0.45	0.2475	0.1361	0.0749	0.0412	0.0226
0.75	0.1875	0.0469	0.0117	0.0029	0.0007

As the table indicates, different values of w cause the weights assigned to previous observations to change. For example, when w equals 0.01, approximately equal weight is given to a value from the most recent observation and to values observed six units earlier. However, when w has the value of 0.75, the weight assigned to previous observations quickly drops, so that values collected six units prior to the current time receive essentially no weight. In a sense, you could say that as the value of w approaches zero, the smoothed average has a longer memory, whereas as w approaches 1, the memory of prior values becomes shorter and shorter.

Forecasting with Exponential Smoothing

Exponential smoothing is often used to forecast the value of the next observation, given the current and prior values. In this situation, you already know the value of y_n and are trying to forecast the next value y_{n+1}. Call the forecast S_n. The formula for S_n is similar to the one we derived for the exponentially weighted moving average; it is

$$S_n = wy_n + w(1 - w)y_{n-1} + w(1 - w)^2 y_{n-2} + \cdots + w(1 - w)^{n-1} y_1 + (1 - w)^n s_0$$

S_n is more commonly written in an equivalent recursive formula, where

$$S_n = wy_n + (1 - w)s_{n-1}$$

so that S_n is equal to the sum of the weighted values of the current observation and the previous forecast. Therefore, to create the forecasted value, an initial forecasted value S_0 is required. One option is to let S_0 equal y_1, the initial observation. Another choice is to let S_0 equal the average of the first few values in the series. The examples in this chapter will use the first option, setting S_0 equal to the first value in the time series.

Once you determine the value of S_0, you can generate the exponentially smoothed values as follows:

$$S_1 = wy_1 + (1 - w)S_0$$

$$S_2 = wy_2 + (1 - w)S_1$$

$$\vdots$$

$$S_n = wy_n + (1 - w)S_{n-1}$$

and then S_n becomes the value you predict for the next observation in the time series.

Assessing the Accuracy of the Forecast

Once you generate the smoothed values, how do you measure their accuracy in forecasting values of the time series? One way is to use exponential smoothing to calculate $\hat{y}_t$, the predicted value of the time series at time t. Then, for each value in the time series, compare $\hat{y}_t$ to the observed value, y_t. The **mean square error (MSE)**, gives the sum of the squared differences between the forecasted values and the observed values. The formula for the MSE is

$$\text{MSE} = \frac{\sum_{t=1}^{n} (y_t - \hat{y}_t)^2}{n}$$

By comparing the MSE of one set of smoothed values to another, one can determine which set does a better job of forecasting the data.

The square root of the MSE gives us the **standard error**, which indicates the magnitude of the typical forecasted error. A standard error of 5 would indicate that the forecasts are typically off by about 5 points.

Another way of measuring the magnitude is to take the sum of the absolute values of the differences between the forecasted and observed values. This measure, called the **mean absolute deviation (MAD)**, has the formula

$$\text{MAD} = \frac{\sum_{t=1}^{n} \left| y_t - \hat{y}_t \right|}{n}$$

One of the differences between the MAD and the MSE is that the MAD does not penalize a forecast as much for very large errors. Because the MSE squares the deviations, large errors become even more prominent.

Another measure is the **mean absolute percent error (MAPE)**, which expresses the accuracy as a percentage of the observed value. The formula for the MAPE is

$$\text{MAPE} = \frac{\sum_{t=1}^{n} \left| (y_t - \hat{y}_t)/y_t \right|}{n} \times 100$$

To help you get a visual image of the impact that differing values of w have on smoothing the data and forecasting the next value in the series, you can open the Exponential Smoothing workbook.

CONCEPT TUTORIALS
One-Parameter Exponential Smoothing

To use the Exponential Smoothing workbook:

1. Open the **Exponential Smoothing** workbook from the Explore folder. Enable the macros in the workbook.
2. Review the contents of the workbook up to the section entitled Explore One-Parameter Exponential Smoothing. The worksheet is shown in Figure 11-12.

Figure 11-12 Exploring one-parameter exponential smoothing

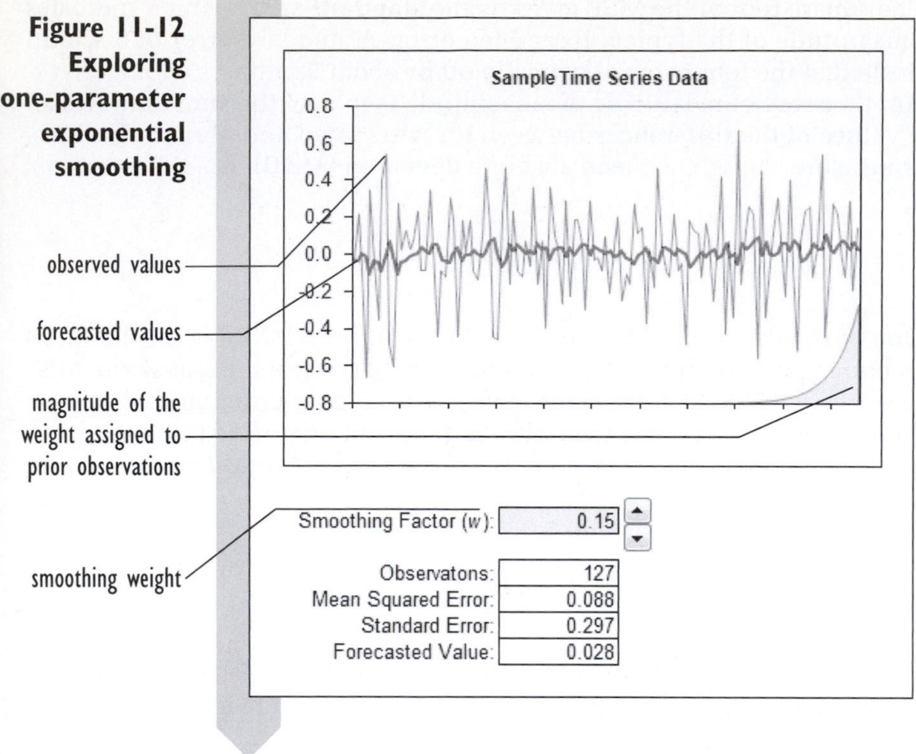

This worksheet shows the observed percentage changes in a sample time series overlaid with the one-parameter exponentially smoothed values. The smoothing factor *w* is set at 0.15. In the lower-right corner, the worksheet contains an area curve indicating the magnitude of the weights assigned to the observations prior to the last value in the series.

The final forecasted value is 0.028. The most recent observation has the most weight in calculating this result, with observations decreasing exponentially in importance. Comparing the curve to the time series tells you that the large drop in the middle of the time series has little weight in estimating the final value. In fact, observations prior to that value have negligible impact.

The mean square error is 0.088 and the standard error is 0.297, showing that if you had used exponential smoothing on this data your typical error in forecasting would have been about 0.297 points.

One way of choosing a value for the smoothing constant is to pick the value that results in the lowest mean square error. Let's see what happens to the mean square error when you decrease the value of the smoothing constant.

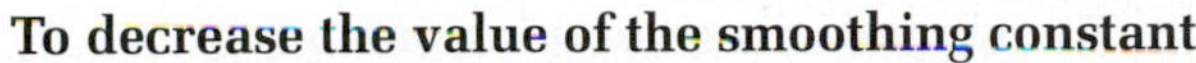

To decrease the value of the smoothing constant:

1 Click the **down spin button** repeatedly to reduce the value of *w* to 0.03. The forecasted values and the weight assigned to prior observations change dramatically. See Figure 11-13.

Figure 11-13 Reducing the *w* value to 0.03

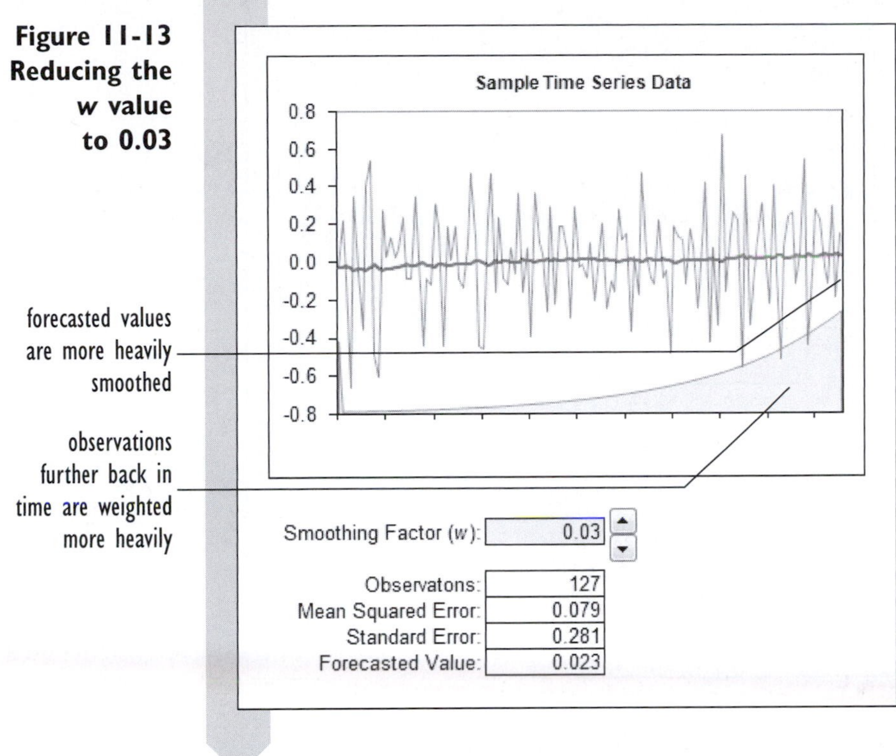

With such a small value for *w*, the smoothed value has a long memory. In fact, the final forecasted value, 0.023, is based in some part on observations spanning the entire time series. A consequence of having such a small value for *w* is that individual events, such as the large drop-off in the middle of the time series, have a minor impact on the smoothed values. The line of forecasted values is practically straight. Note as well that the standard error has declined from 0.293 to 0.281. In that case, the time series data is best estimated by the overall average or smoothed value that has a long memory.

Now increase the value of the smoothing factor to make the forecasts more susceptible to unit-by-unit changes.

To increase the smoothing factor:

1 Click the **up spin button** repeatedly to increase the value of *w* to 0.60. See Figure 11-14.

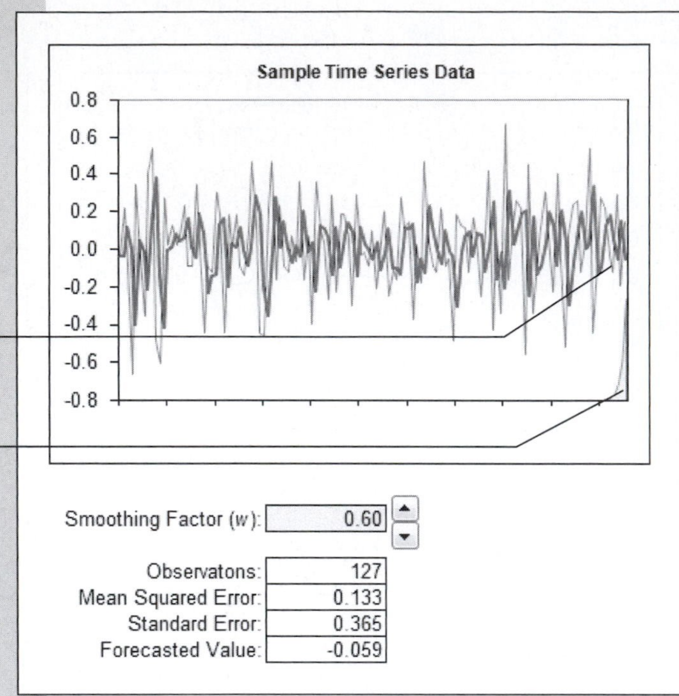

Figure 11-14 Increasing the value of *w* to 0.60

forecasted values are more variable

only recent values are heavily weighted

With a larger value for *w*, the forecasted values are much more variable—almost as variable as the observations themselves. This is a result of so much weight assigned to the value immediately prior to the current value. If one value shows a large upward swing, then the forecasted value for the next value tends to be high. As *w* approaches 1, the forecasted values appear more and more like lag 1 values.

2 Continue trying different values for *w* to see how they affect the smoothed curve and the standard error of the forecasts. Can you find a value for *w* that results in forecasts with the smallest standard error?

3 Close the Exponential Smoothing workbook without saving your changes. You'll return to the workbook later in this chapter.

Choosing a Value for *w*

As you saw in the Exponential Smoothing workbook, you have to choose the value of *w* with care. When choosing a value, keep several factors in mind. Generally, you want the standard error of the forecasts to be low, but this is not the only consideration. The value of *w* that gives the lowest standard error might be very high (such as 0.9) so that the exponential smoothing does not result in very smooth forecasts. If your goal is to simplify the appearance of the data or to spot general trends, you would not want to use such a high value for *w*, even if it produced forecasts with a low standard error. Analysts generally favor values for *w* ranging from 0.01 to 0.3. Choosing appropriate parameter values for exponential smoothing is often based on intuition and experience. Nevertheless, exponential smoothing has proved valuable in forecasting time series data.

The ability to perform exponential smoothing on time series data has been provided for you with StatPlus. Let's smooth the temperature data, using a *w* value of 0.18.

To create exponentially smoothed forecasts of the mean annual temperature:

1. Return to the **Global Temperature Analysis** workbook and go to the Temperature worksheet.
2. Click **Time Series** from the StatPlus menu and then click **Exponential Smoothing**.
3. Click the **Data Values** button and select **Fahrenheit** from the list of range names. Click **OK**.
4. Type **0.18** in the Weight box under General Options.
5. Click the **Output** button and direct the output to a new worksheet named **Smoothed Temperature**. Click **OK**.

The completed dialog box appears in Figure 11-15.

Figure 11-15 The Perform Exponential Smoothing dialog box

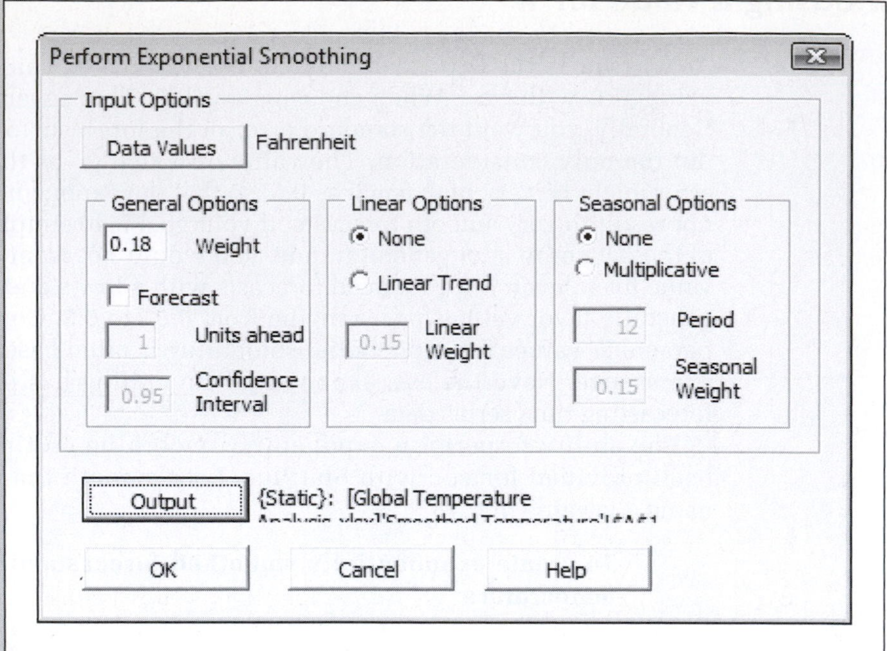

6 Click **OK**.

Excel displays the worksheet shown in Figure 11-16.

Figure 11-16 Smoothed temperatures

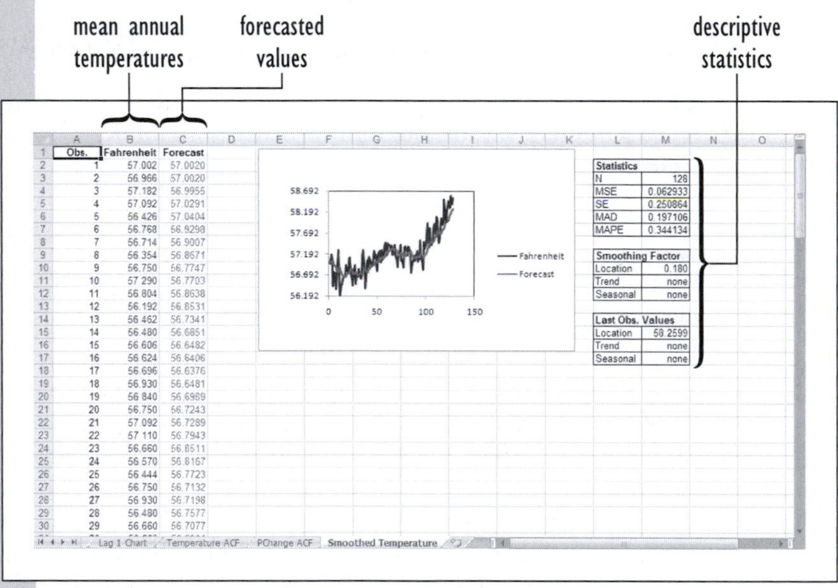

The output shown in Figure 11-16 consists of three columns: the observation numbers, the recorded mean annual temperatures, and the temperatures forecasted for each year based on the smoothing model. The values are then plotted on the chart. It appears that the forecasted values generally underestimated the mean annual temperatures in the last decades of the twentieth century. This may indicate that temperatures are warming faster than expected. The lower forecasted values might also reflect the effect of the slight dip in temperature values that occurred during the middle decades of the century. The standard error of the forecasts, 0.250864, indicates that the typical forecasting error was about 0.25 degrees Fahrenheit points per year.

The one-parameter exponential smoothing only uses weighted averages of previous observations to calculate future results. It does not assume a particular trend for the time series, but it is apparent from the data that the temperature values have been increasing over the time interval being studied. We can insert a trend assumption into our model by using two-parameter exponential smoothing.

EXCEL TIPS

- Excel's Analysis ToolPak also includes a command to perform one-parameter exponential smoothing. To run the command, select Exponential Smoothing from the list of analysis tools in the Data Analysis ToolPak.

Two-Parameter Exponential Smoothing

To explore how to add a trend assumption to exponential smoothing let's first express one-parameter exponential smoothing in terms of the following equation for y_t, the value of the y variable at time t:

$$y_t = \beta_0 + \varepsilon_t$$

where β_0 is the **location parameter** that changes slowly over time, and ε_t is the random error at time t. If β_0 were constant throughout time, you could estimate its value by taking the average of all the observations. Using that estimate, you would forecast values that would always be equal to your estimate of β_0. However, if β_0 varies with time, you weight the more recent observations more heavily than distant observations in any forecasts you make. Such a weighting scheme could involve exponential smoothing. How could such a situation occur in real life? Consider tracking crop yields over time. The average yield could slowly change over time as equipment or soil science technology improved. An additional factor in changing the average yield would be the weather, because a region of the country might go through several years of drought or good weather.

Now suppose the values in the time series follow a linear trend so that the series is better represented by this equation.

$$y_t = \beta_0 + \beta_1 t + \varepsilon_t$$

where β_1 is the **trend parameter**, whose value can also change over time. If β_0 and β_1 were constant throughout time, you could estimate their values using simple linear regression. However, when the values of these parameters change, you can try to estimate their values using the same smoothing techniques you used with one-parameter exponential smoothing (this approach is known as **Holt's method**). This type of smoothing estimates a line fitting the time series, with more weight given to recent data and less weight given to distant data. A county planner might use this method to forecast the growth of a suburb. The planner would not expect the rate of growth to be constant over time. When the suburb was new, it could have had a very high growth rate, which might change as the area becomes saturated with people, as property taxes change, or as new community services are added. In forecasting the probable growth of the community, the planner tends to weight recent growth rates much more heavily than older ones.

Calculating the Smoothed Values

The formulas for two-parameter smoothing are very similar in form to the simple one-parameter equations. Define S_n to be the value of the location parameter for the *n*th observation and T_n to be the trend parameter. Because we have two parameters, we also need two smoothing constants. We'll use the familiar w constant for smoothing the estimates of S_n, and we'll call t the smoothing constant for T_n. Using the same recursive form as was discussed with one-parameter exponential smoothing, we calculate S_n and T_n as follows:

$$S_n = wy_n + (1 - w)(S_{n-1} + T_{n-1})$$

$$T_n = t(S_n - S_{n-1}) + (1 - t)T_{n-1}$$

and the formula for the forecasted value of y_{n+1} is

$$y_{n+1} = S_n + T_n$$

The values of the parameters need not be equal. Although the equations may seem complicated, the idea is fairly straightforward. The value of S_n is a weighted average of the current observation and the previous forecasted value. The value of T_n is a weighted average of the change in S_n and the previous estimate of the trend parameter. As with simple exponential smoothing, you must determine the initial values S_0 and T_0. One method is to fit a linear regression line to the entire series and use the intercept and slope of the regression equation as initial estimates for the location and trend parameters.

CONCEPT TUTORIALS
Two-Parameter Exponential Smoothing

The Exponential Smoothing workbook that you used earlier also contains an interactive tutorial on two-parameter exponential smoothing.

To view the Exponential Smoothing workbook:

1 Return to the **Exponential Smoothing** file in the Explore folder. Enable the macros in the workbook.

2 Scroll through the workbook until you reach "Explore Two-Parameter Exponential Smoothing." See Figure 11-17.

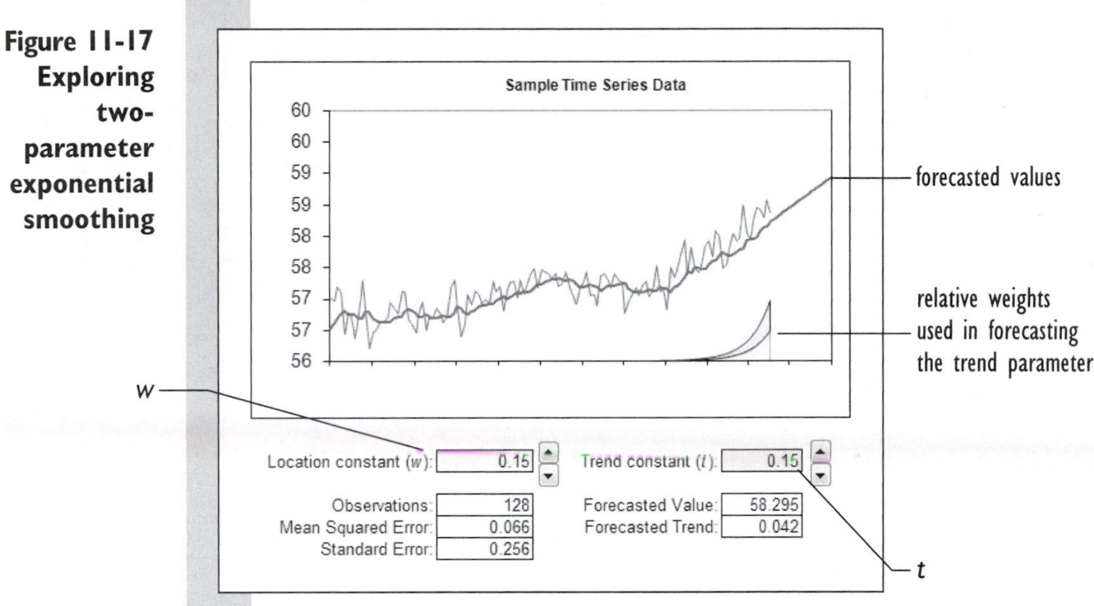

Figure 11-17 Exploring two-parameter exponential smoothing

The worksheet shows data from a sample time series. The smoothing factor for the location is equal to 0.15, as is the smoothing factor for trend. The area curves at the bottom of the chart indicate the relative weights assigned to previous values in calculating the final forecast for the location and trend parameters. For *t* and *w* equal to 0.15, the most prominent observations occur within a few units of the current value. Earlier observations have too little effect to be visible on the chart. On the basis of the two-parameter exponential smoothing estimates, the forecasted value of the time series is projected to be about 58.295, increasing at a rate of 0.042 points per unit of time.

The values chosen for w and t are important in determining what the forecasted value for the time series will be. If we assume that the data will continue to behave as it did for earlier values, smaller values for w and t might be used, because those would result in an estimate that has a longer "memory" of previous values. Let's see what kind of difference this would make by reducing the value of t from 0.15 to 0.05.

To reduce the value of *t*:

1 Repeatedly click the **down spin arrow** next to the Trend constant until the value of t equals **0.05**. See Figure 11-18.

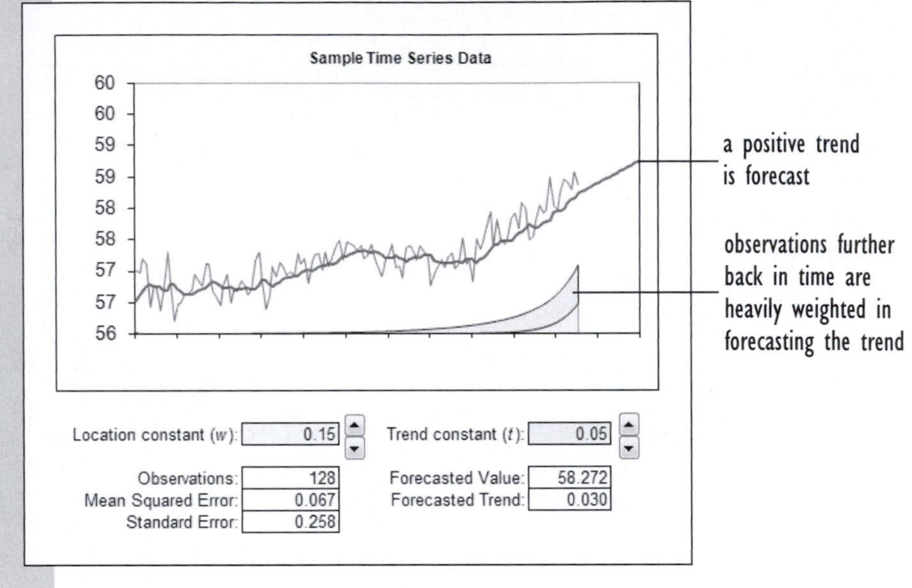

Figure 11-18 Decreasing the value of t to 0.05

With this value of t, the forecasted increase in the time series data changes to 0.030 points per unit, reflecting the assumption that there will be an increase in the data similar to what was observed earlier in the time series. Note that the weights for the trend factor, as shown by the area curve, indicate that older observations are well represented in the forecast. Now let's see what would happen if we increased the value of t, focusing more on short-term trends.

To increase the value of *t*:

1 Repeatedly click the **up spin arrow** next to the Trend constant until the value of t equals **0.40**. See Figure 11-19.

Figure 11-19 Increasing the value of t to 0.40

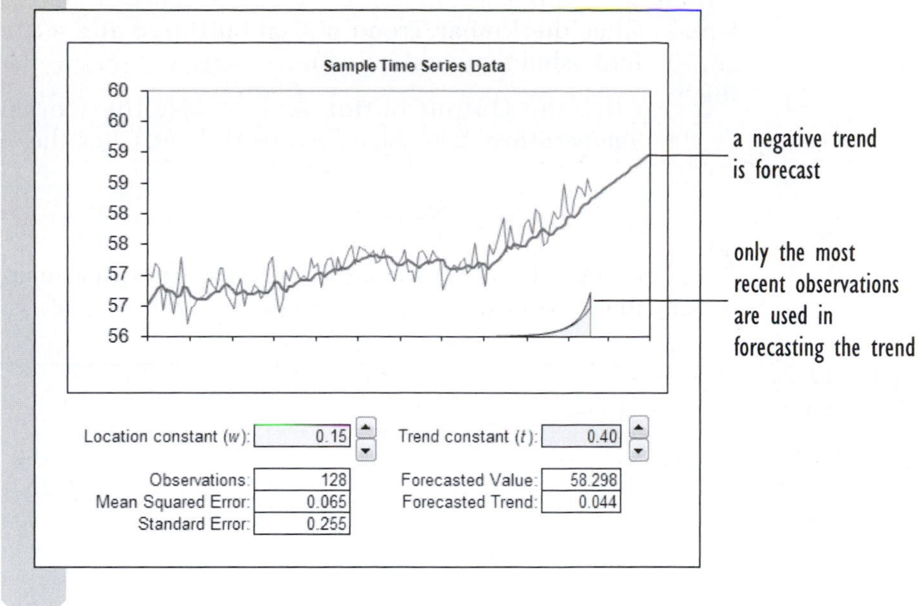

With a higher smoothing constant, the forecasted trend of the time series shows an increase of 0.044 points per unit of time. The area curve indicates that the smoothed trend estimate has a shorter memory; only the most recent observations are relevant in estimating the trend.

Using this worksheet, you can change the values of the smoothing constant for the location and trend parameters. What combinations result in the lowest values for the standard error? When you are finished with your investigations, close the workbook. You do not have to save any of your changes.

Now let's return to the global temperature data. In using one-parameter exponential smoothing the forecasted values underestimated the most recent trend in temperature data. To compensate we'll use two-parameter exponential smoothing in an attempt to "pick up" the most recent trend of increasing temperatures.

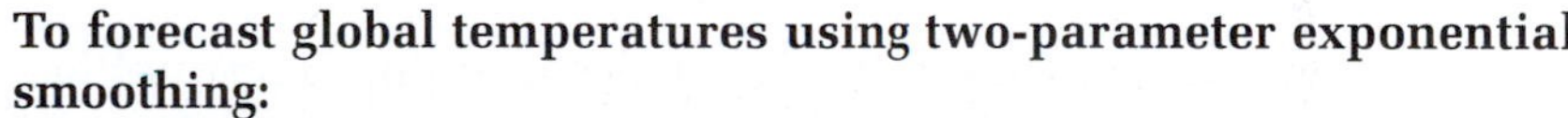

To forecast global temperatures using two-parameter exponential smoothing:

1. Return to the **Global Temperature Analysis** workbook and go to the **Temperature** worksheet.
2. Click **Time Series** from the StatPlus menu and then click **Exponential Smoothing**.
3. Click the **Data Values** button and select **Fahrenheit** from the list of range names. Click **OK**.

4 Click the **Linear Trend** option button to add a linear trend to the forecasted temperature values.

5 Click the **Output** button and specify the worksheet **Smoothed Temperatures 2** as the output worksheet. Click the **OK** button twice.

Figure 11-20 shows the forecasted temperature values using two-parameter exponential smoothing.

Figure 11-20 Forecasting global temperatures with two-parameter exponential smoothing

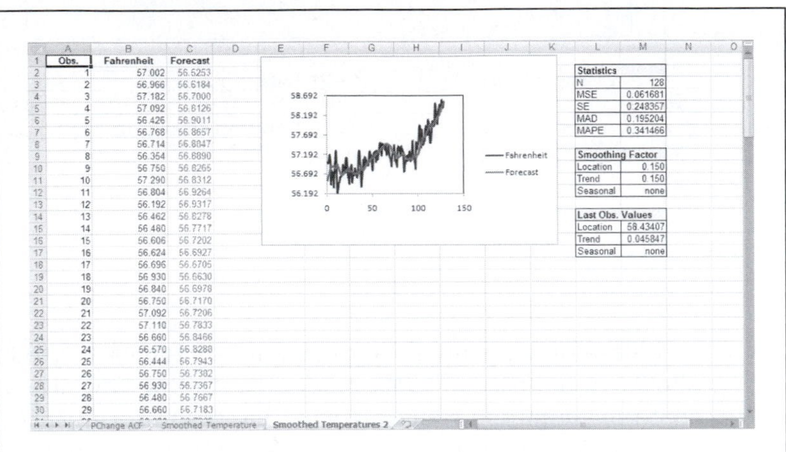

By adding the trend parameter our smoothed values have picked up the recent trend in rising global temperatures indicated in the data. At this point you can close the Global Temperature Analysis workbook.

Seasonality

Often time series are measured on a seasonal basis, such as monthly or quarterly. If the data are sales of ice cream, toys, or electric power, there is a pattern that repeats each year. Ice cream and electric power sales are high in the summer, and toy sales are high in December.

Multiplicative Seasonality

If the sales of some of your products are seasonal, you might want to adjust your sales for the seasonal effect, in order to compare figures from month to month. To compare November and December sales, should you use the difference of the values or the ratio? In many cases seasonal changes are best expressed in ratios, especially if there is substantive growth in yearly sales.

As annual sales increase, the difference between the November and December values should also increase, but the ratio of sales between the two months might remain nearly constant. This is called **multiplicative seasonality**. To quantify the effect of the season on each month's value, we need to assign a multiplicative factor to each month. If the month's sales are equal to the expected yearly average, we'll give it a multiplicative factor of 1. Consequently, months with higher-than-average sales have multiplicative factors greater than 1, and months with lower-than-average sales have multiplicative factors less than 1.

As an example, consider Table 11-4, which shows seasonal sales and multiplicative factors.

Table 11-4 Multiplicative Seasonality

	Jan	Feb	Mar	Apr	May	Jun	Jul	Aug	Sep	Oct	Nov	Dec
Sales	220	310	359	443	374	660	1030	1320	1594	1093	950	610
Factor	0.48	0.58	0.60	0.69	0.59	1.00	1.48	1.69	1.99	1.29	1.02	0.59
Adjusted sales	458.3	534.5	598.3	642.0	633.9	660.0	695.9	781.1	801.0	847.3	931.4	1033.9

The monthly sales figures are shown in the first row of the table. The multiplicative factors based on previous years' sales are shown in the second row. Dividing the sales in each month by the multiplicative factor yields the adjusted sales. Plotting the sales values and adjusted sales values in Figure 11-21 reveals that sales have been steadily increasing throughout the year. This information is masked in the raw sales data by the seasonal effects.

Figure 11-21 Plot of adjusted sales data

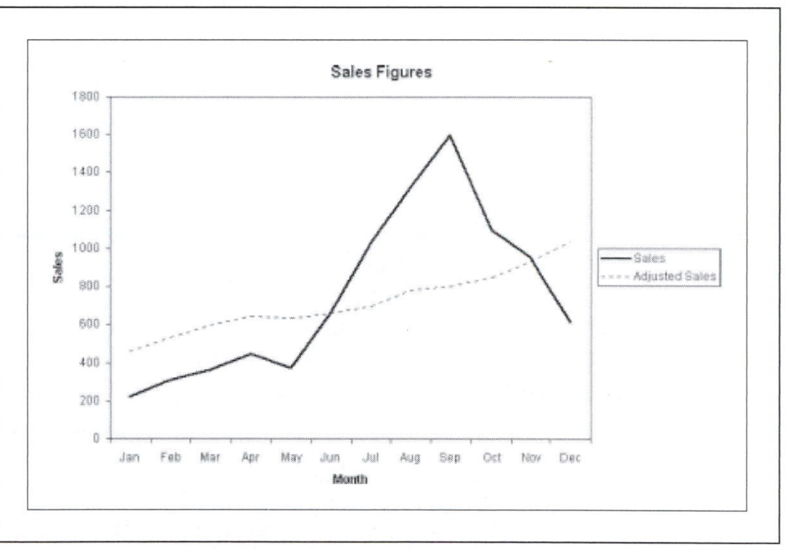

Additive Seasonality

Sometimes the seasonal variation is expressed in additive terms, especially if there is not much growth. If the highest annual sales total is no more than twice the lowest annual sales total, it probably does not matter whether you use differences or ratios. If you can express the month-to-month changes in additive terms, the seasonal variation is called **additive seasonality**. Additive seasonality is expressed in terms of differences from the expected average for the year. In Table 11-5, the seasonal adjustment for December sales is −240, resulting in an adjusted sales for that month of 681. After adjustment for the time of the year, December turned out to be one of the most successful months, at least in terms of exceeding goals.

Table 11-5 Additive Seasonality

	Jan	Feb	Mar	Apr	May	Jun	Jul	Aug	Sep	Oct	Nov	Dec
Sales	298	378	373	443	374	660	1004	1153	1388	904	715	441
Factor	−325	−270	−270	−200	−280	−55	350	450	550	220	70	−240
Adjusted sales	623	648	643	643	654	715	654	703	838	684	645	681

In this chapter you'll work with multiplicative seasonality only, but you should be aware of the principles of additive seasonality.

Seasonal Example: Liquor Sales

Are liquor sales seasonal? The Liquor workbook has monthly liquor store sales in terms of millions of dollars from January 1996 through December 2007. The workbook contains the variables and reference names shown in Table 11-6.

Table 11-6 The Liquor Workbook

Range Name	Range	Description
Year	A2:A145	The year
Month	B2:B145	The month
Year_Month	C2:C145	The year and the month
Sales	D2:D145	The monthly liquor sales in millions of dollars

To open the Liquor workbook:

1 Open the **Liquor** workbook from the Chapter11 data folder.

2 Save the workbook as **Liquor Sales Analysis**. The workbook appears as shown in Figure 11-22.

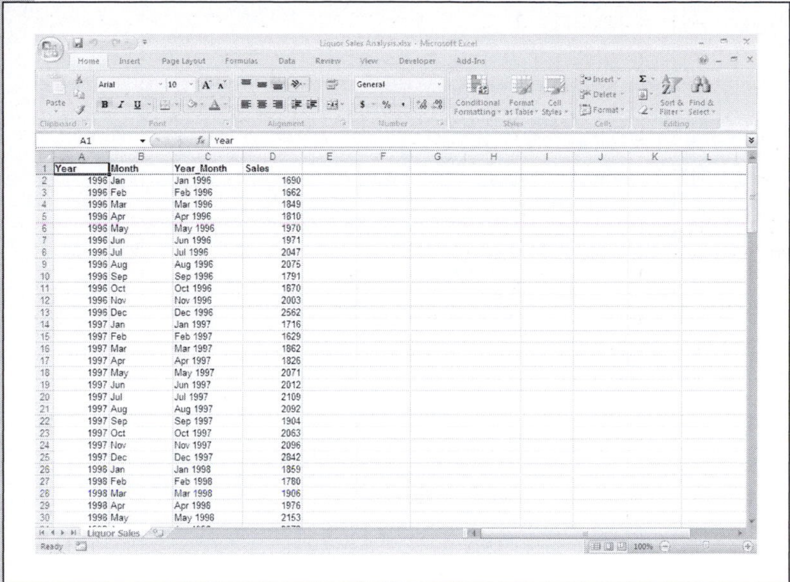

	Year	Month	Year_Month	Sales
2	1996	Jan	Jan 1996	1690
3	1996	Feb	Feb 1996	1662
4	1996	Mar	Mar 1996	1849
5	1996	Apr	Apr 1996	1810
6	1996	May	May 1996	1970
7	1996	Jun	Jun 1996	1971
8	1996	Jul	Jul 1996	2047
9	1996	Aug	Aug 1996	2075
10	1996	Sep	Sep 1996	1791
11	1996	Oct	Oct 1996	1870
12	1996	Nov	Nov 1996	2003
13	1996	Dec	Dec 1996	2562
14	1997	Jan	Jan 1997	1716
15	1997	Feb	Feb 1997	1629
16	1997	Mar	Mar 1997	1862
17	1997	Apr	Apr 1997	1826
18	1997	May	May 1997	2071
19	1997	Jun	Jun 1997	2012
20	1997	Jul	Jul 1997	2109
21	1997	Aug	Aug 1997	2092
22	1997	Sep	Sep 1997	1904
23	1997	Oct	Oct 1997	2063
24	1997	Nov	Nov 1997	2096
25	1997	Dec	Dec 1997	2842
26	1998	Jan	Jan 1998	1859
27	1998	Feb	Feb 1998	1780
28	1998	Mar	Mar 1998	1906
29	1998	Apr	Apr 1998	1976
30	1998	May	May 1998	2153

Figure 11-22 The Liquor Sales Analysis workbook

As a first step in analyzing these data, create a line plot of sales versus year and month.

To create a time series plot:

1 Select the range **C1:D145** and click the **Line** button from the Charts group on the Insert tab. Select the first chart subtype (Line).

2 Move the chart to the chart sheet **Sales Chart**.

3 Enter the chart title **Liquor Sales 1996–2007**, enter **Year** for the *x*-axis title, and enter **Sales ($mil)** for the *y*-axis title. Remove the gridlines and legend from the plot.

4 Click the **Axes** button from the Axes group on the Layout tab of the Chart Tools ribbon. Click **Primary Horizontal Axis** and then click **More Primary Horizontal Axis Options**.

5 Within the Format Axis dialog box, click the **Specify interval unit** option button and enter **24** for the number of units between axis labels. See Figure 11-23.

Figure 11-23 Format Axis dialog box

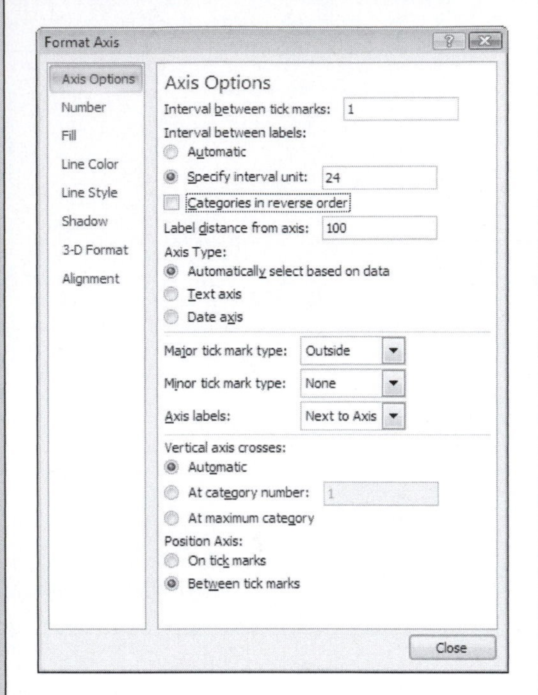

6 Click the **Close** button. Figure 11-24 shows the formatted chart of liquor sales.

Figure 11-24
Liquor sales from
1996 to 2007

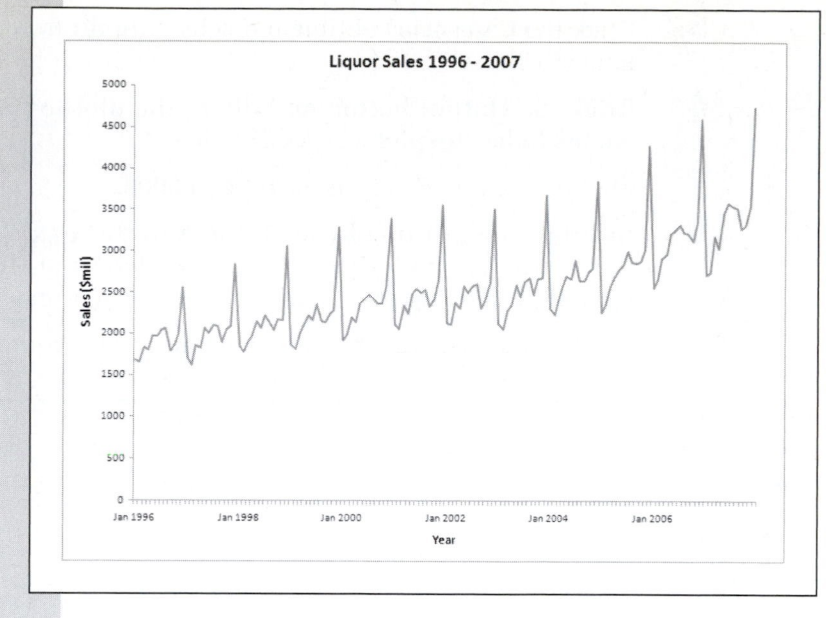

The plot shows that production is seasonal, with peaks occurring each winter (around the holidays). There also appears to be another peak in the summer, perhaps around the fourth of July. In addition to the seasonality of the data, there appears to be a linear trend of increasing sales from 1996 to 2007.

Examining Seasonality with a Boxplot

One way to see the seasonal variation is to make a boxplot, with a box for each of the 12 months. This gives you a picture of the month-to-month variation in liquor sales. The shape of each box tells you how production for that month varied from 1996 to 2007.

To create the boxplot:

1 Click the **Liquor Sales** sheet tab.

2 Click **Single Variable Charts** from the StatPlus menu and click **Boxplots**.

3 Click the **Connect Medians between Boxes** checkbox.

4 Click the **Data Values** button and select **Sales** from the range names list. Click **OK**.

5 Click the **Categories** button and select **Month** from the list of range names. Click **OK**.

6 Click the **Output** button and direct the plot to a new chart sheet named **Sales Boxplot**. Click **OK** twice.

7 Rescale the *y* axis to go from **1000** to **5000**.

8 Insert the chart title **Liquor Sales**. Add the *x*-axis title **Month** and the *y*-axis title **Sales ($mil)**. Edit the labels at the bottom of the boxplot, removing the "Month =" text from each.

Figure 11-25 shows the edited boxplot for the liquor sales data.

Figure 11-25 Liquor sales boxplot

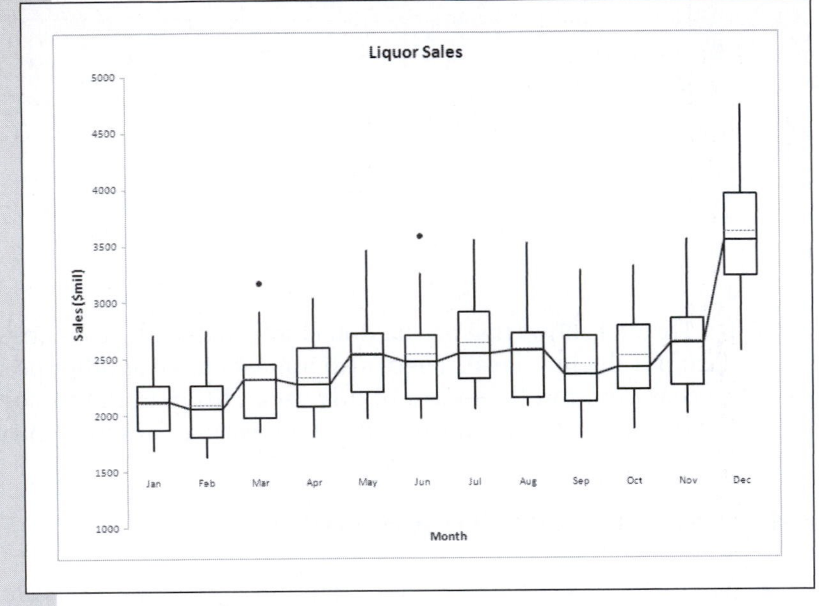

The boxplot in Figure 11-25 shows how monthly liquor sales vary throughout the year. The sales peak in December shows a slight dip in the months of September and October. The boxplot also indicates the range of production levels for each month. There are a couple of outliers in the months of March and June, but there is nothing extreme.

Examining Seasonality with a Line Plot

You can also take advantage of the two-way table to create a line plot of sales versus month for each year of the data set. This is another way to get insight into the monthly sales figures during this time period. You will first have to create a two-way table of the data in the Liquor Sales worksheet.

To create a line plot of liquor sales:

1 Return to the **Liquor Sales** worksheet.

2 Click **Manipulate Columns** from the StatPlus menu and then click **Create Two-way Table.**

3 Select **Sales** for the Data Values variable, **Month** for the Column Levels, and **Year** for the Row Levels.

4 Deselect the **Sort the Column Levels** checkbox.

Note: You want to deselect this checkbox to prevent the two-way table from sorting the columns in alphabetical order, rather than leaving them in time order.

5 Send the two-way table to a new worksheet named **Sales by Year**.

6 Click **OK**.

7 Go to the Sales by Year worksheet.

8 Select the range **B2:M14** and click the **Line** button from the Charts group on the Insert tab. Click the first chart subtype (Line). Move the chart to a new chart sheet named **Liquor Sales Line Plot**.

9 Enter **Liquor Sales versus Month** for the chart title, **Months** for the *x*-axis title, and **Sales ($mil)** for the *y*-axis title. Remove the legend and gridlines from the plot.

Figure 11-26 shows the formatted line chart.

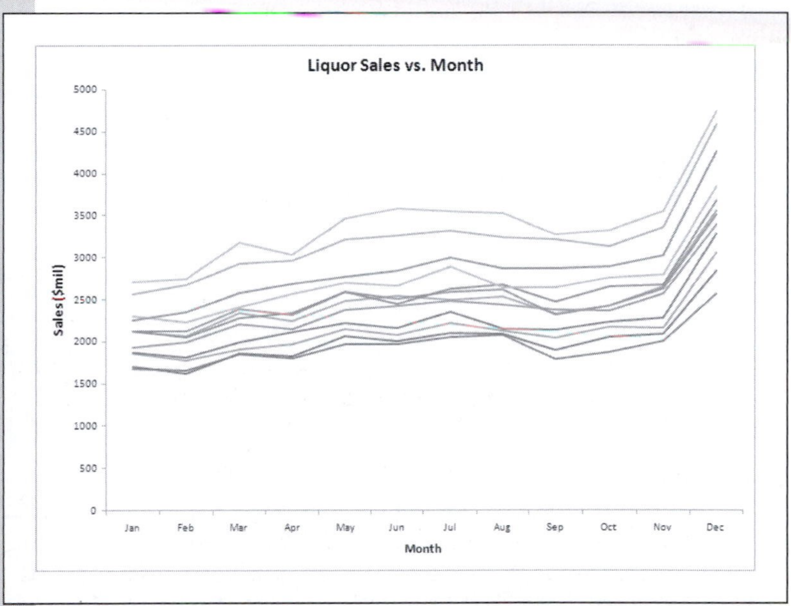

Figure 11-26 Line plot of monthly liquor sales

The line plot in Figure 11-26 demonstrates the seasonal nature of the data and also allows you to observe individual values. Plots like this are sometimes called **spaghetti plots**, for obvious reasons.

Applying the ACF to Seasonal Data

You can also use the autocorrelation function to display the seasonality of the data. For a seasonal monthly series, the ACF should be very high at lag 12, because the current value should be strongly correlated with the value from the same month in the previous year.

To calculate the ACF:

1. Return to the Liquor Sales worksheet.
2. Click **Time Series** from the StatPlus menu and then click **ACF Plot**.
3. Select **Sales** for the Data Values variable.
4. Click the **up spin arrow** to calculate the ACF up through a lag of **24**.
5. Send the output to a new worksheet named **Liquor Sales ACF**.
6. Click **OK**.

Figure 11-27 shows the ACF of the liquor sales data.

Figure 11-27 ACF of the liquor sales data

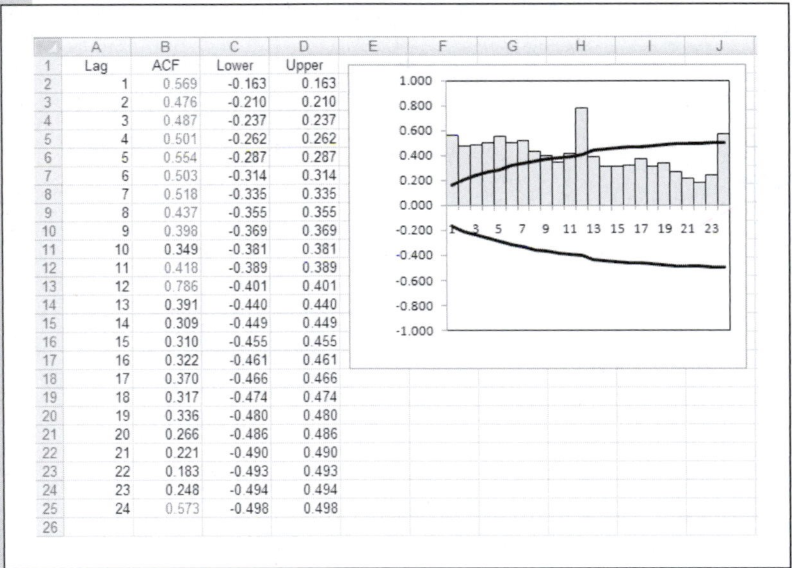

Lag	ACF	Lower	Upper
1	0.569	-0.163	0.163
2	0.476	-0.210	0.210
3	0.487	-0.237	0.237
4	0.501	-0.262	0.262
5	0.554	-0.287	0.287
6	0.503	-0.314	0.314
7	0.518	-0.335	0.335
8	0.437	-0.355	0.355
9	0.398	-0.369	0.369
10	0.349	-0.381	0.381
11	0.418	-0.389	0.389
12	0.786	-0.401	0.401
13	0.391	-0.440	0.440
14	0.309	-0.449	0.449
15	0.310	-0.455	0.455
16	0.322	-0.461	0.461
17	0.370	-0.466	0.466
18	0.317	-0.474	0.474
19	0.336	-0.480	0.480
20	0.266	-0.486	0.486
21	0.221	-0.490	0.490
22	0.183	-0.493	0.493
23	0.248	-0.494	0.494
24	0.573	-0.498	0.498

The autocorrelation holds between adjacent months but the highest autocorrelation exists for sales that are 12 months or one year apart. Also note a second significant autocorrelation occurs at 24 months. So the ACF results do show a seasonal correlation of the sales figures. In other words, the pattern of sales from one year to the next is fairly consistent.

Adjusting for Seasonality

Because the liquor sales data have a seasonal component, it would be useful to adjust the values for the seasonal effect. In this way you can determine whether a drop in production during one month is due to seasonal effects or is a true decline. Adjusting the production data for seasonality also gives you a better indication of the trend in liquor sales over the course of the 10 years of the study. You can use StatPlus to adjust time series data for multiplicative seasonality.

To adjust the liquor sales data:

1. Return to the **Liquor Sales** worksheet.
2. Click **Time series** from the StatPlus menu and then click **Seasonal Adjustment** names list. Click **OK**.
3. Verify that a period of length 12 is entered into the Length of Period box.
4. Click the **Output** button and send the output to a new worksheet named **Adjusted Sales**. Click **OK**.

Your completed dialog box should look like Figure 11-28.

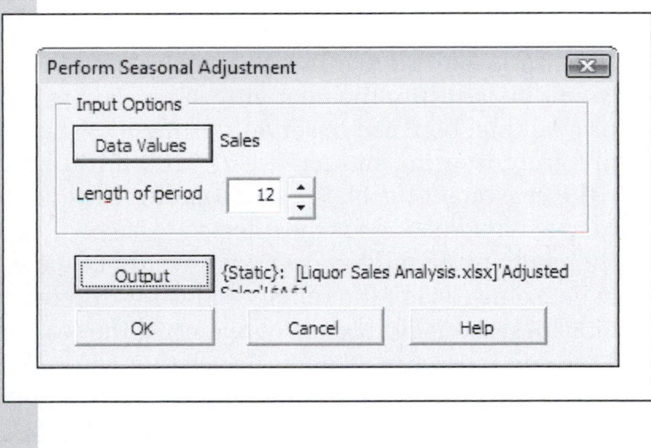

Figure 11-28 The Perform Seasonal Adjustment dialog box

5 Click **OK**.

Excel generates the adjusted sales values shown in Figure 11-29.

Figure 11-29 Liquor sales adjusted for seasonal effects

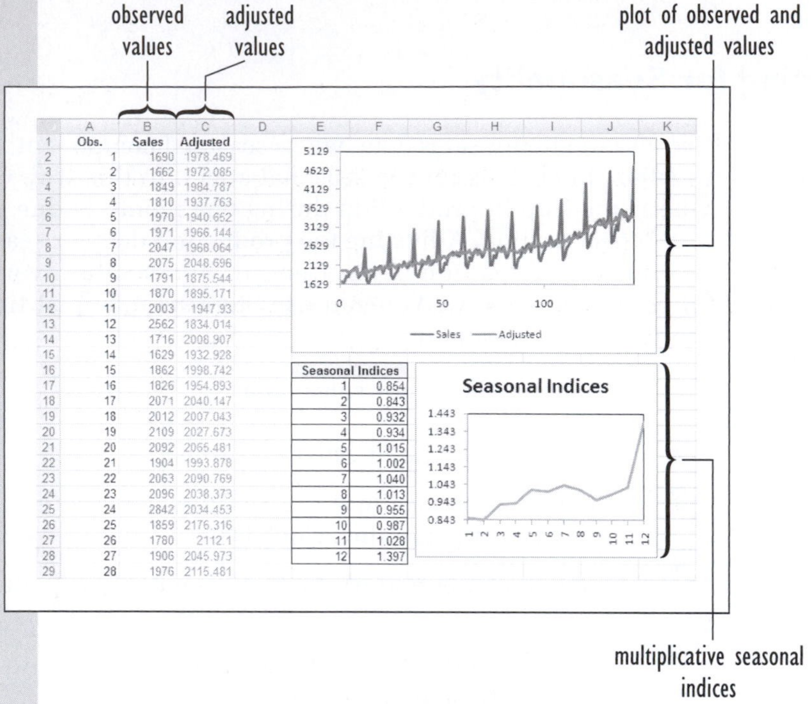

Obs.	Sales	Adjusted
1	1690	1978.469
2	1662	1972.085
3	1849	1984.787
4	1810	1937.763
5	1970	1940.652
6	1971	1966.144
7	2047	1968.064
8	2075	2048.696
9	1791	1875.544
10	1870	1895.171
11	2003	1947.93
12	2562	1834.014
13	1716	2008.907
14	1629	1932.928
15	1862	1998.742
16	1826	1954.893
17	2071	2040.147
18	2012	2007.043
19	2109	2027.673
20	2092	2065.481
21	1904	1993.878
22	2063	2090.769
23	2096	2038.373
24	2842	2034.453
25	1859	2176.316
26	1780	2112.1
27	1906	2045.973
28	1976	2115.481

Seasonal Indices	
1	0.854
2	0.843
3	0.932
4	0.934
5	1.015
6	1.002
7	1.040
8	1.013
9	0.955
10	0.987
11	1.028
12	1.397

The observed production levels are shown in column B, and the seasonally adjusted values are shown in column C. Using the adjusted values can give you some insight into the changing sales values adjusted for seasonal effects. For example, between observations 4 and 5 the sales value increases by 160 units (representing an increase of $160 million); however, when adjusted for the seasonal effects, the increase in sales is about $3 million. In other words, when adjusting for the effects of seasonal variation, the sales increased that month by $3 million over what would be expected in a usual year.

You can get some idea of the relative sales for different months of the year from the table of seasonal indexes. For example, the seasonal index for June is 1.002 and for July it is 1.040. This indicates that you can expect a percentage increase in liquor sales of $(1.040 - 1.002)/1.002 = 0.037546$, or about 3.75%, going from June to July each year. Seasonal indexes for the multiplicative model must add up to the length of the period, in this case 12. You can use this information to tell you that 11.64% of the liquor sales take place in December (because $1.397/12 = 0.1164$). A line plot of the seasonal indexes is provided and shows a profile very similar to the one you saw earlier with the boxplot.

A chart is also included, showing both the production data and the adjusted production values. There is a clear increase in liquor sales in the data set after adjusting for seasonal variation. To further explore this trend, you can smooth the sales data using three-parameter exponential smoothing.

Three-Parameter Exponential Smoothing

You perform exponential smoothing on seasonal data using three smoothing constants. This process is known as **three-parameter exponential smoothing** or **Winters' method**. The smoothing constants in the Winters' method involve location, trend, and seasonality. Winters' method can be used for either multiplicative or additive seasonality, though in this text, we'll assume only multiplicative seasonality. The equation for a time series variable y_t with a multiplicative seasonality adjustment is

$$y_t = (\beta_0 + \beta_1 t) \times I_p + \varepsilon_t$$

and for additive seasonality adjustment the equation is

$$y_t = (\beta_0 + \beta_1 t) + I_p + \varepsilon_t$$

In these equations β_0, β_1, and ε_t once again represent the location, trend, and error parameters of the model and I_p represents the seasonal index at point p in the seasonal data. For example, if we used the multiplicative seasonal indexes shown in Figure 11-29, I_5 would equal 1.015. Once again, these parameters are not considered to be constant but can vary with time. The liquor sales data are an example of such a series. The sales are seasonal, but there is also a time trend to the data such that sales increase from year to year after adjusting for seasonality.

Let's concentrate on smoothing with a multiplicative seasonality factor. The smoothing equations used in three-parameter exponential smoothing are similar to equations you've already seen. For the smoothed location value S_n and the smoothed trend value T_n, from a time series where the length of the period is q, the recursive equations are

$$S_n = w\frac{y_n}{I_{n-q}} + (1 - w)(S_{n-1} + T_{n-1})$$

$$T_n = t(S_n - S_{n-1}) + (1 - t)T_{n-1}$$

Note that the recursive equation for S_n is identical to the equation used in two-parameter exponential smoothing except that the current observation y_n must be seasonally adjusted. Here, I_{n-q} is the seasonal index taken from the index values of the previous period. The recursive equation for T_n is identical to the recursive equation for the two-parameter model.

As you would expect, three-parameter smoothing also smoothes out the values of the seasonal indexes, because these might also change over time. We use a different smoothing constant for these indices. The recursive equation for a seasonal index I_n is

$$I_n = c\frac{y_n}{S_n} + (1 - c)I_{n-q}$$

The value of the smoothed seasonal index is a weighted average of the current seasonal index based on the values of y_n and S_n, and the index value from the previous period. Calculating initial estimates of S_0, T_0, and each initial seasonal index is beyond the scope of this book.

Forecasting Liquor Sales

Let's use exponential smoothing to predict future liquor sales. For the purposes of demonstration, we'll assume a multiplicative model. You will have to decide on values for each of the three smoothing constants. The values need not be the same. For example, seasonal adjustments often change more slowly than the trend and location factors, so you might want to choose a low value for the seasonal smoothing constant, say about 0.05. However, if you feel that the trend factor or location factor will change more rapidly over the course of time, you will want a higher value for the smoothing constant, such as 0.15. As you have seen in this chapter, the values you choose for these smoothing constants depend in part on your experience with the data. Excel does not provide a feature to do smoothing with the Winters' method. One has been provided for you with the Exponential Smoothing command found in StatPlus.

To forecast future liquor sales:

1. Return to the **Liquor Sales** worksheet.
2. Click **Time Series** from the StatPlus menu and then click **Exponential Smoothing**.
3. Select **Sales** for your Data Values variable.
4. Enter **0.15** in the General Options Weight box. This is your value for *w*.
5. Click the **Linear Trend** option button, and enter **0.15** in the Linear Weight box. This is your value for *t*.
6. Click the **Multiplicative** option button, and enter **0.05** in the Seasonal Weight box. This is the value of *c*. Verify that the length of the period is set to **12**.

7 Click the **Forecast** checkbox and enter **12** in the Units ahead box. This will forecast the next year liquor sales.

8 Verify that **0.95** is entered in the Confidence Interval box. This will produce a 95% confidence region around your forecasted values.

9 Click the **Output** button and direct your output to a new sheet named **Forecasted Sales**. Click **OK**.

Your dialog box should look like Figure 11-30.

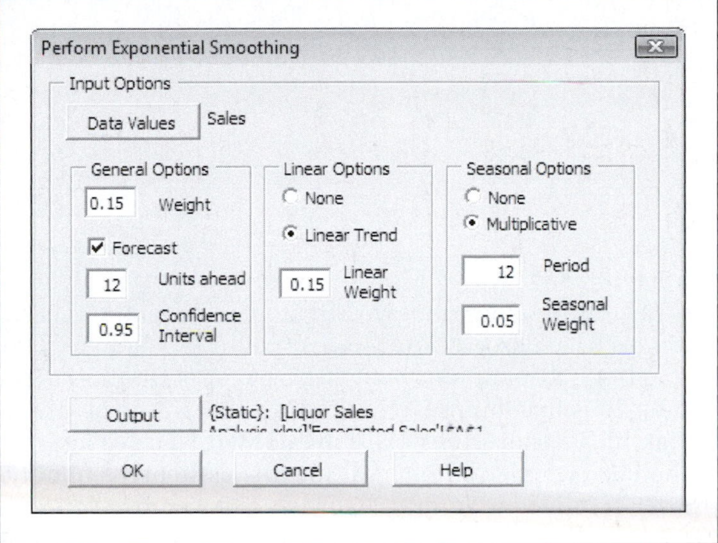

Figure 11-30 The Perform Exponential Smoothing dialog box

10 Click **OK**.

Output from the command appears on the Forecasted Sales worksheet. To view the forecasted values, drag the vertical scroll bar down. See Figure 11-31.

Figure 11-31 Forecasted sales values with 95% confidence region

	A	B	C	D	E
147	Obs.	Forecast	Lower	Upper	
148	145	2,999.4	2,861.3	3,137.5	
149	146	2,972.8	2,833.1	3,112.5	
150	147	3,299.3	3,157.9	3,440.7	
151	148	3,313.7	3,170.5	3,456.9	
152	149	3,613.4	3,468.3	3,758.5	
153	150	3,591.1	3,444.1	3,738.2	
154	151	3,735.6	3,586.5	3,884.8	
155	152	3,651.4	3,500.1	3,802.7	
156	153	3,453.7	3,300.2	3,607.3	
157	154	3,567.9	3,412.1	3,723.7	
158	155	3,739.1	3,580.9	3,897.3	
159	156	5,093.5	4,932.9	5,254.2	
160					

The output does not give the month for each forecast, but you can easily confirm that observation 145 in column A is January 2008 because observation 144 on the Liquor Sales worksheet is December 2007. On the basis of the values in column B, you forecast that in the next year, sales will reach a peak in December (observation 150) with a sales figure of $5,093.5 million. The 95% prediction interval for this estimate is about $4,932.9 million to $5,254.2 million. In other words, in December you would expect sales of not less than 4,932.9 million dollars or more than 5,254.2 million dollars. You could use these estimates to plan your sales strategy for the upcoming year. Before putting much faith in the prediction intervals, you should verify the assumptions for the smoothed forecasts. If the smoothing model is correct, the residuals should be independent (show no discernible ACF pattern) and follow a normal distribution with mean 0. You would find for the liquor sales that these assumptions are met.

Scrolling back up the worksheet, you can view how well the smoothing method forecasted liquor sales in the previous years, as shown in Figure 11-32.

Figure 11-32 Three-parameter exponential smoothing values

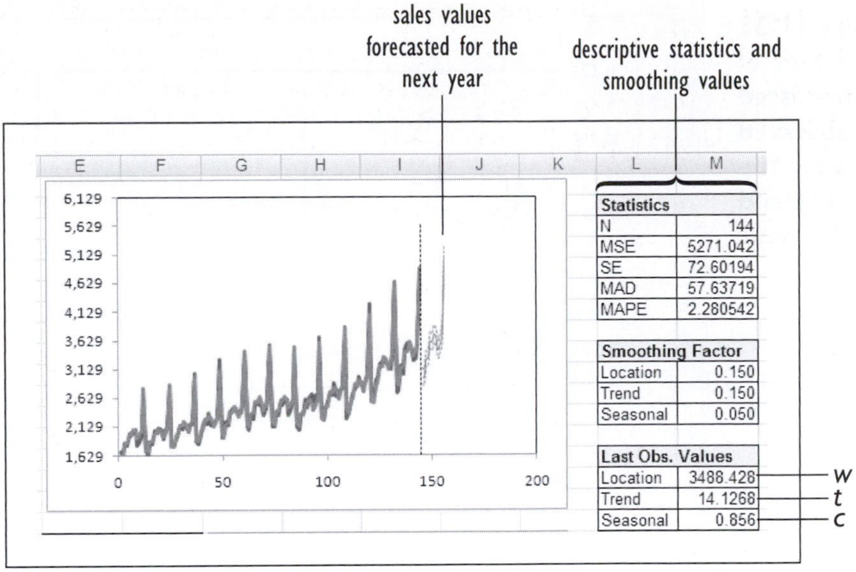

sales values forecasted for the next year

descriptive statistics and smoothing values

Statistics	
N	144
MSE	5271.042
SE	72.60194
MAD	57.63719
MAPE	2.280542

Smoothing Factor	
Location	0.150
Trend	0.150
Seasonal	0.050

Last Obs. Values		
Location	3488.428	w
Trend	14.1268	t
Seasonal	0.856	c

The standard error of the forecast is 72.60194, indicating that the typical forecasting error in the time series is about 73 units (the MAD value is lower with a value of about 57.6 units).

The final estimates for the location and trend values are 3488.428 and 14.1268, respectively. The location value represents the monthly sales after adjusting for seasonal effects. The trend estimate indicates that sales are increasing at a rate of about $14.13 million per month—hardly a large increase given the magnitude of the monthly sales.

The output also includes a scatterplot comparing the observed, smoothed, and forecasted production values. Because of the number of points in the time series, the seasonal curves are close together and difficult to interpret. To make it easier to view the comparison between the observed and forecasted values, rescale the *x* axis to show only the current year and the forecasted year's values.

To rescale the *x* axis:

1. Click the plot to select it.
2. Click the **Axes** button on the Axes group of the Layout tab on the ChartTools ribbon and then click **Primary Horizontal Axis** and **More Primary Horizontal Axes Options**.
3. Click the **Fixed** option button for the Minimum scale value and change it to **130**. Click the **Close** button.

The revised plot appears in Figure 11-33.

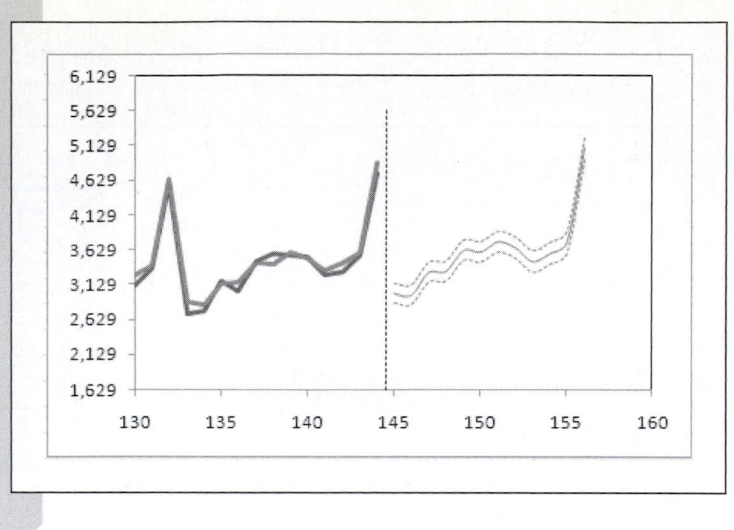

Figure 11-33 Plot of forecasted and observed sales for the current and upcoming year

From the rescaled plot, you would conclude that exponential smoothing has done a good job of modeling the sales data and that the resulting forecasts appear reasonable.

The final part of the exponential smoothing output is the final estimate of the seasonal indexes, shown in Figure 11-34.

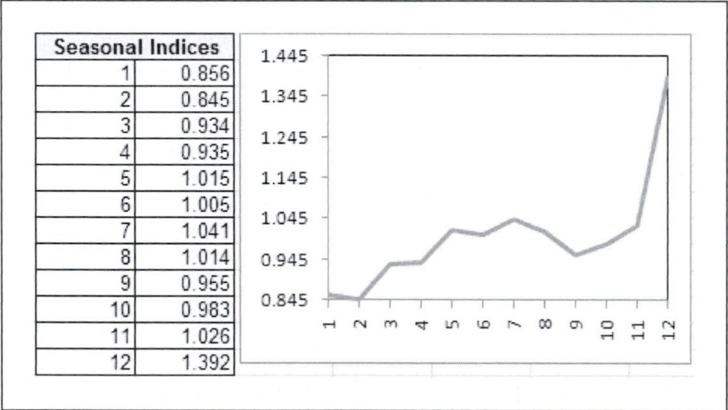

Seasonal Indices	
1	0.856
2	0.845
3	0.934
4	0.935
5	1.015
6	1.005
7	1.041
8	1.014
9	0.955
10	0.983
11	1.026
12	1.392

Figure 11-34 Seasonal indices

The values for the seasonal indexes are very similar to those you calculated using the seasonal adjustment command. The difference is due to the fact that these seasonal indexes are calculated using a smoothed average, whereas the earlier indexes were calculated using an unweighted average.

You're finished with the workbook. You can close it now, saving your changes.

Optimizing the Exponential Smoothing Constant (optional)

As you've seen in this chapter, the choice for the value of the exponential smoothing constant depends partly on the analyst's experience and intuition. Many analysts advocate using the value that minimizes the mean square error. You can use an Excel add-in called the Solver to calculate this value. To demonstrate how this technique works, open the Exponential workbook, which contains a set of time series data.

To open the Exponential workbook:

1. Open **Exponential** workbook from the Chapter11 data folder.
2. Save the file as **Exponential Smoothing**.

The workbook displays the column of sample time series data. Let's create a column of exponentially smoothed forecasts. First we must decide on an initial estimate for the smoothing constant *w*; we can start with any value we want, so let's start with 0.15. From this value, we'll calculate the mean square error.

To calculate the mean square error:

1. Click cell **F1**, type **0.15**, and then press **Enter**.

   Next determine a value for S_0 to be put in cell C2. We'll use the first value in the time series.

2. Click cell **C2**, type **=B2**, and then press **Enter**.

   Now create a column of smoothed forecasts S_n, using the recursive smoothing equation.

3. Select the range **C3:C120**.
4. Type **=$F$1*B2+(1-$F$1)*C2**; then press **Enter**.
5. Click the **Fill** button from the Editing group on the Home tab and click **Down** to fill the formula down the rest of the column.

   Now create a column of squared errors [(forecast – observed)^2].

6. Select the range **D2:D120**.
7. Type **=(C2-B2)^2**, and then press **Enter**.
8. Fill the formula down the rest of the column.

   Finally, calculate the mean square error for this particular value of *w*.

9 Click cell **F2**, type **=SUM(D2:D120)/119**, and then press **Enter**.

Verify that the values in your spreadsheet match the values in Figure 11-35.

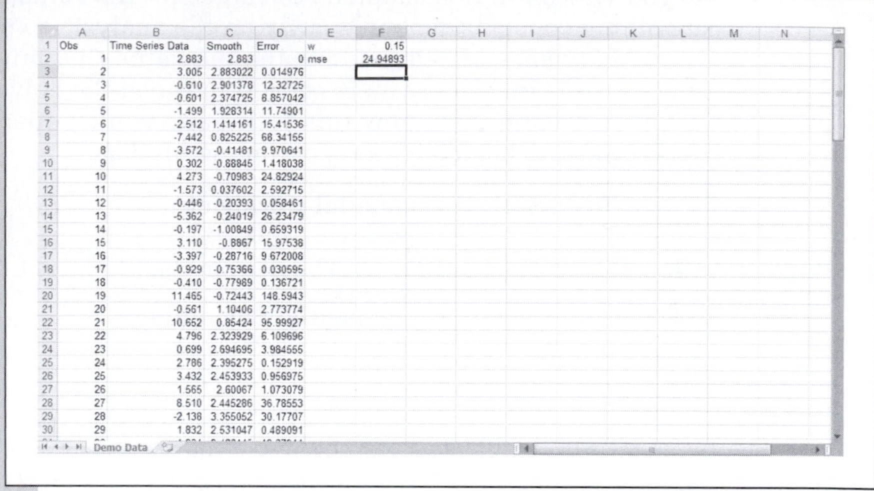

	A	B	C	D	E	F
1	Obs	Time Series Data	Smooth	Error	w	0.15
2	1	2.883	2.883	0	mse	24.94893
3	2	3.005	2.883022	0.014976		
4	3	-0.610	2.901378	12.32725		
5	4	-0.601	2.374725	8.857042		
6	5	-1.499	1.928314	11.74901		
7	6	-2.512	1.414161	15.41536		
8	7	-7.442	0.825225	68.34155		
9	8	-3.572	-0.41481	9.970641		
10	9	0.302	-0.88845	1.418038		
11	10	4.273	-0.70983	24.82924		
12	11	-1.573	0.037602	2.592715		
13	12	-0.446	-0.20393	0.058461		
14	13	-5.362	-0.24019	26.23479		
15	14	-0.197	-1.00849	0.659319		
16	15	3.110	-0.8867	15.97538		
17	16	-3.397	-0.28716	9.672008		
18	17	-0.929	-0.75366	0.030595		
19	18	-0.410	-0.77989	0.136721		
20	19	11.465	-0.72443	148.5943		
21	20	-0.561	1.10406	2.773774		
22	21	10.652	0.85424	95.99927		
23	22	4.796	2.323929	6.109696		
24	23	0.699	2.694695	3.984555		
25	24	2.786	2.395275	0.152919		
26	25	3.432	2.453933	0.956975		
27	26	1.565	2.60067	1.073079		
28	27	8.510	2.445286	36.78553		
29	28	-2.138	3.355052	30.17707		
30	29	1.832	2.531047	0.489091		

Demo Data

Figure 11-35 Exponential smoothing values

You now have everything you need to use Solver.

To open Solver:

1 Click the **Office** button and then click **Excel Options**.

2 Click **Add-Ins** from the list of Excel Options and then click **Go** next to the Manage Excel Add-Ins list box.

3 Click the **Solver Add-In** check box if it is not already selected and then click the **OK** button.

Once the Solver is installed and activated, you can determine the optimal value for the smoothing constant.

To determine the optimal value for the smoothing constant:

1 Click the **Solver** button located on the Analysis group in the Data tab.

2 Type **F2** in the Set Target Cell text box. This is the cell that you will use as a target for the Solver.

3 Click the **Min** option button to indicate that you want to minimize the value of the mean square error (cell F2).

4. Type **F1** in the By Changing Cells text box to indicate that you want to change the value of F1, the smoothing constant, in order to minimize cell F2.

   Because the exponential smoothing constant can take on only values between 0 and 1, you have to add some constraints to the values that the Solver will investigate.

5. Click the **Add** button.

6. Type **F1** in the Cell Reference text box, select **<=** from the Constraint drop-down list, type **1** in the Constraint text box, and then click **Add.**

7. Type **F1** in the Cell Reference text box, select **>=** from the Constraint drop-down list, type **0** in the Constraint text box, and then click **Add.**

8. Click **Cancel** to return to the Solver Parameters dialog text box. The completed Solver Parameters dialog box should look like Figure 11-36.

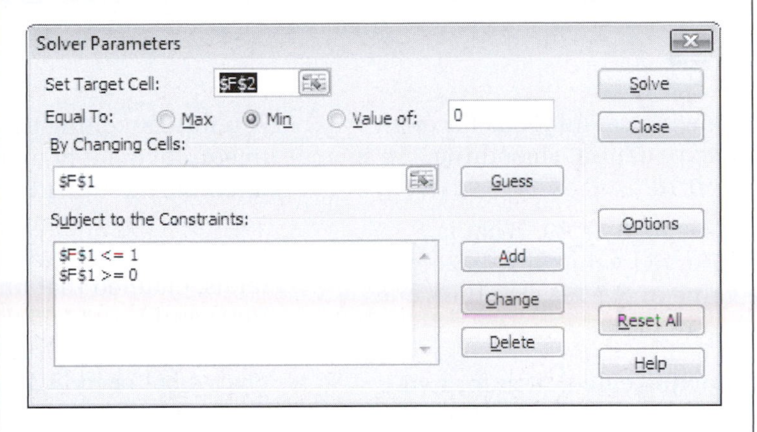

Figure 11-36 The Solver Parameters dialog box

9. Click **Solve.**

   The Solver now determines the optimal value for the smoothing constant (at least in terms of minimizing the mean square error). When the Solver is finished, it will prompt you either to keep the Solver solution or to restore the original values.

10. Click **OK** to keep the solution.

The Solver returns a value of 0.028792 (cell F1) for the smoothing constant, resulting in a mean square error of 23.99456 (cell F2). This is the optimal value for the smoothing constant.

It's possible to set up similar spreadsheets for two-parameter and three-parameter exponential smoothing, but that will not be demonstrated here. The main difficulty in setting up the spreadsheet to do these calculations is in determining the initial estimates of S_0, T_0, and the seasonal indexes.

In the case of two-parameter exponential smoothing, you would use linear regression on the entire time series to derive initial estimates for the location and trend values. Once this is done, you would derive the forecasted values using the recursive equations described earlier in the chapter. You would then apply the Solver to minimize the mean square error of the forecasts by modifying both the location and the trend smoothing constants. Using the Solver to derive the best smoothing constants for the three-parameter model is more complicated because you have to come up with initial estimates for all of the seasonal indexes. The interested student can refer to more advanced texts for techniques to calculate the initial estimates.

You can now save and close the Exponential Smoothing workbook.

Exercises

1. Do the following calculations for one-parameter exponential smoothing, where $w = 0.10$:

a. $S_4 = 23.4$ and $y_5 = 29$. What is S_5?

b. If the observed value of y_6 is 25, what is the value of S_6? Assume the same values as in part a.

2. Do the following calculation for two-parameter exponential smoothing, where $w = 0.10$ and $t = 0.20$:

a. $S_4 = 23.4$, $T_4 = 1.1$, and $y_5 = 29$. What is S_5? What is T_5?

b. If the observed value of y_6 is 25, what are the values of S_6 and T_6? Assume the same values as in part a.

3. If monthly sales are equal to 4,811 units and the seasonal index for that month is 0.85, what is the adjusted sales figure?

4. How can you tell whether a series is seasonal? Mention plots, including the ACF. What is the difference between additive and multiplicative seasonality?

5. A politician citing the latest raw monthly unemployment figures claimed that unemployment had fallen by 88,000 workers. The Bureau of Labor Statistics, however, using seasonally adjusted totals, claimed that unemployment had increased by 98,000. Discuss the two interpretations of the data. Which number gives a better indication of the state of the economy?

6. The Batting Average workbook contains data on the leading major league baseball batting averages for the years 1901 to 2002. Analyze these data.

a. Open the **Batting Average** workbook from the Chapter11 folder and save it as **Batting Average Analysis**.

b. Create a line chart of the batting average versus year. Do you see any apparent trends? Do you see any outliers? Does George Brett's average of 0.390 in 1980 stand out compared with other observations?

c. Insert a trend line smoothing the batting average using a ten-year moving average.

d. Calculate the ACF and state your conclusions (notice that the ACF does not drop off to zero right away, which suggests a trend component).

e. Calculate the difference of the batting averages from one year to the next. Plot the difference series and also compute its ACF. Does the plot show that the variance of the original series is reasonably stable? That is, are the changes roughly the same size at the beginning, middle, and end of the series?

f. Looking at the ACF of the differenced series, do you see much correlation after the first few lags? If not, it suggests that the differenced series does not have a trend, and this is what you would expect. Interpret any lags that are significantly correlated.

g. Perform one-parameter exponential smoothing forecasting one year ahead, using w values of 0.2, 0.3, 0.4, and 0.5. In each case, notice the value predicted for 2003 (observation 103). Which parameter gives the lowest standard error?

h. Save your changes to the workbook and write a report summarizing your observations.

7. The Electric workbook has monthly data on U.S. electric power production, 1978 through 1990. The variable called power is measured in billions of kilowatt hours. The figures come from the 1992 *CRB Commodity Year Book*, published by the Commodity Research Bureau in New York.

a. Open the **Electric** workbook from the Chapter11 folder and save it as **Electric Analysis**.

b. Create a line chart of the power data. Is there any seasonality to the data?

c. Fit a three-parameter exponential model with location, linear, and seasonal parameters. Use a smoothing constant of 0.05 for the location parameter, 0.15 for the linear parameter, and 0.05 for the seasonal parameter. What level of power production do you forecast over the next 12 months?

d. Using the seasonal index, which are the three months of highest power production? Is this in accordance with the plots you have seen? Does it make sense to you as a consumer? By what percentage does the busiest month exceed the slowest month?

e. Repeat the exponential smoothing of part b of Exercise 6 with the smoothing constants shown in Table 11-7.

Table 11-7 Exponential Smoothing Constants

Location	Linear	Seasonal
0.05	0.30	0.05
0.15	0.15	0.05
0.15	0.30	0.05
0.30	0.15	0.05
0.30	0.30	0.05

f. Which forecasts give the smallest standard error?

g. Save your changes to the workbook and report your observations.

8. The Visit workbook contains monthly visitation data for two sites at the Kenai Fjords National Park in Alaska from January 1990 to June 1994. You'll analyze the visitation data for the Exit Glacier site.

a. Open the **Visit** workbook from the Chapter11 data folder and save it as **Visit Analysis**.

b. Create a line plot of visitation for Exit Glacier versus year and month. Summarize the pattern of visitation at Exit Glacier between 1990 and mid-1994.

c. Create two line plots, one showing the visitation at Exit Glacier plotted against year with different lines for different months, and the second showing visitation plotted against month with different lines for different years (you will have to create a two-way table for this). Are there any unusual values? How might the June 1994 data influence future visitation forecasts?

d. Calculate the seasonally adjusted values for visits to the park. Is there a particular month in which visits to the park jump to a new and higher level?

e. Smooth the visitation data using exponential smoothing. Use smoothing constants of 0.15 for both the location and the linear parameters, and use 0.05 for the seasonal parameter. Forecast the visitation 12 months into the future. What are the projected values for the next 12 months?

f. A lot of weight of the projected visitations for 1994–1995 is based on the jump in visitation in June 1994. Assume that this jump was an aberration, and refit two exponential smoothing models with 0.05 and 0.01 for the location parameter (to reduce the effect of the June 1994 increase), 0.15 for the linear parameter, and 0.05 for the seasonal parameter. Compare your results with your first forecasts. How do the standard errors compare? Which projections would you work with and why? What further information would you need to decide between these three projections?

g. What problems do you see with either forecasted value? (*Hint:* Look at the confidence intervals for the forecasts.)

h. Save your changes to the workbook and write a report summarizing your observations.

9. The visitation data in the Visit workbook cover a wide range of values. It might be appropriate to analyze the $\log_{10}$ of the visitation counts instead of the raw counts.

a. Open the **Visit** workbook from the Chapter11 folder and save it as **Visit Log Analysis**.

b. Create a new column in the workbook of the $\log_{10}$ counts of the Exit Glacier data (use the Excel function $\log_{10}$).

c. Create a line plot of $\log_{10}$ (visitation) for the Exit Glacier site from 1990 to mid-1994. What seasonal values does this chart reveal that were hidden when you charted the raw counts?

d. Use exponential smoothing to smooth the $\log_{10}$ (visitation) data. Use a value of 0.15 for the location and linear effects, and use 0.05 for the seasonal effect. Project $\log_{10}$ (visitation) 12 months into the future. Untransform the projections and the prediction intervals by raising 10 to the power of $\log_{10}$ (visitation) [that is, if $\log_{10}$ (visitation) = 1.6, then visitation = $10^{1.6} = 39.8$]. What do you project for the next year at Exit Glacier? What are the 95% prediction intervals? Are the upper and lower limits reasonable?

e. Redo your forecasts, using 0.01 and then 0.05 for the location parameter, 0.15 for the linear parameter, and 0.05 for the seasonal parameter. Which of the three projections results in the smallest standard error?

f. Compare your chosen projections from Exercise 8, using the raw counts, with your chosen projections from this exercise, using the $\log_{10}$ transformed counts. Which would you use to project the 1994–1995 visitations? Which would you use to determine the amount of personnel you will need in the winter months and why?

g. Save your changes to the workbook and write a report summarizing your conclusions.

10. The NFP workbook contains daily body temperature data for 239 consecutive days for a woman in her twenties. Daily temperature readings are one component of natural family planning (NFP) in which a woman uses her monthly cycle with a number of biological signs to determine the onset of ovulation. The file has four columns: Observation, Period (the menstrual period), Day (the day of the menstrual period), and Waking Temperature. Day 1 is the first day of menstruation.

   a. Open the **NFP** workbook from the Chapter11 folder and save it as **NFP Analysis**.
   b. Create a line plot of the daily body temperature values. Do you see any evidence of seasonality in the data?
   c. Create a boxplot of temperature versus day. What can you determine about the relationship between body temperature and the onset of menstruation?
   d. Calculate the ACF for the temperature data up through lag 70. On the basis of the shape of the ACF, what would you estimate as the length of the period in days?
   e. Smooth the data using exponential smoothing. Use 0.15 as the location parameter, 0.01 for the linear parameter (it will not be important in this model), and 0.05 for the seasonal parameter. Use the period length that you estimated in part c of Exercise 9. What body temperature values do you forecast for the next cycle?
   f. Repeat your forecast with values of 0.15 and 0.25 for the seasonal parameters. Which model has the lowest standard error?
   g. Save your changes to the workbook and write a report summarizing your conclusions.

11. The Draft workbook contains data from the 1970 Selective Service draft. Each birth date was given a draft number. Those eligible men with a low draft number were drafted first. One way of presenting the draft number data is through exponential smoothing. The draft numbers vary greatly from day to day, but by smoothing the data, you may be better able to spot trends in the draft numbers. In this exercise, you'll use exponential smoothing to examine the distribution of the draft numbers.

   a. Open the **Draft** workbook from the Chapter11 folder and save it as **Draft Number Analysis**.
   b. Create one-parameter exponential smoothed plots of the number variable on the Draft Numbers worksheet. Use values of 0.15, 0.085, and 0.05 for the location parameter. Which value results in the lowest mean square error?
   c. Examine your plots. Does there appear to be any sort of pattern in the smoothed data?
   d. Test to see whether any autocorrelation exists in the draft numbers. Test for autocorrelation up to a lag of 30. Is there any evidence for autocorrelation in the time series?
   e. Save your changes to the workbook and write a report summarizing your observations.

12. The Oil workbook displays information on monthly production of crude cottonseed oil from 1992 to 1995. The production of cottonseed oil follows a seasonal pattern. Using the data in this workbook, project the monthly values for 1996.

a. Open the **Oil** workbook from the Chapter11 folder and save it as **Oil Forecasts**.
b. Restructure the data in the worksheet into a two-way table. Create a line plot of the production values in the table using a separate line for each year. Describe the seasonal nature of cottonseed oil production.
c. Smooth the production data using a value of 0.15 for all three smoothing factors. Forecast the values 12 months into the future. What are your projections and your upper and lower limits for 1996?
d. Adjust the production data for the seasonal effects. Is there evidence that the adjusted production values have increased over the four-year period? Test your assumption by performing a linear regression of the adjusted values on the month number (1–48). Is the regression significant at the 5% level?
e. Save your changes to the workbook and write a report summarizing your conclusions.

13. The Bureau of Labor Statistics records the number of work stoppages each month that involve 1000 or more workers in the period. Are such work stoppages seasonal in nature? Are there more work stoppages in summer than in winter?

a. Open the **Stoppage** workbook from the Chapter11 folder and save it as **Stoppage Analysis**.
b. Restructure the data in the Work Stoppage worksheet into a two-way table, with each year in a separate row and each month in a separate column.
c. Use the two-way table to create a boxplot and line plot of the work stoppage values. Which months have the highest work stoppage numbers? Do work stoppages occur more often in winter or in summer?
d. Adjust the work stoppage values assuming a 12-month cycle. Is there evidence in the scatterplot that the adjusted number of work stoppages has decreased over the past decade?
e. Smooth adjusted values using one-parameter exponential smoothing. Use a value of 0.15 for the smoothing parameter.
f. Save your changes to the workbook. Summarize your findings regarding work stoppages of 1000 or more workers. Are they seasonal? Have they declined in recent years? Use whatever charts and tables you created to support your conclusions.

14. The Jobs workbook contains monthly youth unemployment rates from 1981 to 1996. Analyze the data in the workbook and try to determine whether unemployment rates are seasonal.

a. Open the **Jobs** workbook from the Chapter11 folder. Save it as **Jobs Analysis**.
b. Restructure the data in the Youth Unemployment worksheet into a two-way table, with each year in a separate row and each month in a separate column.
c. Create a spaghetti plot of the unemployment values.
d. Create a boxplot of youth unemployment rates. Is any pattern apparent in the boxplot?
e. Adjust the unemployment rates assuming a 12-month cycle. Is there evidence in the chart that youth unemployment varies with the season?
f. Save your changes to the workbook and write a report summarizing your observations.

Chapter 12

QUALITY CONTROL

Objectives

In this chapter you will learn to:

- Distinguish between controlled and uncontrolled variation
- Distinguish between variables and attributes
- Determine control limits for several types of control charts
- Use graphics to create statistical control charts with Excel
- Interpret control charts
- Create a Pareto chart

In this chapter you will look at one of the statistical tools used in manufacturing and industry. The proper use of quality control can improve productivity, enhance quality, and reduce production costs. In this chapter, you'll learn about one such tool, the control chart, that is used to determine when a process is out of control and requires human intervention.

Statistical Quality Control

The immediately preceding chapters have been dedicated to the identification of relationships and patterns among variables. Such relationships are not immediately obvious, mainly because they are never exact for individual observations. There is always some sort of variation that obscures the true association. In some instances, once the relationship has been identified, an understanding of the types and sources of variation becomes critical. This is especially true in business, where people are interested in controlling the variation of a process. A **process** is any activity that takes a set of inputs and creates a product. The process for an industrial plant takes raw materials and creates a finished product. A process need not be industrial. For example, another type of process might be to take unorganized information and produce an organized analysis. Teaching could even be considered a process, because the teacher takes uninformed students and produces students capable of understanding a subject (such as statistics!). In all such processes, people are interested in controlling the procedure so as to improve the quality. The analysis of processes for this purpose is called **statistical quality control** (**SQC**) or **statistical process control** (**SPC**).

Statistical process control originated in 1924 with Walter A. Shewhart, a researcher for Bell Telephone. A certain Bell product was being manufactured with great variation in quality, and the production managers could not seem to reduce the variation to an acceptable level. Dr. Shewhart developed the rudimentary tools of statistical process control to improve the homogeneity of Bell's output. Shewhart's ideas were later championed and refined by W. Edwards Deming, who tried unsuccessfully to persuade U.S. firms to implement SPC as a methodology underlying all production processes. Having failed to convince U.S. executives of the merits of SPC, Deming took his cause to Japan, which, before World War II, was renowned for its shoddy goods. The Japanese adopted SPC wholeheartedly, and Japanese production became synonymous with high and uniform quality. In response, U.S. firms jumped on the SPC bandwagon, and many of their products regained market share.

Controlled Variation

The reduction of variation in any process is beneficial. However, you can never eliminate all variation, even in the simplest process, because there are bound to be many small, unobservable, chance effects that influence the process outcome. Variation of this kind is called **controlled variation** and is analogous to the random-error effects in the ANOVA and regression models you studied earlier. As in those statistical models, many individually insignificant random factors interact to have some net effect on the process output. In quality-control terminology, this random variation is said to be "in control," not because the process operator is able to control the factors absolutely, but rather because the variation is the result of normal disturbances, called **common causes**, within the process. This type of variation can be predicted. In other words, given the limitations of the process, each of these common causes is controlled to the greatest extent possible.

Because controlled variation is the result of small variations in the normally functioning process, it cannot be reduced unless the entire process is redesigned. Furthermore, any attempts to reduce the controlled variation without redesigning the process will create more, not less, variation in the process. Endeavoring to reduce controlled variation is called **tampering**; this increases costs and must be avoided. Tampering might occur, for instance, when operators adjust machinery in response to normal variations in the production process. Because normal variations will always occur, adjusting the machine is more likely to harm the process, actually *increasing* the variation in the process, than to help it.

Uncontrolled Variation

The other type of variation that can occur within a process is called uncontrolled variation. **Uncontrolled variation** is due to **special causes**, which are sources of variation that arise sporadically and for reasons outside the normally functioning process. Variation induced by a special cause is usually significant in magnitude and occurs only occasionally. Examples of special causes include differences between machines, different skill or concentration levels of workers, changes in atmospheric conditions, and variation in the quality of inputs.

Unlike controlled variation, uncontrolled variation can be reduced by eliminating its special cause. The failure to bring uncontrolled variation into control is costly.

SPC is a methodology for distinguishing whether variation is controlled or uncontrolled. If variation is controlled, then only improvements in the process itself can reduce it. If variation is uncontrolled, then further analysis is needed to identify and eliminate the special cause.

Table 12-1 summarizes the two types of variation studied in SPC.

Table 12-1 Types of Variation

Variation	Descriptive	Remedy
Controlled	Variation that is native to the process, resulting from normal factors called common causes	Redesign the process to result in a new set of controlled variations with better properties.
Uncontrolled	Variation that is the result of special causes and need not be inherent in the process	Analyze the process to locate the source of the uncontrolled variation and then remove or fix that special cause.

Control Charts

The principal tool of SPC is the control chart. A **control chart** is a graph of the process values plotted in time order. Figure 12-1 shows a sample control chart.

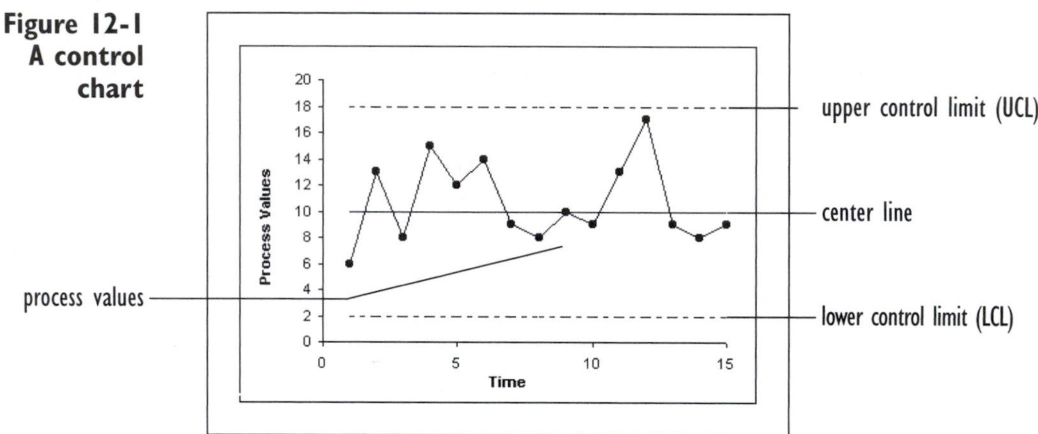

Figure 12-1 A control chart

The chief features of the control chart are the **lower** and **upper control limits** (**LCL** and **UCL**, respectively), which appear as dotted horizontal lines. The solid line between the upper and lower control limits is the **center line** and indicates the expected values of the process.

As the process goes forward, values are added to the control chart. As long as the points remain between the lower and upper control limits, we assume that the observed variation is controlled variation and that the process is **in control** (there are a few exceptions to this rule, which we'll discuss shortly). Figure 12-1 shows a process that is in control. It is important to note that control limits do not represent specification limits or maximum variation targets. Rather, control limits illustrate the limits of normal controlled variation.

In contrast, the process depicted in Figure 12-2 is **out of control**. Both the fourth and the twelfth observations lie outside of the control limits, leading us to believe that their values are the result of uncontrolled variation. At this point a shop manager, or the person responsible for the process, might examine the conditions for those observations that resulted in such extreme values. An analysis of the causes could lead to a better, more efficient, and more stable process.

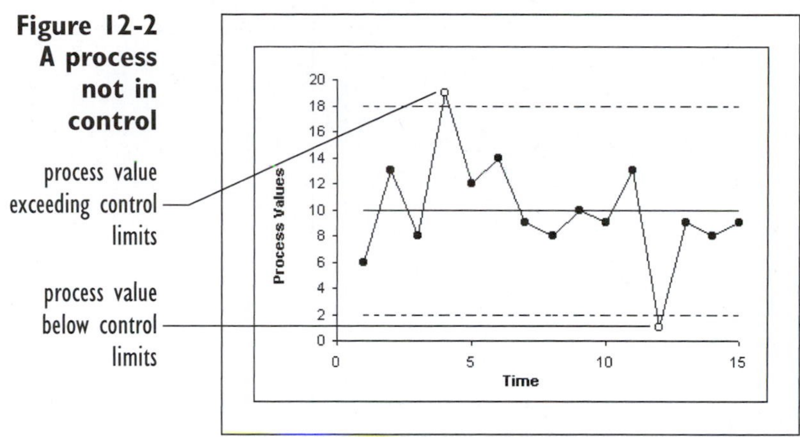

Figure 12-2 A process not in control

Even control charts in which all points lie between the control limits might suggest that a process is out of control. In particular, the existence of a pattern in eight or more consecutive points indicates a process out of control, because an obvious pattern violates the assumption of random variability. In Figure 12-3, for example, the last eight observations depict a steady upward trend. Even though all of the points lie within the control limits, you must conclude that this process is out of control because of the evident trend the data values exhibit.

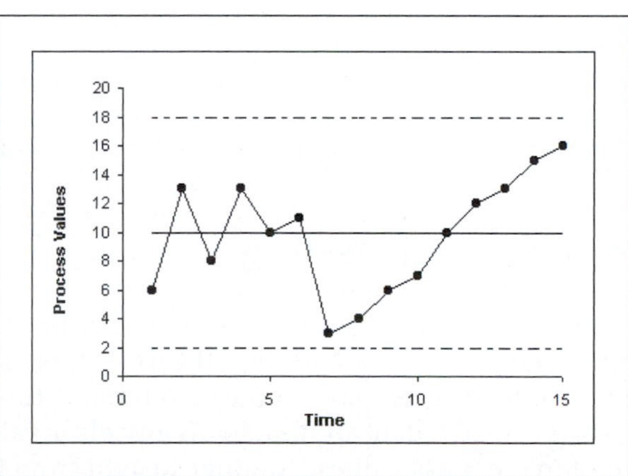

Figure 12-3 A process out of control because of an upward trend

Another common example of a process that is out of control, even though all points lie between the control limits, appears in Figure 12-4. The first eight observations are below the center line, whereas the last seven observations all lie above the center line. Because of prolonged periods where values are either small or large, this process is out of control. One could use the Runs test, discussed in Chapter 8 in the context of examining residuals, to test whether the data values are clustered in a nonrandom way.

Figure 12-4 A process out of control because of a nonrandom pattern

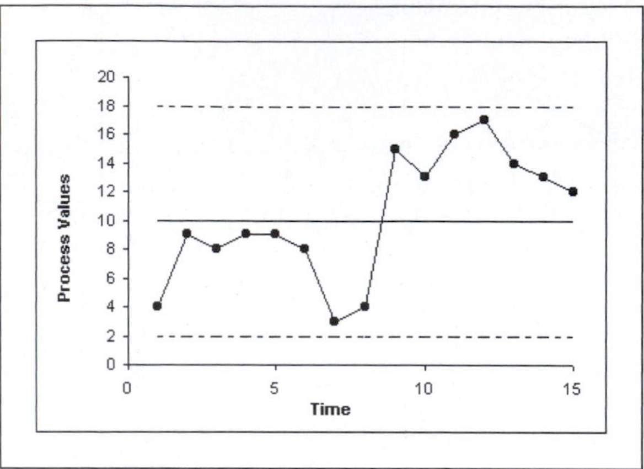

Here are two other situations that may show a process out of control, even though all values lie within the control limits.

- 9 points in a row, all on the same side of the center line
- 14 points in a row, alternating above and below the center line

Other suspicious patterns could appear in control charts. Unfortunately, we cannot discuss them all here. In general, though, any clear pattern in the process values indicates that a process is subject to uncontrolled variation and that it is not in control.

Statisticians usually highlight out-of-control points in control charts by circling them. As you can see, the control chart makes it very easy for you to identify visually points and processes that are out of control without using complicated statistical tests. This makes the control chart an ideal tool for the shop floor, where quick and easy methods are needed.

Control Charts and Hypothesis Testing

The idea underlying control charts should be familiar to you. It is closely related to confidence intervals and hypothesis testing. The associated null hypothesis is that the process is in control; you reject this null hypothesis if any point lies outside the control limits or if any clear pattern appears in the distribution of the process values. Another insight from this analogy is that the possibility of making errors exists, just as errors can occur in standard hypothesis testing. In other words, occasionally a point that lies outside the control limits does not have any special cause but occurs because of normal process variation. On the other hand, there could exist a special cause that is not big enough to move the point outside of the control limits. Statistical analysis can never be 100% certain.

Variable and Attribute Charts

There are two categories of control charts: those which monitor variables and those which monitor attributes. **Variable charts** display continuous measures, such as weight, diameter, thickness, purity, and temperature. As you have probably already noticed, much statistical analysis focuses on the mean values of such measures. In a process that is in control, you expect the mean output of the process to be stable over time.

Attribute charts differ from variable charts in that they describe a feature of the process rather than a continuous variable such as a weight or volume. Attributes can be either discrete quantities, such as the number of defects in a sample, or proportions, such as the percentage of defects per lot. Accident and safety rates are also typical examples of attributes.

Using Subgroups

In order to compare process levels at various points in time, we usually group individual observations together into **subgroups**. The purpose of the subgroup is to create a set of observations in which the process is relatively stable with controlled variation. Thus the subgroup should represent a set of homogeneous conditions. For example, if we were measuring the results of a manufacturing process, we might create a subgroup consisting of values from the same machine closely spaced in time. Once we create the subgroups, we can calculate the subgroup averages and calculate the variance of the values. The variation of the process values within the subgroups is then used to calculate the control limits for the entire set of process values. A control chart might then answer the question Do the averages *between* the subgroups vary more than expected, given the variation *within* the subgroups?

The $\overline{x}$ Chart

One of the most common variable control charts is the $\overline{\mathbf{x}}$ **chart** (the "x bar chart"). Each point in the $\overline{x}$ chart displays the subgroup average against the subgroup number. Because observations usually are taken at regular time intervals, the subgroup number is typically a variable that measures time, with subgroup 2 occurring after subgroup 1 and before subgroup 3. As an example, consider a clothing store in which the owner monitors the length of time customers wait to be served. He decides to calculate the average wait time in half hour increments. The first half hour of customers who were served between 9 and 9:30 a.m. forms the first subgroup, and the owner records the average wait time during this interval. The second subgroup covers customers served from 9:30 to 10:00 a.m., and so forth.

The $\overline{x}$ chart is based on the standard normal distribution. The standard normal distribution underlies the mean chart, because the Central Limit Theorem (see Chapter 5) states that the subgroup averages approximately follow the normal distribution even when the underlying observations are not normally distributed.

The applicability of the normal distribution allows the control limits to be calculated very easily when the standard deviation of the process is known. You might recall from Chapter 5 that 99.74% of the observations in a normal distribution fall within 3 standard deviations of the mean (μ). In SPC, this means that points that fall more than 3 standard deviations from the mean occur only 0.26% of the time. Because this probability is so small, points outside the control limits are assumed to be the result of uncontrolled special causes. Why not narrow the control limits to ±2 standard deviations? The problem with this approach is that you might increase the **false-alarm rate**, that is, the number of times you stop a process that you incorrectly believed was out of control. Stopping a process can be expensive, and adjusting a process that doesn't need adjusting might increase the variability through tampering. For this reason, a 3-standard-deviation control limit was chosen as a balance between running an out-of-control process and incorrectly stopping a process when it doesn't need to be stopped.

You might also recall that the statistical tests you learned earlier in the book differed slightly depending on whether the population standard deviation was known or unknown. An analogous situation occurs with control charts. The two possibilities are considered in the following sections.

Calculating Control Limits When σ Is Known

If the true standard deviation of the process (σ) is known, then the control limits are

$$\text{LCL} = \mu - \frac{3\sigma}{\sqrt{n}}$$

$$\text{UCL} = \mu + \frac{3\sigma}{\sqrt{n}}$$

and 99.74% of the points should lie between the control limits if the process is in control. If σ is known, it usually derives from historical values. Here,

n is the number of observations in the subgroup. Note that in this control chart and the charts that follow, n need not be the same for all subgroups. Control charts are easier to interpret if this is the case, though.

The value for μ might also be known from past values. Alternatively, μ might represent the target mean of the process rather than the actual mean attained. In practice, though, μ might also be unknown. In that case, the mean of all of the subgroup averages $\bar{\bar{x}}$ replaces μ as follows:

$$\text{LCL} = \bar{\bar{x}} - \frac{3\sigma}{\sqrt{n}}$$

$$\text{UCL} = \bar{\bar{x}} + \frac{3\sigma}{\sqrt{n}}$$

The interpretation of the mean chart is the same whether the true process mean is known or unknown.

Here is an example to help you understand the basic mean chart. Students are often concerned about getting into courses with "good" professors and staying out of courses taught by "bad" ones. In order to provide students with information about the quality of instruction provided by different instructors, many universities use end-of-semester surveys in which students rate various professors on a numeric scale. At some schools, such results are even posted and used by students to help them decide in which section of a course to enroll. Many faculty members object to such rankings on the grounds that although there is always some apparent variation among faculty members, there are seldom any significant differences. However, students often believe that variations in scores reflect the professors' relative aptitudes for teaching and are not simply random variations due to chance effects.

$\bar{x}$ Chart Example: Teaching Scores

One way to shed some light on the value of student evaluations of teaching is to examine the scores for one instructor over time. The Teacher workbook provides data ratings of one professor who has taught principles of economics at the same university for 20 consecutive semesters. The instruction in this course can be considered a process, because the instructor has used the same teaching methods and covered the same material over the entire period. Five student evaluation scores were recorded for each of the 20 courses. The five scores for each semester constitute a subgroup. Possible teacher scores run from 0 (terrible) to 100 (outstanding). The range names have been defined in Table 12-2 for the workbook.

Table 12-2 The Teacher Workbook

Range Name	Range	Description
Semester	A2:A21	The semester of the evaluation
Score_1	B2:B21	First student evaluation
Score_2	C2:C21	Second student evaluation
Score_3	D2:D21	Third student evaluation
Score_4	E2:E21	Fourth student evaluation
Score_5	F2:F21	Fifth student evaluation

To open the Teacher workbook:

1. Open the **Teacher** workbook from the Chapter12 data folder.
2. Save the file as **Teacher Control Chart**.
   Figure 12-5 displays the content of the workbook.

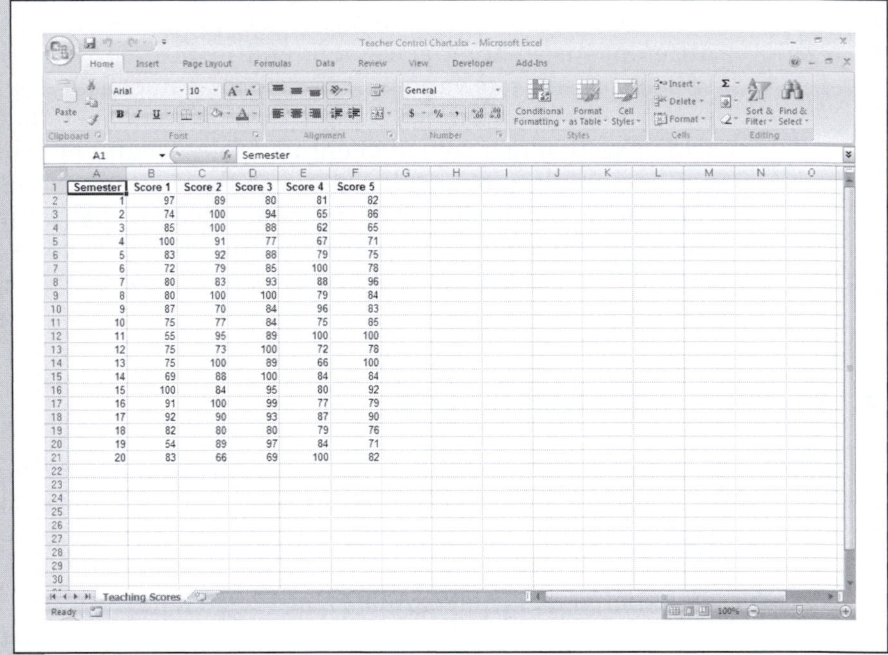

Semester	Score 1	Score 2	Score 3	Score 4	Score 5
1	97	89	80	81	82
2	74	100	94	65	86
3	85	100	88	62	65
4	100	91	77	67	71
5	83	92	88	79	75
6	72	79	85	100	78
7	80	83	93	88	96
8	80	100	100	79	84
9	87	70	84	96	83
10	75	77	84	75	85
11	55	95	89	100	100
12	75	73	100	72	78
13	75	100	89	66	100
14	69	88	100	84	84
15	100	84	95	80	92
16	91	100	99	77	79
17	92	90	93	87	90
18	82	80	80	79	76
19	54	89	97	84	71
20	83	66	69	100	82

Figure 12-5 The Teacher workbook

There is obviously some variation between scores across semesters, with scores varying from a low of 54.0 to a high of 100. Without further analysis, you and your friends might think that such a spread indicates that the professor's classroom performance has fluctuated widely over the course of 20 semesters. Is this interpretation valid?

If you consider teaching to be a process, with student evaluation scores as one of its products, you can use SPC to determine whether the process is in control. In other words, you can use SPC techniques to determine whether the variation in scores is due to identifiable differences in the quality of instruction that can be attributed to a particular semester's course (that is, special causes) or is due merely to chance (common causes).

Historical data from other sources show that σ for this professor is 5.0. Because there are five observations in each subgroup, $n = 5$. You can use StatPlus to calculate the mean scores for each semester and then the average of all 20 mean scores.

To create a control chart of the teacher's scores:

1. Click **QC Charts** from the StatPlus menu and then click **Xbar Chart**.
2. Click the **Subgroups in rows across columns** option button.
3. Click the **Data Values** button and select the range names **Score_1** through **Score_5**. Click **OK**.
4. Click the **Sigma Known** checkbox and type **5** in the accompanying text box.
5. Click the **Output** button and send the control chart to a new chart sheet named **XBar Chart**. Click **OK**.

   Figure 12-6 shows the completed dialog box.

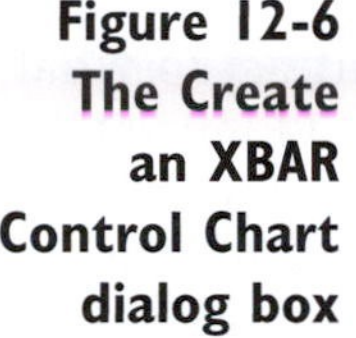

Figure 12-6 The Create an XBAR Control Chart dialog box

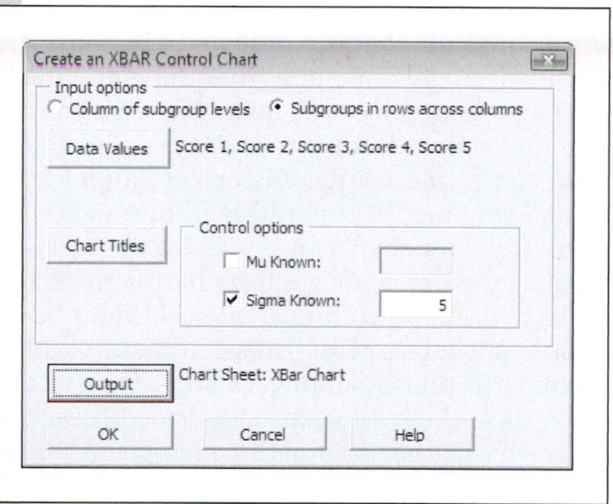

6. Click **OK**.

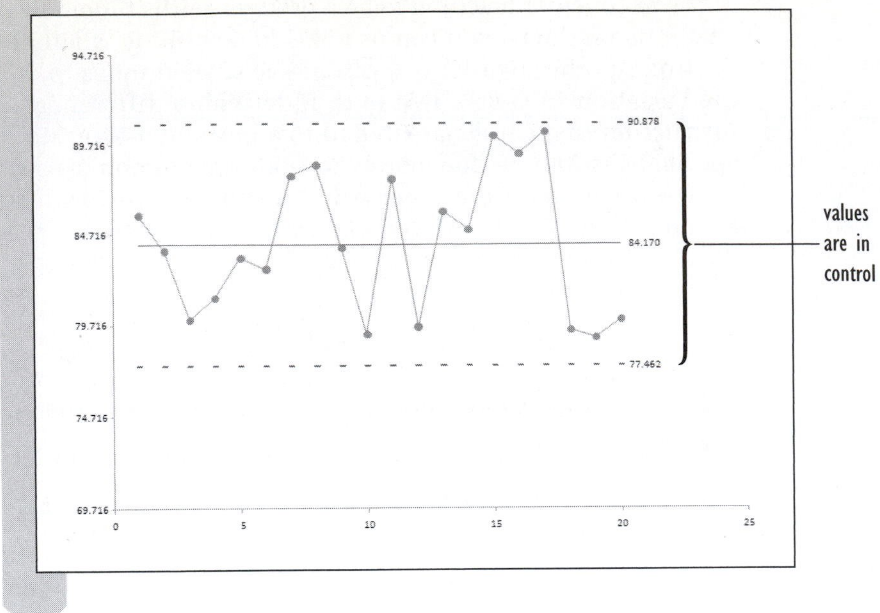

Figure 12-7 Control chart of teacher scores

As you can see from Figure 12-7, no mean score falls outside the control limits. The lower control limit is 77.462, the mean subgroup average is 84.17, and the upper control limit is 90.878. There is no evident trend to the data or nonrandom pattern. You conclude that there is no reason to believe the teaching process is out of control.

Because we conclude that the process is in control, in contrast to what the typical student might conclude from the data, there is no evidence that this professor's performance was better or worse in one semester than in another. The raw scores from the last three semesters are misleading. A student might claim that using a historical value for σ is also misleading, because a smaller value for σ could lead one to conclude that the scores were not in control after all. The exercises at the end of this chapter will examine this issue by redoing the control chart with an unknown value for σ.

One corollary to the preceding analysis should be stated: Because even one professor experiences wide fluctuations in student evaluations over time, apparent differences among various faculty members can also be deceptive. You should use all such statistics with caution.

You can close the Teacher Control Chart workbook now, saving your changes.

Calculating Control Limits When σ Is Unknown

In many instances, the value of σ is not known. You learned in Chapter 6 that the normal distribution does not strictly apply for analysis when σ is unknown and must be estimated. In that chapter, the *t* distribution was used instead of the standard normal distribution. Because SPC is often implemented on the shop floor by workers who have had little or no formal statistical training (and might not have ready access to Excel), the method for estimating σ is simplified and the normal approximation is used to construct the control chart. The difference is that when σ is unknown, the control limits are estimated using the average range of observations within a subgroup as the measure of the variability of the process. The control limits are

$$\text{LCL} = \overline{\overline{x}} - A_2\overline{R}$$

$$\text{UCL} = \overline{\overline{x}} + A_2\overline{R}$$

$\overline{R}$ represents the average of the subgroup ranges, and $\overline{\overline{x}}$ is the average of the subgroup averages. A_2 is a correction factor that is used in quality-control charts. As you'll see, there are many correction factors for different types of control charts. Table 12-3 displays a list of common correction factors for various subgroup sizes n.

Table 12-3 QC Correction Factors

n	A_2	d_2	D_1	D_2	D_3	D_4
2	1.88	1.128	0	3.686	0	3.268
3	1.023	1.693	0	4.358	0	2.574
4	0.729	2.059	0	4.698	0	2.282
5	0.577	2.326	0	4.918	0	2.114
6	0.483	2.534	0	5.078	0	2.004
7	0.419	2.704	0.204	5.204	0.076	1.924
8	0.373	2.847	0.388	5.306	0.136	1.864
9	0.337	2.970	0.547	5.393	0.184	1.816
10	0.308	3.078	0.687	5.469	0.223	1.777
11	0.285	3.173	0.811	5.535	0.256	1.744
12	0.266	3.258	0.922	5.594	0.284	1.717
13	0.249	3.336	1.025	5.647	0.308	1.692
14	0.235	3.407	1.118	5.696	0.329	1.671
15	0.223	3.472	1.203	5.741	0.348	1.652
16	0.212	3.532	1.282	5.782	0.363	1.637
17	0.203	3.588	1.356	5.820	0.378	1.622
18	0.194	3.640	1.424	5.856	0.391	1.608
19	0.187	3.689	1.487	5.891	0.403	1.597
20	0.180	3.735	1.549	5.921	0.415	1.585
21	0.173	3.778	1.605	5.951	0.425	1.575
22	0.167	3.819	1.659	5.979	0.434	1.566
23	0.162	3.858	1.710	6.006	0.443	1.557
24	0.157	3.895	1.759	6.031	0.451	1.548
25	0.153	3.931	1.806	6.056	0.459	1.541

Source: Adapted from "1950 ASTM Manual on Quality Control of Materials," American Society for Testing and Materials, in J. M. Juran, ed., *Quality Control Handbook* (New York: McGraw-Hill, 1974), Appendix II, p. 39. Reprinted with permission of McGraw-Hill.

A_2 accounts for both the factor of 3 from the earlier equations (used when σ was known) and for the fact that the average range represents a proxy for the common-cause variation. (There are other alternative methods for calculating control limits when σ is unknown.) As you can see from the table, A_2 depends only on the number of observations in each subgroup. Furthermore, the control limits become tighter when the subgroup sample size increases. The most typical sample size is 5 because this usually ensures normality of sample means. You will learn to use the control factors in the table later in the chapter.

$\bar{x}$ Chart Example: A Coating Process

The data in the Coats workbook come from a manufacturing firm that sprays one of its metal products with a special coating to prevent corrosion. Because this company has just begun to implement SPC, σ is unknown for the coating process.

To open the Coats workbook:

1 Open the **Coats** workbook from the Chapter12 data folder.

2 Save the workbook as **Coats Control Chart**.

Figure 12-8 shows the contents of the workbook.

Figure 12-8 The Coats workbook

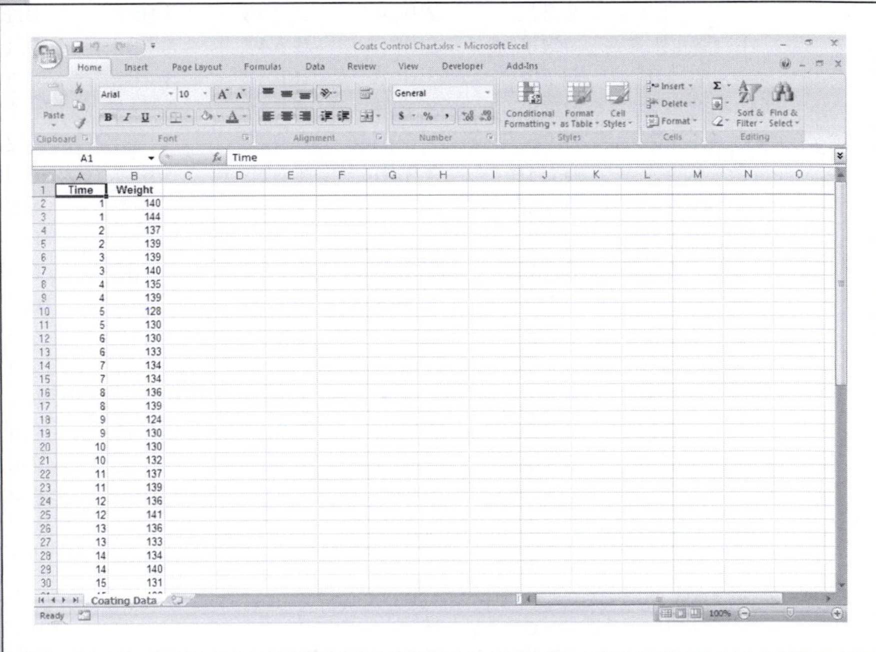

Time	Weight
1	140
1	144
2	137
2	139
3	139
3	140
4	135
4	139
5	128
5	130
6	130
6	133
7	134
7	134
8	136
8	139
9	124
9	130
10	130
10	132
11	137
11	139
12	136
12	141
13	136
13	133
14	134
14	140
15	131

The weight of the spray in milligrams is recorded, with two observations taken at each of 28 times each day. Note that the data are arranged differently, with the Time column indicating the subgroup number. The range names have been defined for the workbook in Table 12-4.

Table 12-4 The Coats Workbook

Range Name	Range	Description
Time	A2:A57	The order of the evaluation (also the subgroup number)
Weight	B2:B57	The weight of the spray in milligrams

As before, you can use StatPlus to create the control chart. Note that because $n = 2$ (there are two observations per subgroup), $A_2 = 1.880$.

To create a control chart of the weight values:

1. Click **QC Charts** from the StatPlus menu and click **Xbar Chart**.
2. Click the **Data Values** button and select the **Weight** range name. Click **OK**.
3. Click the **Subgroups** button and select **Time** from the range names list. Click **OK**.
4. Click the **Output** button and send the control chart to a new chart sheet named **XBar Chart**. Click **OK** twice. See Figure 12-9.

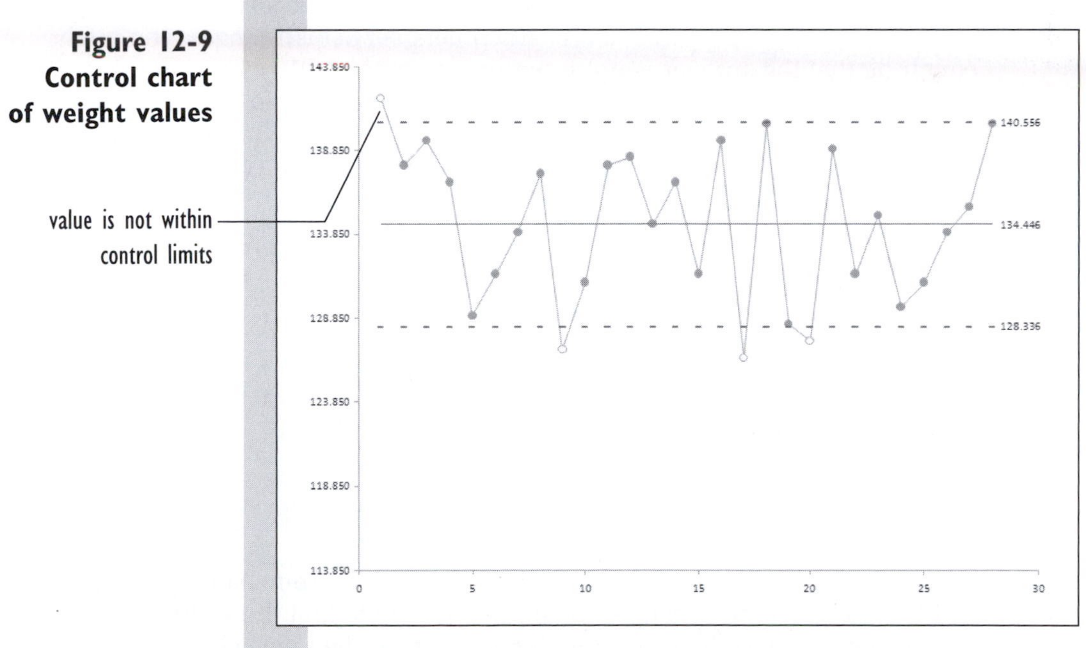

Figure 12-9 Control chart of weight values

The lower control limit is 128.336, the average of the subgroup averages is 134.446, and the upper control limit is 140.556. Note that although most of the points in the mean chart lie between the control limits, four points (observations 1, 9, 17, and 20) lie outside the limits. This process is not in control.

Because the process is out of control, you should attempt to identify the special causes associated with each out-of-control point. Observation 1, for example, has too much coating. Perhaps the coating mechanism became stuck for an instant while applying the spray to that item. The other three observations indicate too little coating on the associated products. In talking with the operator, you might learn that he had not added coating material to the sprayer on schedule, so there was insufficient material to spray.

It is common practice in SPC to note the special causes either on the front of the control chart (if there is room) or on the back. This is a convenient way of keeping records of special causes.

In many instances, proper investigation leads to identification of the special causes underlying out-of-control processes. However, there might be out-of-control points whose special causes cannot be identified.

The Range Chart

The $\overline{x}$ chart provides information about the variation around the average value for each subgroup. It is also important to know whether the range of values is stable from group to group. In the coating example, if some observations exhibit very large ranges and others very small ranges, you might conclude that the sprayer is not functioning consistently over time. To test this, you can create a control chart of the average subgroup ranges, called a **range chart**. As with the $\overline{x}$ chart, the width of the control limits depends on the variability within each subgroup. If σ is known, the control limits for the range chart are

$$\text{LCL} = D_1\sigma$$

$$\text{Center line} = d_2\sigma$$

$$\text{UCL} = D_2\sigma$$

and if σ is not known, the control limits are

$$\text{LCL} = D_3\overline{R}$$

$$\text{UCL} = D_4\overline{R}$$

where d_2, D_1, D_2, D_3, and D_4 are the correction factors from Table 12-3, and $\overline{R}$ is the average subgroup range. It's important to note that the $\overline{x}$ chart is valid only when the range is in control. For this reason the range chart is usually drawn alongside the $\overline{x}$ chart.

Use the information in the Coats workbook to determine whether the range of coating weights is in control.

To create a range chart of the weight values:

1. Return to the Coating Data worksheet.
2. Click **QC Charts** from the StatPlus menu and click **Range Chart**.
3. Select **Weight** as your Data Values variable and **Time** as the Subgroup variable.
4. Verify that the Sigma Known checkbox is *unselected*.
5. Direct the output to a new chart sheet named **Range Chart**. Click **OK** twice.

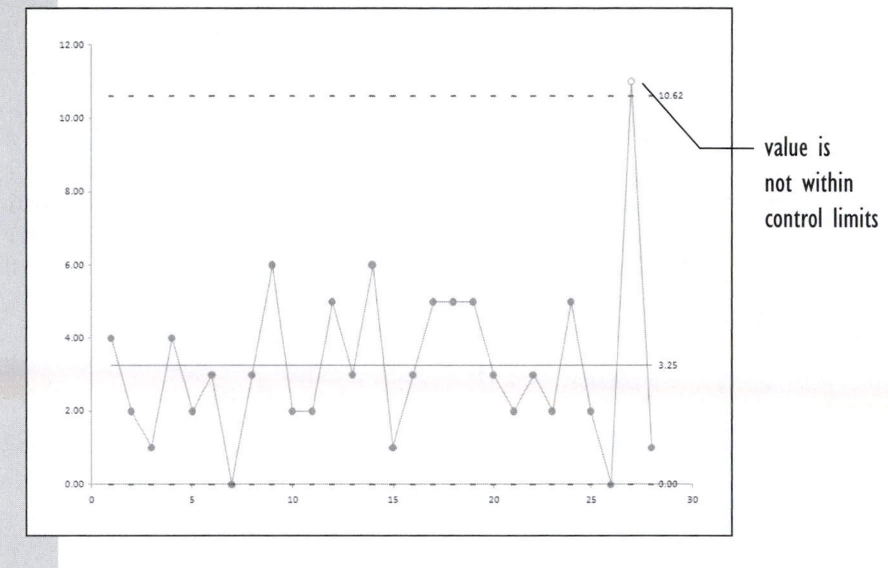

Figure 12-10 The range control chart

Each point on the range chart represents the range within each subgroup. The average subgroup range is 3.25, with the control limits going from 0 to 10.62. According to the range chart shown in Figure 12-10, only the 27th observation has an out-of-control value. The special cause should be identified if possible. However, in discussing the problem with the operator, sometimes you might not be able to determine a special cause. This does not necessarily mean that no special cause exists; it could mean instead that you are unable to determine what the cause is in this instance. It is also possible that there really is no special cause. However, because you are constructing control charts with the width of about 3 standard deviations, an out-of-control value is unlikely unless there is something wrong with the process.

You might have noticed that the range chart identifies as out of control a point that was apparently in control in the $\overline{x}$ chart but does not identify any of the four observations that are out of control in the $\overline{x}$ chart. This is a common occurrence. For this reason, the $\overline{x}$ chart and range charts are often used in conjunction to determine whether a process is in control. In practice, the $\overline{x}$ chart and range chart often appear on the same page because viewing both charts simultaneously improves the overall picture of the process. In this example, you would judge that the process is out of control with both charts but on the basis of different observations.

You can close the Coats Control Chart workbook now, saving your changes.

The C Chart

Both the $\overline{x}$ chart and the range chart measure the values of a particular variable. Now let's look at an attribute chart that measures an attribute of the process. Some processes can be described by counting a certain feature, such as the number of flaws in a standardized section of continuous sheet metal or the number of defects in a production lot. The number of accidents in a plant might also be counted in this manner. A **C chart** displays control limits for the counts attribute. The lower and upper control limits are

$$\text{LCL} = \overline{c} - 3\sqrt{\overline{c}}$$

$$\text{UCL} = \overline{c} + 3\sqrt{\overline{c}}$$

where $\overline{c}$ is the average number of counts in each subgroup. If the LCL is less than zero, by convention it will be set to equal zero, because a negative count is impossible.

C Chart Example: Factory Accidents

The Accidents workbook contains the number of accidents that occurred each month during a period of a few years at a production site. Let's create control charts of the number of accidents per month to determine whether the process is in control.

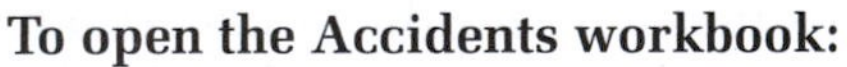

To open the Accidents workbook:

1 Open the **Accidents** workbook from the Chapter12 folder.

2 Save the workbook as **Accidents Control Chart**. See Figure 12-11.

Figure 12-11 The Accidents workbook

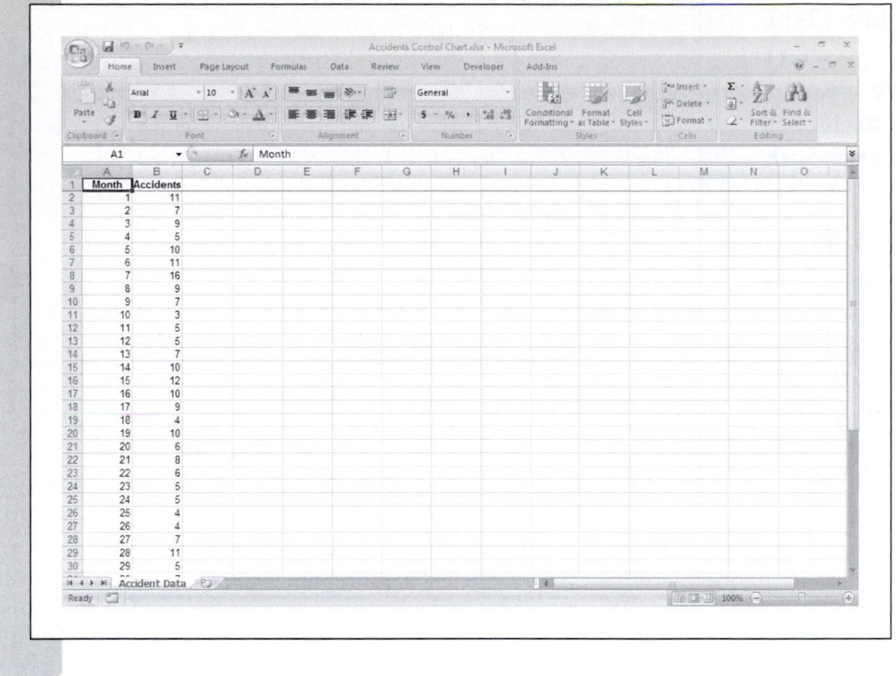

The range names have been defined for the workbook in Table 12-5.

Table 12-5 The Accidents Workbook

Range Name	Range	Description
Month	A2:A45	The month
Accidents	B2:B45	The number of accidents that month

To create a C chart for accidents at this firm:

1. Click **QC Charts** from the StatPlus menu and then click **C Chart**.
2. Select **Accidents** as the Data Values variable.
3. Direct the output to a new chart sheet named **C Chart**.
4. Click **OK**. Excel generates the chart shown in Figure 12-12.

Figure 12-12
C chart
of the number
of accidents
per month

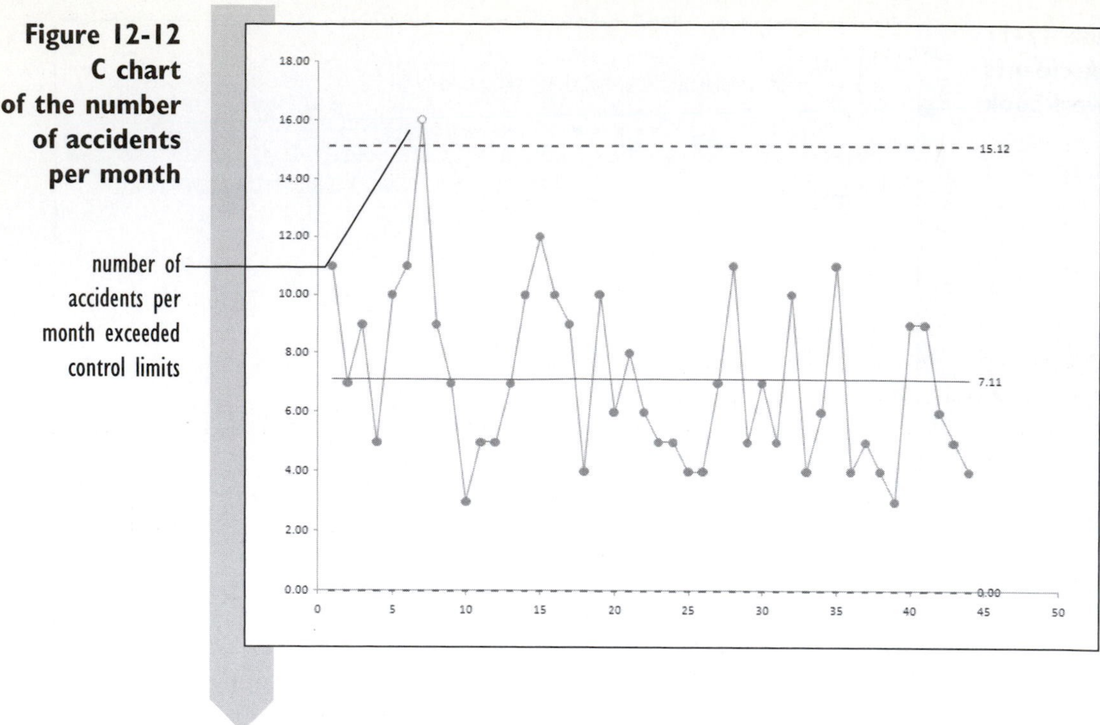

Each point on the C chart in Figure 12-12 represents the number of accidents per month. The average number of accidents per month was 7.11. Only in the seventh month did the number of accidents exceed the upper control limit of 15.12 with 16 accidents. Since then, the process appears to have been in control. Of course, it is appropriate to determine the special causes associated with the large number of accidents in the seventh month. In the case of this firm, the workload was particularly heavy during that month, and a substantial amount of overtime was required. Because employees put in longer shifts than they were accustomed to working, fatigue is likely to have been the source of the extra accidents.

You can close the Accidents Control Chart workbook, saving your results.

The P Chart

Closely related to the C chart is the **P chart**, which depicts the proportion of items with a particular attribute, such as defects. The P chart is often used to analyze the proportion of defects in each subgroup.

Let $\overline{p}$ denote the average proportion of the sample that is defective. The distribution of the proportions can be approximated by the normal distribution, provided that $n\overline{p}$ and $n(1 - \overline{p})$ are both at least 5. If $\overline{p}$ is very close to 0 or 1, a very large subgroup size might be required for the approximation to be legitimate.

The lower and upper control limits are

$$\text{LCL} = \overline{p} - 3\sqrt{\frac{\overline{p}(1 - \overline{p})}{n}}$$

$$\text{UCL} = \overline{p} + 3\sqrt{\frac{\overline{p}(1 - \overline{p})}{n}}$$

P Chart Example: Steel Rod Defects

A manufacturer of steel rods regularly tests whether the rods will withstand 50% more pressure than the company claims them to be capable of withstanding. A rod that fails this test is defective. Twenty samples of 200 rods each were obtained over a period of time, and the number and fraction of defects were recorded in the Steel workbook.

To open the Steel workbook:

1. Open the **Steel** workbook from the Chapter12 data folder.
2. Save your workbook as **Steel Control Chart**. See Figure 12-13.

Figure 12-13 The Steel workbook

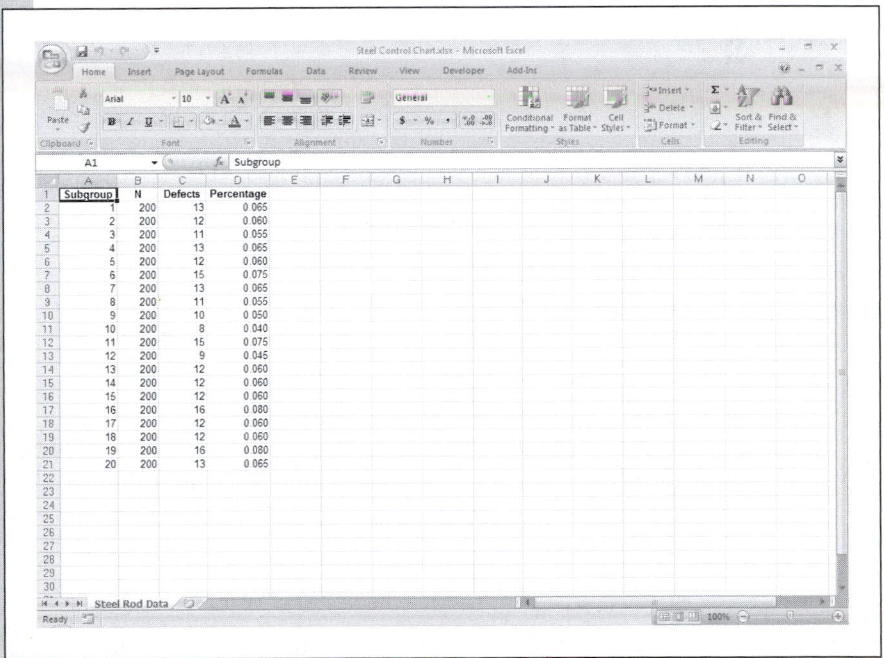

Subgroup	N	Defects	Percentage
1	200	13	0.065
2	200	12	0.060
3	200	11	0.055
4	200	13	0.065
5	200	12	0.060
6	200	15	0.075
7	200	13	0.065
8	200	11	0.055
9	200	10	0.050
10	200	8	0.040
11	200	15	0.075
12	200	9	0.045
13	200	12	0.060
14	200	12	0.060
15	200	12	0.060
16	200	16	0.080
17	200	12	0.060
18	200	12	0.060
19	200	16	0.080
20	200	13	0.065

The range names have been defined for the workbook in Table 12-6.

Table 12-6 The Steel Workbook

Range Name	Range	Description
Subgroup	A2:A21	The subgroup number
N	B2:B21	The size of the subgroup
Defects	C2:C21	The number of defects in the subgroup
Percentage	D2:D21	The fraction of defects in the subgroup

To create a P chart for the percentage of steel rod defects:

1. Click **QC Charts** from the StatPlus menu and click **P Chart**.
2. Click the **Proportions** button and select **Percentage** from the list of range names. Click **OK**.
3. Type **200** in the Sample Size box, because each subgroup has the same sample size.
4. Send the output to a new chart sheet named **P Chart**.
5. Click **OK**.

Excel generates the P chart shown in Figure 12-14.

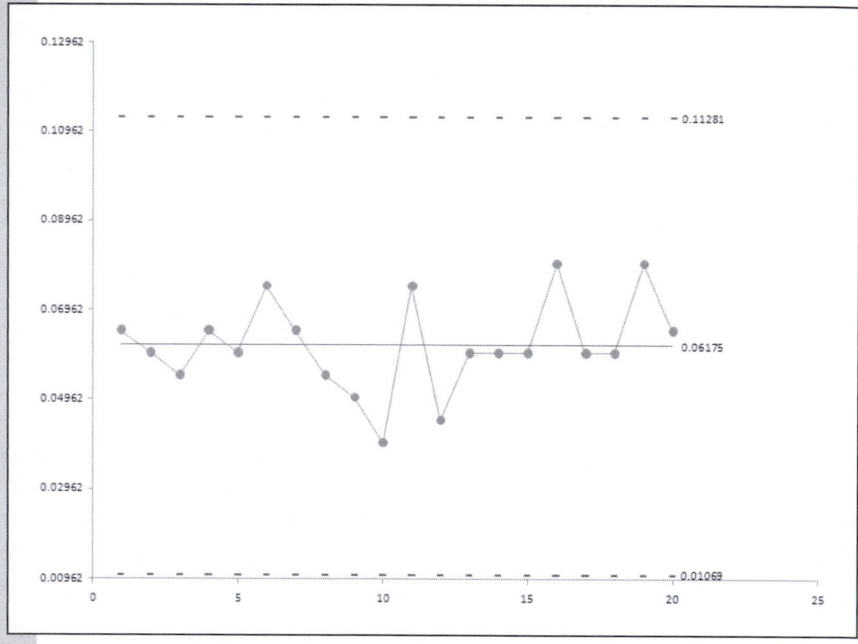

Figure 12-14 P chart of the percentage of steel rod defects

As shown in Figure 12-14, the lower control limit is 0.01069, or a defect percentage of about 1%. The upper control limit is 0.11281, or about 11%. The average defect percentage is 0.06175, about 6%. The control chart clearly demonstrates that no point is anywhere near the 3-σ limits.

Note that not all out-of-control points indicate the existence of a problem. For example, suppose that another sample of 200 rods was taken and that only one rod failed the stress test. In other words, only one-half of 1% of the sample was defective. In this case, the proportion is 0.005, which falls below the lower control limit, so technically it is out of control. Yet you would not be concerned about the process being out of control in this case, because the proportion of defects is so low. Still, you might be inclined to investigate, just to see whether you could locate the source of your good fortune and then duplicate it!

You can save and close the Steel Control Chart workbook now.

Control Charts for Individual Observations

Up to now, we've been creating control charts for processes that can be neatly divided into subgroups. Sometimes it's not possible to group the data into subgroups. This could occur when each measurement represents a single batch in a process or when the measurements are widely spaced in time. With a subgroup size of 1, it's not possible to calculate subgroup ranges. This makes many of the regular formulas impractical to apply.

Instead, the recommended method is to create a subgroup consisting of each consecutive observation and then calculate the moving average of the data. Thus the subgroup variation is determined by the variation from one observation to another, and that variation will be used to determine the control limits for the variation between subgroups. Because we are setting up our subgroups differently, the formulas for the lower and upper control limits are different as well. The LCL and UCL are

$$\text{LCL} = \overline{x} - 3\frac{\overline{R}}{d_2}$$

$$\text{UCL} = \overline{x} + 3\frac{\overline{R}}{d_2}$$

Here $\overline{x}$ is the sample average of all of the observations, $\overline{R}$ is the average range of consecutive values in the data set, and d_2 is the control limit factor shown earlier in Table 12-3. We are using a moving average of size 2, so this will be equal to 1.128. Control charts based on these limits are called **individuals charts**.

We can also create a **moving range chart** of the moving range values, that is, the range between consecutive values. In this case, the lower and upper control limits match the ones used earlier for the range chart.

$$\text{LCL} = D_3\overline{R}$$

$$\text{UCL} = D_4\overline{R}$$

Let's apply these formulas to a workbook recording the tensile strength of 25 steel samples. The values are stored in the Strength workbook.

To open the Strength workbook:

1. Open the **Strength** workbook from the Chapter12 data folder.
2. Save your workbook as Strength Control Chart. See Figure 12-15.

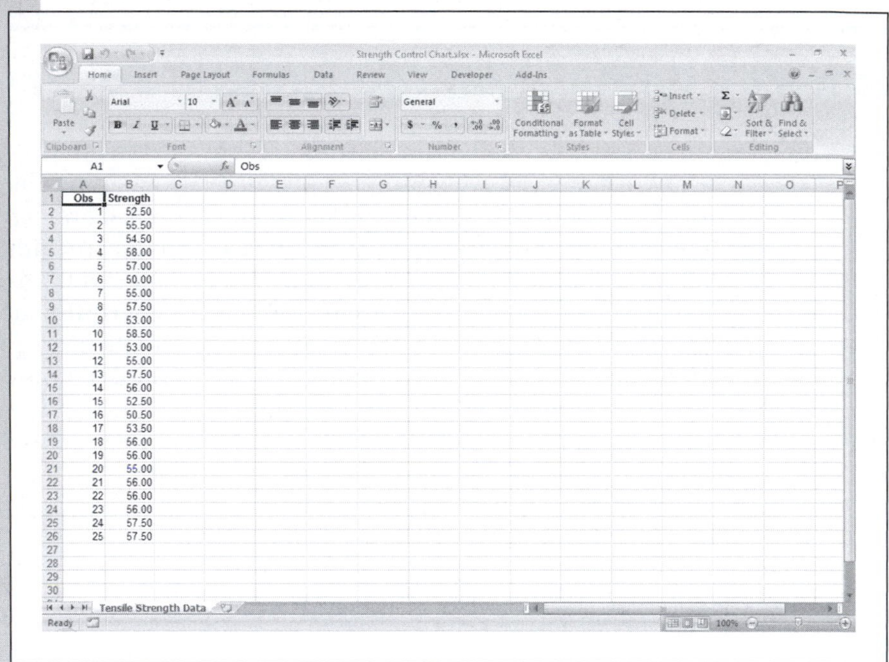

Figure 12-15 The Strength workbook

The range names are shown in Table 12-7.

Table 12-7 The Strength Workbook

Range Name	Range	Description
Obs	A2:A26	The observation number
Strength	B2:B26	The tensile strength of the sample, measured to the nearest 500 pounds in 1,000-pound units

To create an Individuals chart for the steel samples:

1. Click **QC Charts** from the StatPlus menu and click **Individuals Chart**.
2. Select **Strength** for the Data Values variable.
3. Send the output to a new chart sheet named **I-Chart**.
4. Click **OK**. Excel generates the I chart shown in Figure 12-16.

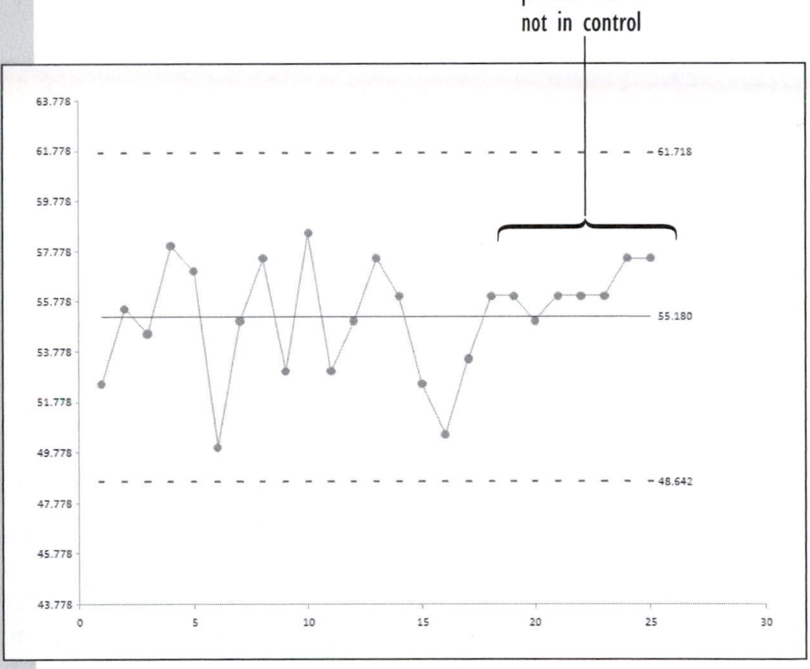

Figure 12-16 The individuals chart for tensile strength samples

The chart shown in Figure 12-16 gives the values of the individual observations (not the moving averages) plotted alongside the upper and lower control limits. No values fall outside the control limits; this leads us to conclude that the process is in control. However, the last eight observations are all either above or near the center line; this might indicate a process going out of control toward the end of the process. This is something that should be investigated further.

We should also plot the moving range chart, to see whether there is any evidence in that plot of an out-of-control process.

To create a moving range chart for the steel samples:

1. Click **QC Charts** from the StatPlus menu and then click **Moving Range Chart**.
2. Select **Strength** for the Data Values variable.
3. Send the output to a new chart sheet named **MR-Chart**.
4. Click **OK**. Excel generates the chart shown in Figure 12-17.

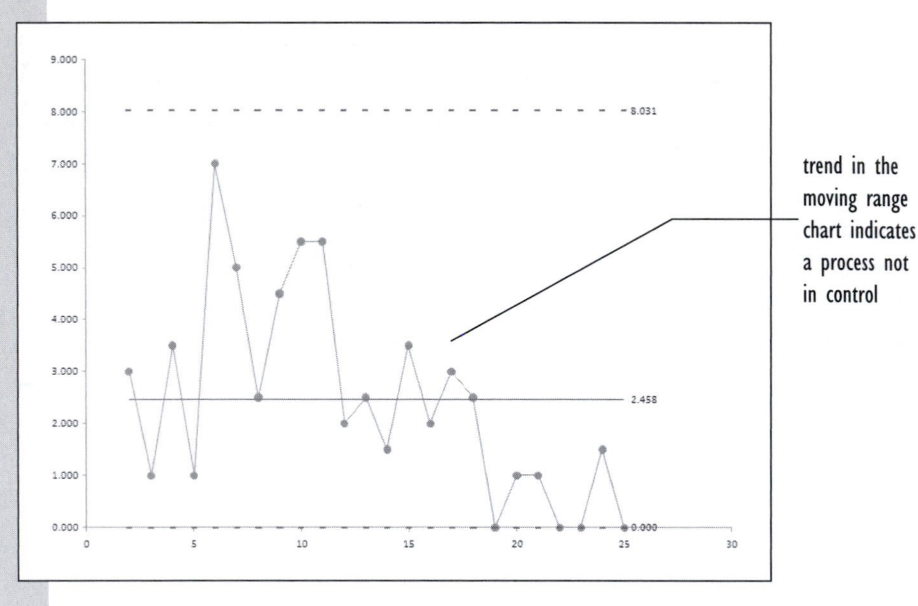

Figure 12-17 The moving range chart for tensile strength samples

The chart in Figure 12-17 shows additional indications of a process that is not in control. The last seven values all fall below the center line, and there appears to be a generally downward trend to the ranges from the sixth observation on. We would conclude that there is sufficient evidence to warrant further investigation and analysis.

You can save and close the Strength Control Chart workbook now.

The Pareto Chart

After you have determined that your process is resulting in an unusual number of problems, such as defects or accidents, the next natural step is to determine what component in the process is causing the problems. This investigation can be aided by a **Pareto chart**, which creates a bar chart of the causes of the problem in order from most to least frequent so that you can focus attention on the most important elements. The chart also includes the cumulative percentage of these components so that you can determine what combination of factors causes a certain percentage of the problems.

The Powder workbook contains data from a company that manufactures baby powder. Part of the process involves a machine called a filler, which pours the powder into bottles to a specified limit. The quantity of powder placed in the bottle varies because of uncontrolled variation, but the final weight of the bottle filled with powder cannot be less than 368.6 grams. Any bottle weighing less than this amount is rejected and must be refilled manually (at a considerable cost in terms of time and labor). Bottles are filled from a filler that has 24 valve heads so that 24 bottles can be filled at one time. Sometimes a head is clogged with powder, and this causes the bottles being filled on that head to receive less than the minimum amount of powder. To gauge whether the machine is operating within limits, you select random samples of 24 bottles (one from each head) at about one-minute intervals over the nighttime shift at the factory. You've been asked to examine the data and determine which part of the filler is most responsible for defective fills.

To open the Powder workbook:

1. Open the **Powder** workbook from the Chapter12 data folder.
2. Save the workbook as **Powder Pareto Chart**. See Figure 12-18.

Figure 12-18
The Powder workbook

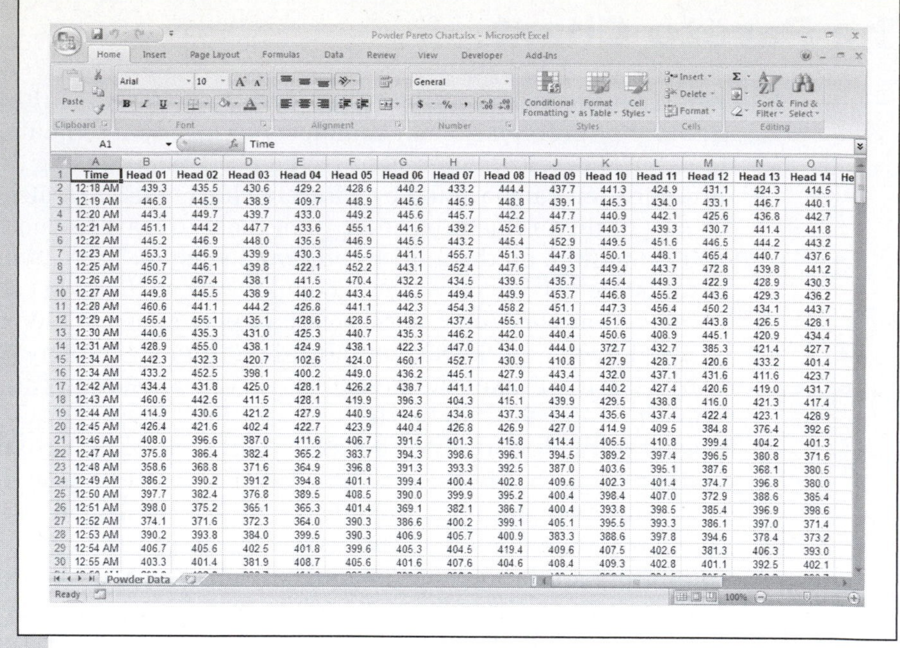

	Time	Head 01	Head 02	Head 03	Head 04	Head 05	Head 06	Head 07	Head 08	Head 09	Head 10	Head 11	Head 12	Head 13	Head 14
2	12:18 AM	439.3	435.5	430.6	429.2	428.6	440.2	433.2	444.4	437.7	441.3	424.9	431.1	424.3	414.5
3	12:19 AM	446.8	445.9	438.9	409.7	448.9	445.6	445.9	448.8	439.1	445.3	434.0	433.1	446.7	440.1
4	12:20 AM	443.4	449.7	439.7	433.0	449.2	445.6	445.7	442.2	447.7	440.9	442.1	425.6	436.8	442.7
5	12:21 AM	451.1	444.2	447.7	433.6	455.1	441.6	439.2	452.6	457.1	440.3	439.3	430.7	441.4	441.8
6	12:22 AM	445.2	446.9	448.0	435.5	446.9	445.5	443.2	445.4	452.9	449.5	451.6	446.5	444.2	443.2
7	12:23 AM	453.3	446.9	439.9	430.3	445.5	441.1	455.7	451.3	447.8	450.1	448.1	465.4	440.7	437.6
8	12:25 AM	450.7	446.1	439.8	422.1	452.2	443.1	452.4	447.6	449.3	449.4	443.7	472.8	439.8	441.2
9	12:26 AM	455.2	467.4	438.1	441.5	470.4	432.2	434.5	439.5	435.7	445.4	449.3	422.9	428.9	430.3
10	12:27 AM	449.8	445.5	438.9	440.2	443.4	446.5	449.4	449.9	453.7	446.8	455.2	443.6	429.3	436.2
11	12:28 AM	460.6	441.1	444.2	426.8	441.1	442.3	454.3	458.2	451.1	447.3	456.4	450.2	434.1	443.7
12	12:29 AM	455.4	455.1	435.1	428.6	428.5	448.2	437.4	455.1	441.9	451.6	430.2	443.8	426.5	428.1
13	12:30 AM	440.6	435.3	431.0	425.3	440.7	435.3	446.2	442.0	440.4	450.6	408.9	445.1	420.9	434.4
14	12:31 AM	428.9	455.0	438.1	424.9	438.1	422.3	447.0	434.0	444.0	372.7	432.7	385.3	421.4	427.7
15	12:34 AM	442.3	432.3	420.7	102.6	424.0	460.1	452.7	430.9	410.8	427.9	428.7	420.6	433.2	401.4
16	12:34 AM	433.2	452.5	398.1	400.2	449.0	436.2	445.1	427.9	443.4	432.0	437.1	431.6	411.6	423.7
17	12:42 AM	434.4	431.8	425.0	428.1	426.2	438.7	441.1	441.0	440.4	440.2	427.4	420.6	419.0	431.7
18	12:43 AM	460.6	442.6	411.5	428.1	419.9	396.3	404.3	415.1	439.9	429.5	438.8	416.0	421.3	417.4
19	12:44 AM	414.9	430.6	421.2	427.9	440.9	424.6	434.8	437.3	434.4	435.6	437.4	422.4	423.1	428.9
20	12:45 AM	426.4	421.6	402.4	422.7	423.9	440.4	426.8	426.9	427.0	414.9	409.5	384.8	376.4	392.6
21	12:46 AM	408.0	396.6	387.0	411.6	406.7	391.5	401.3	415.8	414.4	405.5	410.8	399.4	404.2	401.3
22	12:47 AM	375.8	386.4	382.4	365.2	383.7	394.3	398.6	396.1	394.5	389.2	397.4	396.5	380.8	371.6
23	12:48 AM	358.6	368.8	371.6	364.9	396.8	391.3	393.3	392.5	387.0	403.6	395.1	387.6	368.1	380.5
24	12:49 AM	386.2	390.2	391.2	394.8	401.1	399.4	400.4	402.8	409.6	402.3	401.4	374.7	396.8	380.0
25	12:50 AM	397.7	382.4	376.8	389.5	408.5	390.0	399.9	395.2	400.4	398.4	407.0	372.9	388.6	385.4
26	12:51 AM	398.0	375.2	365.1	365.3	401.4	369.1	382.1	386.7	400.4	393.8	398.5	385.4	396.9	398.6
27	12:52 AM	374.1	371.6	372.3	364.0	390.3	386.6	400.2	399.1	405.1	395.5	393.3	386.1	397.0	371.4
28	12:53 AM	390.2	393.8	384.0	399.5	390.3	406.9	405.7	400.9	383.3	388.6	397.8	394.6	378.4	373.2
29	12:54 AM	406.7	405.6	402.5	401.8	399.6	405.3	404.5	419.4	409.6	407.5	402.6	381.3	406.3	393.0
30	12:55 AM	403.3	401.4	381.9	408.7	405.6	401.6	407.6	404.6	408.4	409.3	402.8	401.1	392.5	402.1

The following range names have been defined for the workbook in Figure 12-18:

Table 12-8 The Powder Workbook

Range Name	Range	Description
Time	A2:A352	The time of the sample
Head_01	B2:B352	Quantity of powder from head 1
Head_02	C2:C352	Quantity of powder from head 2
⋮	⋮	⋮
Head_24	Y2:Y352	Quantity of powder from head 24

Now generate the Pareto chart using StatPlus.

To create the Pareto chart:

1 Click **QC Charts** from the StatPlus menu and then click **Pareto Chart.**

2 Click the **Values in separate columns** option button.

3 Click the **Data Values** button and then select the range names from **Head 01** to **Head 24** in the range names list (do *not* select the Time variable). Click **OK**.

4 Click the Data values represent drop-down list box and select **Error occurs with a value less than.**

5 Type **368.6** in the text box below the drop-down list box.

6 Click the **Output** button and direct the output to a new chart sheet named **Pareto Chart**. Click **OK**.

Figure 12-19 shows the completed dialog box.

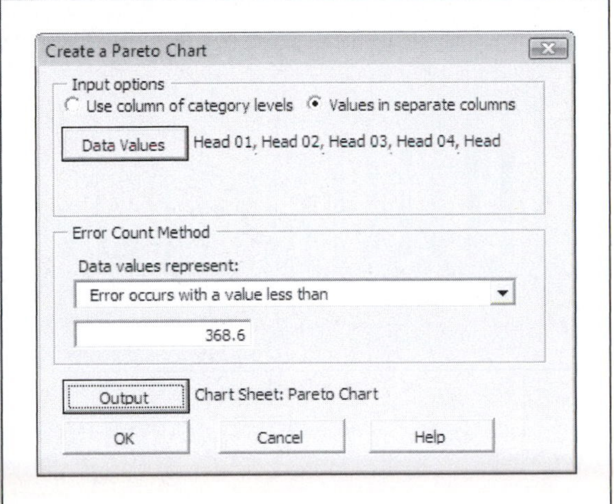

Figure 12-19 The Create a Pareto Chart dialog box

7 Click **OK**.

Excel generates the chart shown in Figure 12-20.

more of the defects come from filler head 18 than from any other filler head

plot of cumulative percentage

Figure 12-20 Pareto chart for the power data

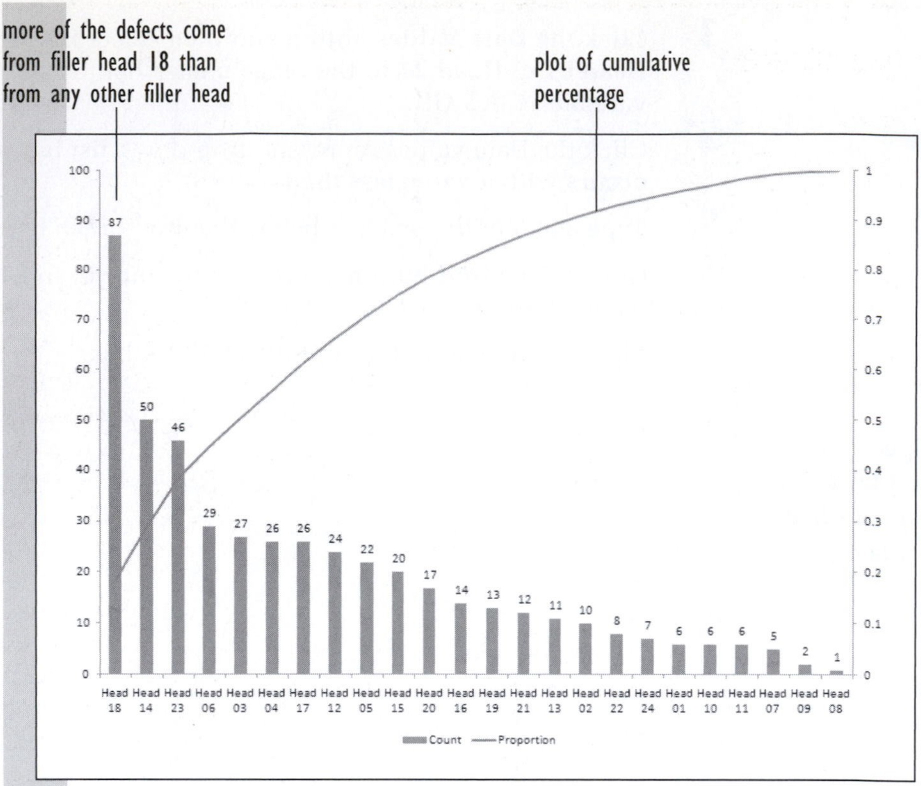

The Pareto chart displayed in Figure 12-20 shows that a majority of the rejects come from a few heads. Filler head 18 accounts for 87 of the defects, and the first three heads in the chart (18, 14, and 23) account for almost 40% of all of the defects. There might be something physically wrong with the heads that made them more liable to clogging up with powder. If rejects were being produced randomly from the filler heads, you would expect that each filler head would produce 1/24, or about 4%, of the total rejects. Using the information from the Pareto chart shown in Figure 12-18, you might want to repair or replace those three heads in order to reduce clogging.

You can close the Powder Pareto Chart workbook now, saving your changes.

Exercises

1. *True or false, and why?* The purpose of statistical process control is to eliminate all variation from a process.

2. *True or false, and why?* As long as the process values lie between the control limits, the process is in control.

3. Calculate the control limits for an $\overline{x}$ chart where
   a. $n = 9$, $\mu = 50$, and $\sigma = 5$.
   b. $n = 9$, $\overline{x} = 50$ $\overline{R} = 8$, and σ is unknown.

4. Calculate the control limits for a range chart where
   a. $n = 4$, $\overline{R} = 10$, and $\sigma = 4$. (What is the value of the center line?)
   b. $n = 4$, $\overline{R} = 10$, and σ is unknown.

5. Calculate the control limits for a C chart where
   a. $\overline{c} = 16$.
   b. $\overline{c} = 22$.

6. Calculate the control limits for a P chart where
   a. $n = 25$ and $\overline{p} = 0.5$.
   b. $n = 25$ and $\overline{p} = 0.2$.

7. Return to the Teacher workbook from this chapter and perform the following analysis:
   a. Open the **Teacher** workbook from the Chapter12 folder and save it as **Teacher Control Chart 2**.
   b. Redo the $\overline{x}$ chart; this time do not assume a value for σ.
   c. Create a range chart of the data; once again, do not assume a value for σ.
   d. Examine your control charts. Is there evidence that the teacher's grades are not in control? Report your conclusions and save your changes to the workbook.

8. A can manufacturing company must be careful to keep the width of its cans consistent. One associated problem is that the metalworking tools tend to wear down during the day. To compensate, the pressure behind the tools is increased as the blades become worn. In the Cans workbook, the width of 39 cans is measured at four randomly selected points. Perform the following analysis on the data:
   a. Open the **Cans** workbook from the Chapter12 folder and save it as **Cans Control Chart**.
   b. Use range and $\overline{x}$ charts to determine whether the process is in control. Does the pressure-compensation scheme seem to correct properly for the tool wear? If not, suggest some special causes that seem still to be present in the process.
   c. Report your results, saving your changes to the workbook.

9. Just-in-time inventory management is an important tool in project management. The OnTime workbook contains data regarding the proportion of on-time deliveries during each month over a two-year period for each of several paperboard products (cartons, sheets, and total). The Total column includes cartons, sheets, and other products. Because sheets were not produced for the entire two-year period, a few data points are missing for that variable. Assume that 1,044 deliveries occurred during each month. Perform the following analysis on the data:
   a. Open the **OnTime** workbook from the Chapter12 folder, saving it as **OnTime Control Chart**.
   b. For each of these products (Cartons, Sheets, and Total), use a P chart to

determine whether the delivery process is in control. If not, suggest some special causes that might exist.

c. Report your results and save your changes to the workbook.

10. A steel sheet manufacturer is concerned about the number of defects, such as scratches and dents, that occur as the sheet is made. In order to track defects, 10-foot lengths of sheet are examined at regular intervals. For each length, the number of defects is counted. Analyze these data to determine whether the process is in control.

    a. Open the **Sheet** workbook from the Chapter12 folder and save the file as **Sheet Control Chart**.
    b. Determine whether the process is in control. If it is not, suggest some special causes that might exist.
    c. Save your changes to the workbook and write a report summarizing your conclusions.

11. A firm is concerned about safety in its workplace. This company does not consider all accidents to be identical. Instead, it calculates a safety index, which assigns more importance to more serious accidents. Examine the data from their study and perform the following analysis:

    a. Open the **Safety** workbook from the Chapter12 folder and save it as **Safety Control Chart**.
    b. Construct a C chart for the data to determine whether safety is in control at this firm.
    c. Save your changes to the workbook and report your conclusions.

12. A manufacturer subjects its steel bars to stress tests to be sure they are up to standard. Three bars were tested in each of 23 subgroups. The amount of stress applied before the bar breaks is recorded by the manufacturer.

    a. Open the **Stress** workbook from the Chapter12 folder and save it as **Stress Control Chart**.
    b. Create a range chart and a $\overline{x}$ chart to determine whether the production process is in control. If it is not, what factors might be contributing to the lack of control?
    c. Report your results, saving your changes to the workbook.

13. A steel rod manufacturer has contracted to supply rods 180 millimeters in length to one of its customers. Because the cutting process varies somewhat, not all rods are exactly the desired length. Five rods were measured from each of 33 subgroups during a week. Analyze these data to determine whether the process is in control.

    a. Open the **Rod** workbook from the Chapter12 folder and save it as **Rod Control Chart**.
    b. Create range and $\overline{x}$ charts to determine whether the cutting process is in statistical control.
    c. Save your changes to the workbook and write a report summarizing your conclusions.

14. An amusement park sampled customers leaving the park over an 18-day period. The total number of customers and the number of customers who indicated they were satisfied with their experience in the park were recorded in an Excel workbook. You've been asked to analyze these data to determine whether the percentage of satisfied customers is in statistical control.

    a. Open the **Satisfy** workbook from the Chapter12 folder and save it as **Satisfy Control Chart**.
    b. Create the appropriate control chart for the percentage of satisfied customers. Is there any indication that the process

is out of control? What factors, if any might have contributed to this?

c. Save your changes to the workbook and write a report of your observations and conclusions.

15. The number of flaws on the surfaces of a particular model of automobile leaving the plant was recorded in the Autos workbook for each of 40 automobiles during a one-week period.

a. Open the **Autos** workbook from the Chapter2 folder and save it as **Autos Control Chart**.

b. Create a control chart of the count of auto flaws. Is this process in control?

c. Save your changes to the workbook and report your results.

16. You've learned in this chapter that filler head 18 is a major factor in the number of defective fills. To investigate further, you decide to look at the head 18 values from the data set to determine at what points in time the head was out of statistical control.

a. Open the **Powder** workbook from the Chapter12 folder and save it as **Powder Control Chart**.

b. Create an Individuals chart and a moving range chart of the Head 18 values. At what times are the head values beyond the control limits?

c. Repeat part b for filler heads 14 and 23.

d. Interpret your findings in light of the fact that a new shift comes in at midnight. Does this fact affect the filler process?

e. Save your changes to the workbook and write a report summarizing your observations.

17. Weather can be considered a process with process variables such as temperature and precipitation and attribute variables such as the number of hurricanes and tornadoes in a given season. One theory of meteorology holds that climatic changes in this process take place over long periods of time, whereas over short periods of time, the process should be stable. On the other hand, concerns have been raised about the effect of CO_2 emissions on the atmosphere, which may lead to major changes in the weather. You've been given the yearly temperature values for northern Illinois from 1895 to 1998, saved in an Excel workbook.

a. Open the **Temp100** workbook from the Chapter12 folder and save it as **Temp100 Control Chart**.

b. Create an Individuals chart and a moving range chart of the average yearly temperature.

c. What is the average yearly temperature? What are the lower and upper control limits? Do the temperature values appear to be in statistical control?

d. Create a moving range chart of the average yearly temperature. Does this chart show any violations of process control?

e. Save your changes to the workbook and write a report summarizing your results.

18. The Rain100 workbook contains the total precipitation for northern Illinois from 1895 to 1998.

a. Open the **Rain100** workbook from the Chapter12 folder and save it as **Rain100 Control Chart**.

b. Create an individuals chart of the total precipitation.

c. Create a moving range chart of the total precipitation.

d. Does the process appear to be in statistical control? Save your workbook and report your conclusions.

19. The Tornado workbook records the number of tornadoes of various levels of severity in Kansas from 1950 to 1999. Tornadoes are rated on the Fujita Tornado Scale, which ranges from minor tornadoes rated at F0 to major tornadoes rated at F5. You've been asked to determine whether the number of tornadoes has changed over this period of time.

   a. Open the **Tornado** workbook from the Chapter12 folder and save it as **Tornado Control Chart**.

   b. Create a C chart of the number of tornadoes each year for each severity level and then for all types of tornadoes.

   c. Which classes of tornadoes show signs of being out of statistical control? Describe the problem.

   d. Techniques in recording and counting tornadoes have improved in the last few decades, especially for minor tornadoes. Explain how this fact may be related to the results you noted in part c.

   e. Save your changes to the workbook and report your results.

Appendix

Excel Reference

Contents

The Excel Reference contains the following:

The Analysis ToolPak add-ins that come with Excel enable you to perform basic statistical analysis. None of the output from the Analysis ToolPak is updated for changing data, so if the source data change, you will have to rerun the command. To use the Analysis ToolPak, you must first verify that it is available to your workbook.

To check whether the Analysis ToolPak is available:

1. Click the **Office** button and click **Excel Options**.
2. Click **Add-Ins** from the list of Excel Options and then click the **Go** button next to the Manage Excel Add-Ins list box.
3. Select the **Analysis ToolPak** checkbox from the Add-Ins dialog box to activate the Analysis ToolPak add-in and click the **OK** button.
4. Verify that the add-in is activated by clicking the Data tab and verifying that the **Data Analysis** button appears in the Analysis group.

The rest of this section documents each Analysis ToolPak command, showing each corresponding dialog box and describing the features of the command.

Output Options

All the dialog boxes that produce output share the following output storage options:

Output Range

Click to send output to a cell in the current worksheet, and then type the cell; Excel uses that cell as the upper left corner of the range.

New Worksheet Ply

Click to send output to a new worksheet; then type the name of the worksheet.

New Workbook

Click to send output to a new workbook.

Anova: Single Factor

The **Anova: Single Factor** command calculates the one-way analysis of variance, testing whether means from several samples are equal.

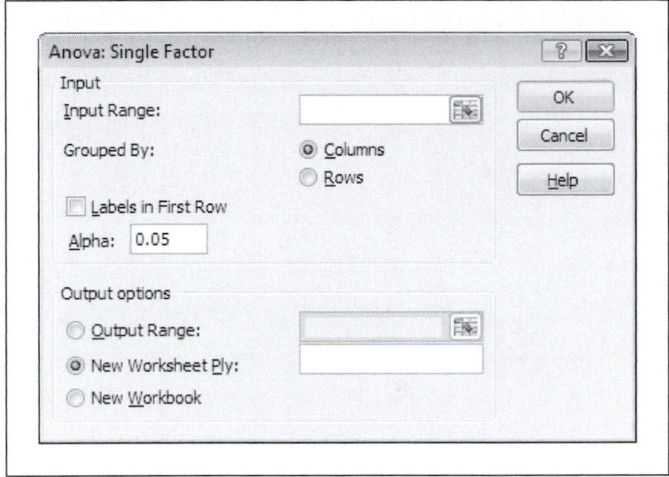

Input Range

Enter the range of worksheet data you want to analyze. The range must be contiguous.

Grouped By

Indicate whether the range of samples is grouped by columns or by rows.

Labels in First Row/Column

Indicate whether the first row (or column) includes header information.

Alpha

Enter the alpha level used to determine the critical value for the *F* statistic.

See "Output Options" at the beginning of this section for information on the output storage options.

Anova: Two-Factor With Replication

The **Anova: Two-Factor With Replication** command calculates the two-way analysis of variance with multiple observations for each combination of the two factors. An analysis of variance table is created that tests for the significance of the two factors and the significance of an interaction between the two factors.

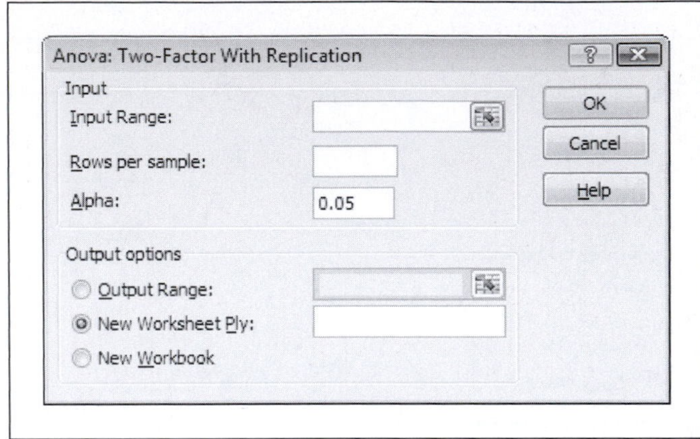

Input Range

Enter the range of worksheet data you want to analyze. The range must be rectangular, the columns representing the first factor and the rows representing the second factor. An equal number of rows are required for each level of the second factor.

Rows per Sample

Enter the number of repeated values for each combination of the two factors.

Alpha

Enter the alpha level used to determine the critical value for the *F* statistic.

See "Output Options" at the beginning of this section for information on the output storage options.

Anova: Two-Factor Without Replication

The **Anova: Two-Factor Without Replication** command calculates the two-way analysis of variance with one observation for each combination of the two factors. An analysis of variance table is created that tests for the significance of the two factors.

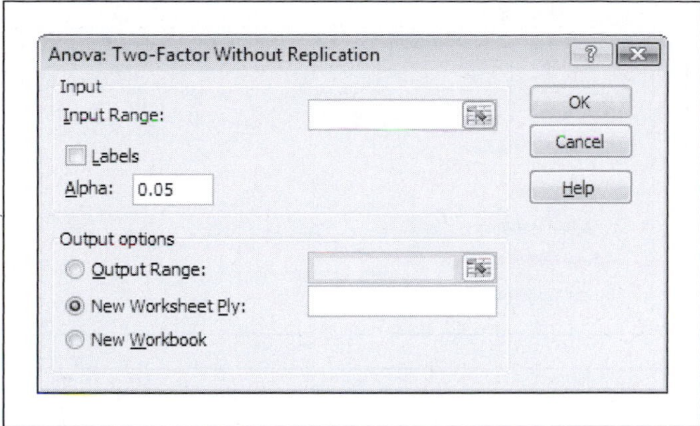

Input Range

Enter the range of worksheet data you want to analyze. The range must be contiguous, with each row and column representing a combination of the two factors.

Labels

Indicate whether the first row (or column) includes header information.

Alpha

Enter the alpha level used to determine the critical value for the *F* statistic.

See "Output Options" at the beginning of this section for information on the output storage options.

Correlation

The **Correlation** command creates a table of the Pearson correlation coefficient for values in rows or columns on the worksheet.

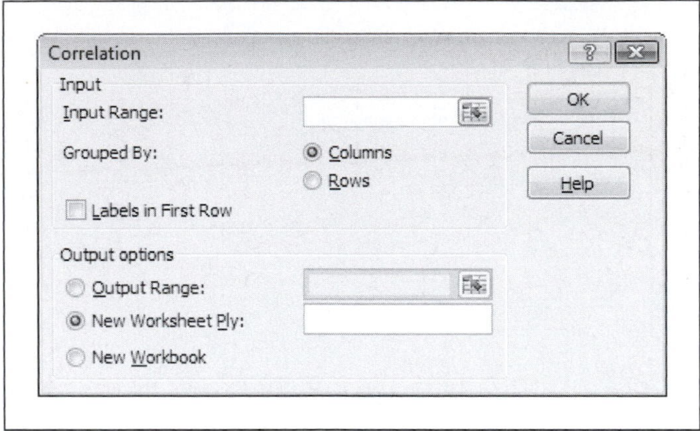

Input Range

Enter the range of worksheet data you want to analyze. The range must be contiguous.

Grouped By

Indicate whether the range of samples is grouped by columns or by rows.

Labels in First Row/Column

Indicate whether the first row (or column) includes header information.

See "Output Options" at the beginning of this section for information on the output storage options.

Covariance

The **Covariance** command creates a table of the covariance for values in rows or columns on the worksheet.

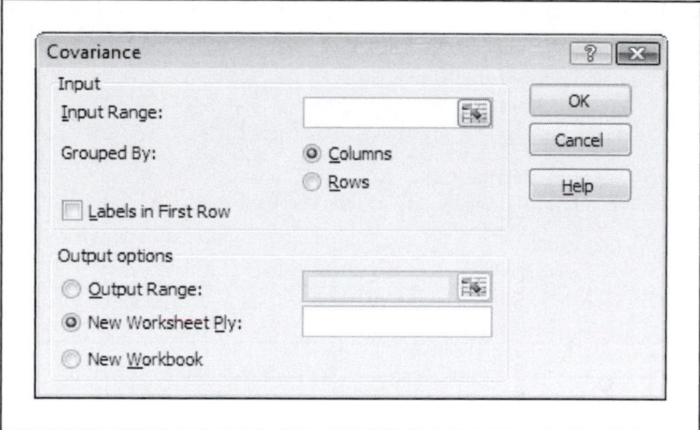

Input Range

Enter the range of worksheet data you want to analyze. The range must be contiguous.

Grouped By

Indicate whether the range of samples is grouped by columns or by rows.

Labels in First Row/Column

Indicate whether the first row (or column) includes header information.

See "Output Options" at the beginning of this section for information on the output storage options.

Descriptive Statistics

The **Descriptive Statistics** command creates a table of univariate descriptive statistics for values in rows or columns on the worksheet.

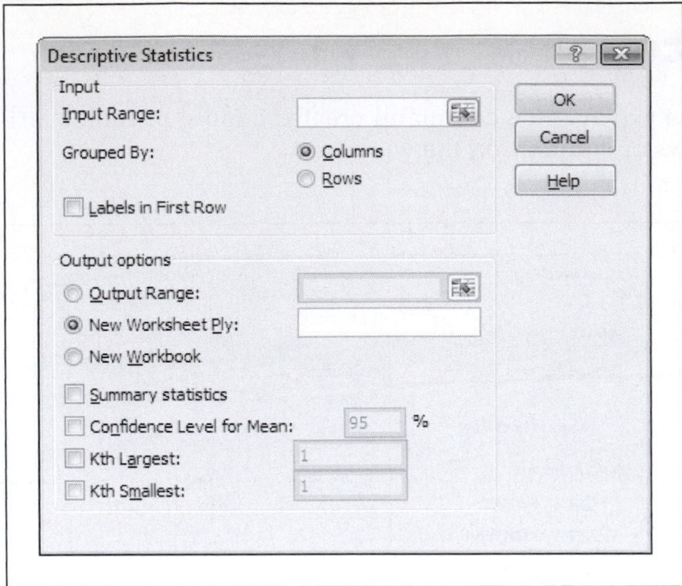

Input Range

Enter the range of worksheet data you want to analyze. The range must be contiguous.

Grouped By

Indicate whether the range of samples is grouped by columns or by rows.

Labels in First Row/Column

Indicate whether the first row (or column) includes header information.

Confidence Level for Mean

Click to print the specified confidence level for the mean in each row or column of the input range.

*K*th Largest

Click to print the *k*th largest value for each row or column of the input range; enter the value for *k* in the corresponding box.

Kth Smallest

Click to print the *k*th smallest value for each row or column of the input range; enter the value for *k* in the corresponding box.

Summary Statistics

Click to print the following statistics in the output range: Mean, Standard Error (of the mean), Median, Mode, Standard Deviation, Variance, Kurtosis, Skewness, Range, Minimum, Maximum, Sum, Count, Largest (#), Smallest (#), and Confidence Level.

See "Output Options" at the beginning of this section for information on the output storage options.

Exponential Smoothing

The **Exponential Smoothing** command creates a column of smoothed averages using simple one-parameter exponential smoothing.

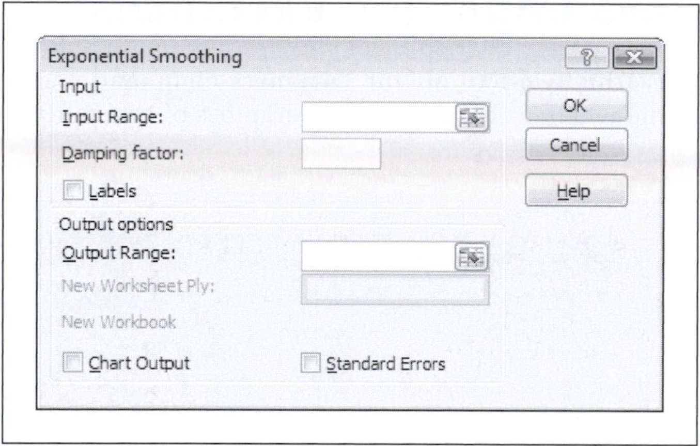

Input Range

Enter the range of worksheet data you want to analyze. The range must be a single row or a single column.

Damping Factor

Enter the value of the smoothing constant. The value 0.3 is used as a default if nothing is entered.

Labels

Indicate whether the first row (or column) includes header information.

Chart Output

Click to create a chart of observed and forecasted values.

Standard Errors

Click to create a column of standard errors to the right of the forecasted column.

Output options

You can send output from this command only to a cell on the current worksheet.

F-Test: Two-Sample for Variances

The **F-Test: Two-Sample for Variances** command performs an *F* test to determine whether the population variances of two samples are equal.

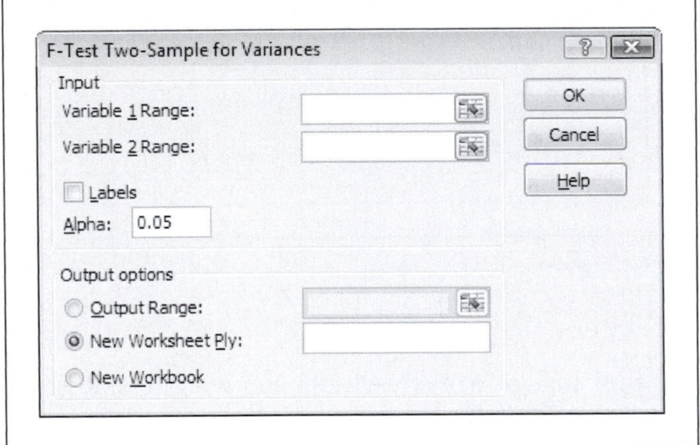

Variable 1 Range

Enter the range of the first sample, either a single row or a single column.

Variable 2 Range

Enter the range of the second sample, either a single row or a single column.

Labels

Indicate whether the first row (or column) includes header information.

Alpha

Enter the alpha level used to determine the critical value for the *F*-statistic.
See "Output Options" at the beginning of this section for information on the output storage options.

Histogram

The **Histogram** command creates a frequency table for data values located in a row, column, or list. The frequency table can be based on default or customized bin widths. Additional output options include calculating the cumulative percentage, creating a histogram, and creating a histogram sorted in descending order of frequency (also known as a Pareto chart).

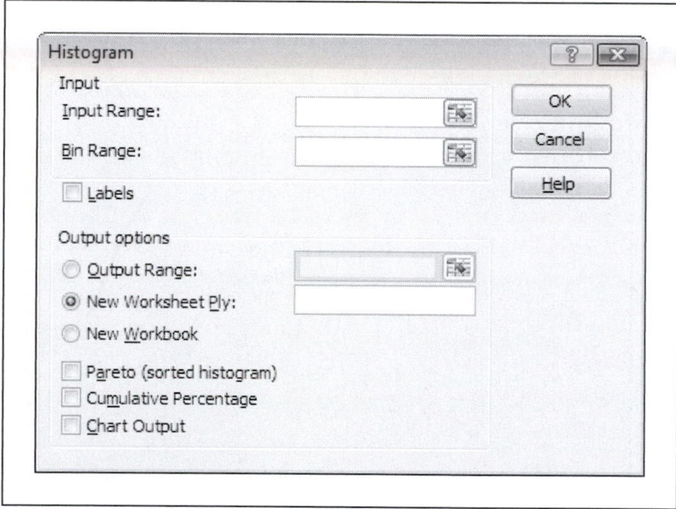

Input Range

Enter the range of worksheet data you want to analyze. The range must be a row, column, or rectangular region.

Bin Range

Enter an optional range of values that defines the boundaries of the bins.

Labels

Indicate whether the first row (or column) includes header information.

Pareto (sorted histogram)

Click to create a Pareto chart sorted by descending order of frequency.

Cumulative Percentage

Click to calculate the cumulative percentages.

Chart Output

Click to create a histogram of frequency versus bin values.

See "Output Options" at the beginning of this section for information on the output storage options.

Moving Average

The **Moving Average** command creates a column of moving averages over the preceding observations for an interval specified by the user.

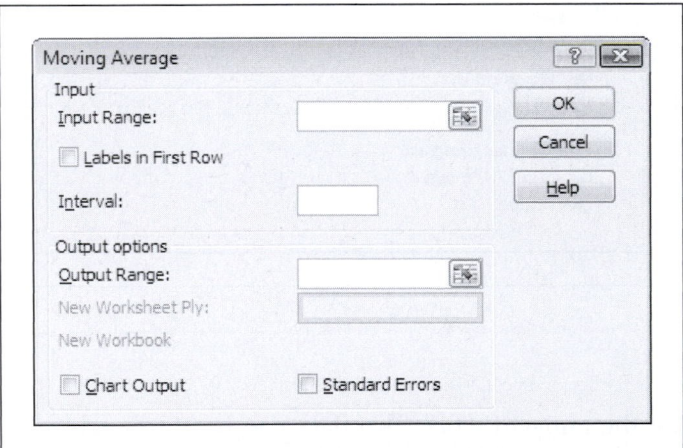

Input Range

Enter the range of worksheet data for which you want to calculate the moving average. The range must be a single row or a single column containing four or more cells of data.

Labels in First Row

Indicate whether the first row (or column) includes header information.

Interval

Enter the number of cells you want to include in the moving average. The default value is three.

Chart Output

Click to create a chart of observed and forecasted values.

Standard Errors

Click to create a column of standard errors to the right of the forecasted column.

Output options

You can only send output from this command to a cell on the current worksheet.

Random Number Generation

The **Random Number Generation** command creates columns of random numbers following a user-specified distribution.

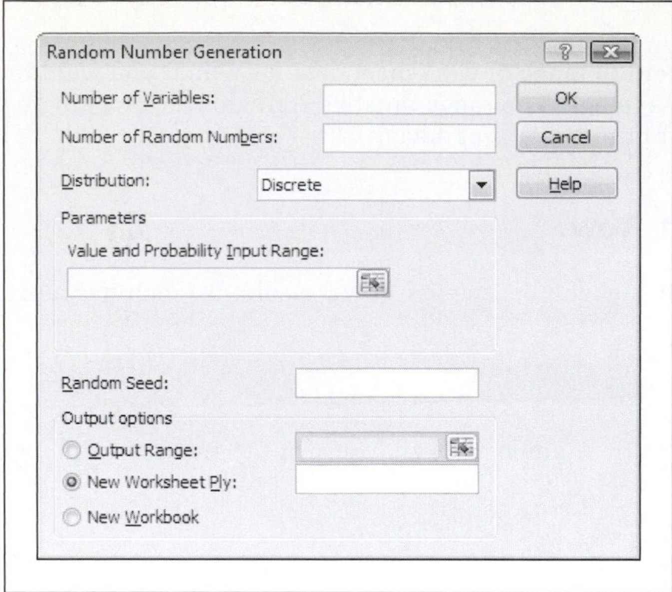

Number of Variables

Enter the number of columns of random variables you want to generate. If no value is entered, Excel fills up all available columns.

Number of Random Numbers

Enter the number of rows in each column of random variables you want to generate. If no value is entered, Excel fills up all available columns. This command is not available for the patterned distribution (see below).

Distribution

Click the down arrow to open a list of seven distributions from which you can choose to generate random numbers and then specify the parameters of that distribution.

Random Seed

Enter an optional value used as a starting point, called a random seed, for generating a string of random numbers. You need not enter a random seed, but using the same random seed ensures that the same string of random numbers will be generated. This box is not available for patterned or discrete random data.

See "Output Options" at the beginning of this section for information on the output storage options.

Rank and Percentile

The **Rank and Percentile** command produces a table with ordinal and percentile values for each cell in the input range.

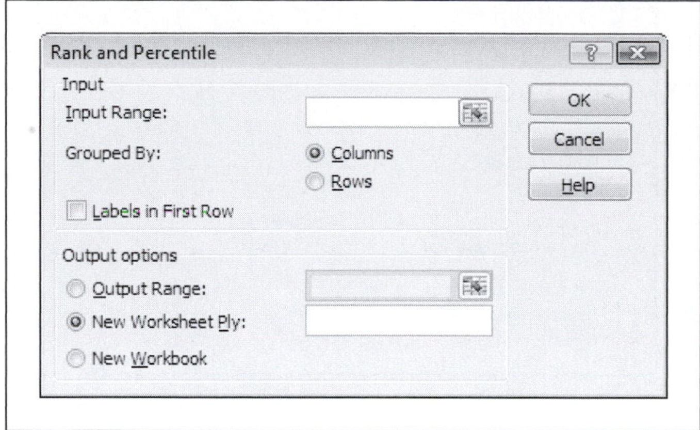

Input Range

Enter the range of worksheet data you want to analyze. The range must be contiguous.

Grouped By

Indicate whether the range of samples is grouped by columns or by rows.

Labels in First Row/Column

Indicate whether the first row (or column) includes header information.

See "Output Options" at the beginning of this section for information on the output storage options.

Regression

The **Regression** command performs multiple linear regression for a variable in an input column based on up to 16 predictor variables. The user has the option of calculating residuals and standardized residuals and producing line fit plots, residuals plots, and normal probability plots.

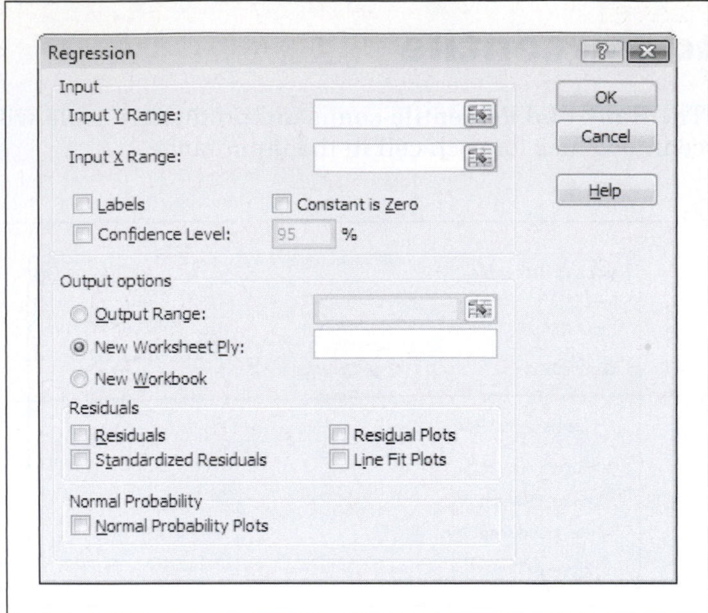

Input Y Range

Enter a single column of values that will be the response variable in the linear regression.

Input X Range

Enter up to 16 contiguous columns of values that will be the predictor variables in the regression.

Labels

Indicate whether the first row of the Y range and that of the X range include header information.

Constant is Zero

Click to include an intercept term in the linear regression or to assume that the intercept term is zero.

Confidence Level

Click to indicate a confidence interval for linear regression parameter estimates. A 95% confidence interval is automatically included; enter a different one in the corresponding box.

Residuals

Click to create a column of residuals (observed—predicted) values.

Residual Plots

Click to create a plot of residuals versus each of the predictor variables in the model.

Standardized Residuals

Click to create a column of residuals divided by the standard error of the regression's analysis of variance table.

Line Fit Plots

Click to create a plot of observed and predicted values against each of the predictor variables.

Normal Probability Plots

Click to create a normal probability plot of the Y variable in the Input Y Range.

See "Output Options" at the beginning of this section for information on the output storage options.

Sampling

The **Sampling** command creates a sample of an input range. The sample can be either random or periodic (sampling values a fixed number of cells apart). The sample generated is placed into a single column.

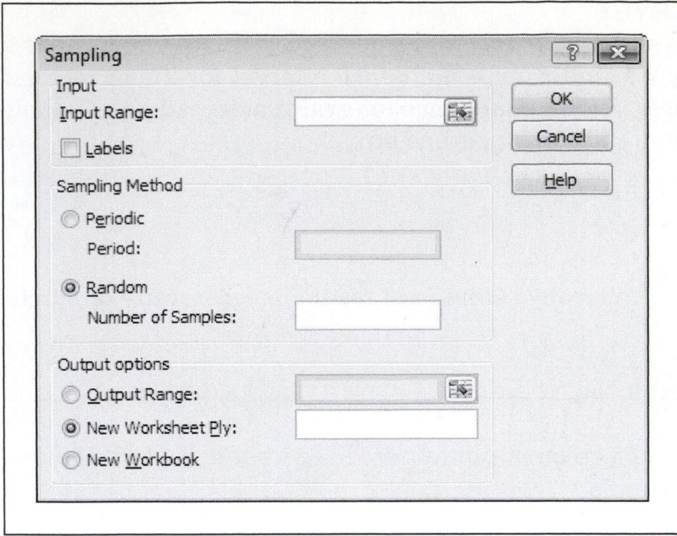

Input Range

Enter the range of worksheet data you want to sample. The range must be contiguous.

Labels

Indicate whether the first row of the Y range and that of the X range include header information.

Sampling Method

Click the sampling method you want.

Periodic

Click to sample values from the input range period cells apart; enter a value for period in the corresponding box.

Random

Click to create a random sample the size of which you enter in the corresponding box.

See "Output Options" at the beginning of this section for information on the output storage options.

t-Test: Paired Two Sample for Means

The ***t*-Test: Paired Two Sample for Means** command calculates the paired two-sample Student's *t*-test. The output includes both the one-tail and the two-tail critical values.

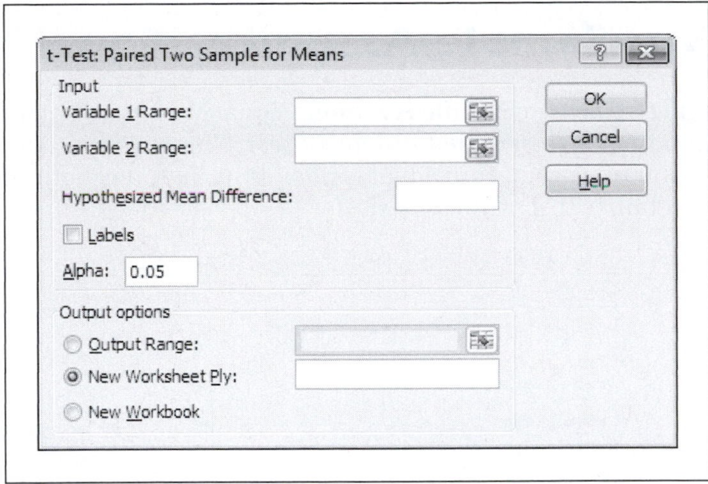

Variable 1 Range

Enter the input range of the first sample; it must be a single row or column.

Variable 2 Range

Enter the input range of the second sample; it must be a single row or column.

Hypothesized Mean Difference

Enter a mean difference value with which to calculate the *t*-test. If no value is entered, a mean difference of zero is assumed.

Labels

Indicate whether the first row of the Y range and that of the X range include header information.

Alpha

Enter an alpha value used to calculate the critical values of the *t* shown in the output.

See "Output Options" at the beginning of this section for information on the output storage options.

t-Test: Two-Sample Assuming Equal Variances

The ***t*-Test: Two-Sample Assuming Equal Variances** command calculates the unpaired two-sample Student's *t*-test. The test assumes that the variances in the two groups are equal. The output includes both the one-tail and the two-tail critical values.

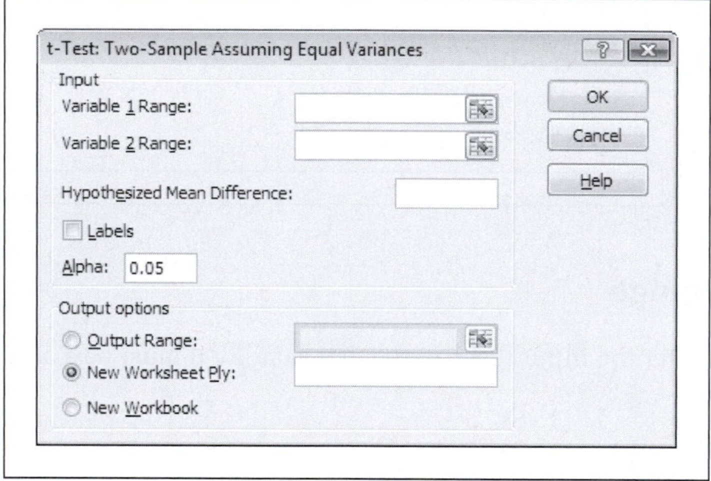

Variable 1 Range

Enter an input range for the first sample; it must be a single row or column.

Variable 2 Range

Enter an input range for the second sample; it must be a single row or column.

Hypothesized Mean Difference

Enter a mean difference with which to calculate the *t*-test. If no value is entered, a mean difference of zero is assumed.

Labels

Indicate whether the first row of the Y range and that of the X range include header information.

Alpha

Enter an alpha value used to calculate the critical values of the *t* shown in the output.

See "Output Options" at the beginning of this section for information on the output storage options.

t-Test: Two-Sample Assuming Unequal Variances

The ***t*-Test: Two-Sample Assuming Unequal Variances** command calculates the unpaired two-sample Student's *t*-test. The test allows the variances in the two groups to be unequal. The output includes both the one-tail and the two-tail critical values.

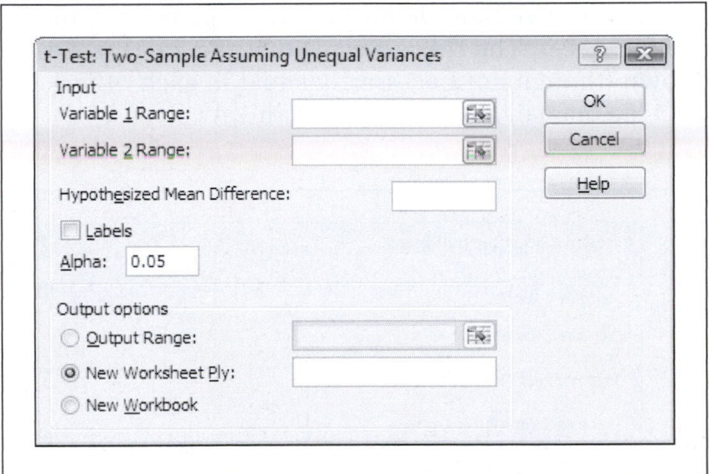

Variable 1 Range

Enter the input range of the first sample; it must be a single row or column.

Variable 2 Range

Enter the input range of the second sample; it must be a single row or column.

Hypothesized Mean Difference

Enter a mean difference with which to calculate the *t* test. If no value is entered, a mean difference of zero is assumed.

Labels

Indicate whether the first row of the Y range and that of the X range include header information.

Alpha

Enter an alpha value used to calculate the critical values of the *t* shown in the output.

See "Output Options" at the beginning of this section for information on the output storage options.

z-Test: Two Sample for Means

The **z-Test: Two Sample for Means** command calculates the unpaired two sample *z* test. The test assumes that the variances in the two groups are known (though not necessarily equal to each other). The output includes both the one-tail and the two-tail critical values.

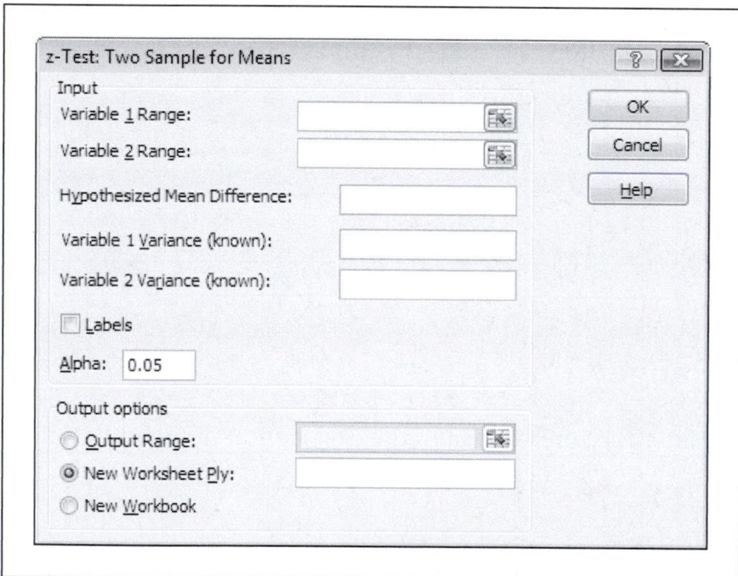

Variable 1 Range

Enter an input range of the first sample; it must be a single row or column.

Variable 2 Range

Enter an input range of the second sample; it must be a single row or column.

Hypothesized Mean Difference

Enter a mean difference with which to calculate the z test. If no value is entered, a mean difference of zero is assumed.

Variable 1 Variance (known)

Enter the known variance s_1^2 of the first sample.

Variable 2 Variance (known)

Enter the known variance s_2^2 of the second sample.

Labels

Indicate whether the first row of the Y range and that of the X range include header information.

Alpha

Enter an alpha value used to calculate the critical values of the z shown in the output.

See "Output Options" at the beginning of this section for information on the output storage options.

Excel's Math and Statistical Functions

This section documents all the functions provided with Excel that are relevant to statistics. So that you can more easily find the function you need, similar functions are grouped together in six categories: Descriptive Statistics for One Variable, Descriptive Statistics for Two or More Variables, Distributions, Mathematical Formulas, Statistical Analysis, and Trigonometric Formulas.

Descriptive Statistics for One Variable

Function Name	Description
AVEDEV	AVEDEV(*number1*, *number2*, . . .) returns the average of the (absolute) deviations of the points from their mean.
AVERAGE	AVERAGE(*number1*, *number2*, . . .) returns the average of the *numbers* (up to 30).
CONFIDENCE	CONFIDENCE(*alpha*, *standarddev*, *n*) returns a confidence interval for the mean.
COUNT	COUNT(*value1*, *value2*, . . .) returns how many numbers are in the *value(s)*.
COUNTA	COUNTA(*value1*, *value2*, . . .) returns the count of nonblank values in the list of arguments.
COUNTBLANK	COUNTBLANK(*range*) returns the count of blank cells in the *range*.
COUNTIF	COUNTIF(*range*, *criteria*) returns the count of nonblank cells in the *range* that meet the *criteria*.
DEVSQ	DEVSQ(*number1*, *number2*, . . .) returns the sum of squared deviations from the mean of the *numbers*.
FREQUENCY	FREQUENCY(*data-array*, *bins-array*) returns the frequency distribution of *data-array* as a vertical array, on the basis of *bins-array*.
GEOMEAN	GEOMEAN(*number1*, *number2*, . . .) returns the geometric mean of up to 30 *numbers*.
HARMEAN	HARMEAN(*number1*, *number2*, . . .) returns the harmonic mean of up to 30 *numbers*.
KURT	KURT(*number1*, *number2*, . . .) returns the kurtosis of up to 30 *numbers*.
LARGE	LARGE(*array*, *n*) returns the *n*th-largest value in *array*.
MAX	MAX(*number1*, *number2*, . . .) returns the largest of up to 30 *numbers*.

MEDIAN	MEDIAN(*number1*, *number2*, . . .) returns the median of up to 30 *numbers*.
MIN	MIN(*number1*, *number2*, . . .) returns the smallest of up to 30 *numbers*.
MODE	MODE(*number1*, *number2*, . . .) returns the value most frequently occurring in up to 30 *numbers* or in a specified array or reference.
PERCENTILE	PERCENTILE(*array*, *n*) returns the *n*th percentile of the values in *array*.
PERCENTRANK	PERCENTRANK(*array*, *value*, *significant-digits*) returns the percent rank of the *value* in the *array*, with the specified number of *significant digits* (optional).
PRODUCT	PRODUCT(*number1*, *number2*, . . .) returns the product of up to 30 *numbers*.
RANK	RANK(*number*, *range*, *order*) returns the rank of the *number* in the *range*. If *order* = 0, then the range is ranked from largest to smallest; if *order* = 1, then the range is ranked from smallest to largest.
QUOTIENT	QUOTIENT(*dividend*, *divisor*) returns the quotient of the numbers, truncated to integers.
SKEW	SKEW(*number1*, *number2*, . . .) returns the skewness of up to 30 *numbers* (or a reference to numbers).
SMALL	SMALL(*array*, *n*) returns the *n*th-smallest number in *array*.
STANDARDIZE	STANDARDIZE(*x*, *mean*, *standard deviation*) normalizes a distribution and returns the *z* score of *x*.
STDEV	STDEV(*number1*, *number2*, . . .) returns the sample standard deviation of up to 30 *numbers*, or of an array of numbers.
STDEVP	STDEVP(*number1*, *number2*, . . .) returns the population standard deviation of up to 30 *numbers* or of an array of numbers.
SUM	SUM(*number1*, *number2*, . . .) returns the sum of up to 30 *numbers* or of an array of numbers.
SUMIF	SUMIF(*range*, *criteria*, *sum range*) returns the sum of the numbers in *range* (optionally in *sum-range*) according to *criteria*.
SUMSQ	SUMSQ(*number1*, *number2*, . . .) returns the sum of the squares of up to 30 *numbers* or of an array of numbers.
TRIMMEAN	TRIMMEAN(*array*, *percent*) returns the mean of a set of values in an *array*, excluding *percent* of the values, half from the top and half from the bottom.

VAR	VAR(*number1, number2, . . .*) returns the sample variance of up to 30 *numbers* (or an array or reference).
VARP	VARP(*number1, number2, . . .*) returns the population variance of up to 30 *numbers* (or an array or reference).

Descriptive Statistics for Two or More Variables

Function Name	**Description**
CORREL	CORREL(*array1, array2*) returns the coefficient of correlation between *array1* and *array2.*
COVAR	COVAR(*array1, array2*) returns the covariance of *array1* and *array2.*
PEARSON	PEARSON(*array1, array2*) returns the Pearson correlation coefficient between *array1* and *array2.*
RSQ	RSQ(*known-y's, known-x's*) returns the square of Pearson's product moment correlation coefficient.
SUMPRODUCT	SUMPRODUCT(*array1, array2, . . .*) returns the sum of the products of corresponding entries in up to 30 *arrays.*
SUMX2MY2	SUMX2MY2(*array1, array2*) returns the sum of the differences of squares of corresponding entries in two *arrays.*
SUMX2PY2	SUMX2PY2(*array1, array2*) returns the sum of the sums of squares of corresponding entries in two *arrays.*
SUMXMY2	SUMXMY2(*array1, array2*) returns the sum of the squares of differences of corresponding entries in two arrays.

Distributions

Function Name	**Description**
BETADIST	BETADIST(*x, alpha, beta, a, b*) returns the value of the cumulative beta probability density function.
BETAINV	BETAINV(*p, alpha, beta, a, b*) returns the value of the inverse of the cumulative beta probability density function.

BINOMDIST	BINOMDIST(*successes*, *trials*, *p*, *type*) returns the probability for the binomial distribution (*type* is TRUE for cumulative distribution function, FALSE for probability mass function).
CHIDIST	CHIDIST(*x*, *df*) returns the probability for the chi-square distribution.
CHIINV	CHIINV(*p*, *df*) returns the inverse of the chi-square distribution.
CRITBINOM	CRITBINOM(*trials*, *p*, *alpha*) returns the smallest value so that the cumulative binomial distribution is greater than or equal to the criterion value, *alpha*.
EXPONDIST	EXPONDIST(*x*, *lambda*, *type*) returns the probability for the exponential distribution (*type* is true for the cumulative distribution function, false for the probability density function).
FDIST	FDIST(*x*, *df1*, *df2*) returns the probability for the *F* distribution.
FINV	FINV(*p*, *df1*, *df2*) returns the inverse of the *F* distribution.
GAMMADIST	GAMMADIST(*x*, *alpha*, *beta*, *type*) returns the probability for the gamma distribution with parameters *alpha* and *beta* (*type* is true for the cumulative distribution function, false for the probability mass function).
GAMMAINV	GAMMAINV(*p*, *alpha*, *beta*) returns the inverse of the gamma distribution.
GAMMALN	GAMMALN(*x*) returns the natural log of the gamma function evaluated at *x*.
HYPGEOMDIST	HYPGEOMDIST(*sample-successes*, *sample-size*, *population-successes*, *population-size*) returns the probability for the hypergeometric distribution.
LOGINV	LOGINV(*p*, *mean*, *sd*) returns the inverse of the lognormal distribution, where the natural logarithm of the distribution is normally distributed with mean *mean* and standard deviation *sd*.
LOGNORMDIST	LOGNORMDIST(*x*, *mean*, *sd*) returns the probability for the lognormal distribution, where the natural logarithm of the distribution is normally distributed with mean *mean* and standard deviation *sd*.
NEGBINOMDIST	NEGBINOMDIST(*failures*, *threshold-successes*, *probability*) returns the probability for the negative binomial distribution.
NORMDIST	NORMDIST(*x*, *mean*, *sd*, *type*) returns the probability for the normal distribution with mean *mean* and standard deviation *sd* (*type* is true for the cumulative distribution function, false for the probability mass function).

NORMINV	NORMINV(*p*, *mean*, *sd)* returns the inverse of the normal distribution with mean *mean* and standard deviation *sd*.
NORMSDIST	NORMSDIST(*number*) returns the probability for the standard normal distribution.
NORMSINV	NORMSINV(*probability*) returns the inverse of the standard normal distribution.
POISSON	POISSON(*x*, *mean*, *type*) returns the probability for the Poisson distribution (*type* is true for the cumulative distribution, false for the probability mass function).
TDIST	TDIST(*x*, *df*, *number-of-tails*) returns the probability for the *t* distribution.
TINV	TINV(*p*, *df)* returns the inverse of the *t* distribution.
WEIBULL	WEIBULL(*x*, *alpha*, *beta*, *type*) returns the probability for the Weibull distribution (*type* is true for the cumulative distribution function, false for the probability mass function).

Mathematical Formulas

Function Name	Description
ABS	ABS(*number*) returns the absolute value of *number* to the point specified.
COMBIN	COMBIN(*x*, *n*) returns the number of combinations of *x* objects taken *n* at a time.
EVEN	EVEN(*number*) returns *number* rounded up to the nearest even integer.
EXP	EXP(*number*) returns the exponential function of *number* with base *e*.
FACT	FACT(*number*) returns the factorial of *number*.
FACTDOUBLE	FACTDOUBLE(*number*) returns the double factorial of *number*.
FLOOR	FLOOR(*number*, *significance*) returns *number* rounded down to the nearest multiple of the *significance* value.
GCD	GCD(*number1*, *number2*, . . .) returns the greatest common divisor of up to 29 *numbers*.
GESTEP	GESTEP(*number*, *step*) returns 1 if *number* is greater than or equal to *step*, 0 if not.
LCM	LCM(*number1*, *number2*, . . .) returns the least common multiple of up to 29 *numbers*.
INT	INT(*number*) truncates *number* to the units place.
LN	LN(*number*) returns the natural logarithm of *number*.

LOG	LOG(*number, base*) returns the logarithm of *number*, with the specified (optional, default is 10) *base*.
LOG10	LOG10(*number*) returns the common logarithm of *number*.
MOD	MOD(*number, divisor*) returns the remainder of the division of *number* by *divisor*.
MULTINOMIAL	MULTINOMIAL(*number1, number2, . . .*) returns the quotient of the factorial of the sum of *numbers* and the product of the factorials of *numbers*.
ODD	ODD(*number*) returns *number* rounded up to the nearest odd integer.
POWER	POWER(*number, power*) returns *number* raised to the *power*.
PERMUT	PERMUT(*x, n*) returns the number of permutations of *x* items taken *n* at a time.
RAND	RAND() returns a randomly chosen number from 0 to but not including 1.
ROUND	ROUND(*number, places*) rounds *number* to a certain number of decimal *places* (if *places* is positive), or to an integer (if *places* is 0), or to the left of the decimal point (if *places* is negative).
ROUNDDOWN	ROUNDDOWN(*number, places*) rounds like ROUND, except always toward 0.
ROUNDUP	ROUNDUP(*number, places*) rounds like ROUND, except always away from 0.
SERIESSUM	SERIESSUM(*x, n, m, coefficients*) returns the sum of the power series $a_1x^n + a_2x^{n+m} + \cdots + a_ix^{n+(i-1)m}$, where $a_1, a_2, \ldots a_i$ are the *coefficients*.
SIGN	SIGN(*number*) returns 0, 1, or −1, the sign of *number*.
SQRT	SQRT(*number*) returns the square root of *number*.
SQRTPI	SQRTPI(*number*) returns the square root of *number*$*\pi$.
TRIM	TRIM(*text*) returns *text* with spaces removed, except for single spaces between words.
TRUNC	TRUNC(*number, digits*) truncates *number* to an integer (optionally, to a number of digits).

Statistical Analysis

Function Name	Description
CHITEST	CHITEST(*observed, expected*) calculates the Pearson chi-square for observed and expected counts.
GROWTH	GROWTH(*known-y's, known-x's, new-x's, constant*) returns the predicted (*y*) values for the *new-x's*, based on exponential regression of the *known-y's* on the *known-x's*.
FISHER	FISHER(*x*) returns the value of the Fisher transformation evaluated at *x*.
FISHERINV	FISHERINV(*y*) returns the value of the inverse Fisher transformation evaluated at *y*.
FORECAST	FORECAST(*x, known-y's, known-x's*) returns a predicted (*y*) value for *x*, based on linear regression of the *known-y's* on the *known-x's*.
FTEST	FTEST(*array1, array2*) returns the *p*-value of the one-tailed *F* statistic, on the basis of the hypothesis that the variances *array1* and *array2* are not significantly different (which is rejected for low *p* values).
INTERCEPT	INTERCEPT(*known-y's, known-x's*) returns the *y* intercept of the linear regression of *known-y's* on *known-x's*.
LINEST	LINEST(*known-y's, known-x's, constant, stats*) returns coefficients in the linear regression of *known-y's* on *known-x's* (*constant* is true if the intercept is forced to be 0, and *stats* is true if regression statistics are desired).
LOGEST	LOGEST(*known-y's, known-x's, constant, stats*) returns the exponential regression of *known-y's* on *known-x's* (*constant* is true if the leading coefficient is forced to be 0, and *stats* is true if regression statistics are desired).
PROB	PROB(*x-values, probabilities, value*) returns the *probability* associated with *value*, given the *probabilities* of a range of values.
PROB	PROB(*x-values, probabilities, lower-limit, upper-limit*) returns the *probability* associated with values between *lower-limit* and *upper-limit*.
SLOPE	SLOPE(*known-y's, known-x's*) returns the slope of a linear regression line.
STEYX	STEYX(*known-y's, known-x's*) returns the standard error of the linear regression.

TREND	TREND(*known-y's, known-x's, new-x's, constant*) returns the *y* values of given input values (*new-x's*) based on the regression of *known-y's* on *known-x's*. If constant = false, the constant value is zero.
TTEST	TTEST(*array1, array2, number-of-tails, type*) returns the *p* value of a *t* test, of *type* paired (1), two-sample equal variance (2), or two-sample unequal variance (3).
ZTEST	ZTEST(*array, x, sigma*) returns the *p* value of a two-tailed *z* test, where *x* is the value to test and *sigma* is the population standard deviation.

Trigonometric Formulas

Function Name	Description
ACOS	ACOS(*number*) returns the arccosine (inverse cosine) of *number*.
ACOSH	ACOSH(*number*) returns the inverse hyperbolic cosine of *number*.
ASIN	ASIN(*number*) returns the arcsine (inverse sine) of *number*.
ASINH	ASINH(*number*) returns the inverse hyperbolic sine of *number*.
ATAN	ATAN(*number*) returns the arctangent (inverse tangent) of *number*.
ATAN2	ATAN2(*x, y*) returns the arctangent (inverse tangent) of the angle from the positive *x* axis.
ATANH	ATANH(*number*) returns the inverse hyperbolic tangent of *number*.
COS	COS(*angle*) returns the cosine of *angle*.
COSH	COSH(*number*) returns the hyperbolic cosine of *number*.
DEGREES	DEGREES(*angle*) returns the degree measure of an *angle* given in radians.
PI	PI() returns π accurate to 15 digits.
RADIANS	RADIANS(*angle*) returns the radian measure of an *angle* given in degrees.
SIN	SIN(*angle*) returns the sine of *angle*.
SINH	SINH(*number*) returns the hyperbolic sine of *number*.
TAN	TAN(*angle*) returns the tangent of *angle*.
TANH	TANH(*number*) returns the hyperbolic tangent of *number*.

StatPlus™ Commands

StatPlus™ is supplied with the textbook *Data Analysis with Microsoft Excel 2007* to perform basic statistical analysis not covered by Excel or the Analysis ToolPak. To use StatPlus, you must first verify that it is available to your workbook.

To check whether StatPlus is available:

1. If the StatPlus menu appears in the Menu Commands group of the Add-Ins tab, StatPlus is loaded and activated on your system.
2. If the menu command does not appear, click **Tools>Add-Ins** from the menu. If the StatPlus option is listed in the Add-Ins list box, click the checkbox. StatPlus is now available to you.
3. If StatPlus is not listed in the Add-Ins list box, you will have to install it from your instructor's disk. See Chapter 1 for more information.

The rest of this section documents each StatPlus Add-In command, showing each corresponding dialog box, and describes the command's options and output.

Creating Data

Bivariate Normal Data

The **StatPlus>Create Data>Bivariate Normal** command creates two columns of random normal data where the standard deviation σ of the data in the first column and the correlation between the columns are specified by the user. The standard deviation of the data in the second column of data is a function of the standard deviation of the data in the first column.

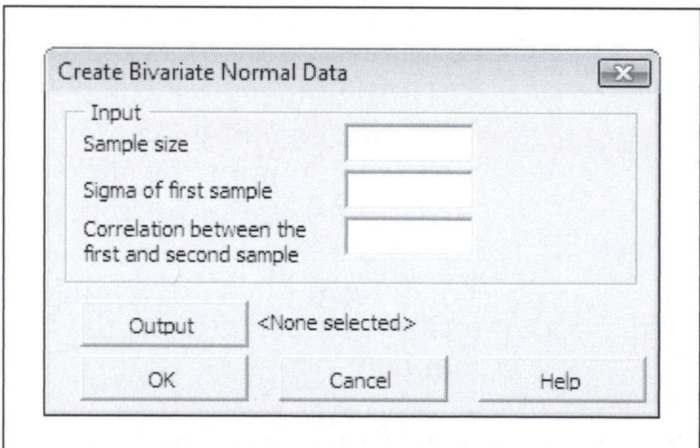

Patterned Data

The **StatPlus>Create Data>Patterned Data** command generates a column of data following a specified pattern. The pattern can be created on the basis of a sequence of numbers or taken from a number sequence entered in a data column already existing in the workbook. The user can specify how often each number in the pattern is repeated and how many times the entire sequence is repeated.

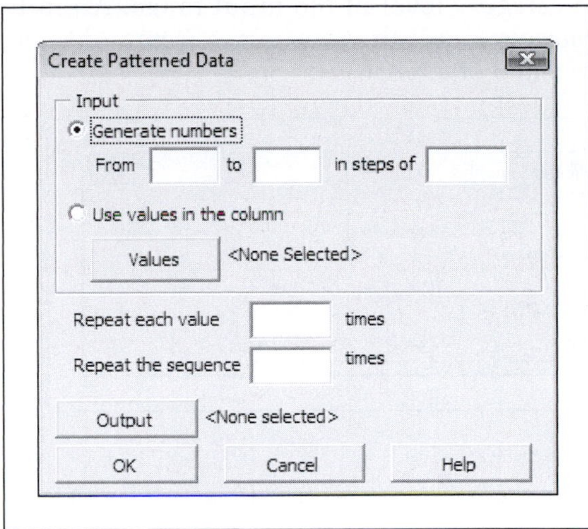

Random Numbers

The **StatPlus>Create Data>Random Numbers** command generates columns of random numbers for a specified probability distribution. The user specifies the number of samples (columns) of random numbers and the sample size (rows) of each sample.

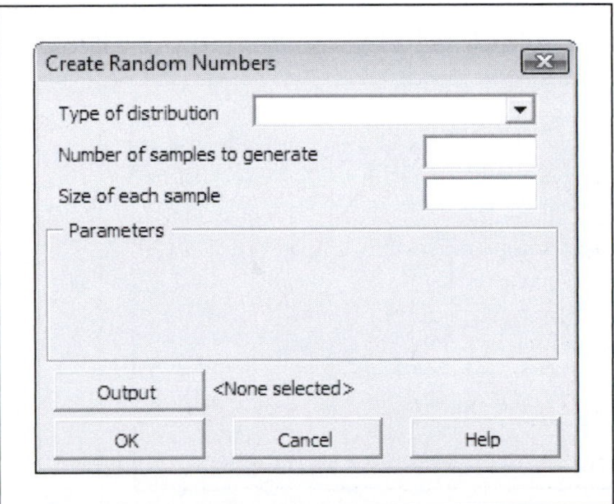

Manipulating Columns

Indicator Columns

The **StatPlus>Manipulate Columns>Create Indicator Columns** command takes a column of category levels and creates columns of indicator variables, one for each category level in the input range. An indicator variable for a particular category = 1 if the row comes from an observation belonging to that category and 0 otherwise.

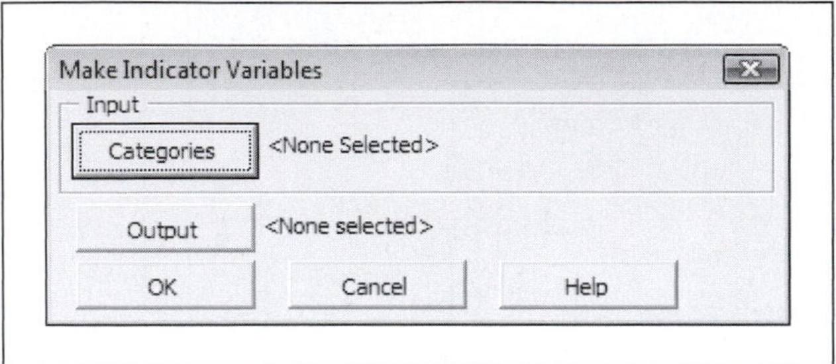

Two-Way Table

The **StatPlus>Manipulate Columns>Create Two-Way Table** command takes data arranged in three columns—a column of values, a column of category levels for one factor, and a second column of category levels for a second factor —and arranges the data into a two-way table. The columns of the table consist of the different levels of the first factor; the rows of the table consist of different levels of the second factor. Multiple values for each combination of the two factors show up in different rows within the table. Output from this command can be used in the Analysis ToolPak's ANOVA commands. The numbers of rows in the three columns must be equal. The user can choose whether to sort the row and column headers of the table.

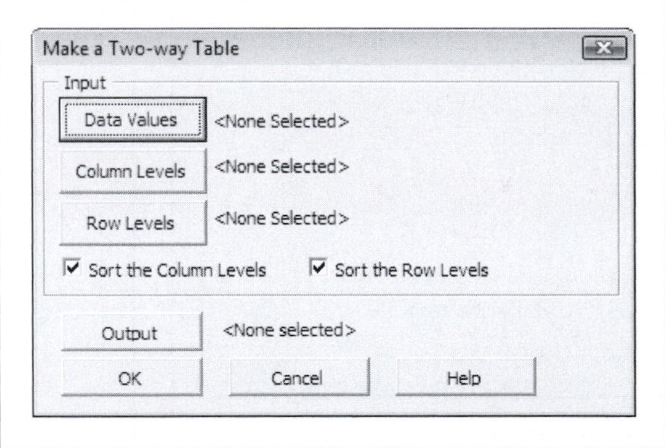

Unstack Column

The **StatPlus>Manipulate Columns>Unstack Column** command takes data found in two columns—a column of data values and a column of categories—and outputs the values into different columns, one column for each category level. The length of the values column and the length of the category column must be equal. The user can choose whether to sort the columns in ascending order of the category variable.

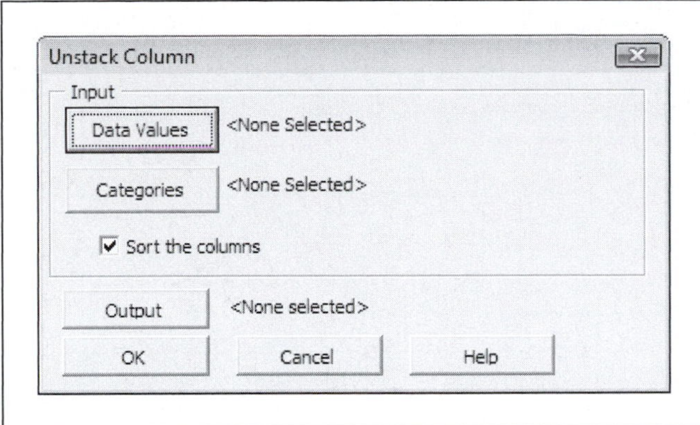

Stack Columns

The **StatPlus>Manipulate Columns>Stack Columns** command takes data that lie in separate columns and stacks the values into two columns. The column to the left contains the values; the column to the right is a category column. Values for the category are found from the header rows in the input columns, or, if there are no header rows, the categories are labeled as Level 1, Level 2, and so forth. The input range need not be contiguous.

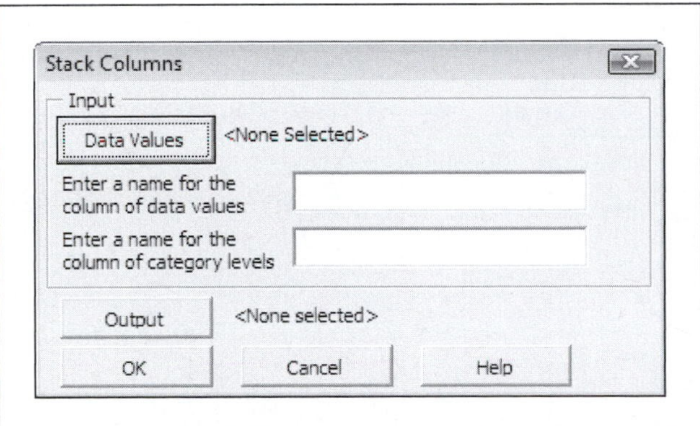

Standardize Data Values

The **StatPlus>Manipulate Columns>Standardize** command standardizes values in a collection of data columns. The user can choose from one of five different standardization methods.

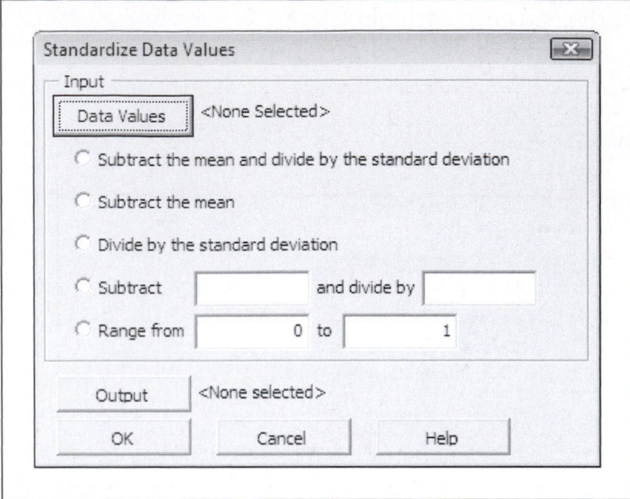

Sampling Data

Conditional Sample

The **StatPlus>Sampling>Conditional Sample** command extracts data values from a collection of columns corresponding to a specified condition.

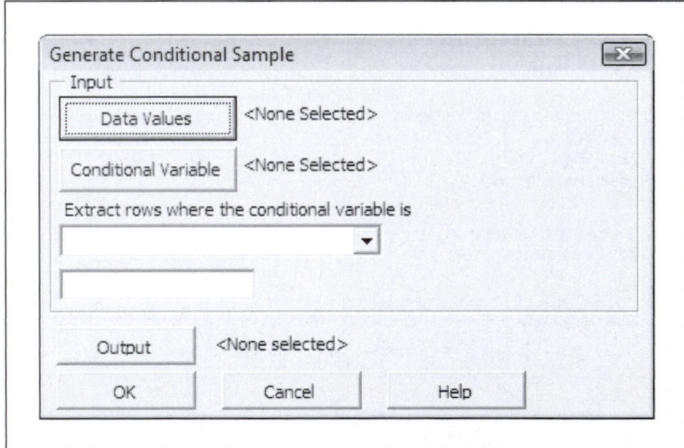

Periodic Sample

The **StatPlus>Sampling>Periodic Sample** command samples data values from a collection of columns starting at a specified row and then extracting every *i* th row, where *i* is specified by the user.

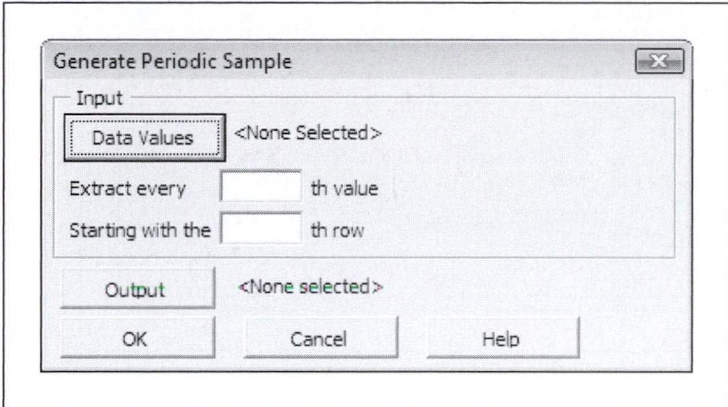

Random Sample

The **StatPlus>Sampling>Random Sample** command extracts a random sample of a given size from a collection of columns. The user can choose whether to sample with replacement or without replacement.

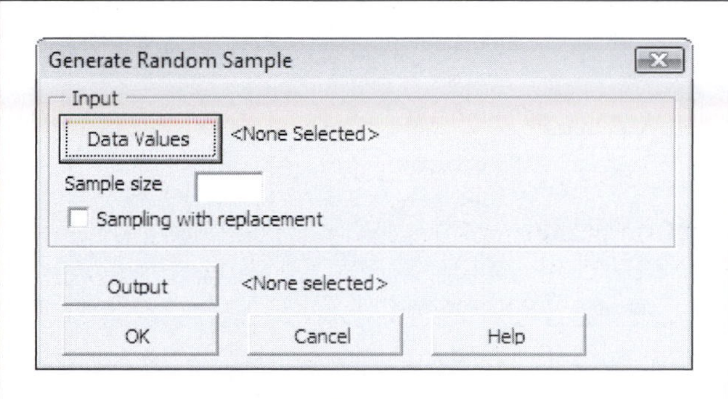

Single-Variable Charts

Boxplots

The **StatPlus>Single Variable Charts>Boxplots** command creates a boxplot. The data values can be arranged either as separate columns or in one column with a category variable. Users can choose to add a dotted line for the sample average and to connect the medians between the boxes. The boxplot can be sent to an embedded chart on a worksheet or to its own chart sheet.

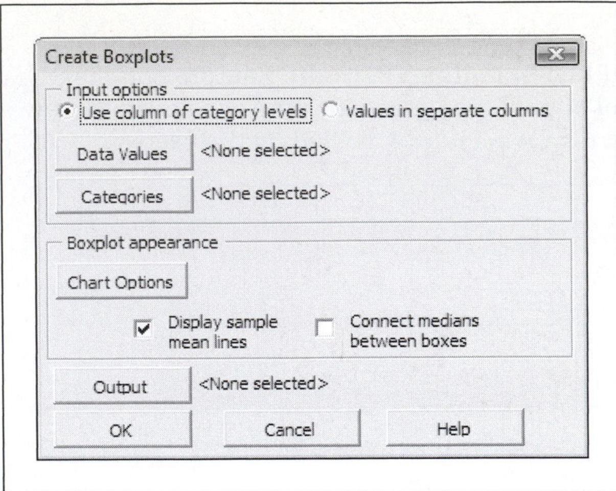

Fast Scatterplot

The **StatPlus>Single Variable Charts>Fast Scatterplot** command creates a quick scatterplot bypassing many of the commands on the Excel ribbon. The scatterplot can be sent to an embedded chart on a worksheet or to its own chart sheet.

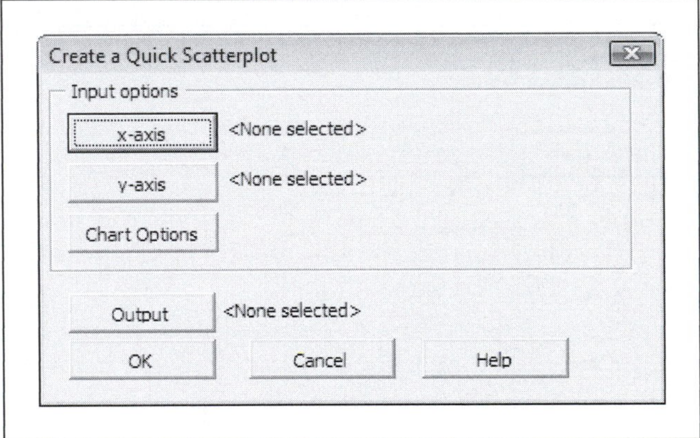

Histogram

The **StatPlus>Single Variable Charts>Histograms** command creates a histogram. The user can specify a frequency, cumulative frequency, percentage, or cumulative percentage chart. Also, the histogram can be broken down into the different levels of a categorical variable. If a categorical variable is used, the histogram bars can be (1) stacked, (2) displayed side by side, or (3) displayed in 3-D. The user can choose to add a normal curve to the histogram, as well as to display the corresponding frequency table. The histogram can be sent to an embedded chart on a worksheet or to its own chart sheet.

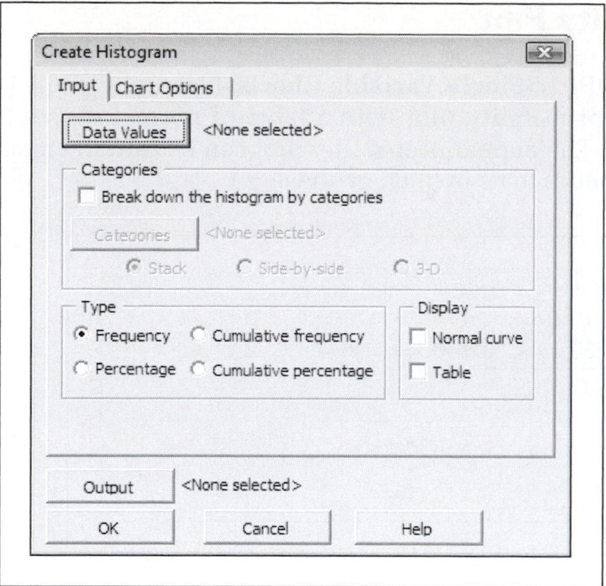

Stem and Leaf Plots

The **StatPlus>Single Variable Charts>Stem and Leaf Plots** command creates a stem and leaf plot. The data values can be arranged either as separate columns or in one column with a category variable. If more than one stem and leaf plot is generated, the user can choose to apply the same stem values to each of the plots and to add a summary stem and leaf plot. The user can also choose to truncate outliers of either moderate or major size. The stem and leaf plot appears as values within a worksheet.

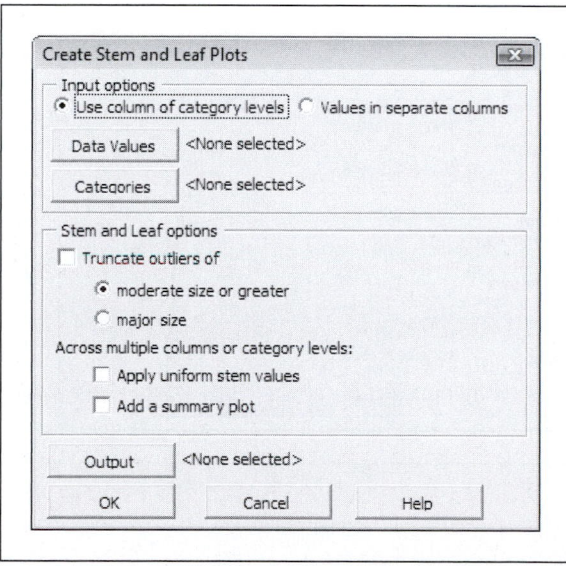

Normal Probability Plot

The **StatPlus>Single Variable Charts>Normal P-plots** command creates a normal probability plot with a table of normal scores for data in a single column. The normal probability plot can be sent to an embedded chart on a worksheet or to its own chart sheet.

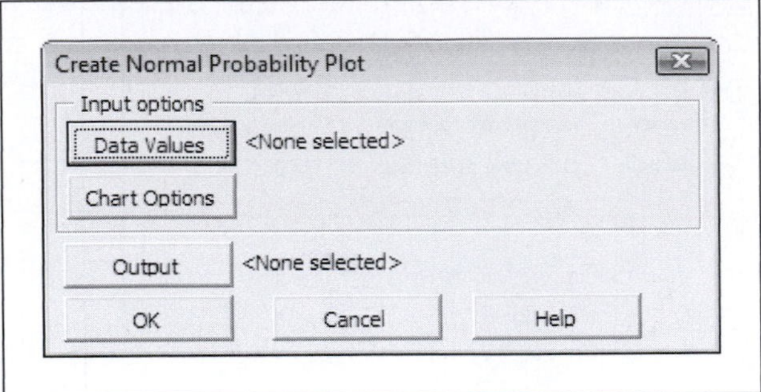

Multivariable Charts

Fast Bubble Plot

The **StatPlus>Multivariable Charts>Fast Bubble Plot** command creates a quick bubble plot for data arranged in different columns. The bubble plot can be sent to a chart sheet or embedded on a worksheet.

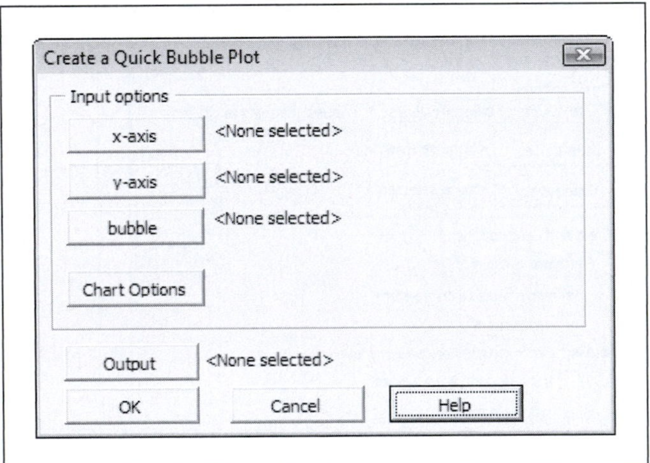

Multiple Histograms

The **StatPlus>Multivariable Charts>Multiple Histograms** command creates stacked histogram charts. The source data can be arranged in separate columns or within a single column along with a column of category values. The user can choose to display frequencies, cumulative frequencies, percentages, or cumulative percentages. A normal curve can also be added to each of the histograms. The histograms have common bin values and are shown in the same vertical-axis scale. The histogram charts are sent to embedded charts on a worksheet.

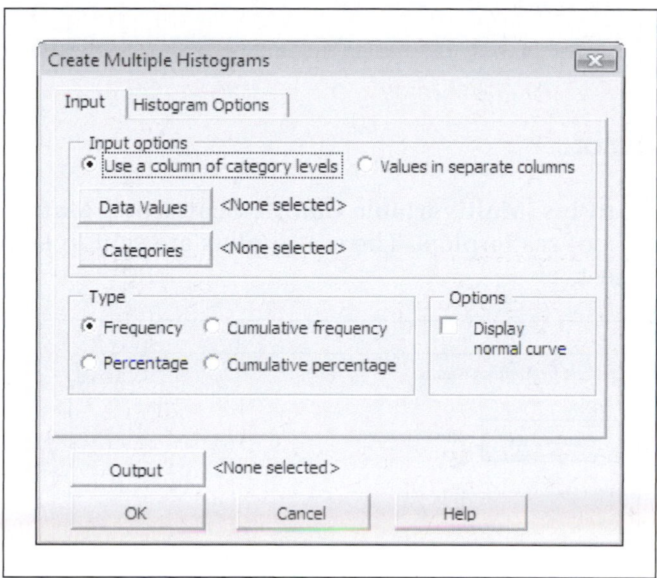

Multiple Normal Probability Plots

The **StatPlus>Multivariable Charts>Normal P-plots** creates a collection of normal probability plots for variables arranged in columns. Normal curves are plotted on a single chart.

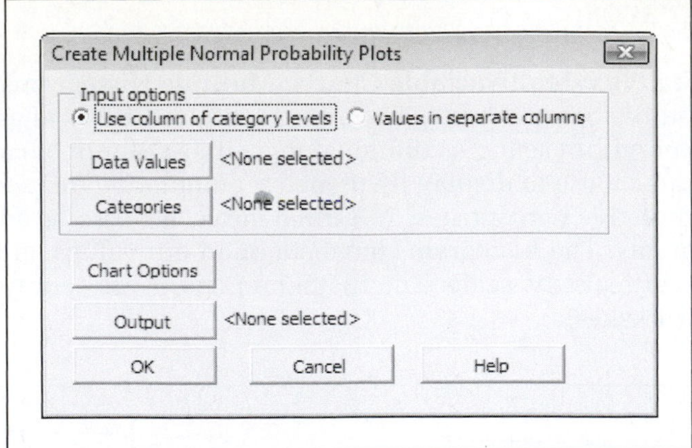

Scatterplot Matrix

The **StatPlus>Multivariable Charts>Scatterplot Matrix** command creates a matrix of scatterplots. The scatterplots are sent to embedded charts on a worksheet.

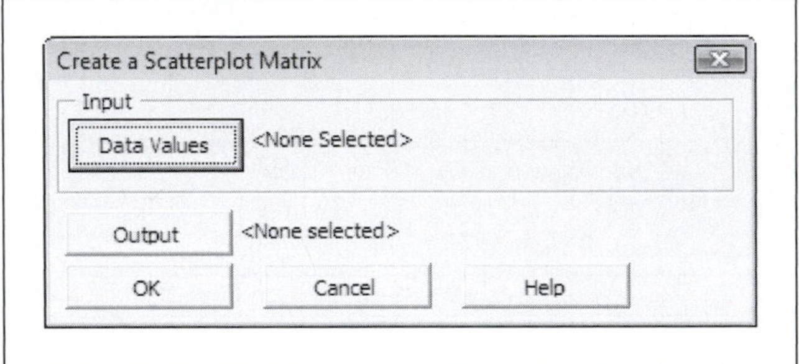

Quality-Control Charts

C-Charts

The **StatPlus>QC Charts>C-Chart** command creates a C-chart (count chart) of quality-control data for a single column of counts (for example, the number of defects in an assembly line). The count chart includes a mean line and lower and upper control limits. The C-chart can be sent to an embedded chart on a worksheet or to its own chart sheet.

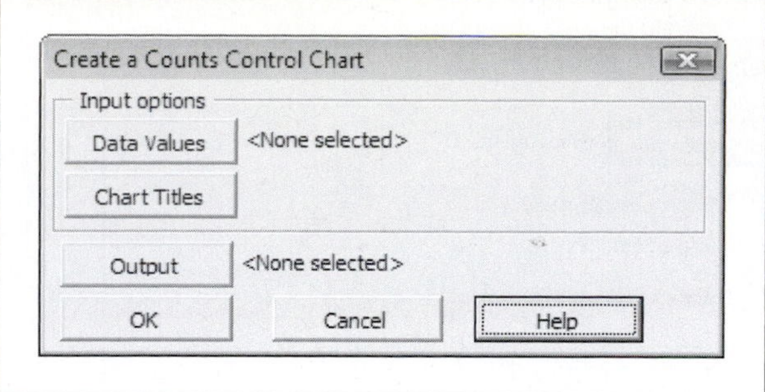

Individuals Chart

The **StatPlus>QC Charts>Individuals Chart** command creates an Individuals chart of quality-control data for a single column of quality-control values, where there is no subgroup available. The Individuals chart can be sent to an embedded chart on a worksheet or to its own chart sheet.

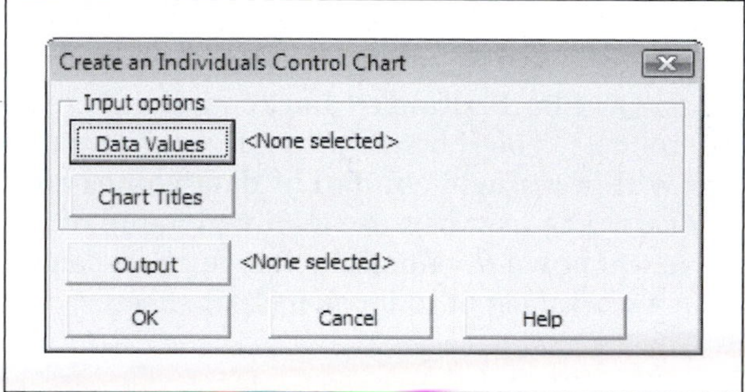

P-Charts

The **StatPlus>QC Charts>P-Chart** command creates a P-chart (proportion chart) of quality-control data. Proportion values are placed in a single column. The user can specify a single sample size for all proportion values or can use a column of sample-size values. The P-chart can be sent to an embedded chart on a worksheet or to its own chart sheet.

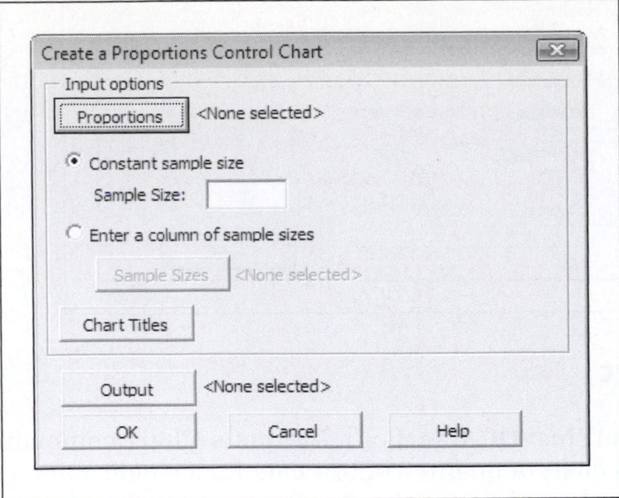

Range Charts

The **StatPlus>QC Charts>Range Chart** command creates a Range chart of quality-control data. The subgroups can be arranged in rows across separate columns or within a single column of data values alongside a column of subgroup levels. The user can use a known value of σ or create the Range chart with an unknown σ value. The Range chart can be sent to an embedded chart on a worksheet or to its own chart sheet.

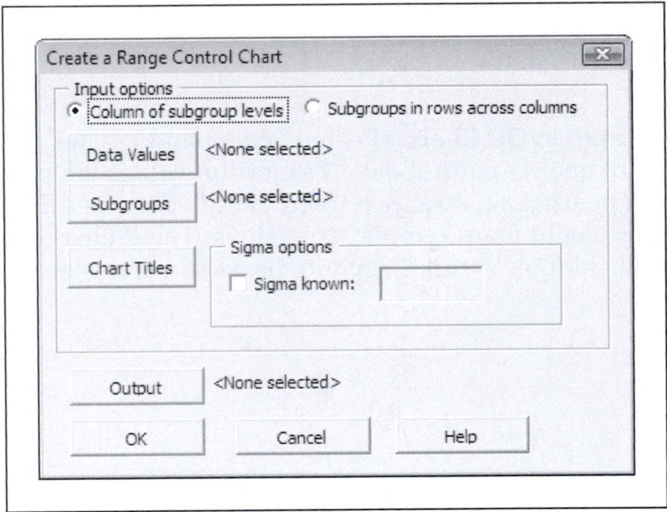

Moving Range Charts

The **StatPlus>QC Charts>Moving Range Chart** command creates a Moving Range chart of quality-control data where there is no subgroup available. The quality-control values must be placed in a single column. The Moving Range chart can be sent to an embedded chart on a worksheet or to its own chart sheet.

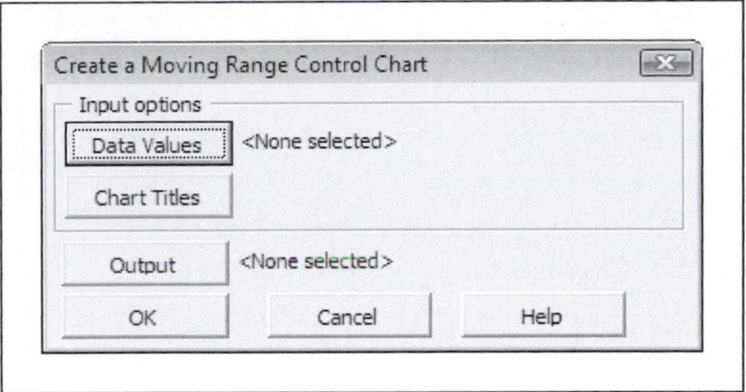

XBAR Charts

The **StatPlus>QC Charts>XBAR Chart** command creates an XBAR chart of quality-control data. The subgroups can be arranged in rows across separate columns or within a single column of data values alongside a column of subgroup levels. The user can use known values of μ and σ or create the XBAR chart with unknown μ and σ values. The XBAR chart can be sent to an embedded chart on a worksheet or to its own chart sheet.

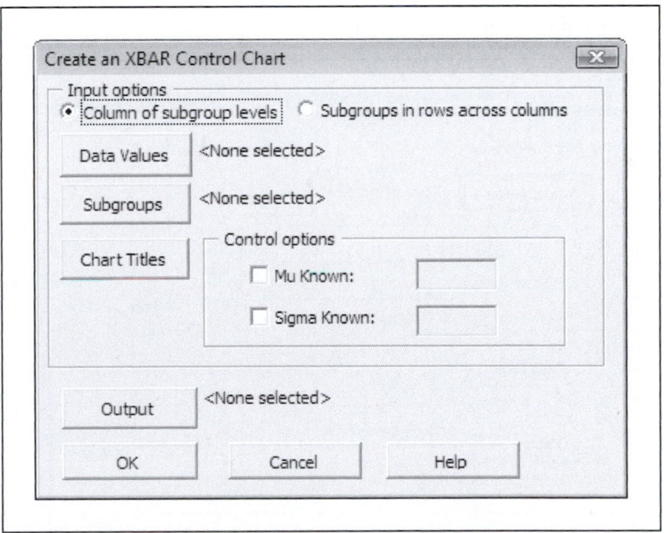

S-Charts

The **StatPlus>QC Charts>S-Chart** command creates a s-chart or sigma-chart of quality control data. The subgroups can be arranged in rows across separate columns or within a single column of data values alongside a column of subgroup levels. The s-chart can be sent to an embedded chart on a worksheet or to its own chart sheet.

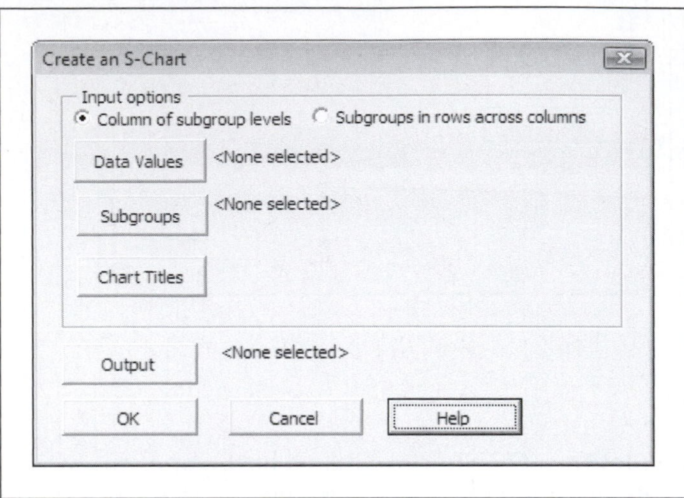

Generic QC Charts

The **StatPlus>QC Charts>Generic QC Chart** command creates a generic quality-control chart where the user specifies the location of the lower control limit (lcl), center line, and upper control limit (ucl). The chart can be sent to an embedded chart sheet on a worksheet or to its own chart sheet.

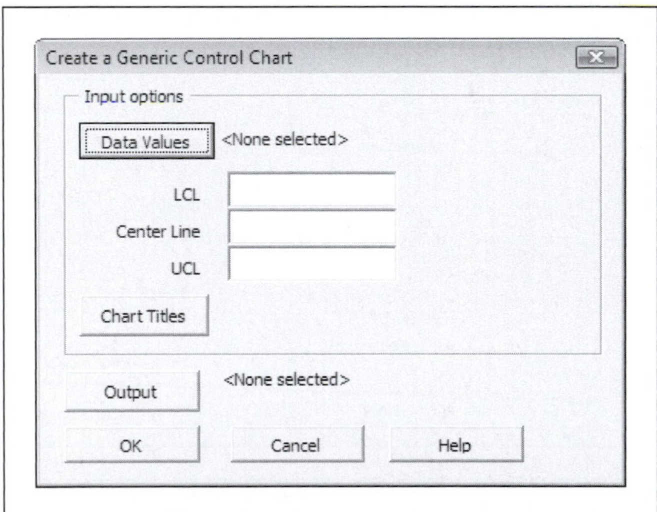

Pareto Chart

The **StatPlus>QC Charts>Pareto Chart** command creates a Pareto chart of quality-control data. Data values can be arranged in separate columns or within a single column along with a column of category values. The user specifies conditions for a defective value. The Pareto chart can be sent to an embedded chart on a worksheet or to its own chart sheet.

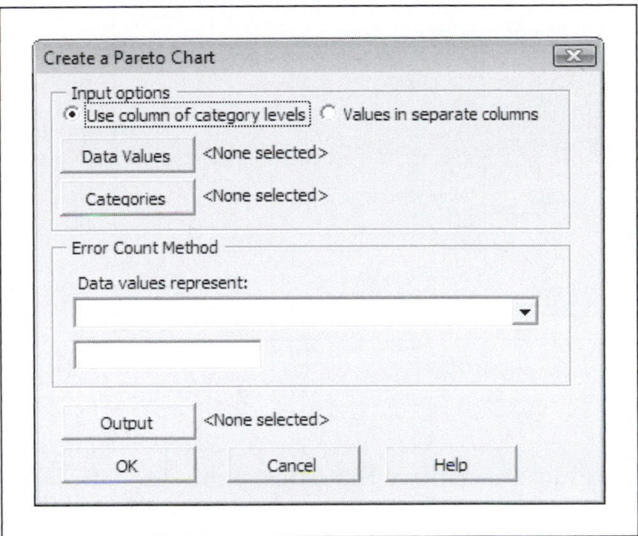

Descriptive Statistics

Frequency Table

The **StatPlus>Descriptive Statistics>Frequency Table** command creates a table containing frequency, cumulative frequency, percentage, and cumulative percentage. The frequency table can either be displayed by discrete values in the data column or by bin values. If bin values are used, the user can specify how the data are counted relative to the placement of the bins. The frequency table can also be broken down by the values of a By variable.

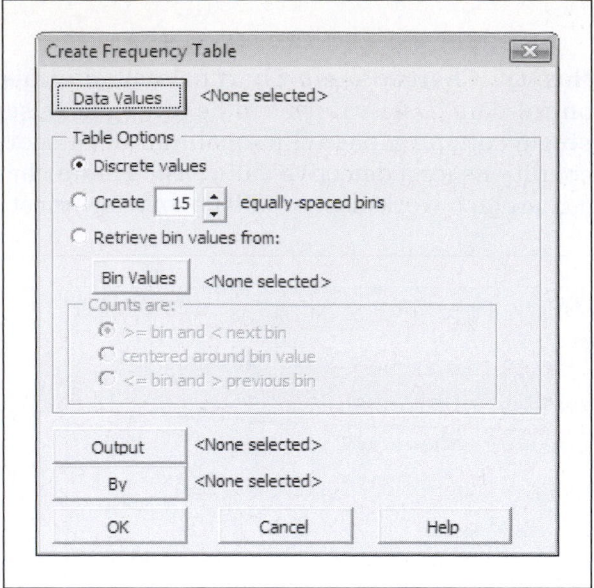

Table Statistics

The **StatPlus>Descriptive Statistics>Table Statistics** command creates a table of descriptive statistics for a two-way cross-classification table. The first column of the table contains the titles of the descriptive statistics, the second column shows their values, the third column indicates the degrees of freedom, and the fourth column shows the *p* value or asymptotic standard error. The user must select the range containing the two-way table, *excluding* the row and column totals but *including* the row and column headers.

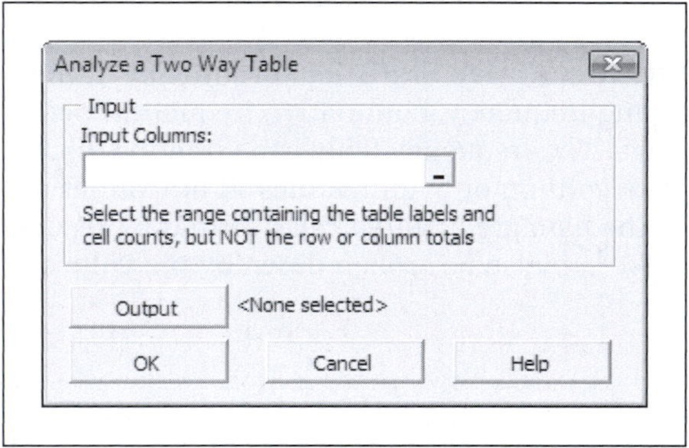

Univariate Statistics

The **StatPlus>Descriptive Statistics>Univariate Statistics** command creates a table of univariate statistics. The user can choose from a selection of 33 different statistics, either by selecting the statistics individually, or selecting entire groups of statistics. Statistics can be displayed in different columns or in different rows. The table can be broken down using a By variable.

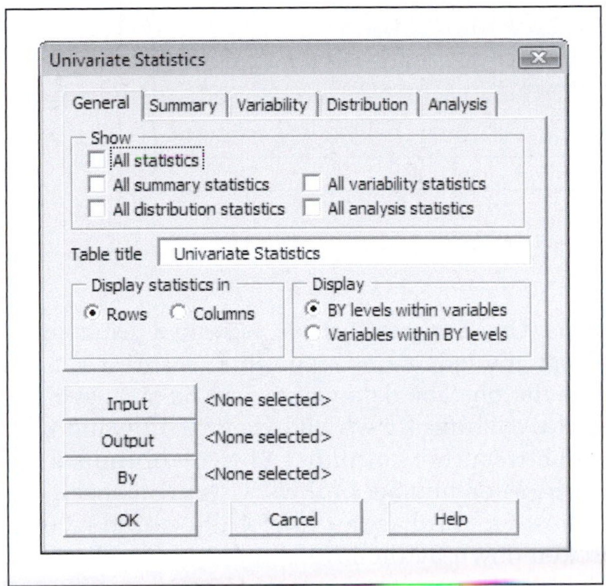

One Sample Tests

One Sample *t*-test

The **StatPlus>One Sample Tests>1 Sample *t*-test** command performs a one-sample *t*-test and calculates a confidence interval. The data values can be arranged either as a single column or as two columns (in which case the command will analyze the paired difference between the columns). If two columns are used, the columns must have the same number of rows. Users can specify the null and alternative hypotheses as well as the size of the confidence interval. The output can be broken down by the levels of a By variable.

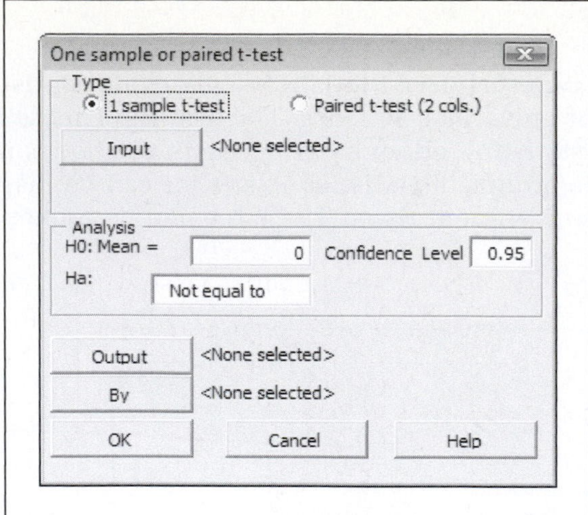

One Sample z test

The **StatPlus>One Sample Tests>1 Sample *z* test** command performs a one sample *z* test and calculates a confidence interval for data with a known standard deviation. The data values can be arranged either as a single column or as two columns (in which case the command will analyze the paired difference between the columns). If two columns are used, the columns must have the same number of rows. Users can specify the null and alternative hypotheses as well as the size of the confidence interval. The output can be broken down by the levels of a By variable. Users must specify the value of the standard deviation.

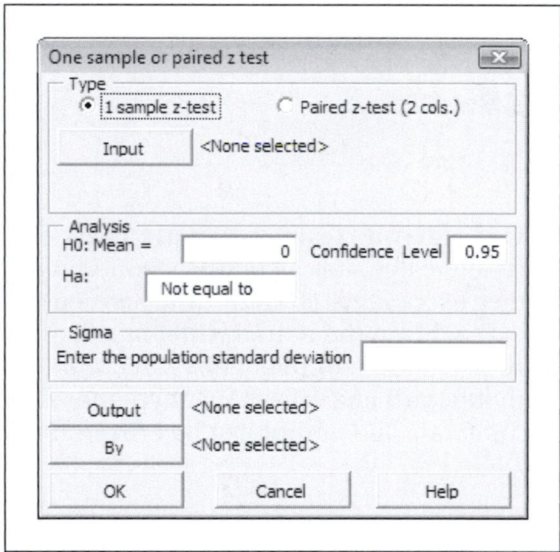

One Sample Sign test

The **StatPlus>One Sample Tests>1 Sample Sign test** command performs a one-sample Sign test and calculates a confidence interval. The data values can be arranged either as a single column or as two columns (in which case the command will analyze the paired difference between the columns). If two columns are used, the columns must have the same number of rows. Users can specify the null and alternative hypotheses as well as the size of the confidence interval. For confidence intervals, the user specifies that the calculated interval be approximately, at least, or at most the size of the specified interval. The output can be broken down by the levels of a By variable.

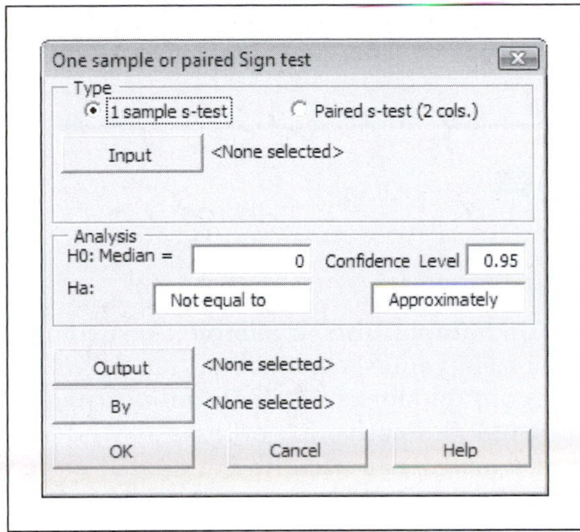

One Sample Wilcoxon Signed Rank test

The **StatPlus>One Sample Tests>1 Sample Wilcoxon Signed Rank test** command performs a one-sample Wilcoxon Signed Rank test and calculates a confidence interval. The data values can be arranged either as a single column or as two columns (in which case the command will analyze the paired difference between the columns). If two columns are used, the columns must have the same number of rows. Users can specify the null and alternative hypotheses as well as the size of the confidence interval. The output can be broken down by the levels of a By variable.

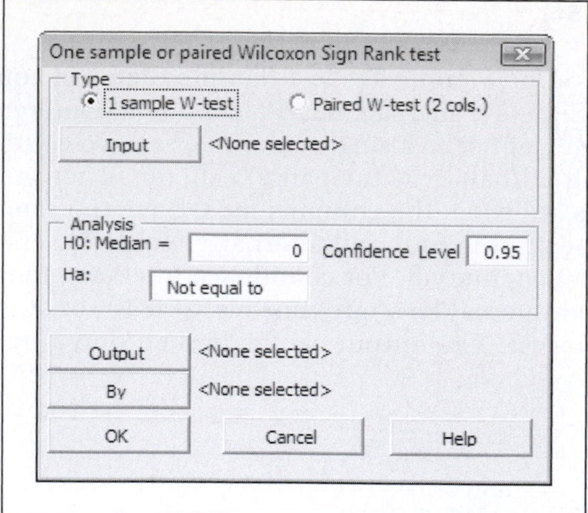

Two Sample Tests

Two Sample *t*-test

The **StatPlus>Two Sample Tests>2 Sample *t*-test** command performs a two sample *t*-test for data values, arranged either in two separate columns or within a single column alongside a column of category levels. Users can specify the null and alternative hypotheses as well as the size of the confidence interval. The test can use either a pooled or an unpooled variance estimate. The output can be broken down by the levels of a By variable.

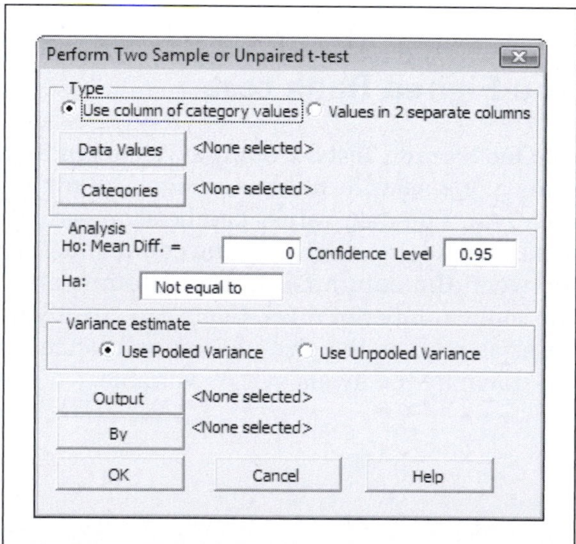

Two Sample z test

The **StatPlus>Two Sample Tests>2 Sample *z*-test** command performs a two sample *z* test for data values, arranged either in two separate columns or within a single column alongside a column of category levels. Users can specify the null and alternative hypotheses as well as the size of the confidence interval. Users must enter the standard deviation for each sample. The output can be broken down by the levels of a By variable.

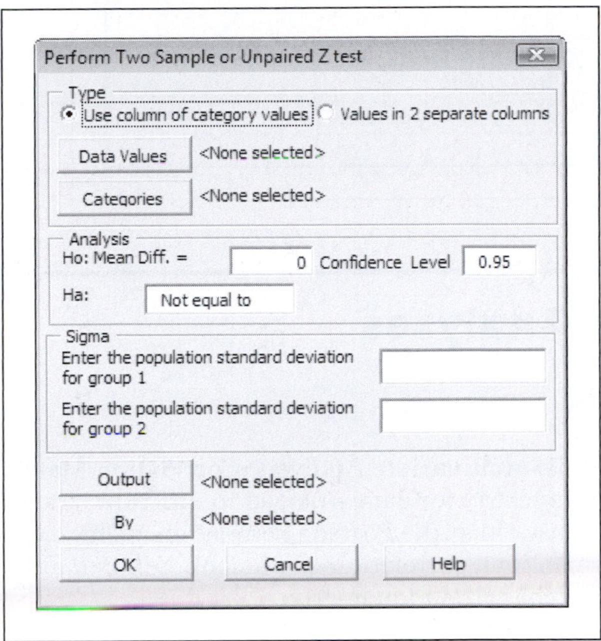

Two Sample Mann-Whitney test

The **StatPlus>Two Sample Tests>2 Sample Mann-Whitney Rank test** command performs a two sample Mann-Whitney Rank test for data values, arranged either in two separate columns or within a single column alongside a column of category levels. Users can specify the null and alternative hypotheses as well as the size of the confidence interval. The output can be broken down by the levels of a By variable.

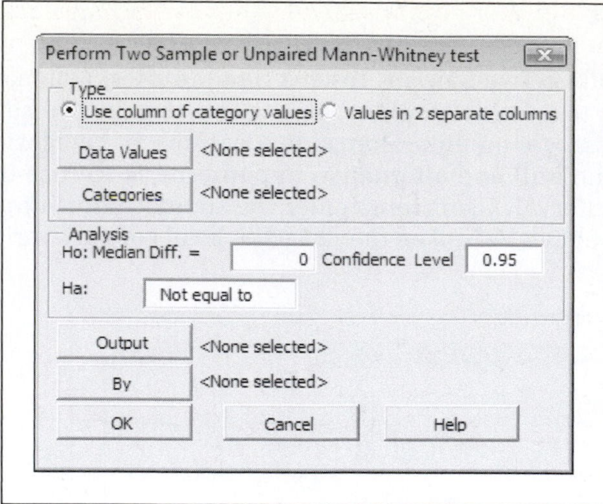

Multivariate Analyses

Correlation Matrix

The **StatPlus>Multivariate Analysis>Correlation Matrix** command creates a correlation matrix for data arranged in different columns. The correlation matrix can use either the Pearson correlation coefficient or the nonparametric Spearman rank correlation coefficient. You can also output a matrix of *p* values for the correlation matrix.

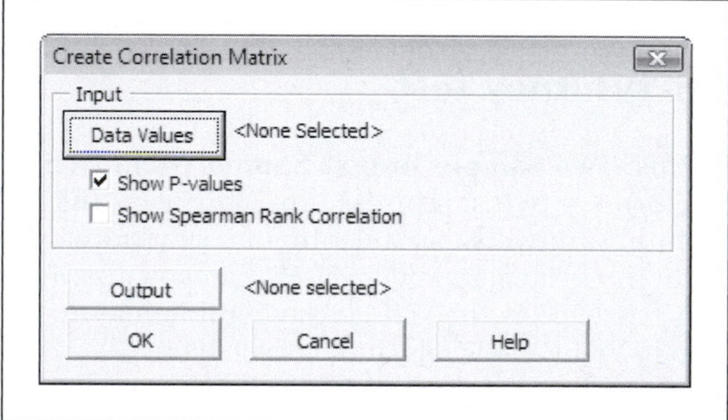

Means Matrix

The **StatPlus>Multivariate Analysis>Means Matrix** command creates a matrix of pairwise mean differences for data. The data values can be arranged in separate columns or within a single column alongside a column of category levels. The output includes a matrix of *p* values with an option to adjust the *p* value for the number of comparisons using the Bonferroni correction factor.

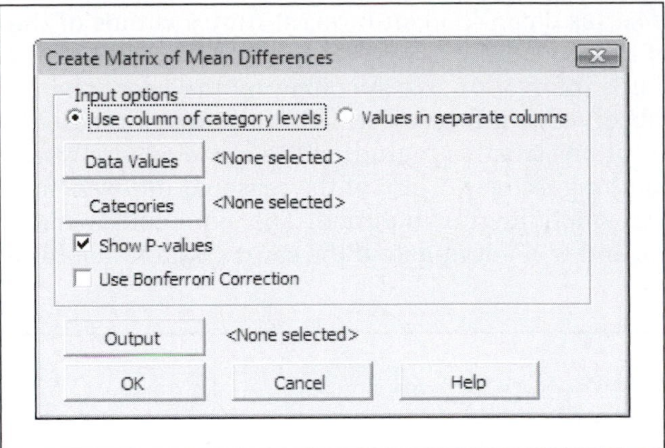

Time Series Analyses

ACF Plot

The **StatPlus>Time Series>ACF Plot** command creates a table of the autocorrelation function and a chart of the autocorrelation function, for time series data arranged in a single column. The first column in the output table contains the lag values up to a number specified by the user, the second column contains the auto correlation, the third column of the table contains the lower 95% confidence boundary, and the fourth column contains the upper 95% confidence boundary. Autocorrelation values that lie outside the 95% confidence interval are shown in red. The chart shows the autocorrelations and the 95% confidence width.

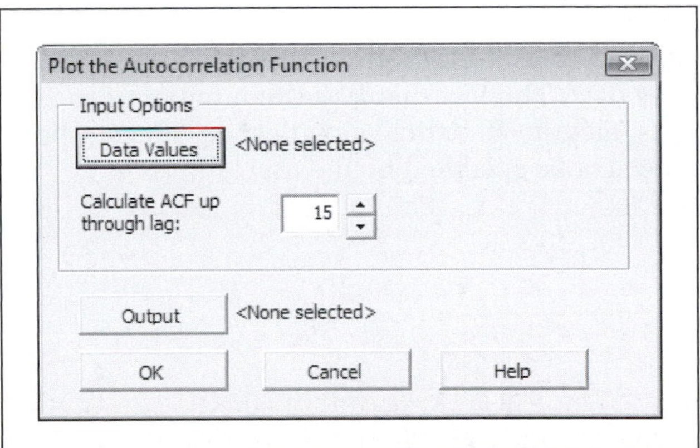

Exponential Smoothing

The **StatPlus>Time Series>Exponential Smoothing** command calculates one-, two-, or three-parameter exponential smoothing models for a single column of time series data. You can forecast future values of the time series on the basis of the smoothing model for a specified number of units and include a confidence interval of size specified by the user. The output includes a table of observed and forecasted values, future forecasted values, and a table of descriptive statistics including the mean square error and final values of the smoothing factors. A plot of the seasonal indexes (for three-parameter exponential smoothing) is included. The exponential smoothing output is not dynamic and will not update if the source data in the input range change.

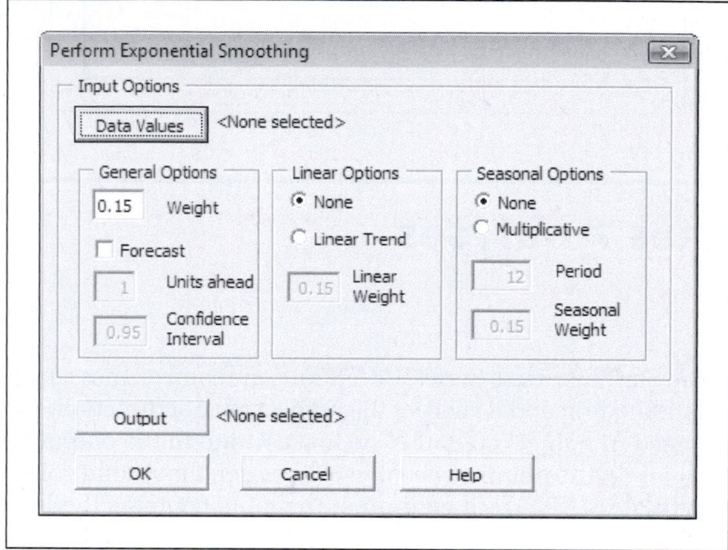

Runs Test

The **StatPlus>Time Series>Runs Test** command performs a Runs test on time series data. The test displays the number of runs, the expected number of runs, and the statistical significance. The cut point can either be the sample mean or be specified by the user.

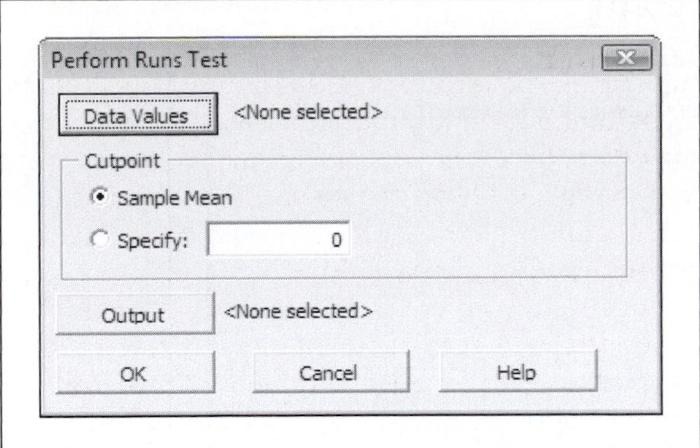

Seasonal Adjustment

The **StatPlus>Time Series>Seasonal Adjustment** command creates a column of seasonally adjusted time series values that show periodicity, and creates a plot of unadjusted and adjusted values. A plot of the seasonal indexes is included in the output (multiplicative seasonality is assumed). The seasonal adjustment output is not dynamic and will not update if the source data in the input range change.

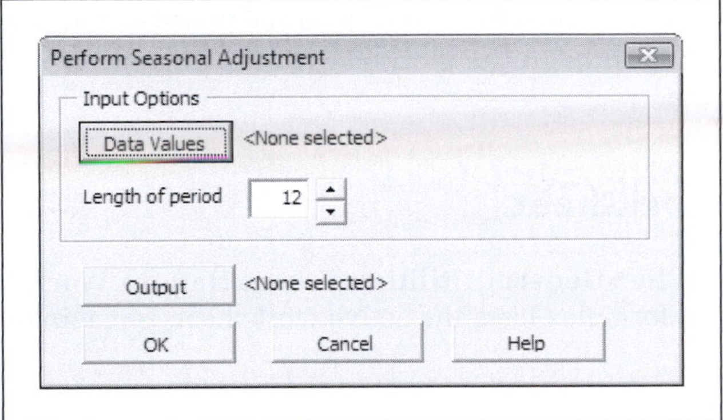

StatPlus Options

The **StatPlus>StatPlus Options** command allows the user to specify the default input and output options for the different StatPlus modules. One can also specify how StatPlus should handle hidden data.

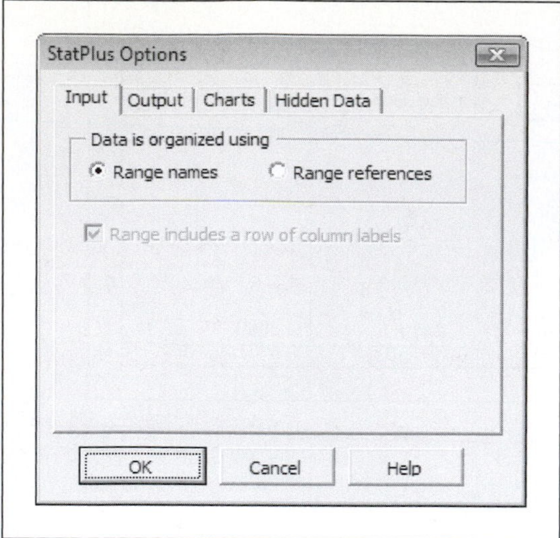

General Utilities

Resolve StatPlus Links

The **StatPlus>General Utilities>Resolve StatPlus Links** command redirects all StatPlus links in the workbook to the current location of the StatPlus add-in file.

Freeze Data in Worksheet

The **StatPlus>General Utilities>Freeze Data in Worksheet** command removes all formulas from the active worksheet, replacing them with values.

Freeze Hidden Data

The **StatPlus>General Utilities>Freeze Hidden Data** command removes all formulas from the StatPlus hidden worksheet, replacing them with values.

Freeze Data in Workbook

The **StatPlus>General Utilities>Freeze Data in Workbook** command removes all formulas from the active workbook, replacing them with values.

View Hidden Data

The **StatPlus>General Utilities>View Hidden Data** command unhides all of the hidden worksheets created by StatPlus.

Rehide Hidden Data

The **StatPlus>General Utilities>Rehide Hidden Data** command hides all of the hidden worksheets created by StatPlus.

Remove Unlinked Hidden Data

The **StatPlus>General Utilities>Remove Unlinked Hidden Data** command removes any hidden data created by StatPlus that are no longer linked to a worksheet in the active workbook.

Unload Modules

The **StatPlus>Unload Modules** command unloads StatPlus modules. Select the individual modules to unload from the list of loaded modules.

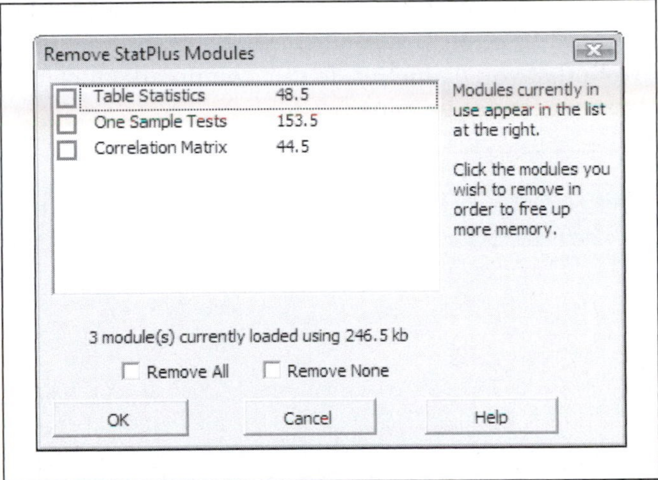

About StatPlus

The **StatPlus>About StatPlus** command provides information about the installation and version number of the StatPlus add-in.

Chart Commands

Label Chart Points

The **StatPlus>Label Series points** command can be run when a chart is the active object in the workbook. You can link the labels to a cell range in the workbook and copy the cell format. You can also replace the points in the scatterplot with the labels.

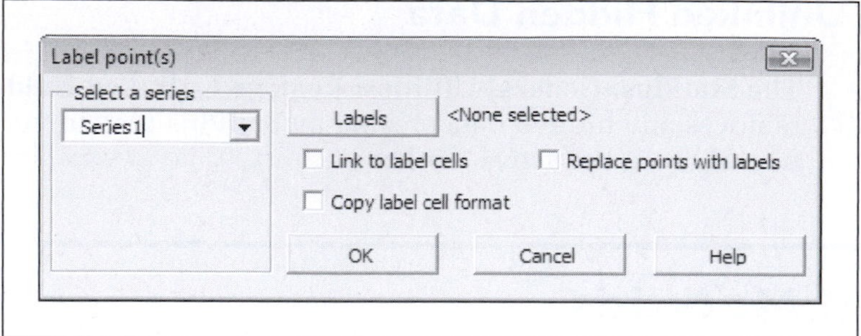

Display Chart Series by Category

The **StatPlus>Display series by category** command can be run when a chart is the active object in the workbook. The command divides the chart series into several different series on the basis of the levels of the category variable. Note that you cannot undo this command. Once the chart series is broken down, it cannot be joined again.

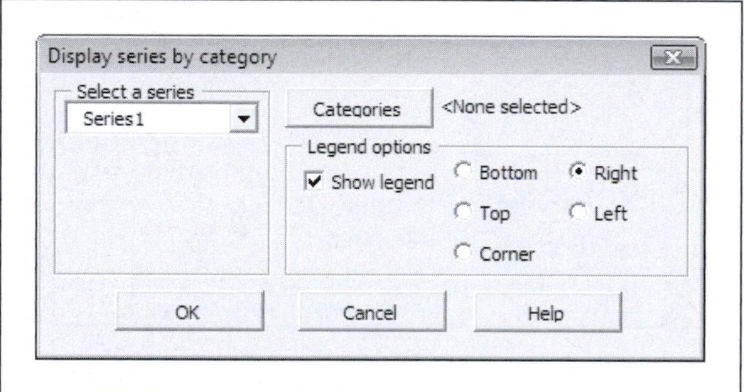

Select Row from Chart Series

The **StatPlus>Select Row** command can be run when a chart is the active object in the workbook. The command selects the row in the worksheet corresponding to the point you selected.

StatPlus™ Math and Statistical Functions

The following functions are available in Excel when StatPlus™ is loaded:

Descriptive Statistics for One Variable

Function Name	Description
COUNTBETW	COUNTBETW(*range, lower, upper, boundary*) returns the count of nonblank cells in the *range* that lie between the *lower* and *upper* values. The *boundary* variable determines how the end points are used. If *boundary* = 1, the interval is > the lower value and < the upper value. If *boundary* = 2, the interval is ≥ the lower value and < the upper value. If *boundary* = 3, the interval is > the lower value and ≤ the upper value. If *boundary* = 4, the interval is ≥ lower value and ≤ the upper value.
IQR	IQR(*range*) calculates the interquartile range for the data in *range*.
MODEVALUE	MODEVALUE(*range*) calculates the mode of the data in *range*. The data are assumed to be in one column.
NSCORE	NSCORE(*number, range*) returns the normal score of *number* (or cell reference to *number*) from a *range* of values.
RANGEVALUE	RANGEVALUE(*range*) calculates the difference between the maximum and minimum values from a *range* of values.
RANKTIED	RANKTIED(*number, range, order*) returns the rank of the *number* in *range*, adjusting the rank for ties. If *order* = 0, then the range is ranked from largest to smallest; if *order* = 1, the range is ranked from smallest to largest.
RUNS	RUNS(*range*, [*center*]) returns the number of runs in the data column *range*. The center = 0 unless a *center* value is entered.
SE	SE(*range*) calculates the standard error of the values in *range*.
SIGNRANK	SIGNRANK(*number, range*) returns the sign rank of the *number* in *range*, adjusting the rank for ties. Values of zero receive a sign rank of 0. If *order* = 0, then the range is ranked from largest to smallest in absolute value; if *order* = 1, the range is ranked from smallest to largest in absolute value.

Descriptive Statistics for Two or More Variables

Function Name	Description
CORRELP	CORRELP(*range1*, *range2*) returns the *p* value for the Pearson coefficient of correlation between *range1* and *range2*. (Note: Range values must be in two columns.)
MEDIANDIFF	MEDIANDIFF (*range*, *range2*) calculates the pairwise median difference between values in two separate columns.
MWMedian	MWMedian(*range*, *range2*) calculates the median of the Walsh averages between two columns of data.
MWMedian2	MWMedian2(*range*, *range2*) calculates the median of the Walsh averages for data values in one column (*range*) with category levels in a second column (*range2*). There can be only two levels in the categories column.
PEARSONCHISQ	PEARSONCHISQ(*range*) returns the Pearson chi-square test statistic for data in *range*.
PEARSONP	PEARSONP(*range*) returns the *p* value for the Pearson chi-square test statistic for data in *range*.
SPEARMAN	SPEARMAN(*range*) returns the Spearman nonparametric rank correlation for values in *range*. (Note: Range values must be in one column only.)
SPEARMANP	SPEARMANP(*range*) returns the *p* value for the Spearman nonparametric rank correlation for values in *range*. (Note: Range values must be in one column only.)

Distributions

Function Name	Description
NORMBETW	NORMBETW(*lower, upper, mean, stdev*) calculates the area under the curve between the *lower* and *upper* limits for a normal distribution with $\mu = mean$ and $\sigma = stdev$.
TDF	TDF(*number, df, cumulative*) calculates the area under the curve to the left of *number* for a *t* distribution with degrees of freedom *df*, if *cumulative* = true. If *cumulative* = false, this function calculates the probability density function for *number*.

Mathematical Formulas

Function Name	Description
IF2FUNC	IF2FUNC(*Fname, IFRange1, IFValue1, IFRange2, IFValue2, RangeAnd*, [*Arg1, Arg2, . . .*]) calculates the value of the Excel function *Fname*, for rows in a data set where the values of *IFRange1* are equal to *IFValue1* and the values of *IFRange2* are equal to *IFValue2*. Parameters of the *Fname* function can be inserted as *Arg1, Arg2,* and so forth. If *RangeAnd* = true, an AND clause is assumed between the two values. If *RangeAnd* = false, an OR clause is assumed.
IFFUNC	IFFUNC(*Fname, IFRange, IFValue,* [*Arg1, Arg2, . . .*]) calculates the value of the Excel function *Fname*, for rows in a data set where the values of *IFRange* are equal to *IFValue*. Parameters of the *Fname* function can be inserted as *Arg1*, *Arg2*, and so forth.
RANDBERNOULLI	RANDBERNOULLI(*prob*) returns a random number from the Bernoulli distribution with probability = *prob*.
RANDBETA	RANDBETA(*alpha*, *beta*, [*a*], [*b*]) returns a random number from the Beta distribution with parameters *alpha*, *beta*, and (optionally) *a* and *b* where *a* and *b* are the end points of the distribution.
RANDBINOMIAL	RANDBINOMIAL(*prob, trials*) returns a random number from the binomial distribution with probability = *prob* and number of trials = *trial.*
RANDCHISQ	RANDCHISQ(*df*) returns a random number from the chi-square distribution with degrees of freedom *df.*
RANDDISCRETE	RANDDISCRETE(*range, prob*) returns a random number from a discrete distribution where the values of the distribution are found in the cell range *range*, and the associated probabilities are found in the cell range *prob*.
RANDEXP	RANDEXP(*lambda*) returns a random number from the exponential distribution where λ = *lambda*.
RANDF	RANDF(*df1, df2*) returns a random number from the *F* distribution with numerator degrees of freedom *df1* and denominator degrees of freedom *df2*.

RANDGAMMA	RANDGAMMA(*alpha, beta*) returns a random number from the gamma distribution with parameters *alpha* and *beta*.
RANDINTEGER	RANDINTEGER(*lower, upper*) returns a random integer from a discrete uniform distribution with the lower boundary = *lower* and the upper boundary = *upper*.
RANDLOG	RANDLOG(*mean, stdev*) returns a random number from the log normal distribution with $\mu = mean$ and $\sigma = stdev$.
RANDNORM	RANDNORM(*mean, stdev*) returns a random number from the normal distribution with $\mu = mean$ and $\sigma = stdev$.
RANDPOISSON	RANDPOISSON(*lambda*) returns a random number from the Poisson distribution where $\lambda = lambda$.
RANDT	RANDT(*df*) returns a random number from the *t* distribution with degrees of freedom *df*.
RANDUNI	RANDUNI(*lower, upper*) returns a random number from the uniform distribution where the lower boundary = *lower* and the upper boundary = *upper*.

Statistical Analysis

Function Name	Description
ACF	ACF(*range*, *lag*) calculates the autocorrelation function for values in *range* for lag = *lag*. *Note:* Range values must lie within one column.
Bartlett	Bartlett(*range*, *range2*, . . .) calculates the *p* value for the Bartlett test assuming that the data are arranged in multiple columns.
Bartlett2	Bartlett2(*range*, *range2*) calculates the *p* value for the Bartlett test assuming one column of data values and one column of category values.
DW	DW(*range*) calculates the Durbin-Watson statistics for data in a single column.
FTest2	FTest2(*range*, *range2*) calculates the *p* value for the *F* test assuming one column of data values and one column of category values.
Levene	Levene(*range*, *range2*, . . .) calculates the *p* value for Levene test assuming that the data are arranged in multiple columns.
Levene2	Levene2(*range*, *range2*) calculates the *p* value for the Levene test assuming one column of data values and one column of category values.

MannW MannW(*range, range2*, [*median*]) calculates the Mann-Whitney test statistic for data values in two columns. The median difference is assumed to be 0, unless a *median* value is specified.

MannWp MannWp(*range, range2*, [*median*], [*Alt*]) calculates the *p* value of the Mann-Whitney test statistic for data values in two columns. The median difference is assumed to be 0, unless a *median* value is specified. The *p* value is for a two-sided alternative hypothesis unless *Alt* = 1, in which case a one-sided test is performed.

MannW2 MannW2(*range, range2*, [*median*]) calculates the Mann-Whitney test statistic for data values in one column (*range*) and category values in a second column (*range2*). There can be only two levels in the categories column. The median difference is assumed to be 0, unless a *median* value is specified.

MannWp2 MannWp2(*range, range2*, [*median*]) calculates the *p* value of the Mann-Whitney test statistic for data values in one column (*range*) and category values in a second column (*range2*). There can be only two levels in the categories column. The median difference is assumed to be 0, unless a *median* value is specified. The *p* value is for a two-sided alternative hypothesis unless *Alt* = 1, in which case a one-sided test is performed.

Oneway Oneway(*range, range2*) calculates the *p* value of the one-way ANOVA for data arranged in two columns.

RUNSP RUNSP(*range*, [*center*]) calculates the *p* value of the Runs test for values in the data column *range*. Center = 0 unless a *center* value is entered.

TSTAT TSTAT(*range*, [*mean*]) calculates the one-sample t-test statistic for values in the data column *range*. The mean value under the null hypothesis is assumed to be 0, unless a *mean* value is specified.

TSTATP TSTATP(*range*, [*mean*], [*Alt*]) calculates the *p* value for the one-sample *t* test statistic for values in the data column *range*. The mean value under the null hypothesis is assumed to be 0, unless a *mean* value is specified. A two-sided alternative hypothesis is assumed unless *Alt* = −1, in which case the "less than" alternative hypothesis is assumed, or *Alt* = 1, in which case the "greater than" alternative hypothesis is assumed.

WILCOXON	WILCOXON(*range*, [*median*]) calculates the Wilcoxon Signed Rank statistic for values in the data column *range*. The median value under the null hypothesis is assumed to be 0, unless a *median* value is specified.
WILCOXONP	WILCOXONP(*range*, [*median*], [*Alt*]) calculates the *p* value of the Wilcoxon Signed Rank statistic for values in the data column *range*. The median value under the null hypothesis is assumed to be 0, unless a *median* value is specified. A two-sided alternative hypothesis is assumed unless $Alt = -1$, in which case the "less than" alternative hypothesis is assumed, or $Alt = 1$, in which case the "greater than" alternative hypothesis is assumed.
ZSTAT	ZSTAT(*range, sigma,* [*mean*]) calculates the *z*-test statistic for values in the data column *range* with a standard deviation *sigma*. The mean value under the null hypothesis is assumed to be 0, unless a *mean* value is specified.
ZSTATP	ZSTATP(*range, sigma,* [*mean*], [*Alt*]) calculates the *p* value for the *z* test statistic for values in the data column *range* with a standard deviation *sigma*. The mean value under the null hypothesis is assumed to be 0, unless a *mean* value is specified. A two-sided alternative hypothesis is assumed unless $Alt = -1$, in which case the "less than" alternative hypothesis is assumed, or $Alt = 1$, in which case the "greater than" alternative hypothesis is assumed.

Bibliography

Bliss, C. I. (1964). *Statistics in Biology*. New York: McGraw-Hill.

Booth, D. E. (1985). Regression methods and problem banks. *Umap Modules: Tools for Teaching 1985*. Arlington, MA: Consortium for Mathematics and Its Applications, pp.179–216.

Bowerman, B.L., and O'Connell, R.T. (1987). *Forecasting and Time Series, An Applied Approach*. Pacific Grove, CA: Duxbury Press.

Cushny, A. R., and Peebles, A. R. (1905). The action of optical isomers, II: Hyoscines. *Journal of Physiology* 32: 501–510.

D'Agostino, R. B., Chase, W., and Belanger A. (1988). The appropriateness of some common procedures for testing the equality of two independent binomial populations. *The American Statistician* 42: 198–202.

Deming, W. E. (1982). *Quality, Productivity, and Competitive Position*. Cambridge, MA: M.I.T. Center for Advanced Engineering Study.

Deming, W. E. (1982). *Out of the Crisis*. Cambridge, MA: M.I.T. Center for Advanced Engineering Study.

Donoho, D., and Ramos, E. (1982). PRIMDATA: Data Sets for Use with PRIM-H (DRAFT). FTP stat library at Carnegie Mellon University.

Edge, O. P., and Friedberg, S. H. (1984). Factors affecting achievement in the first course in calculus. *Journal of Experimental Education* 52: 136–140.

Fosback, N. G. (1987) *Stock Market Logic*. Fort Lauderdale, FL: Institute for Econometric Research.

Halio, M. P. (1990). Student writing: Can the machine maim the message? *Academic Computing*, January 1990, 16–19, 45.

Juran, J. M., ed.(1974) *Quality Control Handbook*. New York: McGraw-Hill.

Lea, A. J. (1965). "Relationship Between Environmental Temperature and the Death Rate from Neoplasms of the Breast", *British Medical Journal* i: 488.

Longley, J. W. (1967). An appraisal of least squares programs for the electronic computer from the point of view of the user. *Journal of the American Statistical Association* 62: 819–831.

Milliken, G., and Johnson, D. (1984). *Analysis of Messy Data*, Volume 1: Designed Experiments, Princeton, NJ: Van Nostrand.

Neave, H. R. (1990). *The Deming Dimension*. Knoxville, TN: SPC Press.

Rosner, B., and Woods, C. (1988). Autoregressive modeling of baseball performance and salary data. *1988 Proceedings of the Statistical Graphics Section*, American Statistical Association, pp. 132–137.

Shewhart, W. A. (1931). *Economic Control of Quality of Manufactured Product*. Princeton, NJ: Van Nostrand.

Tukey, J. W. (1977). *Exploratory Data Analysis*. Reading, MA: Addison-Wesley.

Weisberg, S. (1985). *Applied Linear Regression*, 2nd ed. New York: Wiley.

Index

E

F

G

H

S

T

U

V

Quick Reference Guide

for Berk & Carey's *Data Analysis with Microsoft® Excel: Updated for Office 2007®*.

Objective	Steps	Refer to
Patterned data, create	Click **StatPlus > Create Data > Patterned Data**. Specify the data pattern. *Requires StatPlus.*	Chapter 7
PivotTable, create	Click the **PivotTable** button from the Tables group on the Insert tab.	
PivotTable, grouping categories in	Select cells from the PivotTable and then click the **Group Selection** button from the Group group on the Options tab of the PivotTable Tools ribbon.	Chapter 7
PivotTable, remove categories from	Click and drag the row or column label off of the pivot table.	Chapter 7
Random numbers, create	Click the **Data Analysis** button from the Analysis group on the Data tab and then click **Random Number Generation**.	
Random numbers, create	Click **StatPlus > Create Data > Random Numbers**. Select the probability distribution, number of samples, and sample size. *Requires StatPlus.*	Chapter 5
Range control chart, create	Click **StatPlus > QC Charts > Range Chart**. Select the control data and specify the control chart options. *Requires StatPlus.*	Chapter 12
Range names, create from column labels	Select the cell range and then click the **Create from Selection** button from the Defined Names group on the Formulas tab; then click the **Top Row** check box and click **OK**.	Chapter 2
Regression analysis, perform	Click the **Data Analysis** button from the Analysis group on the Data tab and then click **Regression.** *Requires the Analysis ToolPak*	Chapter 8
Regression line, add to scatterplot	Right-click the chart series and click **Add Trendline**. Select the regression type from the Trendline Options dialog sheet.	Chapter 8
Runs test, perform	Click **StatPlus > Time Series > Runs test**. Enter the options of the test. *Requires StatPlus.*	Chapter 8
Sample, create a conditional	Click **StatPlus > Sampling > Conditional Sample**. Enter the sampling conditions. *Requires StatPlus.*	
Sample, create a periodic	Click **StatPlus > Sampling > Periodic Sample**. Enter the sampling conditions. *Requires StatPlus.*	
Sample, create a random	Click **StatPlus > Sampling > Random Sample**. Enter the sampling conditions. *Requires StatPlus.*	
Scatterplot matrix, create	Click **StatPlus > Multi-variable Charts > Scatterplot Matrix**. Enter the variables in the scatterplot matrix. *Requires StatPlus.*	Chapter 8
Scatterplot, break into categories	Select the chart and click **StatPlus > Display by Category**. Specify the categorical variable to use. *Requires StatPlus.*	Chapter 3
Scatterplot, create quickly	Click **StatPlus > Single Variable Charts > Fast Scatterplot**. Enter the data columns for the *x* and *y* axes. *Requires StatPlus.*	Chapter 11
Seasonal adjustment, perform	Click **StatPlus > Time Series > Seasonal Adjustment**. Select the data column and the period of the season. *Requires StatPlus.*	Chapter 11
Standardize data	Click **StatPlus > Manipulate Columns > Standardize**. Enter the data columns and select the method of standardization. *Requires StatPlus.*	
StatPlus modules, unloading	Click **StatPlus > Unload Modules**, select the module's checkbox and click **OK**. *Requires StatPlus.*	Chapter 1

Quick Reference Guide

for Berk & Carey's *Data Analysis with Microsoft® Excel: Updated for Office 2007®.*

Objective	Steps	Refer to
Histogram, create	Click the **Data Analysis** button from the Analysis group on the Data tab and then click **Histogram**.	
Histogram, create	Click **StatPlus > Single Variable Charts > Histograms**. Select the histogram options. *Requires StatPlus.*	Chapter 4
Histograms, create multiple	Click **StatPlus > Multi-variable Charts > Multiple Histograms**. Select the variables and enter the histogram options. *Requires StatPlus.*	Chapter 10
Indicator variables, create	Click **StatPlus > Manipulate Columns > Create Indicator Columns**. Select the data column. *Requires StatPlus.*	Chapter 10
Individuals control chart, create	Click **StatPlus > QC Charts > Individuals Chart**. Select the control data and specify the control chart options. *Requires StatPlus.*	Chapter 12
Mann-Whitney Rank test, perform	Click **StatPlus > Two Sample Tests > 2 Sample Mann-Whitney Rank test**. Enter the null and alternative hypotheses. *Requires StatPlus.*	Chapter 6
Means matrix, create	Click **StatPlus > Multivariate Analysis > Means Matrix**. Select the columns to display in the means matrix. *Requires StatPlus.*	Chapter 10
Moving average, add to scatterplot	Right-click the chart series and click **Add Trendline**. Select Moving Average from the Trendline Options tab.	Chapter 11
Moving Range control chart, create	Click **StatPlus > QC Charts > Moving Range Chart**. Select the control data and specify the control chart options. *Requires StatPlus.*	Chapter 12
Multiple regression analysis, perform	Click the **Data Analysis** group from the Analysis group on the Data tab and click **Regression**. *Requires the Analysis ToolPak.*	Chapter 9
Normal probability plot, create	Click **StatPlus > Single Variable Charts > Normal P-plots**. *Requires StatPlus.*	Chapter 5
One-sample Sign test, perform	Click **StatPlus > One Sample Tests > 1 Sample Sign test**. Enter the null and alternative hypotheses. *Requires StatPlus.*	
One-sample *t* test, perform	Click **StatPlus > One Sample Tests > 1 Sample t-test**. Enter the null and alternative hypotheses. *Requires StatPlus.*	Chapter 6
One-sample *z* test, perform	Click **StatPlus > One Sample Tests > 1 Sample z-test**. Enter the null and alternative hypotheses. *Requires StatPlus.*	
One-parameter exponential smoothing, perform	Click **StatPlus > Time Series > Exponential Smoothing**. Select the options for the one-parameter model. *Requires StatPlus.*	Chapter 11
One-way analysis of variance, perform	Click the **Data Analysis** group from the Analysis group on the Data tab and click **ANOVA: Single Factor**. *Requires the Analysis ToolPak.*	Chapter 10
P control chart, create	Click **StatPlus > QC Charts > P-Chart**. Select the control data and specify the control chart options. *Requires StatPlus.*	Chapter 12
Paired *t* test, perform	Click **StatPlus > One Sample Tests > 1 Sample t-test**. Enter the null and alternative hypotheses. *Requires StatPlus.*	Chapter 6
Pareto chart, create	Click **StatPlus > QC Charts > Pareto Chart**. Select the control data and specify the options for the Pareto chart. *Requires StatPlus.*	Chapter 12

Quick Reference Guide

for Berk & Carey's *Data Analysis with Microsoft® Excel: Updated for Office 2007®*.

Objective	Steps	Refer to
StatPlus, hidden data viewing	Click **StatPlus > General Utilities > View Hidden Data**. *Requires StatPlus.*	Chapter 1
StatPlus, installing	Access the online installation program from the website and follow the instructions on the Installation Wizard.	
StatPlus, set options	Click **StatPlus > StatPlus Options**. *Requires StatPlus.*	Chapter 1
StatPlus, update links	Click **StatPlus > General Utilities > Resolve StatPlus Links**.	Chapter 1
Stem and Leaf plot, create	Click **StatPlus > Single Variable Charts > Stem and Leaf**. Select the Stem and Leaf options. *Requires StatPlus.*	Chapter 4
Table statistics, calculate	Click **StatPlus > Descriptive Statistics > Table Statistics**. Select the cell range containing the cell counts and table labels but *not* the column and row totals. *Requires StatPlus.*	Chapter 7
Three-parameter exponential smoothing, perform	Click **StatPlus > Time Series > Exponential Smoothing**. Select the options for the three-parameter model. *Requires StatPlus.*	Chapter 11
Two-sample *t* test, perform	Click **StatPlus > Two Sample Tests > 2 Sample t-test**. Enter the null and alternative hypotheses. *Requires StatPlus.*	Chapter 6
Two-sample *z* test, perform	Click **StatPlus > Two Sample Tests > 2 Sample z-test**. Enter the null and alternative hypotheses. *Requires StatPlus.*	
Two-parameter exponential smoothing, perform	Click **StatPlus > Time Series > Exponential Smoothing**. Select the options for the two-parameter model. *Requires StatPlus.*	Chapter 11
Two-way analysis of variance with replication, perform	Click the **Data Analysis** button from the Analysis Group on the Data tab and click **ANOVA: Two-Factor with Replication**. *Requires the Analysis ToolPak.*	Chapter 10
Two-way analysis of variance without replication, perform	Click the **Data Analysis** button from the Analysis group on the Data tab and click **ANOVA: Two-Factor without Replication.** *Requires the Analysis ToolPak.*	
Two-way table, create	Click **StatPlus > Manipulate Columns > Create Two-Way Table**. Select the columns for the table. *Requires StatPlus.*	Chapter 10
Univariate statistics, display	Click the **Data Analysis** button from the Analysis group on the Data tab and click **Descriptive Statistics**. *Requires the Analysis ToolPak.*	
Univariate statistics, display	Click **StatPlus > Descriptive Statistics > Univariate Statistics**. Select the statistics to display. *Requires StatPlus.*	Chapter 4
Unpaired *t* test, perform	Click **StatPlus > Two Sample Tests > 2 Sample t-test**. Enter the null and alternative hypotheses. *Requires StatPlus.*	Chapter 6
Wilcoxon Signed Rank test, perform	Click **StatPlus > One Sample Tests > 1 Sample Wilcoxon Signed Rank test**. Enter the null and alternative hypotheses. *Requires StatPlus.*	Chapter 6
XBar control chart, create	Click **StatPlus > QC Charts > XBar Chart**. Select the control data and specify the control chart options. *Requires StatPlus.*	Chapter 12

Quick Reference Guide

for Berk & Carey's *Data Analysis with Microsoft® Excel: Updated for Office 2007®*.

Objective	Steps	Refer to
Add-Ins, installing	Click the **Office** button, click the **Excel Options** button, click **Add-Ins** from the list of Excel options, click the **Go** button, and click the **Browse** button from within the Add-Ins dialog box to locate and load the add-in file.	Chapter 1
Autocorrelation, plot	Click **StatPlus > Time Series > ACF Plot**. Select the data column and range of lag values. *Requires StatPlus.*	Chapter 11
Bivariate Normal data, create	Click **StatPlus > Create Data > Bivariate Normal**. Specify the parameters of the bivariate distribution. *Requires StatPlus.*	
Boxplot, create	Click **StatPlus > Single Variable Charts > Boxplots**. Select the boxplot options. *Requires StatPlus.*	Chapter 4
Bubble plots, create	Click the **Other Charts** button from the Charts group on the Insert tab and then click a bubble chart subtype.	Chapter 3
C control chart, create	Click **StatPlus > QC Charts > C-Chart**. Select the control data and specify the control chart options. *Requires StatPlus.*	Chapter 12
Chart axis, reformat	With a chart selected, click the **Axes** button from the Axes group on the Layout tab of the ChartTools ribbon and then select whether to reformat the horizontal or vertical axis.	
Chart background, format	With the chart selected, click the **Plot Area** button from the Background group on the Layout tab of the ChartTools ribbon and select the background option.	Chapter 3
Chart point labels, create	With the chart selected, click **StatPlus > Label series points**. *Requires StatPlus.*	Chapter 3
Chart points, format	With the chart selected, click any data point in the chart and click **Format Selection** from the Current Selection group on the Format tab of the ChartTools ribbon.	Chapter 3
Columns, stack	Click **StatPlus > Manipulate Columns > Stack**. Select the columns to stack. *Requires StatPlus.*	
Columns, unstack	Click **StatPlus > Manipulate Columns > Unstack**. Select the columns to unstack. *Requires StatPlus.*	Chapter 10
Correlation matrix, create	Click **StatPlus > Multivariate Analysis > Correlation Matrix**. Enter the variables in the correlation matrix. *Requires StatPlus.*	Chapter 8
Data, enter from keyboard	Click the cell and type the data values.	Chapter 2
Data, import from a database	Click the **Get External Data** button on the Data tab; then select the **From Other Sources** button and select the data source.	Chapter 2
Data, import from text files	Click the **Office** button, click the **Excel Options** button, click **Add-Ins** from the list of Excel options, click the **Go** button, and click the **Browse** button from within the Add-Ins dialog box to locate and load the add-in file.	Chapter 2
Data, query with advanced filter	Enter the query conditions in the worksheet, and click the **Advanced** button from the Sort & Filter group on the Data tab.	Chapter 2
Data, query with AutoFilter	Select the cell range and click the **Filter** button from the Sort & Filter group on the Data tab.	Chapter 2
Data, sort	Select the cell range and then click the **Sort** button from the Sort & Filter group on the Data tab.	Chapter 2
Frequency table, create	Click **StatPlus > Descriptive Statistics > Frequency Tables**. Select the frequency table options. *Requires StatPlus.*	Chapter 4

BROOKS/COLE
CENGAGE Learning™

www.cengage.com/brookscole